In recent years, Rydberg atoms have been the subject of intense study, becoming the testing ground for several quantum mechanical problems. This book provides a comprehensive description of the physics of Rydberg atoms, highlighting their remarkable properties by reference to their behavior in a wide range of physical situations.

Beginning with a brief historical overview, the basic properties, creation and detection of Rydberg atoms are described. The effects of blackbody radiation are discussed, as are optical excitation in static electric fields, ionization by pulsed electric fields, Rydberg spectroscopy in high magnetic fields, and microwave excitation and ionization. The collisions of Rydberg atoms with neutral atoms and molecules, charged particles, and other Rydberg atoms are dealt with in detail. The powerful method of multichannel quantum defect theory is presented, and used in the description of autoionizing Rydberg states, interseries interactions and double Rydberg states.

In addition to providing a clear introduction to the basic properties of Rydberg atoms, experimental and theoretical research in this extensive field is reviewed. The books will therefore be valuable to both graduate students and established researchers in physics and physical chemistry.

**Cambridge Monographs on
Atomic, Molecular, and Chemical Physics
3**

General editors: A. Dalgarno, P. L. Knight, F. H. Read, R. N. Zare

RYDBERG ATOMS

Cambridge Monographs on
Atomic, Molecular, and Chemical Physics

RYDBERG ATOMS

THOMAS F. GALLAGHER

Jesse W. Beams Professor of Physics
Department of Physics, University of Virginia

CAMBRIDGE
UNIVERSITY PRESS

CAMBRIDGE UNIVERSITY PRESS
Cambridge, New York, Melbourne, Madrid, Cape Town, Singapore, São Paulo

Cambridge University Press
The Edinburgh Building, Cambridge CB2 2RU, UK

Published in the United States of America by Cambridge University Press, New York

www.cambridge.org
Information on this title: www.cambridge.org/9780521385312

© Cambridge University Press 1994

First published 1994
This digitally printed first paperback version 2005

A catalogue record for this publication is available from the British Library

Library of Congress Cataloguing in Publication data

Gallagher, Thomas F.
Rydberg atoms / Thomas F. Gallagher.
p. cm.
ISBN 0-521-38531-8
1. Rydberg states. I. Title.
QC454.A8S27   1994
539.7—dc20   93-37132   CIP

ISBN-13 978-0-521-38531-2 hardback
ISBN-10 0-521-38531-8 hardback

ISBN-13 978-0-521-02166-1 paperback
ISBN-10 0-521-02166-9 paperback

Contents

Preface

My intent in writing this book is to present a unified description of the many properties of Rydberg atoms. It is intended for graduate students and research workers interested in the properties of Rydberg states of atoms or molecules. In many ways it is similar to the excellent volume *Rydberg States of Atoms and Molecules* edited by R. F. Stebbings and F. B. Dunning just over a decade ago. It differs, however, in covering more topics and in being written by one author. I have attempted to focus on the essential physical ideas. Consequently the theoretical developments are not particularly formal, nor is there much emphasis on the experimental details.

The constraints imposed by the size of the book and my energy have forced me to limit the topics covered in this book to those of general interest and those about which I already knew something. Consequently, several important topics which might well have been included by another author are not included in the present volume. Two examples are molecular Rydberg states and cavity quantum electrodynamics.

Finally, it is a great pleasure to acknowledge the fact that this book would never have been written without the efforts of many people. First I would like to acknowledge the help of my colleagues in the Molecular Physics Laboratory of SRI International (originally Stanford Research Institute). They had the confidence that our initial experiments would develop into a productive research program, and they completed my education as a physicist. My colleagues at the University of Virginia have continued to provide both a critical audience and the encouragement necessary to undertake the writing of this book.

My collaborators have contributed substantially to my understanding of Rydberg atoms, and it is a pleasure to acknowledge the contributions of L. A. Bloomfield, W. E. Cooke, S. A. Edelstein, F. Gounand, R. M. Hill, R. Kachru, R. R. Jones, D. J. Larson, D. C. Lorents, L. Noordam, P. Pillet, K. A. Safinya, W. Sandner, and R. C. Stoneman. In addition I have been the beneficiary of the insights of my students, post doctoral fellows, and visitors. Without all of their contributions this book could not have been written.

I am indebted to Tammie Shifflett, Bessie Truzy, Warrick Liu, and Sibyl Hale for their careful typing of the manuscript, and it is a pleasure to acknowledge the encouragement of James Deeny, Rufus Neal, and Philip Meyler of the Cambridge

University Press. Finally, the gentle prodding of my mother, Margaret Gallagher, and the patience of my wife, Betty, played important roles in the completion of this book.

Charlottesville, Virginia   *T. F. Gallagher*
November, 1993

1

Introduction

Rydberg atoms, atoms in states of high principal quantum number, n, are atoms with exaggerated properties. While they have only been studied intensely since the nineteen seventies, they have played a role in atomic physics since the beginning of quantitative atomic spectroscopy. Their role in the early days of atomic spectroscopy is described by White.[1]

The first appearance of Rydberg atoms is in the Balmer series of atomic H. Balmer's formula, from 1885, for the wavelengths of the visible series of atomic H, is given by[1]

$$\lambda = \frac{bn^2}{n^2 - 4},$$

(1.1)

where $b = 3645.6$ Å. We now recognize Eq. (1.1) as giving the wavelengths of the Balmer series of transitions from the $n = 2$ states to higher lying levels.

While the H atom was the first atom to be understood in a quantitative way, other atoms played a crucial role in unravelling the mysteries of atomic spectroscopy. For example, Liveing and Dewar[2] made the important observation that the observed spectral lines of Na could be grouped into different series. Specifically they observed the Na ns–3p and nd–3p emissions from an arc. Up to the 9s and 8d states they observed doublets due to the fine structure splitting of the 3p state, but for the 10s and 9d states they were unable to resolve the 3p doublet. The ns–3p doublets appeared as sharp lines while the nd–3p doublets appeared as diffuse lines. Most important, they recognized that these sharp and diffuse doublets were members of two series of related lines, although they were unsuccessful in their attempts to relate the wavelengths of the observed transitions.

Knowing, as we do today, that the Na d states are almost degenerate with the higher angular momentum states of the same n, we are not surprised that the series of transitions to the d Rydberg series is diffuse. They are much more likely to suffer pressure broadening. In contrast to Na, Liveing and Dewar noted that in the K spectrum the s and d series were not particularly different, both being "more or less diffuse".[2] We now realize that the difference between Na and K is due to the fact that both the s and the d Rydberg states of K are energetically well removed from the high angular momentum states.

A crucial advance was made by Hartley, who in his studies of the spectra of Mg, Zn, and Cd, first realized the significance of the frequency of a transition as opposed to the experimentally measured wavelength.[1,3] Hartley observed that the splittings of series of multiplets always had the same wavenumber splitting, irrespective of the wavelengths of the transitions. The wavenumber, v, is the inverse of the wavelength in vacuum. The importance of this realization is hard to overstate and is immediately apparent if we rewrite Balmer's formula in terms of the wavenumber of the observed lines instead of the wavelength,

$$v = \left(\frac{1}{4b}\right)\left(\frac{1}{4} - \frac{1}{n^2}\right). \tag{1.2}$$

It is evident that it simply reflects the energy difference between the $n = 2$ and higher lying states.

Following the precedent of Liveing and Dewar, Rydberg began to classify the spectra of other atoms, notably alkali atoms, into sharp, principal, and diffuse series of lines.[4] Each series of lines has a common lower level and a series of ns, np, or nd levels, respectively, as the upper levels of the sharp, principal, and diffuse series. He realized that the wavenumbers of the series members were related and that the wavenumbers of the observed lines could be expressed as[1,4]

$$v_s = v_{\infty s} - \frac{Ry}{(n - \delta_s)^2},$$

$$v_p = v_{\infty p} - \frac{Ry}{(n - \delta_p)^2}, \tag{1.3}$$

$$v_d = v_{\infty d} - \frac{Ry}{(n - \delta_d)^2},$$

where the constants $v_{\infty s}$, $v_{\infty p}$, and $v_{\infty d}$ and δ_s, δ_p, and δ_d are the series limits and quantum defects of the sharp, principal, and diffuse series. The constant Ry is a universal constant, which can be used to describe the wavenumbers of the transitions, not only for different series of the same atom, but for different atoms as well. This constant, Ry, is now known as the Rydberg constant. Defining the wavenumber as the inverse of the wavelength in air, Rydberg[4] assigned Ry the value $109721.6 \, \text{cm}^{-1}$. Recognition of the generality of the Rydberg constant is one of Rydberg's two major accomplishments. The other was to show that all the spectral lines from an atom are related. Specifically, he realized that $v_{\infty s}$ and $v_{\infty d}$ were the same to within experimental error. For example, for the series originating in the Na $3p_{3/2}$ and $3p_{1/2}$ levels he assigned the values 24485.9 and 24500.5 cm^{-1} for $v_{\infty s}$, and 24481.8 and 24496.4 cm^{-1} for $v_{\infty d}$. In addition, he observed that $v_{\infty p} - v_{\infty s}$ was equal to the wavenumber of the first member of the principal series. He reasoned that it should be possible to see other series which did not have the first term of the principal series in common, but some other term instead. This notion led him to a general expression for the wavenumbers of lines connecting different series. The wavenumbers of lines connecting the s and p series, for example, are given by[4]

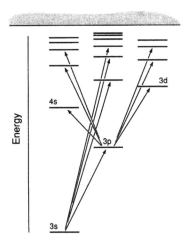

Fig. 1.1 Energy level diagram for Na showing the sharp 3p–ns, principal 3s–np, and diffuse 3p–nd series.

$$\pm \nu = \frac{Ry}{(m - \delta_s)^2} - \frac{Ry}{(n - \delta_p)^2}. \qquad (1.4)$$

In Eq. (1.4) the + sign and constant n describe a sharp series of s states and the minus sign and a constant m describe a principal series of p states. If we consider the special case $\delta_s = \delta_p = 0$ and $m = 2$ we arrive at Balmer's formula for the H transitions from $n = 2$.

From the early spectroscopic work it is possible to construct an energy level diagram for Na, as shown in Fig. 1.1. From this figure it is apparent that the difference in the principal and sharp series limits is the wavenumber of the 3s–3p transition. It is also apparent that the Rydberg states, states of high principal quantum number n, lie close to the series limit.

The physical significance of high n did not become clear until Bohr proposed his model of the H atom in 1913.[1] In spite of its shortcomings, the Bohr atom contains the properties of Rydberg atoms which make them interesting. The basis of the Bohr atom is an electron moving classically in a circular orbit around an ionic core. To the currently accepted ideas of classical physics Bohr added two notions; that the angular momentum was quantized in integral units of $\hbar$, Planck's constant divided by 2π, and that the electron did not radiate continuously in a classical manner, but gave off radiation only in making transitions between states of well defined energies. The first was an assumption, and the second was forced upon him by the experimental observations.

Rydberg atoms

The Bohr theory can be summarized as follows. An electron of charge $-e$ and mass m in a circular orbit of radius r about an infinitely heavy positive charge of Ze obeys Newton's law for uniform circular motion[5]

$$\frac{mv^2}{r} = \frac{kZe^2}{r^2},$$ (1.5)

where $k = 1/4\pi\varepsilon_0$, ε_0 being the permittivity of free space. Adding Bohr's requirement of the quantization of angular momentum yields

$$mvr = n\hbar.$$ (1.6)

Combining Eqs. (1.5) and (1.6) leads immediately to an expression for the radius r of the orbit

$$r = \frac{n^2\hbar^2}{Ze^2mk}.$$ (1.7)

In other words, the size of the orbit increases as the square of the principal quantum number, and states of high n have very large orbits. The energy, W, of a state is obtained by adding its kinetic and potential energies,

$$W = \frac{mv^2}{2} - \frac{kZe^2}{r} = \frac{-k^2Z^2e^4m}{2n^2\hbar^2}.$$ (1.8)

The energies are negative, i.e. the electron is bound to the proton, and the binding energy decreases as $1/n^2$. The allowed transition frequencies are the differences in the energies. Explicitly

$$W_2 - W_1 = \frac{k^2Z^2e^4m}{2\hbar^2}\left(\frac{1}{n_1^2} - \frac{1}{n_2^2}\right).$$ (1.9)

Comparing this expression to the general form of Rydberg's formula we can see that

$$Ry = \frac{k^2Z^2e^4m}{2\hbar^2}.$$ (1.10)

Historically, the most important aspect of the Bohr atom was that it related the spectroscopic Rydberg constant to the mass and charge of the electron. From our present point of view, the Bohr atom physically defines Rydberg atoms and shows us why they are interesting. In high n, or Rydberg, states the valence electrons have binding energies which decrease as $1/n^2$ and orbital radii which increase as n^2. In other words, in a Rydberg atom the valence electron is in a large, loosely bound orbit. Although Rydberg atoms of principal quantum number well over 100 have been studied in the laboratory, it is instructive to consider a relatively low Rydberg state, $n = 10$, and compare it to a ground state atom. The ground state of H is bound by 1 Ry, 13.6 eV, and has an orbital radius of $1a_0$ and a geometric cross section of a_0^2. In contrast, the $n = 10$ state has a binding energy of 0.01 Ry, an

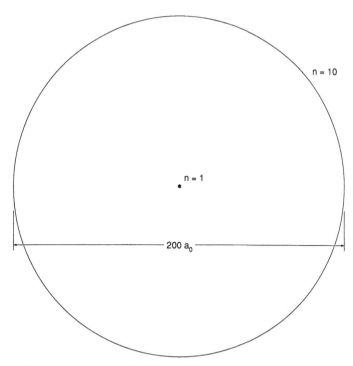

Fig. 1.2 Bohr orbits of $n = 1$ and $n = 10$. The solid dark circle at the center is the $n = 1$ orbit of radius $1a_0$. The $n = 10$ orbit has radius $100a_0$. $1a_0 = 0.53$ Å.

orbital radius of $100a_0$, and a geometric cross section of $10^4a_0^2$. Fig. 1.2 is a drawing of the Bohr orbits for $n = 1$ and $n = 10$, which demonstrates graphically the difference between a Rydberg atom and a ground state atom. The binding energy of the $n = 10$ atom is comparable to thermal energies, and the geometric cross section is orders of magnitude larger than gas kinetic cross sections.

Once the Bohr theory was formulated it was apparent that Rydberg atoms should have bizarre properties. Why then were Rydberg atoms not studied more extensively until the nineteen seventies? There are two reasons. First, the most pressing problem at the time was not the properties of atoms near the series limit, but the development of the quantum theory. Second, there was simply no means of making Rydberg atoms efficiently enough to study them in the detail that is now possible. Nonetheless, the most sophisticated tool available at the time, high resolution absorption spectroscopy, was used in the first experiments to demonstrate the bizarre properties of Rydberg atoms. By examining the fine details of the spectra, the splittings, shifts, and broadenings, it was possible to learn a great deal about Rydberg states.

One of the first properties to be explored was the sensitivity of Rydberg atoms to external electric fields, the Stark effect. While ground state atoms are nearly immune to electric fields, relatively modest electric fields not only perturb the Rydberg energy levels, but even ionize Rydberg atoms, as was shown in early

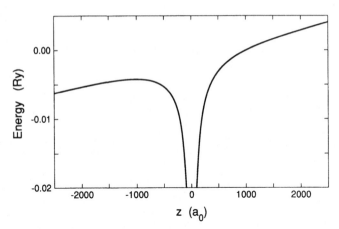

Fig. 1.3 Combined Coulomb–Stark potential along the z axis for a field of $5.14\,kV/cm$ in the z direction.

studies of the Balmer series in electric fields of up to $10^6\,V/cm$.[6,7] The Balmer lines exhibit a splitting which is approximately linear in the electric field, and the components actually disappear at well defined values of the electric field due to field ionization of the Rydberg state.

While the Bohr atom is of no help in understanding the splittings of the Balmer lines, using it we can calculate the field at which a state is ionized by an electric field. Consider a H atom with its nucleus at the origin in the presence of an electric field in the z direction. The potential experienced by an electron moving along the z axis is given by

$$V = -\frac{k}{|z|} + Ez \tag{1.11}$$

and is shown in Fig. 1.3. Only electrons with energies lower than the local maximum of the potential, at $z = -1000a_0$, are classically bound. In the real three dimensional potential the local maximum at $z = -1000a_0$ is a saddle point, and electrons with energies above the saddle point of the potential are ionized by the field. The potential at the saddle point is given by

$$V_s = -2\sqrt{kE}. \tag{1.12}$$

If this energy is equated to the energy of the Rydberg state, $-Ry/n^2$, we find the classical field for ionization of a state of principal quantum number n

$$E_c = \frac{Ry^2}{4kn^4}. \tag{1.13}$$

This formula is usually encountered in atomic units, in which it reads

$$E_c = 1/16n^4. \tag{1.14}$$

In the next chapter we shall introduce atomic units. Although we have derived Eq. (1.13) with no regard for tunneling or the shifts of the energy levels in an electric field, it is quite a useful expression, as it is, in a practical sense, nearly exact for nonhydrogenic atoms.

The size of a Rydberg atom is large, the geometric cross section scaling as n^4, and, not surprisingly, early experiments were focused on this aspect of Rydberg atoms. A classic experiment was undertaken by Amaldi and Segre to observe the energy shifts of the high lying members of the Rydberg series of K in the presence of high pressures of rare gas atoms.[8] Specifically, they observed the shifts of the 4s–np absorption lines due to the added rare gas. The expectation at the time was that a red shift of the lines would be observed due to the fact that the space between the Rydberg electron and the K^+ ion is filled with a polarizable dielectric, the rare gas atoms, instead of vacuum. While the pressure shifts for Ar and He were to the red, the pressure shift for Ne was to the blue, an observation completely incompatible with the dielectric model. Shortly after the experiment Fermi explained the pressure shifts as arising predominantly from the short range scattering of the Rydberg electron from the rare gas atom, not from the anticipated dielectric effect.[9] This experiment was perhaps the first of many in which the result was diametrically opposed to expectations.

Due to their large size, Rydberg atoms also exhibit large diamagnetic energy shifts. Since the Zeeman and Paschen–Back effects are proportional to the angular momentum, they are not particularly striking in the optically accessible low ℓ Rydberg states. The diamagnetic shifts on the other hand depend on the area of the orbit and thus scale as n^4. The diamagnetic shifts are very difficult to see in low n states, but in high n states they are quite evident. The diamagnetic shifts were first observed by Jenkins and Segre in the absorption spectrum of K taken in the 27 kG field of the Berkeley cyclotron.[10] As n increases the field first mixes states of different ℓ, then mixes states of different n. Only within the last 20 years have experiments progressed substantially beyond these first experiments of Jenkins and Segre. Ironically, the study of Rydberg atoms in magnetic fields, one of the first topics studied, is still a topic of active research 50 years later.

The interest in Rydberg atoms waned until it became apparent that they might play an important role in real physical systems. In particular it became apparent that they were important astrophysically.[11] The radio recombination lines detected by radio astronomers are the emissions from transitions between Rydberg states subsequent to radiative recombination. A particularly favorable frequency regime in which to search for radio signals is 2.4 GHz, which corresponds to the $n = 109$ to $n = 108$ transition. While it is not surprising that Rydberg atoms might be found in interstellar space, where the density is very low, they also play an important role in astrophysical and laboratory plasmas.[12,13] Radiative recombination of low energy electrons and ions results in Rydberg atoms, and dielectronic recombination of higher energy electrons and ions often proceeds by the capture of the electrons into autoionizing Rydberg states and the subsequent radiative decay to bound Rydberg states. Furthermore, the microscopic electric fields in a plasma cut off the Rydberg series at or near the classical ionization limit, as described by Eq. (1.4). Precisely how the Rydberg series is terminated is a matter of some importance, for it determines the thermodynamic properties of the plasma.[12]

The single event which probably contributed the most to the resurgence of interest in Rydberg atoms was the development of the tunable dye laser.[14,15] With a tunable dye laser and very simple apparatus it became possible to excite large numbers of atoms to a single, well defined Rydberg state. This ability allowed one to study a wide range of properties in astonishing detail. Many properties familiar from low lying excited states, such as the radiative lifetimes, energy level spacings, and collision cross sections were measured systematically, and in many, but not all, cases the properties followed the expected extrapolations from the low lying states. Perhaps more interesting were the investigations of the properties which are in a practical sense peculiar to Rydberg atoms. A good example of such properties is the effect of an electric field. While the gross effects of electric fields on Rydberg atoms had been observed long before in optical spectra, and worked out theoretically for some cases, they had not been understood in a systematic way, mainly because there appeared to be no pressing reason to do so. Not only do electric fields produce dramatic effects on Rydberg atoms, but they also afford us novel techniques for the detection and manipulation of Rydberg atoms. To exploit fully these techniques has required a systematic study of the static and dynamic effects of electric fields.

The most fascinating opportunities presented by Rydberg atoms are those which take advantage of their exaggerated properties to carry out experiments that would be impossible in other systems. One example of this is the interaction of atoms with radiation fields. In the low intensity limit, Rydberg atoms provide an ideal system with which to test the interaction of an atom with the vacuum, particularly with the structured vacuum produced by a resonant cavity. Outstanding examples are the one atom maser[16] and the two photon maser.[17] Rydberg atoms have two properties which make them ideal for such investigations. First, the characteristic frequencies of Rydberg atom transitions are low, and the wavelengths correspondingly long, making the construction of resonant or near resonant cavities a straightforward matter. Second, the dipole moments of Rydberg atoms are so large that they have usable radiative decay rates in spite of the low transition frequencies. At the other intensity extreme, advances in laser technology have made atoms in strong radiation fields, fields comparable in magnitude to the coulomb field, a topic of both practical and fundamental interest.[18–20] While it is apparent that any practical applications will involve intense lasers and ground state atoms and molecules, a Rydberg atom in a microwave field is an ideal system in which to carry out quantitative measurements of the properties of atoms in strong radiation fields.[21]

All the properties we have discussed thus far are those of atoms with one Rydberg electron. Double Rydberg atoms, atoms in which there are two highly excited electrons, exhibit pronounced correlation between the motions of the two electrons and have properties which differ in a qualitative way from those of normal Rydberg atoms.[22] While some experiments have been done, they have only scratched the surface, and much is yet to be learned about these exotic atoms.

References

1. H. E. White, *Introduction to Atomic Spectra* (McGraw-Hill, New York, 1934).
2. G. D. Liveing and J. Dewar, *Proc. Roy. Soc. Lond.* **29**, 398 (1879).
3. W. N. Hartley, *J. Chem. Soc.* **43**, 390 (1883)
4. J. R. Rydberg, *Phil. Mag.* 5th Ser. **29**, 331 (1890)
5. H. Semat and J. R. Albright, *Introduction to Atomic and Nuclear Physics* (Holt, Rinehart, and Winston, New York, 1972).
6. H. A. Bethe and E. A. Salpeter, *Quantum Mechanics of One and Two Electron Atoms* (Academic Press, New York, 1957).
7. H. Rausch v. Traubenberg, *Z. Phys.* **54** 307 (1929).
8. E. Amaldi and E. Segre, *Nuovo Cimento* **11**, 145 (1934).
9. E. Fermi, *Nuovo Cimento* **11**, 157 (1934).
10. F. A. Jenkins and E. Segre, *Phys. Rev.* **55**, 59 (1939).
11. A. Dalgarno, in *Rydberg States of Atoms and Molecules*, eds. R.F. Stebbings and F.B. Dunning (Cambridge University Press, Cambridge, 1983).
12. D. G. Hummer and D. Mihalis, *Astrophys. J.* **331**, 794, (1988).
13. V. L. Jacobs, J. Davis, and P. C. Kepple, *Phys. Rev. Lett.* **37**, 1390, (1976).
14. P. P. Sorokin and J. R. Lankard, *IBM J. Res. Dev.* **10**, 162, (1966).
15. T. W. Hansch, *Appl. Opt.* **11**, 895, (1972).
16. D. Meschede, H. Walther, and G. Müller, *Phys. Rev. Lett.* **54**, 551, (1985).
17. M. Brune, J. M. Raimond, P. Goy, L. Davidovitch, and S. Haroche, *Phys. Rev. Lett.* **59**, 1899, (1987).
18. A. L'Huillier, L.A. Lompre, G. Mainfray, and C. Manus, *Phys. Rev. Lett.* **48**, 1814, (1982).
19. T. S. Luk, H. Pommer, K. Boyer, M. Shahidi, H. Egger, and C. K. Rhodes, *Phys. Rev. Lett.* **51**, 110, (1983).
20. R. R. Freeman, P. H. Bucksbaum, H. Milchberg, S. Darack, D. Schumacher, and M. E. Geusic, *Phys. Rev. Lett.* **59**, 1092, (1987).
21. J. E. Bayfield and P. M. Koch, *Phys. Rev. Lett.* **33**, 258, (1974).
22. U. Fano, *Rep. Prog. Phys.* **46**, 97, (1983).

2
Rydberg atom wavefunctions

To calculate the properties of Rydberg atoms we need wavefunctions to describe them. In this chapter we use the methods of quantum defect theory to develop the wavefunctions for an electron in a coulomb potential. This approach may be applied to any single valence electron Rydberg atom, including the H atom as a special case. From a practical point of view it is an important special case, because the analytic properties of the hydrogenic wavefunctions are well known and are useful in obtaining analytic solutions to a variety of problems. We also present numerical methods for generating accurate coulomb wavefunctions for use in later problems. Finally, we show how scaling laws of many properties of Rydberg atoms are easily obtained from the wavefunctions.

If we consider Rydberg states of H and Na, as shown in Fig. 2.1, they are essentially similar. The only difference between the two is that the Na$^+$ core, while having charge 1, has a finite size due to its being made up of a nucleus of charge $+11$ and ten electrons. When the Rydberg electron is far from the Na$^+$ core it is only sensitive to the net charge. Since the Rydberg electron spends most

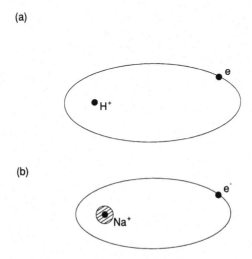

Fig. 2.1 Rydberg atoms of (a) H and (b) Na. In H the electron orbits around the point charge of the proton. In Na it orbits around the $+11$ nuclear charge and ten inner shell electrons. In high ℓ states Na behaves identically to H, but in low ℓ states the Na electron penetrates and polarizes the inner shell electrons of the Na$^+$ core.

of its time near its classical outer turning point, where the differences between Na and H are minimal, we expect the properties of all Rydberg atoms to be similar. On the other hand, when the Rydberg electron comes near the Na^+ core the precise charge distribution of the Na^+ core does play a role. In particular, the Rydberg electron can both polarize and penetrate the Na^+ core, and both polarization and penetration change the wavefunctions and energies of a Na Rydberg state from their hydrogenic counterparts.

If we consider Na and H atoms in states of high orbital angular momentum, for example in circular Bohr orbits, the differences are negligible, for the Na electron never comes close enough to the Na^+ core to detect that it is anything other than a point charge. On the other hand, if the electron is in a highly elliptic low ℓ orbit, on each orbit the Rydberg electron comes close to the core, where there is a significant difference between Na and H. In fact, there are very interesting differences between the low ℓ and high ℓ states of Na, the most obvious being the depression of the energies of the low ℓ states below the H levels, due to core polarization and penetration. When the Na Rydberg electron penetrates the cloud of the inner ten electrons, it is exposed to the unshielded +11 nuclear charge, and its binding energy is increased, or, equivalently, its total energy decreased. Similarly, polarization of the Na^+ core produces a decrease in the energy of a Rydberg state. In Fig. 2.2 we show the energy levels of H and Na. As expected, the H and high ℓ Na levels are degenerate on the scale of Fig. 2.2, but the low ℓ states of Na are depressed in energy.[1] Their energies are given by

$$W = \frac{-Ry}{(n - \delta_\ell)^2}, \tag{2.1}$$

where δ_ℓ is an empirically observed quantum defect for the series of orbital angular momentum ℓ. The depression of the energies of the low ℓ states of Na below the H values is perhaps the most obvious difference between Na and H. However the alteration in the wavefunction required to produce these energy differences leads to many other differences as well.

In the calculation of atomic properties it is convenient to introduce atomic units which are defined so that all the relevant parameters for the ground state of H have magnitude one. The atomic units most useful for our purposes are given in Table 2.1.[2] An extensive list is given by Bethe and Salpeter.[2] Throughout the book atomic units will be used for all calculations, with conversions to other units to facilitate comparisons to experiment.

In developing Rydberg atom wavefunctions we begin with the Schroedinger equation for the H atom, which, in atomic units, may be written as

$$\left(-\frac{\nabla^2}{2} - \frac{1}{r} \right) \psi = W\psi, \tag{2.2}$$

where r is the distance of the electron from the proton, assumed to be infinitely massive, and W is its energy. Unless stated otherwise, we shall consider only

Rydberg atoms

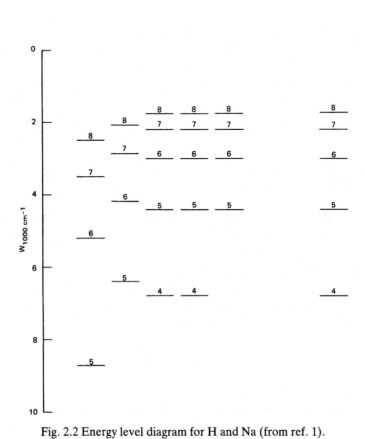

Fig. 2.2 Energy level diagram for H and Na (from ref. 1).

Table 2.1. *Atomic units.*

Quantity	Atomic unit	Definition
Mass	Electron mass	9.1×10^{-28} g
Charge	Electron charge e	1.6×10^{-19} C
Energy	Twice the ionization potential of hydrogen	27.2 eV
Length	Radius of a_0, the first Bohr orbit	0.529 Å
Velocity	Velocity of the first Bohr orbit	2.19×10^8 cm/s
Electric field	Field at the first Bohr orbit	5.14×10^9 V/cm

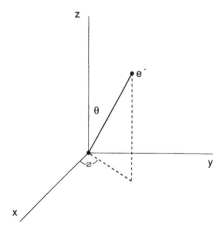

Fig. 2.3 Geometry of the atom.

neutral atoms in which an electron far from the ionic core sees a charge of +1. In spherical coordinates

$$\nabla^2 = \frac{\partial}{\partial r^2} + \frac{2}{r}\frac{\partial}{\partial r} + \frac{1}{r^2 \sin\theta}\frac{\partial}{\partial\theta}\left(\sin\theta\frac{\partial}{\partial\theta}\right) + \frac{1}{r^2 \sin^2\theta}\frac{\partial^2}{\partial\phi^2}, \tag{2.3}$$

where θ and ϕ are the polar angle relative to the quantization axis and the azimuthal angle, respectively, as shown in Fig. 2.3. If we assume that Eq. (2.2) is separable and write ψ as a product of radial and angular functions, i.e. $\psi = Y(\theta,\phi)\,R(r)$, Eq. (2.2) becomes

$$\left[\frac{\partial^2 R}{\partial r^2} + \frac{2}{r}\frac{\partial R}{\partial r} + 2\left(W + \frac{1}{r}\right)R\right]Y +$$

$$\left[\frac{1}{r^2 \sin\theta}\frac{\partial}{\partial\theta}\left(\sin\theta\frac{\partial Y}{\partial\theta}\right) + \frac{1}{r^2 \sin^2\theta}\frac{\partial^2 Y}{\partial\phi^2}\right]R = 0. \tag{2.4}$$

Dividing by RY/r^2 yields the separated form

$$\frac{r^2}{R}\left[\frac{\partial^2 R}{\partial r^2} + \frac{2}{r}\frac{\partial R}{\partial r} + 2\left(W + \frac{1}{r}\right)R\right] + \frac{1}{Y}\left[\frac{1}{\sin\theta}\frac{\partial}{\partial\theta}\left(\sin\theta\frac{\partial Y}{\partial\theta}\right) + \frac{1}{\sin^2\theta}\frac{\partial Y}{\partial^2\phi}\right] = 0. \tag{2.5}$$

The two terms, which depend only on R and Y, respectively, are independent of each other and therefore must be separately equal to $\pm\lambda$, where λ is a constant.

If we further assume that we can write the angular function as $Y(\theta,\phi) = \Theta(\theta)\Phi(\phi)$, we may write the angular equation as

$$\frac{1}{\sin\theta}\frac{\partial}{\partial\theta}\left(\sin\theta\frac{\partial\Theta}{\partial\theta}\right)\Phi + \frac{\Theta}{\sin^2\theta}\frac{\partial^2\Phi}{\partial^2\phi} = -\lambda\Phi\Theta. \tag{2.6}$$

The solution to Eq. (2.6) is the normalized spherical harmonic,[2]

Table 2.2. *Normalized spherical
harmonics for s, p, and d states.*[a]

$$Y_{00} = \frac{1}{\sqrt{4\pi}}$$

$$Y_{10} = \sqrt{\frac{3}{4\pi}} \cos\theta$$

$$Y_{1\pm1} = \pm\sqrt{\frac{3}{8\pi}} \sin\theta\, e^{\pm i\phi}$$

$$Y_{20} = \sqrt{\frac{5}{4\pi}} \left(\frac{3}{2}\cos^2\theta - \frac{1}{2}\right)$$

$$Y_{2\pm1} = \pm\sqrt{\frac{15}{8\pi}} \sin\theta \cos\theta\, e^{\pm i\phi}$$

$$Y_{2\pm2} = \sqrt{\frac{15}{32\pi}} \sin^2\theta\, e^{\pm i2\phi}$$

[a] (from ref. 2)

$$\Theta(\theta)\Phi(\phi) = Y_{\ell m}(\theta,\phi) = \sqrt{\frac{(\ell - m)!}{(\ell + m)!}} \frac{2\ell + 1}{4\pi} P_\ell^m(\cos\theta)e^{im\phi}, \qquad (2.7)$$

where $P_\ell^m(x)$ is the unnormalized associated Legendre polynomial, ℓ is zero or a
positive integer, m takes integral values from $-\ell$ to ℓ, and $\lambda = \ell(\ell + 1)$. The
spherical harmonics are normalized,

$$\int_0^1 \sin\theta\, d\theta \int_0^{2\pi} d\phi Y_{\ell m}^*(\theta,\phi)Y_{\ell m}(\theta,\phi) = 1. \qquad (2.8)$$

In Table 2.2 we have listed the first few spherical harmonics, for the s, p, and d
states. It is worth noting that some authors introduce a factor of $[-1]^m$ in defining
the associated Legendre polynomials, producing a corresponding difference in
the spherical harmonics.[2] There are ℓ-m nodes in the θ coordinate, and none in
the ϕ coordinate.

The angular equation requires $\lambda = \ell(\ell + 1)$ where ℓ is a positive integer. Using
this value for λ the radial equation of Eq. (2.5) can be written as

$$\frac{\partial^2 R}{\partial r^2} + \frac{2}{r}\frac{\partial R}{\partial r} + \left[2W + \frac{2}{r} - \frac{\ell(\ell + 1)}{r^2}\right]R = 0. \qquad (2.9)$$

We now convert the radial equation to the standard form of the coulomb problem
by introducing $\rho(r)$, which is defined by[3] $R(r) = \rho(r)/r$. With this substitution the
radial equation of Eq. (2.9) may be written as

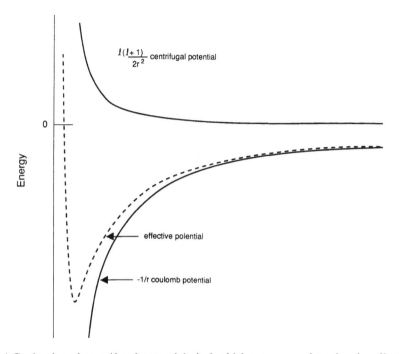

Fig. 2.4 Coulomb and centrifugal potentials (—) which are summed to give the effective radial potential (– –) in which an electron of angular momentum ℓ moves.

$$\frac{\partial^2 \rho}{\partial r^2} + \left[2W + \frac{2}{r} - \frac{\ell(\ell + 1)}{r^2} \right] \rho = 0. \tag{2.10}$$

The terms in the square brackets of Eq. (2.10) have the significance shown in Fig. 2.4: $2/r$ is twice the coulomb potential of the electron, $[\ell(\ell+1)]/r^2$ is twice the centrifugal potential, and $2W$ is twice the total energy. The sum in the brackets is $2T$, where T is the radial kinetic energy of the electron. In the classically allowed region $T > 0$, and Eq. (2.10) has two independent oscillatory solutions much like sines and cosines. On the other hand, in the classically forbidden region $T < 0$, and the solutions to Eq. (2.10) are increasing and decreasing exponential functions.

Since Eq. (2.10) is a second order equation it has two independent solutions. The two solutions of most use in solving our problem are[4]

$$\rho(r) = f(W, \ell, r)$$

and $\qquad\qquad\qquad\qquad\qquad\qquad\qquad\qquad\qquad\qquad$ (2.11)

$$\rho(r) = g(W, \ell, r).$$

The f and g functions are commonly termed the regular and irregular coulomb functions.[4,5] In the classically allowed region the f and g functions are real oscillatory functions with a phase difference of 90 degrees. In the classically forbidden regions they are increasing or decreasing exponentials.

To find the solutions of the radial equation the forms of the f and g functions as $r \to 0$ and $r \to \infty$ are useful. For $W > 0$ and $W < 0$, as $r \to 0$

$$f(W,\ell,r) \propto r^{\ell+1} \tag{2.12a}$$

and

$$g(W,\ell,r) \propto r^{-\ell}. \tag{2.12b}$$

From Fig. 2.4 it is apparent that the small r behavior is not very energy dependent, because T is very large, except for very high angular momentum. For $W < 0$ we introduce the effective quantum number ν, defined by $W = -1/2\, \nu^2$. For $r \to \infty$ the f and g functions may be expressed in terms of increasing and decreasing exponential functions $u(\nu,\ell,r)$ and $v(\nu,\ell,r)$. As $r \to \infty$, $u \to \infty$ and $v \to 0$, and f and g are given by[4]

$$f \to u(\nu,\ell,r) \sin \pi\nu - v(\nu,\ell,r)e^{i\pi\nu}, \tag{2.13a}$$

$$g \to -u(\nu,\ell,r) \cos \pi\nu + v(\nu,\ell,r)\, e^{i\pi\,(\nu\,+\,1/2)}. \tag{2.13b}$$

For $W > 0$ and $r \to \infty$ the f and g functions are given by the oscillatory functions

$$f \to (2/k\pi)^{1/2} \sin\left[kr - \pi\ell/2 + \frac{1}{k}\ell n\,(2kr) + \sigma_\ell \right], \tag{2.14a}$$

$$g \to -(2/k\pi)^{1/2} \cos\left[kr - \pi\ell/2 + \frac{1}{k}\ell n\,(2kr) + \sigma_\ell \right], \tag{2.14b}$$

where $k = (2W)^{1/2}$ and σ_ℓ is the coulomb phase, $\sigma_\ell = \arg \Gamma(\ell + 1 - i/k)$.

To find bound state solutions, $W < 0$, for the H atom we apply the $r = 0$ and $r = \infty$ boundary conditions. Specifically we require that ψ be finite as $r \to 0$ and that $\psi \to 0$ as $r \to \infty$. We can see from Eqs. (2.12) that only the f functions are allowed due to the $r = 0$ boundary condition. As $r \to \infty$ we require that $\psi \to 0$, and, as indicated by the asymptotic form of the f function, this requirement is equivalent to requiring that $\sin \pi\nu$ be zero or that ν be an integer. Combining the angular function of Eq. (2.7) with the f radial function yields the bound H wavefunction

$$\psi_{n\ell m}(\theta,\phi,r) = \frac{Y_{\ell m}(\theta,\phi)f(W,\ell,r)}{r}, \tag{2.15}$$

where the radial function is assumed to be normalized. Normalization procedures will be discussed shortly. Since $W = -1/2\nu^2$ and ν must be a positive integer n, we recover the familiar expression for the allowed energies,

$$W = -\frac{1}{2n^2}. \tag{2.16}$$

If the radial equation is solved by using the more conventional power series solution,[2] the two mathematically allowed solutions for $R(r)$ have leading terms

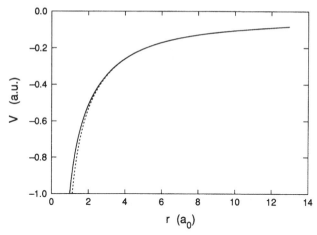

Fig. 2.5 Schematic coulomb potential of H (——) and $V_{Na}(r)$, the effective one-electron potential of Na, (– – –) which is different from the coulomb potential only at small r (from ref. 1).

proportional to r^ℓ and $r^{-\ell-1}$. The $r^{-\ell-1}$ solution, the irregular coulomb function, is discarded since it diverges at the origin and does not satisfy the $r=0$ boundary condition. The $r \to \infty$ boundary condition imposes the requirement that the series terminate, which in turn requires that $W = -1/2n^2$, where n is an integer. The power series solution also shows that, the radial function $R_{n\ell}(r)$ has $n - \ell$ nodes. If we choose solutions for $R_{n\ell}(r) \propto r^\ell$ at $r \sim 0$, the outer lobe of the wavefunction changes sign as either n or ℓ is increased by one.

The above treatment of H, based on coulomb waves, may be easily extended to generate wavefunctions for single valence electron atoms with spherical ionic cores. This approach, quantum defect theory,[5] enables us to generate wave functions which are good when the valence electron is outside the ionic core. We first recall the difference between a Na atom and a H atom, the replacement of the point charge of the proton by the finite sized ionic core of the Na atom. Far from the Na^+ core the potential experienced by the valence electron is indistinguishable from the coulomb potential due to a proton. However, at small orbital radii, where the outer electron can penetrate the ten electron cloud of the Na^+ core, the Na potential is deeper than the coulomb potential. Since the Na^+ core is spherically symmetric, and we assume it to be frozen in place, the effective potential, V_{Na} seen by the valence electron is spherically symmetric and depends only on r. When the coulomb $-1/r$ potential of Eq. (2.2) is replaced by $V_{Na}(r)$, Eq. (2.2) is still separable, and the angular equation, Eq. (2.6), and its solutions, Eq. (2.7), are unchanged. Consequently the Na wavefunction analogous to the H wavefunction of Eq. (2.15) differs only in its radial function.

In Fig. 2.5 we show in a schematic way the coulomb $-1/r$ potential and the effective potential $V_{Na}(r)$ seen by the Rydberg electron in the Na atom. The effect

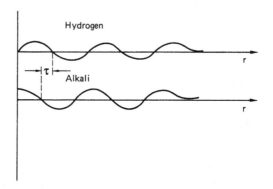

Fig. 2.6 Wavefunctions of H and alkali atoms. The lower potential of the alkali introduces the phase shift τ, as shown, leading to the depression of the energies of alkali, low ℓ states relative to those of H (from ref. 1).

of the lower potential at small r seen by the valence electron in Na is to increase its kinetic energy and decrease the wavelength of the radial oscillations relative to H. In Na all the nodes of the radial wavefunction are pulled closer to the origin than in H, as shown by Fig. 2.6. For $r > r_0$, the radius of the Na$^+$ core, the Na potential is equal to the coulomb potential. Thus in this region the wavefunction is simply shifted in phase from the H wavefunction. The magnitude of the radial phase shift is the difference in momentum of an electron of energy W in Na and H integrated from $r = 0$ to r_0. For an s electron, the phase shift τ relative to H can be written as

$$\tau = \int_0^{r_0} \left\{ [W - V_{Na}(r)]^{1/2} - \left[W + \frac{1}{r}\right]^{1/2} \right\} \sqrt{2} \, dr. \tag{2.17}$$

The terms in the square brackets are the kinetic energies of the electron in Na and H. If we replace $V_{Na}(r)$ by $-1/r - V_d(r)$, where V_d is the difference between the Na and H potentials, and take advantage of the fact that for $r < r_0$ the kinetic energy is large so that $1/r \gg W$ and $V_d(r)$, we can expand Eq. (2.17). Keeping the first order term in the expansion leads to

$$\tau = \int_0^{r_0} V_d(r) \left(\frac{r}{2}\right)^{1/2} dr. \tag{2.18}$$

Thus τ, the radial phase shift, is independent of the energy W as long as $|W| \ll 1/r_0$. A phase shift of τ in the radial wave from H implies that for $r > r_0$, the pure f wave of the H wavefunction of Eq. (2.15) is replaced by $f(W,\ell,r) \cos \tau - g(W,\ell,r) \sin \tau$, and the bound state radial wavefunctions are given by

$$\rho(r) = f(W,\ell,r) \cos \tau - g(W,\ell,r) \sin \tau. \tag{2.19}$$

The hydrogenic requirement that the wavefunction be finite at the origin has been replaced by the requirement that at $r \geq r_0$ the wavefunction be shifted in phase from the hydrogenic wavefunction by τ. This new boundary condition necessitates g functions as well as the regular f functions in the coulomb $r > r_0$

region. We note that the wavefunction that we generate is only good for $r > r_0$. We have no knowledge about the Na potential for $r < r_0$ save the fact that it produces at $r = r_0$ a wave phase shifted from the analogous H solution by τ, as is shown in Fig. 2.6. The requirement that $\psi \rightarrow 0$ as $r \rightarrow \infty$ is equivalent to requiring that the coefficient of $u(\nu, \ell, r)$ be zero. This requirement leads to

$$\cos \tau \sin \pi \nu + \sin \tau \cos \pi \nu = 0, \qquad (2.20)$$

or $\sin(\pi \nu + \tau) = 0$. Thus $\pi \nu + \tau$ must be an integral multiple of π, and $\nu = n - \tau/\pi$ where n is an integer. The effective quantum number ν of a Na state of angular momentum ℓ differs from an integer by an amount τ/π which we recognize as the quantum defect δ_ℓ. Since τ is independent of W, the quantum defect δ_ℓ of a series of ℓ states is a constant. Due to the $\ell(\ell + 1)/2r^2$ centrifugal potential, in the higher ℓ states the electron does not sample the small r part of the potential as much, and the quantum defects are smaller. As the radial wavefunctions for nonhydrogenic atoms obey the same $r \rightarrow \infty$ boundary condition they are similar to hydrogenic wavefunctions at large r, but they do have different energies.

Replacing the f radial function of Eq. (2.15) by the radial function of Eq. (2.19) leads to a wavefunction for a nonhydrogenic atom valid for $r > r_0$. Explicitly,

$$\psi(\theta, \phi, r) = \frac{Y_{\ell m}(\theta, \phi, r)[f(W, \ell, r) \cos \pi \delta_\ell - g(W, \ell, r) \sin \pi \delta_\ell]}{r}, \qquad (2.21a)$$

which has the allowed energies

$$W = -\frac{1}{2(n - \delta_\ell)^2}, \qquad (2.21b)$$

where n is an integer. A point which is worth keeping in mind is that we have not considered core polarization in obtaining the wavefunction of Eq. (2.21a). Due to the long range of a polarization potential, it is not possible to define a value of r_0 such that for $r > r_0$ the potential is a coulomb potential.

An important property of the wavefunction is its normalization, and we have yet to normalize the radial coulomb radial functions. Following the approach of Merzbacher, we can find an approximate WKB radial wavefunction, good in the classically allowed region, given by[6]

$$\rho(r) = \frac{N \sin \int_{r_i}^{r} k \, dr'}{\sqrt{k}}, \qquad (2.22)$$

where N is a normalization constant, r_i is the inner classical turning point and $k = \sqrt{2T}$, the square root of twice the kinetic energy. The factor of $\sqrt{k}$ in the denominator reflects the fact that the amplitude of the wavefunction is larger where the electron is moving more slowly. In the classically allowed region between the inner and outer turning points, r_i and r_o,

$$k = \sqrt{2W + \frac{2}{r} - \frac{\ell(\ell + 1)}{r^2}}. \qquad (2.23)$$

We can use the wavefunction of Eq. (2.22) to obtain an approximate normalization for the bound Rydberg states. Specifically we require that

$$\int_0^{2\pi} \int_0^\pi \int_{r_i}^{r_o} \psi \cdot \psi r^2 dr \sin \theta d\theta d\phi = 1,$$ (2.24)

where

$$r_{i,o} = n^2 \pm n \sqrt{n^2 - \ell(\ell + 1)},$$ (2.25)

the plus and minus signs corresponding to r_o and r_i, respectively. Since the angular part of the wavefunction is normalized, Eq. (2.24) reduces to

$$\int_{r_i}^{r_o} \rho^2(r) dr = 1.$$ (2.26)

Explicitly, Eq. (2.26) can be written as

$$N^2 \int_{r_i}^{r_o} \frac{\sin^2 \int_{r_i}^r k dr' dr}{\sqrt{2W + \dfrac{2}{r} - \dfrac{\ell(\ell + 1)}{r^2}}} = 1.$$ (2.27)

Approximating $\sin^2 \int_{r_i}^r k dr'$ by its average value 1/2 reduces the integral over r to one which is easily done, with the result

$$N^2 = \frac{2}{\pi n^3}$$ (2.28a)

or

$$N = \sqrt{\frac{2}{\pi n^3}}.$$ (2.28b)

The normalization constant decreases as $1/n^{3/2}$. At small r, the wavefunction's only dependence on energy is through the normalization constant.

For $W > 0$, the Rydberg electron is no longer in a bound state but in the continuum. In developing a wavefunction for the continuum we realize that the $r \to \infty$ boundary condition has been removed; only the $r \to 0$ boundary condition remains. For H, the $r=0$ boundary condition excludes the g function so

$$\rho(W,\ell,r) = f(W,\ell,r).$$ (2.29)

For Na or any other atom, the $r \to 0$ boundary condition of H is replaced by the requirement that at $r \geq r_0$ the wavefunction is a wave phase shifted from the coulomb wave by τ;

$$\rho(W,\ell,r) = f(W,\ell,r) \cos \tau - g(W,\ell,r) \sin \tau.$$ (2.30)

As $r \to \infty$ the wavefunctions are oscillatory sine and cosine functions, as shown by Eqs. (2.14). For small r the wavefunctions of the continuum are functionally identical to the bound wavefunctions, differing only in their normalization. Since continuum waves extend to $r = \infty$ they cannot be normalized in the same way as a bound state wavefunction. We shall normalize the continuum waves per unit energy. This is one of several options, and we shall return to the reasons for this choice. If we again take advantage of the fact that the angular part of the wavefunction is normalized, the continuum normalization requirement is[2]

$$\int_0^\infty \rho_W^*(r)\mathrm{d}r \int_{W-\Delta W}^{W+\Delta W} \rho_{W'}(r)\,\mathrm{d}W' = 1. \tag{2.31}$$

It is easier to work out the normalization integral for normalization per unit wavenumber[2]

$$\int_0^\infty \rho_k^*(r)\mathrm{d}r \int_{k-\Delta k}^{k+\Delta k} \rho_{k'}(r)\mathrm{d}k' = 1. \tag{2.32}$$

The normalization integral is dominated by the contribution from $r \rightarrow \infty$ where k is a constant. To determine the normalization we use the WKB form of a continuum wave, also given by Eq. (2.21). If we take advantage of the fact that the dominant contribution to the integral comes from large r, where k is independent of r, we can express Eq. (2.32) as

$$N_k^2 \int_0^\infty \frac{\sin kr}{\sqrt{k}}\,\mathrm{d}r \int_{k-\Delta k}^{k+\Delta k} \frac{\sin k'r}{\sqrt{k'}}\,\mathrm{d}k' = 1, \tag{2.33}$$

where we have written the wavefunction normalized per unit wavenumber as

$$\rho_k(r) = N_k \frac{\sin kr}{\sqrt{k}}, \tag{2.34}$$

N_k being the normalization constant to be determined. The equivalent wavefunction normalized per unit energy is

$$\rho_W(r) = N_W \frac{\sin kr}{\sqrt{k}}. \tag{2.35}$$

The integral of Eq. (2.33) is easily reduced to

$$N_k^2 \cdot \frac{\pi}{2k} = 1. \tag{2.36}$$

The energy and wavenumber normalization constants are related by the derivative dW/dk. Explicitly,[2]

$$N_k^2 = N_W^2 \frac{\mathrm{d}W}{\mathrm{d}k} = N_W^2 k. \tag{2.37}$$

Using Eqs. (2.36) and (2.37) we calculate N_W, yielding the wavefunction normalized per unit energy. Explicitly,

$$\rho_W(r) = \sqrt{\frac{2}{\pi k}} \sin \int_{r_i}^{r} k\,\mathrm{d}r'. \tag{2.38}$$

We note that this wavefunction differs from the bound state wavefunctions by a factor of $n^{-3/2}$. The factor of $n^{-3/2}$ is, in essence, the factor obtained by converting from normalization per unit energy to normalization per unit state. The conversion of the normalization integral requires the factor $dW/dn = d(-1/2n^2)/dn = n^{-3}$. Explicitly,

$$N_n^2 = N_W^2\left(\frac{dW}{dn}\right) = N_W^2 n^{-3}. \tag{2.39}$$

Many of the properties of Rydberg atoms depend on the wavefunction in the $r > r_0$ region where the potential is simply a coulomb potential. In this region we can very easily calculate the wavefunction numerically using the Numerov method, which can be applied to an equation of the form[7]

$$\frac{d^2Y}{dx^2} = g(x)Y(x). \tag{2.40}$$

If x increases in steps of size h, then the basic equation of the Numerov method is[7]

$$[1 - T(x + h)]Y(x + h) + [1-T(x - h)]Y(x - h) = [2 + 10T(x)]Y(x) + O(h^6), \tag{2.41}$$

where $T(x) = h^2 g(x)/12$. If we ignore the terms of order h^6 and if we know $Y(x)$ and $Y(x - h)$ we can calculate $Y(x + h)$, since everything else is known.

Zimmerman *et al.*[8] calculated bound state radial functions and matrix elements starting from Eq. (2.9). To remove the first derivative they made the substitutions

$$x = \ell n\,(r) \tag{2.42a}$$

and

$$Y(x) = R\sqrt{r}, \tag{2.42b}$$

yielding a radial equation in the form of Eq. (2.40) with

$$g(x) = 2e^{2x}\left(-\frac{1}{r} - W\right) + \left(\ell + \frac{1}{2}\right)^2. \tag{2.43}$$

In addition to transforming the radial equation to the Numerov form, the transformation also changes the scale to one more nearly matching the frequency of the oscillations of the radial function. For bound functions, the wavelength of the wavefunction increases with r, and the substitution $x = \ell n(r)$ keeps the number of points per lobe closer to being constant.

We start in the classically forbidden region, $r > 2n^2$, and decrease r to r ~ 0. We follow this procedure since we know that as $r \to \infty$ the physically interesting solution to the coulomb equation is a decaying exponential. Successive values of r are given by $r_j = r_s e^{-jh}$, where r_s is the starting radius (typically $r_s = 2n(n + 15)$) and h is the logarithmic step size, which is usually chosen to be 0.01. As the index j is increased r decreases. As can be seen from Eq. (2.41), we need to specify the values of Y at two adjacent points, corresponding to $j = 0$ and 1. At $r = r_s$ the wavefunction is a factor of 10^{10} smaller than at the classical turning point, so a good estimate for $Y(0)$, corresponding to $j = 0$, is 10^{-10}. The value of $Y(h)$, corresponding to $j = 1$, must only be larger than the value of $Y(0)$ to approximate the decaying exponential. The orthogonal, nonphysical solution to the coulomb equation diverges as $r \to \infty$. Thus as we decrease r, toward $r = 0$, the wrong solution is exponentially damped. The wavefunction can be calculated to the inner turning point or to the radius of the ion core whichever is encountered sooner. The calculation must be stopped at this point, since for all nonzero quantum defects the g part of the wave function will begin to exhibit its small r, $r^{-\ell(\ell+1)}$ behavior and diverge. The above procedure generates a wavefunction which is not normalized. Its normalization integral is given by[8]

$$N^2 = \int_0^\infty r^2 R^2(r) \mathrm{d}r = \sum_k Y_k^2 r_k^2. \tag{2.44}$$

With this normalization integral it is possible to generate either a normalized wavefunction, by converting $Y(x)$ back to $R(r)$, or matrix elements between two bound wavefunctions. The Numerov technique described above assumes that the outermost lobe is positive, which gives wavefunctions for $n - \ell$ even with the signs reversed from the usual convention that the functional form at small r is independent of n.

Matrix elements of r^σ between two wavefunctions, i.e. $\langle 1|r^\sigma|2 \rangle$, may be calculated by carrying out the Numerov procedure for two radial wavefunctions at the same time. The procedure must be started at the appropriate starting point for the higher energy of the two wavefunctions. For concreteness, let us say $|1\rangle$ has the higher energy. The normalization integral $N_1^2 = \Sigma Y_{1k}^2 r_k^2$ is accumulated until the appropriate starting point for the second wavefunction is reached. At this point we begin to accumulate the second normalization integral $N_2^2 = \Sigma Y_{2k}^2 r_k^2$ as well as the matrix element $R_{12}^\sigma = \Sigma Y_{1k} Y_{2k} r_k^{2+\sigma}$. The matrix element is thus given by[8]

$$\langle 1 | r^\sigma | 2 \rangle = \frac{R_{12}^\sigma}{N_1 N_2} = \frac{\Sigma Y_{1k} Y_{2k} r_k^{2+\sigma}}{(\Sigma Y_{1k}^2 r_k^2)^{1/2} (\Sigma Y_{2k}^2 r_k^2)^{1/2}}. \tag{2.45}$$

An alternative, physically appealing way of calculating the bound wavefunctions was used by Bhatti *et al.*[9] In the classically allowed region, the kinetic energy of the electron is given approximately by $1/r$, thus the wavelength of the wavefunction varies as $\sqrt{r}$. If we make the substitution $x = \sqrt{r}$ and increment x by a uniform step size h, we should always have nearly the same number of steps in each lobe of the wavefunction. Making the substitutions

$$x = \sqrt{r} \tag{2.46a}$$

and

$$Y(x) = r^{3/4} R(r) \tag{2.46b}$$

leads to the following form of the radial equation

$$\left[-\frac{\mathrm{d}^2}{\mathrm{d}x^2} - 8Wx^2 + \frac{\left(2\ell + \frac{1}{2}\right)\left(2\ell + \frac{3}{2}\right)}{x^2} \right] Y(x) = 8Y(x). \tag{2.47}$$

In this case $g(x)$ is given by

$$g(x) = -8 - 8Wx^2 + \frac{\left(2\ell + \frac{1}{2}\right)\left(2\ell + \frac{3}{2}\right)}{x^2}, \tag{2.48}$$

and the solution has nearly the same number of points in each lobe of the radial wavefunction.

Using either of the above approaches it is straightforward to calculate bound wavefunctions and bound–bound matrix elements. Occasionally, though, we need continuum functions, which we cannot generate in the same way. As $r \rightarrow \infty$ the physically significant linear combination of f and g waves, i.e. $f \cos \tau - g \sin \tau$ is not a decaying exponential, but a sinusoidally varying wave phase shifted from the hydrogenic f wave by τ. Therefore the easiest way to calculate these wavefunctions is to begin at the origin with an approximation to the regular f function, which is proportional to $r^{\ell+1}$, and calculate the wavefunction out to a point where the coulomb potential is negligible compared to the kinetic energy. At this point the wavefunction is simply a sine wave of constant amplitude. Introducing a phase shift of τ towards the origin produces the desired coulomb wave at large r which may then be numerically propagated to small r as r is decreased to zero.[10] The wavefunction is normalized by requiring that the amplitude of the sine wave be $\sqrt{\pi/2k}$. Explicitly, we start with the radial equation for ρ, Eq. (2.10), which is already in the Numerov form. The Numerov substitutions for Eq. (2.40) which give the same number of points per lobe as $r \rightarrow \infty$ are

$$x = r \tag{2.49a}$$

and

$$Y(x) = \rho, \tag{2.49b}$$

yielding

$$g(x) = -\left[2W + \frac{2}{r} - \ell(\ell + 1)\right]. \tag{2.50}$$

Using the procedures outlined above we may calculate bound and continuum wavefunctions as well as matrix elements of r^σ, for $\sigma \geq 0$. These wavefunctions are often called coulomb wavefunctions, and properties calculated using them are said to be obtained in the coulomb approximation. In addition, we can calculate matrix elements of inverse powers of r for H. We cannot calculate with confidence matrix elements of inverse powers of r for anything but H since the inverse r matrix elements weight $r \sim 0$ heavily and the results can be highly dependent on the radius at which we truncate the sums of Eq. (2.45).

From their wavefunctions, we can infer some of the properties of Rydberg atoms. We begin by estimating the expectation values of r^σ where σ is a positive or negative integer. The expectation values of r^σ, $\sigma > 0$, are determined mostly by the location of the outer classical turning point, $r = 2n^2$. Since the electron spends most of its time there, a good estimate for the expectation values of r^σ, $\sigma > 0$, is

Table 2.3. *Expectation values of* r^σ *for H.*[a]

$$\langle r \rangle = \frac{1}{2}[3n^2 - \ell(\ell + 1)]$$

$$\langle r^2 \rangle = \frac{n^2}{2}[5n^2 + 1 - 3\,\ell(\ell + 1)]$$

$$\langle 1/r \rangle = 1/n^2$$

$$\langle 1/r^2 \rangle = \frac{1}{n^3\,(\ell + 1/2)}$$

$$\langle 1/r^3 \rangle = \frac{1}{n^3\,(\ell + 1)(\ell + 1/2)\ell}$$

$$\langle 1/r^4 \rangle = \frac{3n^2 - \ell(\ell + 1)}{2n^5(\ell + 3/2)(\ell + 1)(\ell + 1/2)\ell(\ell - 1/2)}$$

$$\langle 1/r^6 \rangle = \frac{35\,n^4 - 5n^2[6\ell(\ell + 1) - 5] + 3(\ell + 2)(\ell + 1)\ell(\ell - 1)}{8n^7(\ell + 5/2)(\ell + 2)(\ell + 3/2)(\ell + 1)(\ell + 1/2)\ell(\ell - 1/2)(\ell - 1)(\ell - 3/2)}$$

[a](from refs. 2, 11)

Table 2.4. *Properties of Rydberg atoms.*[a]

Property	n dependence	Na(10d)
Binding energy	n^{-2}	0.14 eV
Energy between adjacent n states	n^{-3}	0.023 eV
Orbital radius	n^2	147 a_0
Geometric cross section	n^4	68 000 a_0^2
Dipole moment $\langle nd|er|nf \rangle$	n^2	143 ea_0
Polarizability	n^7	0.21 MHz cm^2/V^2
Radiative lifetime	n^3	1.0 μs
Fine-structure interval	n^{-3}	-92 MHz

[a](from ref. 1)

$$\langle r^\sigma \rangle \sim n^{2\sigma}. \tag{2.51}$$

If we are interested in inverse powers of r, $\sigma < -1$, it is the behavior of the small r part of the wavefunction which is important, and therefore the precise value of the quantum defect is critical. However, at small r, for $n \gg \ell$, the only energy, or ν, dependence is on the normalization, and thus

$$\langle r^{-\sigma} \rangle \propto n^{-3} \tag{2.52}$$

While the n^{-3} scaling of the expectation values of inverse powers of r is evident, the precise values are not. For future use we give in Table 2.3 expectation values for H of $\langle r^\sigma \rangle$ which have been determined analytically.[2,11]

Using combinations of radial matrix elements and energy separations it is possible to deduce the n scaling of many properties of Rydberg atoms. For example, the polarizability is proportional to the sum of squares of electric dipole matrix elements divided by energy denominators, and it is dominated by the contributions from a few nearby levels. The dipole matrix elements between neighboring levels scale as the orbital radius, as n^2, while the energy differences scale as n^{-3}, yielding polarizabilities that scale as n^7. Analogous reasoning allows us to develop scaling laws for many of the properties of Rydberg atoms. Table 2.4 is a short list of representative properties which shows several ways in which Rydberg atoms differ substantially from ground state atoms.

References

1. T. F. Gallagher, *Rep. Prog. Phys.* **51**, 143 (1988).
2. H. A. Bethe and E. A. Salpeter, *Quantum Mechanics of One and Two Electron Atoms* (Academic Press, New York, 1957).
3. M. Abromowitz and I. A. Stegun, *Handbook of Mathematical Functions* (Dover, New York 1972).
4. U. Fano *Phys. Rev. A* **2**, 353 (1970).
5. M. J. Seaton, *Rep. Prog. Phys.* **46**, 167 (1983).
6. E. Merzbacher, *Quantum Mechanics* (Wiley, New York, 1961).
7. J. M. Blatt, *J. Comput. Phys.* **1**, 382 (1967).
8. M. L. Zimmerman, M. G. Littman, M. M. Kash, and D. Kleppner, *Phys. Rev. A* **20**, 2251 (1979).
9. S. A. Bhatti, C. L. Cromer, and W. E. Cooke, *Phys. Rev. A* **24**, 161 (1981).
10. W. P. Spencer, A. G. Vaidyanathan, D. Kleppner, and T. W. Ducas, *Phys. Rev. A* **26**, 1490 (1982).
11. B. Edlen, in *Handbuch der Physik* (Springer-Verlag, Berlin, 1964).

3
Production of Rydberg atoms

The original method of studying Rydberg atoms was by absorption spectroscopy, and it remains a generally useful technique. For example, the quasi-Landau resonances were first observed by Garton and Tomkins using absorption spectroscopy.[1] Many variants have been developed in which a dye laser is used to populate an excited state preferentially, allowing the absorption spectrum of the atoms in the excited state to be recorded.[2–4] However, most of the modern work on Rydberg atoms has been carried out using methods in which the Rydberg atoms themselves are detected. Rydberg atoms may be detected in two general ways. First, since they are optically excited, they decay radiatively to low lying states, emitting visible fluorescence, which is conveniently detected using photomultiplier tubes. Second, in a Rydberg atom the valence electron is in a large, weakly bound orbit, and as a result it is very easy to ionize the atom by either collisional ionization or field ionization, after which either the ion or the electron produced is easily detected.

Excitation to the Rydberg states

Rydberg atoms have been produced by charge exchange, electron impact, and photoexcitation, the processes

$$A^+ + B \rightarrow A\ n\ell + B^+ \tag{3.1a}$$

$$e^- + A \rightarrow A\ n\ell + e^- \tag{3.1b}$$

$$h\nu + A \rightarrow A\ n\ell, \tag{3.1c}$$

as well as by techniques in which both collisional and optical excitation are used. Although combining collisional and optical excitation has proven to be very effective, we shall for the moment consider the three processes of Eqs. (3.1) separately. Charge exchange has been used to convert ion beams to fast beams of Rydberg atoms, and many of the techniques are described by Koch.[5] Thermal beams of Rydberg atoms have been made by electron impact or optical excitation of beams of ground state atoms, and many of the relevant considerations are described by Ramsey.[6] It is also possible to produce thermal Rydberg atoms in cells. Cells of glass or quartz are useful for containing permanent gases or low densities ($\sim 10^{-5}$ Torr) of alkali atoms, and cells similar to optical pumping cells described by Happer[7] have often been used in fluorescence experiments. To contain higher densities ($\sim$ 1 Torr) of alkali or alkaline earth atoms heat pipe ovens, as described by Vidal and Cooper,[8] are often used.

The cross sections for all three of the processes of Eqs. (3.1) scale as n^{-3}, which may be easily understood. Consider, for example, electron impact excitation of He. If an electron of energy W_0 impinges upon a ground state He atom it has a total cross section, $\sigma_I(W_0)$, for exciting the He atom from the ground state. Subsequent to excitation the final state of the He atom is an excited He atom or a He^+ ion and a free electron. The probability of these outcomes is represented by $d\sigma_I(W_0)/dW$, where W is the energy of the electron ejected from the He atom. $W < 0$ corresponds to the "ejected" electron remaining in a bound state of He, and $W > 0$ corresponds to ionization of the He. In general, $d\sigma_I/dW$ is a smooth function of W. More important, there is no fundamental difference between $W > 0$ and $W < 0$. The collision process occurs when the electrons are within a short range, <10Å, of the He^+ core, and at these distances there is almost no difference between the kinetic energies of a Rydberg electron and a free electron. As a result the differential cross section $d\sigma_I/dW$ passes smoothly from $W > 0$ to $W < 0$.

In general σ_I depends on the energy, W_0, of the incident electron. However, once W_0 is fixed it is a good approximation to assume that $d\sigma_I(W_0)/dW$ is constant for $W \approx 0$ and given by

$$\frac{d\sigma_I(W_0)}{dW} = \sigma_I'(W_0). \tag{3.2}$$

Below the limit, $W < 0$, the energy is no longer a continuous variable, and it is more useful to write the cross section per principal quantum number

$$\sigma(n, W_0) = \frac{d\sigma_I(W_0)}{dn} = \frac{d\sigma_I(W_0)}{dW} \cdot \frac{dW}{dn} = \frac{\sigma_I'(W_0)}{n^3}, \tag{3.3}$$

showing the n^{-3} dependence of the cross section for the production of Rydberg states.

Another way of seeing the n scaling is the following. Prior to the collision the He atom is compact, having a radius of approximately 1 Å. Immediately after the collision the Rydberg electron is in its large orbit, which has a density at the origin scaling as n^{-3}. Thus when the initial state is projected onto the final state in the excitation process, it is hardly surprising that the cross section has an n^{-3} dependence.

Similar arguments apply to charge exchange and photoexcitation, and the basic result is the same; the cross section for the production of Rydberg atoms is the continuation below the limit of the ionization cross section, leading to an n^{-3} dependence of the excitation cross section.

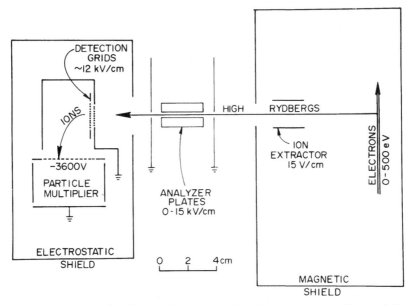

Fig. 3.1 Apparatus for electron beam excitation of rare gas atoms (from ref. 9).

Electron impact

Schiavone *et al.* have measured the cross sections for the production of Rydberg states of rare gases using the apparatus shown in Fig. 3.1.[9,10] Rydberg states of rare gas atoms are formed by electron impact by a pulsed electron beam in the source filled with 10^{-5} Torr of rare gas. Some of the Rydberg atoms formed travel towards the detector, which ionizes all atoms which can be ionized by a 12 kV/cm field and detects the resulting ions. The distribution of Rydberg states reaching the detector is affected by several processes, the excitation from the ground state, ℓ changing collisions produced by the electron beam, and radiative decay of the Rydberg atoms during the flight from the exciting electron beam to the detector. To extract the excitation cross section they first measured the number of Rydberg atoms reaching the detector as a function of ionizing field by varying the voltage on the analyzer plates shown in Fig. 3.1. The observed Rydberg atom signals as a function of field were converted to n distributions by making several assumptions. First they assumed that the excitation cross section scales as n^{-3}. Second, they assumed the radiative lifetime to be given by $\tau_1 n^3$, τ_1 being a constant. Finally, ionization was assumed to occur at the classical field,

$$E = 1/16n^4. \tag{3.4}$$

Using these assumptions they could then fit their data to determine excitation cross sections, $\sigma(n, W_0)$, for the excitation of a Rydberg state of principal quantum number n by an electron of energy W_0. They found the values of $\sigma_1'(W_0)$ given in

Table 3.1. *Electron impact cross sections
for the excitation of rare gas Rydberg
states for the electron energy at the peak
of the cross section and 100 eV.*[a]

Rare gas	Electron energy W_0 (eV)	$\sigma'_1(W_0)$ (Å^2)
He	70	0.77
He	100	0.67
Ne	60	0.63
Ne	100	0.61
Ar	28	6.5
Ar	100	1.5
Kr	20	4.0
Kr	100	2.0
Xe	20	10.0
Xe	100	4.6

[a](from ref. 10).

Table 3.1 for the rare gas atoms.[10] Our $\sigma'_1(W_0)$ corresponds to $\sigma^{ex}(n = 1, E)$ in the notation of Schiavone *et al.*[10]. For a typical value of $\sigma'(W_0) = 1$ Å, the cross section $\sigma(20, W_0)$ for exciting an $n = 20$ state is 1.25×10^{-4} Å^2.

An interesting aspect of electron impact excitation noted by Schiavone *et al.*[10] is that the electrons play two roles. First, they excite the ground state atoms to Rydberg states. Second, they collisionally redistribute the atoms initially excited to low ℓ states over all ℓ states, including the long lived high ℓ states. The production of high ℓ states allows the detection of n states which would normally not live long enough to reach the detector. The cross section for ℓ changing collisions by electrons is so large, ˜10^6 Å^2 , that only at low currents does the observed signal depend on the electron current in an obviously nonlinear way. At higher electron currents the ℓ mixing is so saturated that the signal is linear in the electron current. Electron impact excitation has also been used in a long series of experiments by Wing and MacAdam[11] to measure the $\Delta\ell$ intervals of He Rydberg states by radio frequency spectroscopy. The fact that there are observable population differences makes it clear that electron collisions do not produce a thermal population distribution on the time scale of the radiative decay of the Rydberg states in these experiments.

In sum, electron impact excitation has the advantages of relative simplicity and generality, and it has the disadvantages of being inefficient and nonselective, with nearly all energetically possible states being produced.

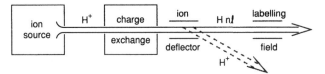

Fig. 3.2 The fast beam approach of Bayfield and Koch (ref. 13). H^+ ions of roughly 10 keV energy pass through a charge exchange cell forming a fast beam of H Rydberg atoms. Down-stream from the charge exchange cell the ions are deflected from the beam and a band of n states is selected by a square wave modulated ionization field.

Charge exchange

Charge exchange of H^+ and He^+ with various target gases has been used to produce fast beams of Rydberg atoms since the first experiments of Riviere and Sweetman.[12] The method of producing fast Rydberg H atoms used by Bayfield and Koch[13] in their initial experiment to study the microwave ionization of Rydberg atoms is shown schematically in Fig. 3.2. The 10 keV protons from an ion source pass through a charge exchange cell filled with Xe, where some of the protons are converted to fast Rydberg atoms.

The distribution of product states of charge exchange of ions with incident energy W_0 is described by $d\sigma_L(W_0)/dW$, where W is the energy of the Rydberg electron in the neutral product atom and $\sigma_L(W_0)$ is the total electron loss cross section. As in Eq. (3.3), we can write the charge exchange cross section for the population of a specific n state as

$$\sigma_{CE}(n, W_0) = \frac{\sigma'_L(W_0)}{n^3}, \tag{3.5}$$

where $\sigma'_L(W_0) = d\sigma_L(W_0)/dW|_{W=0}$.

In Table 3.2 we give representative values of $\sigma'_L(W_0)$.[14–16] Only a small fraction of the incident ions which undergo collisions with the gas in the charge exchange cell will be left in any particular Rydberg state. For $n = 10$ the fraction is $<10^{-3}$, and for $n = 30$ it is $<10^{-4}$. If the number density of the charge exchange gas is N and the length of the cell is L, in principle the charge exchange cell can be operated with $NL\sigma_L \sim 1$, so that nearly all the ions in the beam undergo charge exchange. In fact it is usually not possible to use such a thick target since it will scatter the fast neutral atoms into large enough angles to remove them from the beam. Since the cross section for scattering atoms out of the beam is $\sim 1 \text{ Å}^2$, the cell must be operated at a density length product low enough that the Rydberg atoms produced are not scattered out of the beam.

The Rydberg states produced in the cell by charge exchange can be collisionally depopulated by subsequent collisions with the target gas. In the charge exchange collision, it is presumably low ℓ states which are populated, by virtue of the overlap of their wavefunctions with the ground state wavefunction at the origin. In

Table 3.2. *Values of the charge exchange cross section for protons on several vapor targets.*

Target	Incident energy (keV)	$\sigma'_L(W_0)$ (Å^2)
Xe[a]	20	10
Kr[a]	20	9
Ar[a]	20	4
Ar[b]	30	3.3
He[a]	20	0.4
He[c]	60	0.6
H$_2$[c]	60	1.1
N$_2$[c]	60	2.5
CO$_2$[c]	60	3.8

[a] (see ref. 14)
[b] (see ref. 15)
[c] (see ref. 16)

thermal collisions it has been observed that the cross sections for ℓ and m changing collisions are large,[17,18] with values for Xe reaching 10^5 Å^2 at $n = 20$, and minimum values >20 Å^2. These cross sections are far larger than σ_L, so it is likely that efficient charge exchange is accompanied by collisions which populate high ℓ and m states. Thermal energy cross sections for changing n are smaller than the ℓ changing cross sections, with typical values of ~ 10 Å^2. Since virtually all n states are populated, some further redistribution over n has no significant effect.

In sum it is reasonable to expect a distribution of $n\ell m$ states from charge exchange, with n^{-3} scaling of the population in each n. Such a distribution of states must be filtered in some way before it is useful in an experiment. A good way is the method used by Bayfield and Koch and shown in Fig. 3.2.[13] The Rydberg atoms pass between two field plates, which produce a modulated field to select a band of n states by field ionization. The field required to ionize a state of principal quantum number n is given by an expression similar to Eq. (3.4). If the field is switched between E_1 and E_2 with $E_2 < E_1$, atoms in states of $n > n_1$ or n_2 are ionized by the field. Thus the difference in signal obtained with $E = E_1$ and $E = E_2$ must be due to atoms with $n_1 \leq n \leq n_2$. Using a field switched between 28.5 and 41.0 V/cm, Bayfield and Koch selected the band $63 \leq n \leq 69$ for the first experiments with microwave ionization.[13]

Optical excitation

Optical excitation differs from collisional excitation in a fundamental way; the exciting photon is absorbed by the target atom. As a result, specifying the energy

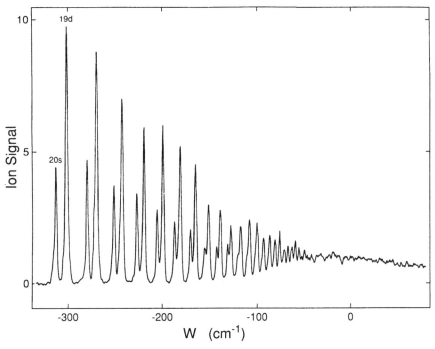

Fig. 3.3 Excitation spectrum of the Na *n*s and *n*d states from the Na $3p_{3/2}$ state (from ref. 19).

of the absorbed photon specifies the Rydberg state produced. In contrast, specifying the energy of an incoming electron does not specify the energy of the Rydberg state produced because there is no way to control how the energy is shared between the incident electron and the electron which is excited to the Rydberg state.

In spite of the above difference from collisional excitation, photoexcitation has the similarity of having a cross section with an n^{-3} dependence. This cross section is simply the continuation of the photoionization cross section from above the limit, $W > 0$, to below the limit, $W < 0$. This point is demonstrated experimentally by Fig. 3.3, the excitation spectrum of the Na *n*s and *n*d states from the $3p_{3/2}$ state.[19] The signal, or the cross section, is apparently the same just above and just below the limit, in spite of the fact that the electron is photoionized in the former case and excited to a bound Rydberg state in the latter. Only when $n < 40$ is the instrumental resolution adequate for resolved Rydberg series to appear in the spectrum, and for $n < 30$ the *n*s and *n*d Rydberg states are clearly resolved and increase in intensity with decreasing n. A useful way to describe photoexcitation is to specify σ_{PI}, the photoexcitation cross section in the vicinity of the ionization limit, $W = 0$. Above the limit the photoionization cross section is given by σ_{PI}. Below the limit, the cross section averaged over an integral number of n states equals σ_{PI}. The cross section $\sigma(n)$ for exciting a resolved Rydberg state is given by

Table 3.3. *Photoionization cross sections*
at threshold (W = 0).[a]

Atom	Wavelength (Å)	σ_{PI} (10^{-18} cm^2)
H	912	6.3
Li	2299	1.8
Na	2412	0.125
K	2856	0.007
Rb	2968	0.10
Cs	3184	0.20
Mg	1621	1.18
Ca	2028	0.45
Sr	2177	3.6

[a](from ref. 20)

$$\sigma(n) = \frac{\sigma_{PI}}{\Delta W n^3}, \tag{3.6}$$

where ΔW is the energy resolution of the excitation. The cross section for exciting the Rydberg state is the photoionization cross section times the ratio of the Δn energy interval to the experimental resolution.

In most optical excitations the resolution is determined by the Doppler effect or the finite linewidth of the light source. The Doppler effect gives a typical frequency width of 1 GHz, and the width of the light source can be anywhere from 1 kHz to 30 GHz. We assume that these widths are larger than the radiative width. The photoionization cross sections from the ground states of H, alkali, and the alkaline earth atoms are given in Table 3.3. [20]

The most widely used method of optically exciting Rydberg atoms is to use a pulsed dye laser pumped by a N_2 or Nd:YAG laser. It is straightforward to generate visible laser pulses 5 ns long with $100\,\mu J$ energies and 1 cm^{-1} bandwidths. Such a pulse contains $\sim 3 \times 10^{14}$ photons, and, if collimated into a beam of cross sectional area 10^{-2} cm^2, it has an integrated photon flux of 10^{16} cm^{-2}. At $n = 20$ the Δn spacing is ~ 28 cm^{-1}, and with a resolution of 1 cm^{-1}, the cross section of Eq. (3.6) is a factor of 28 larger than the photoionization cross section. Using 10^{-18} cm^2 as the photoionization cross section we find a probability of 10% for exciting atoms exposed to a photon flux of 10^{16} cm^{-2} to the $n = 20$ state.

A typical example of the pulsed dye laser excitation is the beam experiment shown in Fig. 3.4 in which Na atoms in a thermal beam are excited in two steps from the ground state 3s to the 3p state with a yellow dye laser photon and from the 3p state to a high lying ns or nd state with a second, blue photon.[21]

Using continuous wave (cw) laser excitation it is possible to excite atoms with substantially higher efficiency than using pulsed lasers. For example a single mode laser of 1 MHz linewidth has a resolution 3×10^4 better than the pulsed laser

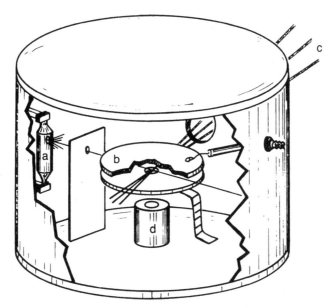

Fig. 3.4 Apparatus for the study of Rydberg states of alkali metal atoms: a, the atomic beam source; b, the electric field plates; c, the pulsed laser beams; and d, the electron multiplier (from ref. 21).

described above. Accordingly, for single photon excitation the resolution limited cross section can be 3×10^4 times larger if the excitation is done so as to avoid Doppler broadening. Examples of such an excitation are the experiments of Zollars *et al.*[22] using a frequency doubled cw dye laser to excite a beam of Rb ground state atoms to the *np* states and the experiments of Fabre *et al.*[23] in which they used a cw dye laser and diode lasers to excite Na nf states.

Collisional–optical excitation

Purely optical excitation is possible for alkali and alkaline earth atoms. For most other atoms the transition from the ground state to any other level is at too short a wavelength to be useful. To produce Rydberg states of such atoms a combination of collisional and optical excitation is quite effective. A good example is the study of the Rydberg states of Xe by Stebbings *et al.*[24] As shown in Fig. 3.5, a thermal beam of Xe atoms is excited by electron impact, and a reasonable fraction of the excited atoms is left in the metastable state. Downstream from the electron excitation the atoms in the metastable state are excited to a Rydberg state by pulsed dye laser excitation.

   Metastable atoms have also been used as the starting point for laser excitation in cell experiments. Devos *et al.*[25] used metastable He atoms in the stationary

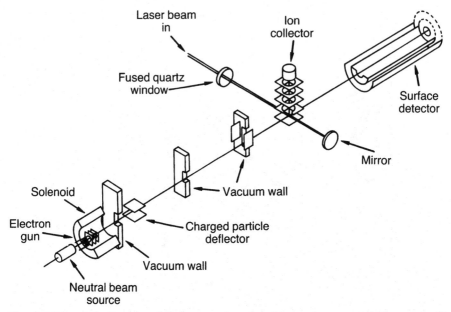

Fig. 3.5 Schematic diagram of the apparatus using electron impact excitation to the Xe
metastable state followed by laser excitation to Rydberg states (from ref. 24).

afterglow of a pulsed discharge as the starting point for pulsed laser excitation to
high lying np states. They observed the time and wavelength resolved fluorescence to determine the collision rate constants of the np states. An example in
which selective detection of the Rydberg atoms is not required is the optogalvanic
spectroscopy of Ba Rydberg states of Camus *et al.*[26] They optically excited the
atoms from metastable levels produced in a discharge to high lying Rydberg
states. The Rydberg atoms are collisionally ionized and the resulting ions and
electrons alter the discharge current, a readily detectable signature of the
excitation of a Rydberg state.

In fast beams optical excitation has proven to be most useful. Since the fast
beams are low in intensity, but continuous, cw lasers have been used. Usually,
fixed frequency lasers have been used since fine tuning can be done using the Stark
shift or the Doppler shift of the fast beam. The Doppler shift can be used either by
changing the angle at which the laser beam and fast beam cross, or by altering the
velocity of the fast beam. An early example was the use of the uv line of an Ar laser
to drive transitions from the metastable H 2s state to the $40 < n < 55$ np states.[27] In
this particular case the velocity of the beam was changed to tune different np states
into resonance.

The most commonly used technique has been to use a CO_2 laser which has
about 20 lines at intervals of 1.3 cm^{-1} in each of the 9.6 and 10.6 μm bands. These
lines match the transition frequencies of $n = 10$ to $n \sim 30$ transitions,[28] and allow
efficient selective population of single states of n $\sim$ 30. The essential notion is to

populate the Rydberg levels of the fast beam by charge exchange, remove all atoms in $n > 10$ states by field ionization, and then repopulate the selected level of $n \approx 30$ using the CO_2 laser. This approach, or a variant using a second CO_2 laser to drive $n = 7$ to $n = 10$ transitions, has become a standard approach in the study of Rydberg atoms using fast beams.[28]

References

1. W. R. S. Garton and F. S. Tomkins, *Astrophys. J.* **158**, 839 (1969).
2. D. J. Bradley, P. Ewart, J. V. Nicholas, and J. R. D. Shaw, *J. Phys.B* **6**, 1594, (1973).
3. J. R. Rubbmark, S. A. Borgstrom, and K. Bockasten, *J Phys. B* **10**, 421 (1977).
4. E. Amaldi and E. Segre, *Nuovo Cimento* **11**, 145 (1943).
5. P. M. Koch, in *Rydberg States of Atoms and Molecules*, eds R.F. Stebbings and F.B. Dunning (Cambridge University. Press, Cambridge, 1983).
6. N. F. Ramsey, *Molecular Beams* (Oxford University Press, London, 1956).
7. W. Happer, *Rev. Mod. Phys.* **44**, 169 (1972).
8. C. R. Vidal and J. Cooper, *J. Appl. Phys.* **40**, 3370 (1969).
9. J. A. Schiavone, D. E. Donohue, D. R. Herrick, and R. S. Freund, *Phys. Rev. A* **16**, 48 (1977).
10. J. A. Schiavone, S. M. Tarr, and R. S. Freund, *Phys. Rev. A* **20**, 71 (1979).
11. W. A. Wing and K. B. Mac Adam, in *Progress in Atomic Spectroscopy A*, eds W. Hanle and H. Kleinpoppen (Plenum, New York, 1978).
12. A. C. Riviere and D. S. Sweetman, in *Atomic Collision Processes*, ed. M. R. C. McDowell (North Holland, Amsterdam, 1964).
13. J. E. Bayfield and P. M. Koch, *Phys. Rev. Lett.* **33**, 258 (1974).
14. R. F. King and C. J. Latimer, *J. Phys B.* **12**, 1477 (1979).
15. J. E. Bayfield, G. A. Khayrallah, and P. M. Koch, *Phys. Rev. A* **9**, 209(1974).
16. R. N. Iℓ' in, B. Kikiani, V. A. Oparin, E. S. Solov'ev, and N. V. Fedorenko, *Sov. Phys. JETP* **20**, 835(1965). [*J. Exptl. Theor. Phys. USSR* **47**, 1235 (1964)].
17. R. Kachru, T. F. Gallagher, F. Gounand, K. A. Safinya, and W. Sandner, *Phys. Rev. A* **27**, 795 (1983).
18. M. Hugon, F. Gounand, P. R. Fournier, and J. Berlande, *J. Phys. B* **12**, 2707 (1979).
19. W. R. Anderson, Q. Sun, and M. J. Renn private communication.
20. G. V. Marr, *Photoionization Processes in Gases* (Academic Press, New York, 1967).
21. D. Kleppner, M. G. Littman, and M. L. Zimmerman, in *Rydberg States of Atoms and Molecules*, eds R. F. Stebbings and F. B. Dunning (Cambridge University Press, Cambridge, 1983).
22. B. G. Zollars, C. Higgs, F. Lu, C. W. Walter, L. G. Gray, K. A. Smith, F. B. Dunning, and R. F. Stebbings, *Phys. Rev. A* **32**, 3330 (1985).
23. C. Fabre, Y. Kaluzny, R. Calabrese, L. Jun, P. Goy, and S. Haroche, *J. Phys. B* **17**, 3217 (1984).
24. R. F. Stebbings, C. J. Latimer, W. P. West, F. B. Dunning, and T. B. Cook, *Phys. Rev. A* **12**, 1453 (1975).
25. F. Devos, J. Boulmer and J. F. Delpech, *J. Phys. (Paris)* **40**, 215 (1979).
26. P. Camus, M. Dieulin, and C. Morillon, *J. Phys. Lett. (Paris)* **40**, L513 (1979).
27. J. E. Bayfield, L. D. Gardner, and P. M. Koch, *Phys. Rev. Lett.* **39**, 76 (1977).
28. P. M. Koch and D. R. Mariani, *J. Phys. B* **13**, L645 (1980).25

4

Oscillator strengths and lifetimes

In Chapter 3 we considered briefly the photoexcitation of Rydberg atoms, paying particular attention to the continuity of cross sections at the ionization limit. In this chapter we consider optical excitation in more detail. While the general behavior is similar in H and the alkali atoms, there are striking differences in the optical absorption cross sections and in the radiative decay rates. These differences can be traced to the variation in the radial matrix elements produced by nonzero quantum defects. The radiative properties of H are well known, and the radiative properties of alkali atoms can be calculated using quantum defect theory.

Oscillator strengths

A convenient way of expressing the strengths of transitions is to use the oscillator strength. The oscillator strength $f_{n'\ell'm',n\ell m}$ from level $n\ell m$ to level $n'\ell'm'$ is defined by[1]

$$f_{n'\ell'm',n\ell m} = 2\frac{m}{\hbar}\omega_{n'\ell',n\ell}|\langle n'\ell'm\,|x|\,n\ell m\rangle|^2, \qquad (4.1)$$

where $\omega_{n'\ell',n\ell} = (W_{n'\ell'} - W_{n\ell})/\hbar$. We could equally well have used the matrix elements of y or z in Eq. (4.1), but in any case, $f_{n'\ell'm',n\ell m}$ is m dependent. Since it is evident that the radiative decay of an atom in free space cannot depend upon m, it is useful to introduce the average oscillator strength $\bar{f}_{n'\ell',n\ell}$ which does not depend upon m. It may be written as[1]

$$\bar{f}_{n'\ell',n\ell} = \frac{2}{3}\omega_{n'\ell',n\ell}\frac{\ell_{max}}{2\ell+1}|\langle n'\ell'|r|n\ell\rangle|^2, \qquad (4.2)$$

where ℓ_{max} is the larger of ℓ and ℓ'. If we reverse the roles of ℓ and ℓ', it is straightforward to show that

$$\bar{f}_{n'\ell',n\ell} = -\frac{2\ell'+1}{2\ell+1}\bar{f}_{n\ell,n'\ell'}. \qquad (4.3)$$

The usefulness of the oscillator strengths stems in part from the fact that they satisfy several sum rules. The Thomas–Reiche–Kuhn sum rule is given by[1]

$$\sum_{n'\ell'm'} f_{n'\ell'm',n\ell m} = Z, \qquad (4.4)$$

where Z is the number of electrons in the atom and the sum implicitly includes the accessible continua. When applied to Na, this sum rule leads to $\Sigma f_{n'\ell'm',n\ell m} = 11$, which tells us nothing about the oscillator strength of transitions of the outer electron alone. More useful sum rules, valid for one electron in a central potential, are[1]

$$\sum_{n'} \bar{f}_{n'\ell-1,n\ell} = -\frac{1}{3}\frac{\ell(2\ell-1)}{2\ell+1} \tag{4.5}$$

and

$$\sum_{n'} \bar{f}_{n'\ell+1,n\ell} = \frac{1}{3}\frac{(\ell+1)(2\ell+3)}{2\ell+1}, \tag{4.6}$$

from which it is apparent that

$$\sum_{n'\ell'} \bar{f}_{n'\ell',n\ell} = 1, \tag{4.7}$$

where $\ell' = \ell\pm1$. Eqs. (4.5)–(4.7) apply to, for example, the transitions of the valence electron of Na, in spectral regions where the transitions of the Na^+ core are unimportant. Since the lowest lying excited states of Na^+ lie approximately 25 eV above the ground state of Na^+, and 30 eV above the ground state of Na, Eqs. (4.5)–(4.7) can be used to described the transitions of the valence electron of Na due to photons of energy less than 10 eV.

The attractions of the oscillator strengths are that they are dimensionless and they sum to 1. Thus the determination of one oscillator strength is often enough to give a good indication of the overall distribution of oscillator strengths. We shall see that the effect of nonzero quantum defects is to redistribute the alkali oscillator strengths relative to those of H. If we return to Eqs. (4.5) and (4.6) we can see that the strongest $\ell \to \ell$-1 transitions are to lower energies and that the strongest $\ell \to \ell$+1 transitions are to higher energies.

Finally, we introduce the Einstein A coefficient, $A_{n'\ell',n\ell}$, which defines the spontaneous decay rate of the $n\ell$ state to the lower lying $n'\ell'$ state. Explicitly,[1]

$$A_{n'\ell',n\ell} = \frac{4e^2\omega_{n\ell,n'\ell'}^3}{3\hbar c^3}\frac{\ell_{max}}{2\ell+1}|\langle n'\ell'|r|n\ell\rangle|^2. \tag{4.8}$$

In terms of the average oscillator strength we can write the A coefficient as

$$A_{n'\ell',n\ell} = \frac{-2e^2\omega_{n'\ell',n\ell}^2}{\hbar c^3}\bar{f}_{n'\ell',n\ell}. \tag{4.9}$$

The minus sign comes from the fact that $\bar{f}_{n'\ell',n'\ell'} < 0$. The radiative lifetime, $\tau_{n\ell}$, of the $n\ell$ state is simply the inverse of the total radiative decay rate, which is obtained by summing $A_{n'\ell',n\ell}$ over all lower lying $n'\ell'$ states. Explicitly

$$\tau_{nl} = \left[\sum_{n'\ell'} A_{n'\ell',n\ell}\right]^{-1}. \tag{4.10}$$

From Eq. (4.8) we see that the A values contain a factor of ω^3, which generally means that the transition with the highest frequency contributes the most to the radiative decay rate and therefore dominates the overall dependence on n.

For all Rydberg $n\ell$ states save the s states the highest frequency decay transition is to the lowest lying state of $\ell - 1$. For an s state it is the transition to the lowest p state. In either case, for a high $n\ell$ state, as $n \to \infty$ the frequency of the highest frequency transition approaches a constant. Thus, in the limit of high n, the A value depends only on the radial matrix element between the Rydberg state and the low lying state. Only the part of the Rydberg state wavefunction which spatially overlaps the wavefunction of the low lying state contributes to the matrix element in Eq. (4.8). As a result, the squared radial matrix element of Eq. (4.8) exhibits an n^{-3} scaling due to the normalization of the Rydberg state wavefunction. Correspondingly,

$$\tau_{n\ell} \propto n^3. \tag{4.11}$$

For $\ell \approx n$ states the same reasoning does not apply. For the $\ell = n - 1$ state the only allowed transition is to the $n' = n - 1$, $\ell' = n - 2$ state. In this case the frequency of the transition is not constant but is given by $1/n^3$. When cubed, this frequency contributes an n^{-9} scaling to the decay rate. Offsetting this scaling is the fact that the radial matrix element represents the size of the atom in the n and $n - 1$ states. Since $\langle n - 1\, n - 2 | r | n\, n - 1 \rangle \sim n^2$ the lifetimes of the highest ℓ states scale as n^5, i.e.

$$\tau_{n\ell=n-1} \propto n^5. \tag{4.12}$$

The average lifetime for a statistical mixture of ℓm states of the same n is given by[1,2]

$$\bar{\tau}_n \propto n^{4.5}. \tag{4.13}$$

Finally, it is useful to extend the definition of oscillator strength above the ionization limit using

$$\frac{d\bar{f}_{\varepsilon'\ell',n\ell}}{dW} = \frac{2}{3}\, \omega_{\varepsilon'\ell',n\ell}\, \frac{\ell_{max}}{2\ell+1}\, |\langle \varepsilon\ell' | r | n\ell \rangle|^2, \tag{4.14}$$

where the continuum wavefunction is normalized per unit energy.

The photoionization cross section of a statistical mixture of m levels of the $n\ell$ state to the $\varepsilon'\ell'$ continuum is then given by

$$\sigma = \frac{2\pi^2}{c} \cdot \frac{d\bar{f}_{\varepsilon'\ell',n\ell}}{cdW}. \tag{4.15}$$

It is interesting to examine the dependence on n and ℓ of the oscillator strength and lifetimes for the H and alkali atoms. First we show in Fig. 4.1 a plot of the oscillator strengths of the H $1s \to np$ transitions using the format of Fano and Cooper.[3] Above the ionization limit $d\bar{f}/dW$ is plotted directly. Below the ionization limit the oscillator strength to the np level is plotted as a rectangular block of area $\bar{f}_{np,1s}$ from $W = -1/2(n - 1/2)^2$ to $W = -1/2(n + 1/2)^2$, i.e. it is approximately $1/n^3$ wide, $\bar{f}_{np,1s}n^3$ high, and centered at energy $W = -1/2n^2$. Thus the area of each block corresponds to the oscillator strength for the excitation of one np state. Plotting the oscillator strength in this fashion brings out clearly the fact that

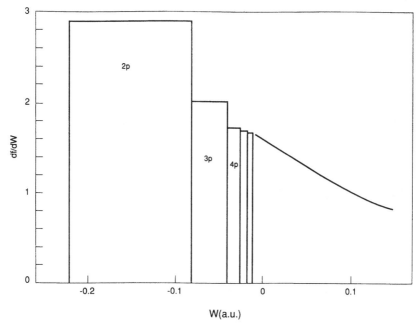

Fig. 4.1 Oscillator strength distribution from the H 1s state to the excited p states. Below the ionization limit the area of each block corresponds to the average oscillator strength from the 1s state to an np state.

oscillator strength per unit energy passes smoothly from the bound to the continuous region of the spectrum. From Fig. 4.1 it is also apparent that at the limit the oscillator strength is slowly varying, and thus $\bar{f}_{np,1s} \propto n^{-3}$ as expected from the normalization of the np wavefunctions near the origin. Fig. 4.1 and direct calculation show that a reasonable fraction, about one half, of the oscillator strength lies above the ionization limit.

In the excited states of H the lifetimes exhibit a predictable behavior. For states of a given n the p states decay most rapidly, followed by the d,f,g,s,h . . . states in that order.[4] Putting aside the s states, the dominant decay mechanism for each $n\ell$ series is to the lowest $n'\ell - 1$ state. The frequency dependence of this transition upon ℓ leads to the observed ℓ ordering of the lifetimes. The ns states have long lifetimes due to the fact that the radial matrix elements to the lower lying p states are small. For the high ns states of H the first two lobes of the ns radial wavefunction overlap the 2p wavefunction, and their contributions cancel to some extent. In contrast, only one lobe, the innermost, of the nd radial wavefunction overlaps the 2p radial wavefunction. The nd wavefunction also has a larger amplitude than the ns wavefunction at the position of the 2p wavefunction because of the centrifugal potential, which ensures that the nd atom has less kinetic energy in radial motion. In general, matrix elements involving decreases in n accompanied by a decrease in ℓ by one are larger than matrix elements involving

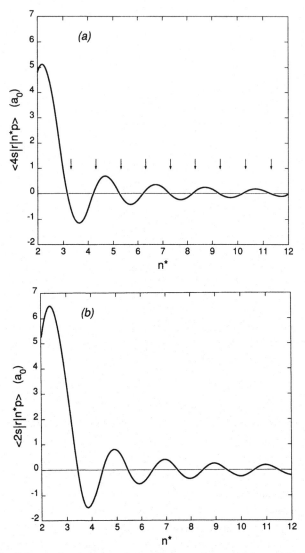

Fig. 4.2 (a) Radial matrix element of the K 4s → n*p transition with n* a continous variable. The quantum defect of the 4s state is 2.23. The quantum defects of the np states are 1.71, so the np states fall at the locations shown by the arrows near where the matrix element crosses zero. (b) H 2s–n*p radial matrix element as a function of n*. Note that the maximum amplitudes of the matrix element occur at integer values of n*.

decreases in n with an increase of ℓ by one. With the exception of the ns states, all states decay predominantly by Δℓ = -1 transitions.[1,4]

Now let us consider alkali atoms. We begin by considering the K 4s–np excitations analogous to the excitation from the ground state of H. In K the situation is in fact quite different, as shown by Fig. 4.2(a), a plot of the radial

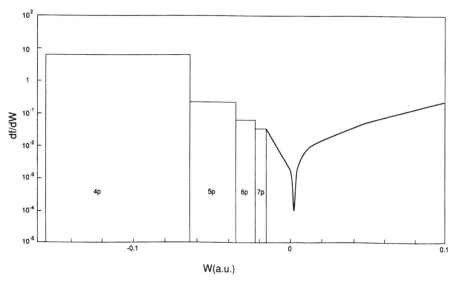

Fig. 4.3 Oscillator strength distribution from the K 4s state to the np states and εp continuum. Note that the vertical scale is logarathmic, not linear as in Fig. 4.1. The oscillator strengths to the high np states are orders of magnitude smaller than in H.

matrix element $\langle n^*p|r|4s \rangle$ for K with n^* a continuous variable. The K 4s state has a quantum defect of 2.23, and the high lying p states have quantum defects of 1.71, so that $n^* = 0.29(\text{mod } 1)$. In Fig. 4.2 we have shown the locations of the np states. It is quite apparent that the 4s → 4p transition has a large matrix element, but the matrix elements to higher n states fall near the zero crossings in the matrix element and are therefore very small. The analogous plot of the matrix element between the H 2s and np states is shown in Fig. 4.2(b). The H $\langle 2p|r|2s \rangle$ matrix element is larger than the K 4s–4p matrix element due to the smaller binding energy of the H 2s state, and all the matrix elements between the 2s and the higher np states are further away from the zero crossings and hence larger than the K $\langle np|r|4s \rangle$ matrix elements.

If we convert the squares of the matrix elements shown in Fig. 4.2 to average oscillator strengths, by multiplying by $2\omega/3$, we find that in H the 2s → 2p transition has vanishing oscillator strength due to the 2s–2p degeneracy. All the oscillator strength must come from the $n \geq 3$ 2s → np and 2s → εp transitions. On the other hand, in K the 4s → 4p transition has a significant oscillator strength since the 4p state lies halfway between the 4s state and the ionization limit. In fact, it has an oscillator strength of very nearly 1, consuming nearly all of the available K 4s–np oscillator strength. In Fig. 4.3 we show a plot of $d\bar{f}/dW$ for K. The values for the bound oscillator strengths and for the continuum more than 0.03 au above the limit are taken from Marr and Creek.[5,6] The values of $d\bar{f}/dW$ near the minimum just above the limit are taken from the relative measurements of Sandner *et al.*[7] normalized to the bound state oscillator strengths given by Marr and Creek.[5] In K the variation in $d\bar{f}/dW$ is so great that Fig. 4.3 is plotted on a logarithmic scale. As shown, $d\bar{f}/dW$ declines throughout the bound spectrum reaching a minimum just above the ionization limit. This minimum, usually termed the Cooper minimum, is found in the continua of all the alkalis save Li.

The origin of the minimum is most easily understood by considering a spinless K atom. In a spinless K atom the 4s–np radial matrix elements decrease with increasing energy, as implied by Fig. 4.3. The decrease with increasing energy of the np state occurs because the nodes of the bound np, or continuum εp, wavefunction become increasingly close together with increasing energy. Stated another way, the 4s–np or 4s–εp radial matrix element is an integral of the product of a fixed function, $r^3 R_{4s}(r)$, and a function the wavelength of which decreases with energy, $R_{np}(r)$ or $R_{\varepsilon p}(r)$. In general the resulting integral oscillates about zero as a function of energy. In K the first change in sign is just above the ionization limit, leading to $d\bar{f}/dW = 0$ and a vanishing photoionization cross section. Having neglected spin we predict a zero cross section instead of the nonzero minimum of Fig. 4.3. The energy at which the minimum occurs is determined by the radial phases of the ground state and εp wavefunctions. In Na, K, Rb, and Cs these phases differ by $\sim\pi/2$, since the ground state and np quantum defects differ by 1/2. For these atoms the Cooper minimum occurs in the continuum not far above the ionization limit. In Li the quantum defects of the ground state and the np states differ by 1/4, and $d\bar{f}/dW$ has its minimum below the ionization limit.

As shown by Fig. 4.3, the cross section does not actually vanish, but only decreases to a nonzero minimum. As pointed out by Seaton,[8] when the spin orbit interaction is taken into account there are two εp continuum waves, $\varepsilon p_{1/2}$ and $\varepsilon p_{3/2}$, which have slightly different radial phases corresponding to the slightly different quantum defects of the bound $np_{1/2}$ and $np_{3/2}$ series. As a result of having different phases the $\varepsilon p_{1/2}$ and $\varepsilon p_{3/2}$ radial matrix elements cross zero at slightly different energies. Each of these continuum channels has its own oscillator strength, and the total cross section is never zero. While the location of the minimum is a function of the difference in the quantum defects of the ground state and the excited np states, the minimum cross section and the width of the minimum are proportional to the np fine structure splitting, or the difference in the $np_{1/2}$ and $np_{3/2}$ quantum defects. Consequently the minimum cross section and the width of the Cooper minimum increase from Na to Cs.

Since the radial matrix elements from the ground state to the $np_{1/2}$ and $np_{3/2}$ states are not the same, due to the differing radial phases, the oscillator strengths and intensity ratios of the transitions to the $np_{1/2}$ and $np_{3/2}$ states cannot be in the 1:2 ratio predicted by the angular factors.[8] The largest deviations occur in Cs, for the Cs $6s_{1/2}$–εp$_{1/2}$ matrix element vanishes just above the ionization limit.[9] As a result, just below the ionization limit large intensity ratios have been measured.[10–13] For example, at $n=30$ Raimond *et al.* measured a ratio of 1:1170.[13] The measurements agree with calculations of Norcross[14] which predict that the intensity ratio at $n=\infty$ should be roughly 1:12,000. While the Cooper minimum in K falls much closer to the ionization limit than it does in Cs, both the 4s–$\varepsilon p_{1/2}$ and 4s–$\varepsilon p_{3/2}$ matrix elements cross zero at nearly the same energy, and in the bound states the maximum $np_{1/2}$:$np_{3/2}$ intensity ratio is only 1:4.[15]

Radiative lifetimes

Both the absorption cross sections and the lifetimes of low ℓ states of alkali atoms are different from the hydrogenic values. The most glaring examples occur the alkali np states. In H the np states have very short radiative lifetimes, but in an alkali atom they have very long lifetimes, due to their small oscillator strengths to the ground state. For example, as shown in Figs. 4.1 and 4.3, the oscillator strengths from the ground states of H and K to the high, $n \approx 10$, np states differ by a factor of 100. This factor, when multiplied by the ratio of the squares of the frequencies, 16, leads to K $np \rightarrow$ 4s Einstein A coefficients roughly 1000 times smaller than their H $np \rightarrow$ 1s counterparts. The K np lifetimes are not a thousand times longer than the H np lifetimes because decay to states other than the ground state plays a significant role in the total decay rate.

The radiative lifetimes of many excited states of Li,[16] Na,[17–21] K,[22] Rb,[23–25] and Cs,[26–28] have been measured using a variety of techniques, the most common being time resolved laser induced fluorescence, which is typically carried out using a cell, as shown in Fig. 4.4. In all cases, the observed lifetimes are in reasonable agreement with values calculated in the coulomb approximation, corrected for the decrease due to black body radiation.[29] In the following chapter we show that if the 0 K lifetime of a state is τ that its lifetime at a finite temperature T is given by[24,30]

$$\frac{1}{\tau^T} = \frac{1}{\tau} + \frac{1}{\tau^{bb}}. \tag{4.16}$$

where $1/\tau^{bb}$ is the black body radiation induced decay rate.

Systematic lifetime measurements have been made for Rb[23–25] and Na.[17–21] In both cases the lifetimes of the s, p, d, and f states have been measured, and we discuss the Na lifetimes as a representative example. The lifetimes of Na $n < 15$ s, p, d, and f states have been measured using the time and wavelength resolved resolved laser induced fluorescence arrangement shown in Fig. 4.4. Na atoms at a pressure of 10^{-6} Torr in a pyrex cell are excited from ground state to the 3p state and then to the ns or nd states using two dye laser pulses of 5 ns duration. The ns and nd lifetimes are easily measured by observing the time resolved $ns \rightarrow$ 3p or $nd \rightarrow$ 3p fluorescence with a photomultiplier[17,18]. These fluorescent decays have large branching ratios and are easily detected in spite of the fact that they are at the

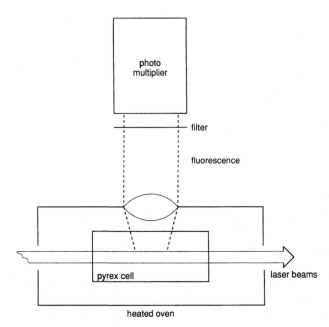

Fig. 4.4 Schematic diagram of a typical arrangment for laser induced fluorescence measurements of lifetimes. A pulsed laser beam (or beams) passes through a heated glass cell containing alkali vapor and the time and wavelength resolved fluorescence is detected at a right angle.

same wavelength as the second exciting laser. For rapidly decaying low lying states the fluorescence is typically stronger than the scattered laser light, and for slowly decaying high lying states the phototube can be gated on after the laser pulse. The longer lived Na np lifetimes can be observed by exciting the nd state which quickly decays, leaving an observable fraction, ~10%, of the initial nd population in the much longer lived np state.[19] The time resolved $4p \rightarrow 3s$ fluorescence can be directly detected, but for higher np states the $np \rightarrow 3s$ fluorescence has too short a wavelength to pass through a glass cell. Instead, the time resolved $ns \rightarrow 3p$ fluorescence, the second step of the $np \rightarrow ns \rightarrow 3p$ cascade is observed. The long time behavior of the observed signal reflects the lifetime of the np state, which is much longer than the lifetime of either the ns or nd states. Finally, the lifetimes of several Na nf states have been measured using a resonant microwave field to equalize the populations in the nd and nf levels.[20] Observing the time resolved fluorescence from either the nf or nd state gives a decay rate indistinguishable from the nd state decay rate, indicating that the nf lifetime is very nearly the same as the nd lifetime. Had the nf lifetimes been much longer than the nd lifetimes, this method would have been relatively insensitive to the nf lifetimes.

The fluorescence methods described above work well for states of principal quantum numbers less than 15. For higher principal quantum numbers they do not work so well for three reasons. First, the amplitude of the fluorescence signal

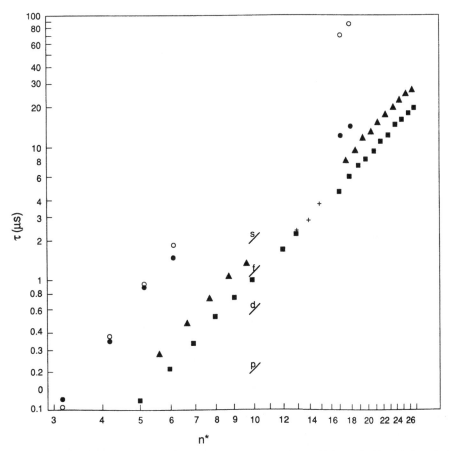

Fig. 4.5 Na radiative lifetimes vs n^*. Experimental values for ns (▲), np (—), nd (■) and nf (+) states are shown. The lifetimes states below $n = 15$ have been measured by fluorescence techniques, at temperatures of approximately 400 K. The lifetimes of $n > 15$ states have been measured by field ionization, the ns and nd states at 30 K and the np states at 300 K. The theoretical 0 K np lifetimes are also shown (○). They are far above the measured values at high n due to black body radiation. Finally, the hydrogenic lifetimes are shown by the line segments (—) (from refs. 4, 18–22, 30).

decreases as n^{-6}, second, the atoms pass out of the field of view due to their longer lifetimes, and, third, black body radiation can appreciably alter the lifetimes.[8] The time delayed field ionization method employed by Spencer *et al.*[21] to measure the Na ns and nd lifetimes circumvents all of these problems. They passed a well collimated Na atomic beam through an interaction region cooled to 30 K. The atoms were excited by two pulsed dye laser beams collinear with the atomic beam and tuned to the 3s → 3p and 3p → ns or 3p → nd transitions. At a variable time after the laser excitation a field ionization pulse was applied to atoms in the interaction region and the resulting ions detected. The atoms which passed out of the interaction region between the laser pulse and the field ionization pulse were

Table 4.1. *Lifetime parameters for alkali atoms.*

		s	p	d	f
Li[a]	τ_0 (ns)	1.39	5.69	0.59	0.96
	α	2.80	2.78	2.92	3.06
Na[b]	τ_0 (ns)	1.38	8.35	0.96	1.13
	α	3.00	3.11	2.99	2.96
K[b]	τ_0 (ns)	1.32	6.78	5.94	0.83
	α	3.00	2.78	2.82	2.95
Rb[b]	τ_0 (ns)	1.43	2.76	2.09	0.76
	α	2.94	3.02	2.85	2.95
Cs[b]	τ_0 (ns)	1.43	4.42	0.96	0.69
	α	2.96	2.94	2.93	2.94

[a] (computed from values in ref. 29)
[b] (from ref. 31)

replaced by atoms which had been excited further upstream by the laser. Using this approach they were able to make measurements at delay times of up to 25 μs, corresponding to about 2 cm of travel of the atoms. It is useful to note that if the interaction region is cooled the only criterion for the field ionization pulse is that it be strong enough to ionize the Rydberg state in question. If significant black body redistribution occurs, as it does at 300 K, the field ionization pulse must be chosen with more care, as described in Chapter 5.

In Fig. 4.5 we show measured values of Na ns, np, nd, and nf lifetimes vs n^{*3} on logarithmic scales. The scales differ by a factor of 3, so an n^{*3} dependence appears as a 45° line. As shown by Fig. 4.5 the n^{*3} dependence suggested by Eq. (4.11) is obtained. The values for $n \leq 15$ have all been obtained using fluorescence detection from atoms in a cell at about 425 K. For all but the np states the black body radiation is a negligible effect. For the np states we also show the calculated 0 K values of Theodosiou, and for the 7p state, the measured value is about 30% below the calculated 0 K value and consistent with the calculated 425 K value of Theodosiou.[29] For $n>15$ all values have been obtained using field ionization. The ns and nd values are those obtained by Spencer *et al.*[21] at 30 K and are clearly in reasonably good agreement with an extrapolation of the values obtained at lower n using fluorescence detection. For the 18p and 19p lifetimes measured at 300 K, however, black body radiation reduces the lifetimes far below an n^{*3} extrapolation of the lifetimes of the lower np states.[30]

We also show in Fig. 4.5 short 45° lines at $n = 10$ corresponding to the lifetimes of the hydrogenic ns, np, nd, and nf series. The Na np and nd lifetimes are respectively factors of 50 and 2 longer than their hydrogenic counterparts, while the Na ns lifetimes are about 30% shorter than the H ns lifetimes. The Na and H nf lifetimes are the same, which is not surprising as the Na nf states and the Na nd states to which they predominantly decay all have quantum defects of nearly zero.

The lifetimes of alkali Rydberg states can be calculated accurately, as shown by Theodosiou,[29] who has made extensive comparisons between observed lifetimes and calculated values. From this work is is also apparent that the black body radiation correction of Eq. (4.16) is important for high lying states. In light of these two points, it seems that the most compact and consistent way of presenting the alkali lifetimes is by a fit of the calculated 0 K values to the form[31]

$$\tau = \tau_0(n^*)^\alpha \qquad \qquad . (4.17)$$

The values of τ_0 and α are presented in Table 4.1. As shown by Table 4.1, the values of α are all near 3, as expected from Eq. (4.11). The biggest discrepancies occur in the Li and K np series which have their Cooper minima nearest the ionization limit. It is also apparent that the np lifetimes are the longest, as expected from Figs. 4.1 and 4.2. In using the values of Table 4.1 it is worth remembering that for high n states the black body decay rate is ℓ independent, so the lifetimes of the longest lived states are the most affected by black body radiation.[32]

References

1. H. A. Bethe and E. A. Salpeter, *Quantum Mechanics of One and Two Electron Atoms* (Academic Press, New York, 1957).
2. E. S. Chang, *Phys. Rev. A* **31**, 495 (1985).
3. U. Fano and J. W. Cooper, *Rev. Mod. Phys.* **40**, 441 (1968).
4. A. Lindgard and S. E. Nielsen, *Atomic Data and Nuclear Data Tables* **19**, 533 (1977).
5. G. V. Marr and D. M. Creek, *Proc. Phys. Soc. (London) A* **304**, 233 (1968).
6. G. V. Marr and D. M. Creek, *Proc. Roy. Soc. (London) A* **304**, 233 (1968).
7. W. Sandner, T. F. Gallagher, K. A. Safinya, and F. Gounand, *Phys. Rev. A* **23**, 2732 (1981).
8. M. J. Seaton, *Proc. Roy. Soc. (London) A* **208**, 408 (1951).
9. G. Baum, M. S. Lubell, and W. Raith, *Phys. Rev. A* **5**, 1073 (1972).
10. R. J. Exton, J. Quant, *Spectrosc. Radiat. Transfer* **16**, 309 (1975).
11. G. Pichler, J. Quant, *Spectrosc. Radiat. Transfer* **16**, 147 (1975).
12. C. J. Lorenzen and K. Niemax, *J. Phys. B* **11**, L723 (1978).
13. J. M. Raimond, M. Gross, C. Fabre, S. Haroche and H. H. Stroke, *J. Phys. B* **11**, L765 (1978).
14. D. W. Norcross, *Phys. Rev. A* **20**, 1285 (1979).
15. C. M. Huang and C.W. Wang, *Phys. Rev. Lett.* **46**, 1195 (1981).
16. W. Hansen, *J. Phys. B.* **16**, 933 (1983).
17. D. Kaiser, *Phys. Lett.* **51A**, 375 (1975).
18. T. F. Gallagher, S. A. Edelstein and R. M. Hill, *Phys. Rev. A* **11**, 1504 (1975).
19. T. F. Gallagher, S. A. Edelstein and R. M. Hill, *Phys. Rev. A* **14**, 2360 (1976).
20. T. F. Gallagher, W. E. Cooke, and S. A. Edelstein, *Phys. Rev. A* **17**, 904 (1978).
21. W. P. Spencer, A. G. Vaidyanathan, D. Kleppner, and T. W. Ducas, *Phys. Rev. A* **24**, 2513 (1981).
22. T. F. Gallagher and W. E. Cooke, *Phys. Rev. A* **20**, 670 (1980).
23. F. Gounand, P. R. Fournier, J. Cuvellier, and J. Berlande, *Phys. Lett.* **59A**, 23 (1976).
24. F. Gounand, M. Hugon, and P. R. Fournier, *J. Phys. (Paris)* **41**, 119 (1980).
25. M. Hugon, F. Gounand, and P. R. Fournier, *J. Phys. B* **11**, L605 (1978).
26. H. Lundberg and S. Svanberg, *Phys. Lett.* **56A**, 31 (1976).
27. K. Marek and K. Niemax, *J. Phys. B* **9**, L483 (1976).
28. J. S. Deech, R. Luypaert, L. R. Pendrill, and G. S. Series, *J. Phys. B* **10**, L137 (1977).
29. C. E. Theodosiou, *Phys. Rev. A* **30**, 2881 (1984).
30. T. F. Gallagher and W. E. Cooke, *Phys. Rev. Lett.* **42**, 835 (1979).
31. F. Gounand, *J. Phys. (Paris)* **40**, 457 (1979).
32. W. E. Cooke and T. F. Gallagher, *Phys. Rev. A* **21**, 588 (1980).

Black body radiation

Lying ≥ 4 eV above the ground state, Rydberg states are not populated thermally, except at very high temperatures. Accordingly, it is natural to assume that thermal effects are negligible in dealing with Rydberg atoms. However, Rydberg atoms are strongly affected by black body radiation, even at room temperature. The dramatic effect of thermal radiation is due to two facts. First, the energy spacings ΔW between Rydberg levels are small, so that $\Delta W < kT$ at 300 K. Second, the dipole matrix elements of transitions between Rydberg states are large, providing excellent coupling of the atoms to the thermal radiation. The result of the strong coupling between Rydberg atoms and the thermal radiation is that population initially put into one state, by laser excitation for example, rapidly diffuses to other energetically nearby states by black body radiation induced dipole transitions.[1-3] Both the redistribution of population and the implicit increase in the radiative decay rates are readily observed. Although the above mentioned effects on level populations are the most obvious effects, the fact that a Rydberg atom is immersed in the thermal radiation field increases its energy by a small amount, 2 kHz at 300 K. While the radiation intensity is vastly different in the two cases, this effect is the same as the ponderomotive shift of the ionization limit in high intensity laser experiments.

Black body radiation

The most familiar way of representing black body radiation is to use the Planck radiation law for the energy density, or the square of the electric field, $\rho(v)$. Explicitly[4]

$$\rho(v)\,dv = \frac{8\pi h v^3}{c^3(e^{hv/kT}-1)},\tag{5.1}$$

where k is the Boltzmann constant, h is Planck's constant, v is the frequency of the black body radiation, and T is the temperature.

In Fig. 5.1 we show $\rho(v)$ vs v at 300 K using both frequency and wavenumber as abscissae. A typical optical transition from the ground state of an atom has frequency $v = 3 \times 10^{14}$ Hz, and a transition between two Rydberg states has frequency $v \sim 3 \times 10^{11}$ Hz. Thus, it is apparent from Fig. 5.1 that to a ground state atom the black body radiation appears as a slowly varying, nearly static field, whereas to a Rydberg atom it appears to be a rapidly varying field.

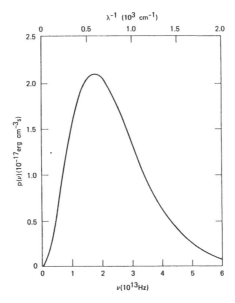

Fig. 5.1 Energy density $\rho(\nu)$ as a function of frequency ν at 300 K (from ref. 5).

While the representation of black body radiation given in Eq. (5.1) and Fig. 5.1 is the most familiar, it is not the most useful for the most important effect of black body radiation on Rydberg atoms, inducing transitions between neighboring levels.

To calculate the rates of these transitions it is more convenient to express the radiation field in terms of the number of photons per mode of the radiation field, i.e. the photon occupation number $\bar{n}$ of each mode. The photon occupation number $\bar{n}$ is given by[4]

$$\bar{n} = \frac{1}{e^{h\nu/kT} - 1}. \qquad (5.2)$$

At low frequencies, for which $h\nu \ll kT$, Eq. (5.2) reduces to

$$\bar{n} \simeq \frac{kT}{h\nu} \qquad (5.3)$$

Fig. 5.2 shows the dependence of $\bar{n}$ on ν for $T = 300$ K, again with both frequency (Hz) and wavenumber (cm^{-1}) used as abscissae. It is useful to recall that at 300 K, $kT/h \approx 6 \times 10^{12}$ Hz or $kT/hc \approx 200$ cm^{-1}. This point is apparent in Fig. 5.2, since $\bar{n} = 1$ at $\lambda^{-1} \cong 200$ cm^{-1}. Since the vacuum fluctuations, which lead to spontaneous emission, are given by $\bar{n} = 1/2$, black body radiation at frequencies greater than $kT/h\bar{n}$, where $\bar{n} \ll 1$, does not lead to significant effects. For an atom in its ground state with transitions at $\sim 10^4$ cm^{-1}, black body induced transitions are unimportant, since $\bar{n} \ll 1$. However, for a Rydberg state with transitions at 10 cm^{-1}, where $\bar{n} \sim 10$, the black body induced transition rates can

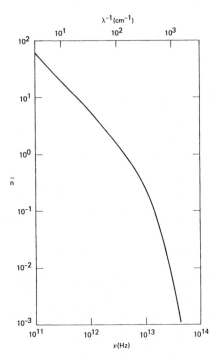

Fig. 5.2 Occupation number $\bar{n}$ as a function of frequency ν at 300 K (from ref. 5).

be an order of magnitude larger than the spontaneous emission rates, and the effect of black body radiation is significant.

Eqs. (5.1)–(5.3) present black body radiation in a familiar form. Both to conform to the general use of atomic units in this book and to simplify the calculation of the Rydberg atoms' response to the radiation we reexpress Eqs. (5.1)–(5.3) in atomic units.[5]

We reexpress Eq. (5.1) as

$$\rho(\omega)d\omega = \frac{2\alpha^3\omega^3 d\omega}{\pi\left(e^{\omega/kT} - 1\right)}, \tag{5.4}$$

where α is the fine structure constant and ω is the energy in au. Similarly, we may reexpress Eq. (5.2) as

$$\bar{n} = \frac{1}{e^{\omega/kT} - 1} \tag{5.5}$$

and Eq. (5.3) for $\omega \ll kT$ as

$$\bar{n} \approx \frac{kT}{\omega}. \tag{5.6}$$

We now consider the transition rates for absorption of and stimulated emission induced by the black body radiation and compare these rates to the spontaneous emission rates. This comparison provides a reasonable indication of the importance of the effects of black body radiation induced transitions for Rydberg atoms.[6]

Black body induced transitions

As shown in the previous chapter, the spontaneous decay rate of the $n\ell$ state to the lower lying n'ℓ' state is given by the Einstein A coefficient $A_{n'\ell',n\ell}$.[7] In the presence of thermal radiation the stimulated emission rate $K_{n'\ell',n\ell}$, is simply $\bar{n}$ times as large as the spontaneous rate. Explicitly,

$$K_{n'\ell',n\ell} = \bar{n} A_{n'\ell',n\ell}, \tag{5.7}$$

where $\bar{n}$ is implicitly evaluated at the frequency $\omega_{n'\ell',n\ell}$. It is convenient to reexpress Eq. (5.7) in terms of the average oscillator strength,

$$K_{n'\ell',n\ell} = -2\bar{n}\alpha^3\omega^2_{n'\ell',n\ell}\bar{f}_{n'\ell',n\ell}, \tag{5.8}$$

where α is the fine structure constant, $\omega_{n'\ell',n\ell}$ is the energy difference $W_{n'\ell'} - W_{n\ell}$, and

$$\bar{f}_{n'\ell',n\ell} = \frac{2}{3}\omega_{n'\ell',n\ell} \frac{\ell_{max}}{2\ell+1} |\langle n'\ell'|r|n\ell\rangle|^2, \tag{5.9}$$

where ℓ_{max} is the larger of ℓ and ℓ'.

The black body photons can also be absorbed as the atoms in the $n\ell$ state make the transition to a higher lying $n'\ell'$ state. Both the stimulated emission and absorption rates are given by

$$K_{n'\ell',n\ell} = 2\bar{n}\alpha^3\omega^2_{n'\ell',n\ell}|\bar{f}_{n'\ell',n\ell}|. \tag{5.10}$$

Because of the variation of $\bar{n}$ with frequency, the frequency dependence of $K_{n'\ell',n\ell}$ is quite different from that of $A_{n'\ell',n\ell}$, and these two processes favor different final states. This point is made graphically in Fig. 5.3, which is a plot of $A_{n'p,18s}$ and $K_{n'p,18s}$ vs n' for the Na 18s state and $T = 300$ K. As shown in Fig. 5.3, black body radiation favors transitions to nearby states and spontaneous emission favors transitions to the lowest lying states.

In addition to driving transitions to discrete states the black body radiation can also photoionize the Rydberg atoms. The photoionization rate, $1/\tau^P_{n\ell}$ is given by[8]

$$\frac{1}{\tau^P_{n\ell}} = \frac{2\pi}{3} \int_{1/2n^2}^{\infty} \left\{ \frac{\ell}{2\ell+1} |\langle n\ell|r|\varepsilon\ell-1\rangle|^2 + \frac{(\ell+1)}{2\ell+1} |\langle n\ell|r|\varepsilon\ell+1\rangle|^2 \right\} \rho(\omega)d\omega. \tag{5.11}$$

The matrix elements are the radial matrix elements of r between the initial $n\ell$ state and continuum $\ell - 1$ and $\ell + 1$ waves which are normalized per unit energy.

Just as we previously summed the spontaneous emission rates over the possible final states to obtain the total spontaneous decay rate $1/\tau_{n\ell}$, we can sum the black

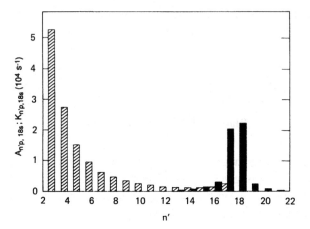

Fig. 5.3 Spontaneous emission rates of the 18s state to lower lying n'p states, $A_{n'p,\,18s}$, ($\square$) and the 300 K black body transition rates of the 18s state to higher and lower lying n'p states, $K_{n'p,\,18s}$, ($\blacksquare$) as a function of n' (from ref. 5).

body radiation transition rates to compute the total black body radiation induced decay rate $1/\tau_{n\ell}^{bb}$. Explicitly,

$$\frac{1}{\tau_{n\ell}^{bb}} = 2\alpha^3 \sum_{n'} \bar{n}\omega_{n'\ell',n\ell}^2 \left|f_{n'\ell',n\ell}\right| + \frac{1}{\tau_{n\ell}^{P}}, \tag{5.12}$$

and the total decay rate $1/\tau_{n\ell}^{T}$ is given by

$$\frac{1}{\tau_{n\ell}^{T}} = \frac{1}{\tau_{n\ell}} + \frac{1}{\tau_{n\ell}^{bb}}. \tag{5.13}$$

From Fig. 5.3 it is apparent that the most important contributions to the black body radiation decay rate are from transitions to nearby states for which $|\omega_{n'\ell',n\ell}| < kT$. In this case we may substitute Eq. (5.6) for $\bar{n}$ and write $1/\tau_{n\ell}^{bb}$ as

$$\frac{1}{\tau_{n\ell}^{bb}} = 2\alpha^3 kT \sum_{n'} \omega_{n'\ell',n\ell} \bar{f}_{n'\ell',n\ell}. \tag{5.14}$$

In Eq. (5.14) the summation over n' implicitly includes the continuum. For a one-electron atom we may use the sum rule[7]

$$\sum \omega_{n'\ell',n\ell} \bar{f}_{n'\ell',n\ell} = \frac{2}{3n^2}, \tag{5.15}$$

which we may use to write

$$\frac{1}{\tau_{n\ell}^{bb}} = \frac{4\alpha^3 kT}{3n^2}. \tag{5.16}$$

For $T = 300$ K, Eq. (5.16) is accurate to 30% for $n > 15$, and as n is increased Eq. (5.16) becomes an increasingly good approximation. More important, it brings

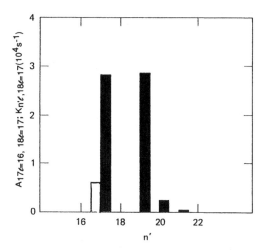

Fig. 5.4 Spontaneous emission rate of the $n = 18$, $\ell = 17$ state to the $n = 17$, $\ell = 16$ state, $A_{17\ell=16,\,18\ell=17}$, ($\square$) and the 300 K black body transition rates of the $n = 18$, $\ell = 17$ state to $n'\ell'$ states of $\ell' = \ell \pm 1$, $K_{n'\ell',\,18\ell=17}$, ($\blacksquare$) as a function of n' (from ref. 5).

out two interesting points. First, the total black body decay rate $1/\tau_{n\ell}^{\mathrm{bb}}$ does not depend on ℓ, only on n. Since the radiative decay rate decreases rapidly as ℓ is increased,[9] for high ℓ states the black body induced decay rate is often much larger than the spontaneous emission rate. This point is shown graphically in Fig. 5.4 in which the spontaneous emission rates and black body radiation emission and absorption rate are shown for the $n = 18$, $\ell = 17$ state. In contrast to the 18s state, for which $1/\tau^{\mathrm{bb}} \sim 0.2/\tau$, the $n = 18$, $\ell = 17$ state has a black body decay rate which is ten times larger than its spontaneous decay rate. Second, because of the $1/n^2$ scaling of the black body radiation decay rate, it is evident that even for low ℓ states that at high enough n the black body radiation rate exceeds the spontaneous emission rate, which scales as $1/n^3$. Even at $n = 20$, the black body radiation rate is at least 20% of the spontaneous emission rate for all ℓ states, and in many experiments it is a significant source of population redistribution.

Black body energy shifts

That part of the black body spectrum coincident with the atomic transition frequencies leads to the transitions which redistribute the population. In contrast, all the energy of the black body radiation contributes to the shift of the energy levels. The energy shift is a second order ac Stark shift, and for state n the shift ΔW_{b} is given by[10]

$$\Delta W_{n\ell} = \sum_{n'\ell'} \int_0^\infty \frac{|\langle n'\ell|r|n\ell\rangle|^2 \, E_{\omega_b}^2 \omega_{n\ell,n'\ell'}}{2(\omega_{n\ell,n'\ell'}^2 - \omega_b^2)} \, d\omega_b \tag{5.17}$$

where E_{ω_b} is the electric field of the black body radiation in a bandwidth $d\omega_b$ at frequency ω_b. Since $E_{\omega_b}^2$ is proportional to the energy density, examining Fig. 5.2 we can see that at room temperature the peak of the black body radiation energy density is at ~ 500 cm^{-1}, far above the strong $n \sim 20$ transitions which have wavenumbers of ~ 20 cm^{-1}. In fact, for atoms in Rydberg states it is often true that the frequencies of the strong transitions are much lower than the frequency of the black body radiation, i.e. $\omega_{n\ell,n'\ell'} \ll \omega_b$, in which case we can ignore $\omega_{n\ell,n'\ell'}$ in the denominator of Eq. (5.17) and rewrite $\Delta W_{n\ell}$ as

$$\Delta W_{n\ell} = \left(\sum_{n'\ell'} \bar{f}_{n'\ell',n\ell} \right) \left(\int_0^\infty \frac{E_{\omega_b}^2 \, d\omega_b}{4\omega_b^2} \right). \tag{5.18}$$

Using the oscillator strength sum rule[7]

$$\sum_{n'\ell'} \bar{f}_{n'\ell',n\ell} = 1 \tag{5.19}$$

and integrating over ω_b in Eq. (5.18), we may express Eq. (5.18) as

$$\Delta W_{n\ell} = \frac{\pi}{3} \alpha^3 (kT)^2. \tag{5.20}$$

Eq. (5.20) is accurate to $\sim 10\%$ for $n > 15$ at $T = 300$ K. Eq. (5.20) indicates that all Rydberg states experience the same energy or frequency shift, which is $+2.2$ kHz at 300 K. Eq. (5.20) corresponds to the energy shift of a free electron in an oscillating electric field. This point is more apparent if we simply use the oscillator strength sum of Eq. (5.19) and apply it to the case of a monochromatic field of angular frequency ω_b for which $\int_0^\infty E_b^2 d\omega_b = E^2$. In this case Eq. (5.18) can be written

$$\Delta W_{n\ell} = \frac{E^2}{4\omega^2}, \tag{5.21}$$

which is the average kinetic energy of a free electron in a field $E \cos \omega t$. The equality of the energy shift of a free electron and an electron in a Rydberg state is not surprising since the physical effect of the ac black body field, at a frequency high compared to the orbital frequency of the electron, is to superimpose a fast, $\sim \omega_b$, wiggle on the orbital motion of the electron. The energy of the wiggle motion is independent of the much slower orbital velocity.

While the black body radiation energy is mostly at frequencies high compared to the Rydberg state frequencies, it is low compared to the frequencies of transitions of low lying states. Explicitly, $\omega_{n'\ell',n\ell} \gg \omega_b$, and for low lying states we can ignore ω_b in the denominator of Eq. (5.18). In this case the Stark shift is equal to that produced by a static field and can be expressed as

$$\Delta W_{n\ell} = \sum_{n'\ell'} -\frac{\bar{f}_{n'\ell',n\ell}}{\omega_{n'\ell',n\ell}^2} \int_0^\infty \frac{E_{\omega b}^2 \, d\omega_b}{4}. \tag{5.22}$$

For a one-electron Rydberg atom we may use the hydrogenic sum rule[11]

$$\sum_{n'} \frac{\overline{f}_{n'\ell',n\ell}}{\omega^2_{n'\ell',n\ell}} = \frac{9}{2} \tag{5.23}$$

so that we may write Eq. (5.22) as

$$\Delta W_n = -\int \frac{9}{8} E^2_{\omega_b} d\omega_b = -\frac{3}{5}(\alpha\pi)^3(kT)^4. \tag{5.24}$$

Evaluation of Eq. (5.24) gives a shift of -0.036 Hz at 300 K, which is negligible for all practical purposes.

Finally, we must remember that for some states it will be true that $\omega_{n'\ell',n\ell} \sim \omega_b$, in which case neither of the above approximations is valid and Eq. (5.17) must be evaluated explicitly as has been done by Farley and Wing.[12] At 300 K $\omega_{n\ell,(n+1)\ell'} \sim \omega_b$ for $n \sim 8$.

If we confine our attention to Rydberg states of $n > 15$, for which Eq. (5.20) is valid, it is apparent that all the Rydberg states and the ionization limit are shifted up in energy by an amount which is proportional to the square of the temperature. If the atoms are exposed to a monochromatic radiation field satisfying the same frequency criteria, the Rydberg state energies and the ionization limit are shifted according to Eq. (5.21), which is often termed the ponderomotive energy shift.[13]

In this chapter we have implicitly assumed the Rydberg atom to be a one electron-atom. In the perturbed Rydberg series of, for example, alkaline earth atoms, Rydberg states can have mixed valence–Rydberg character. In such states the black body effects are reduced by a factor equal to the fractional Rydberg character.[14]

Initial verification

The experimental observations which called attention to the effects of room temperature black body radiation on Rydberg atoms were made using Rydberg states of Xe and Na.[1-3] However, Pimbert had earlier noted the effect of higher temperatures.[15] Two kinds of experiments were performed, measurements of population redistribution and increased decay rates. Both of the effects measured could conceivably arise from collisions, and this possibility was systematically ruled out in the experiments. In the Xe experiment atoms in a metastable beam were excited to the 26f state by a pulsed laser and exposed to the thermal radiation for periods of 1.5, 7.5, and 15.5 μs, after which a ramped field was applied to the atoms. This procedure leads to the field ionization signals shown in Fig. 5.5.[1] For each time delay there is a clear progression of peaks in the field ionization spectra which are assigned the n values shown. At longer delay times the intensities of the

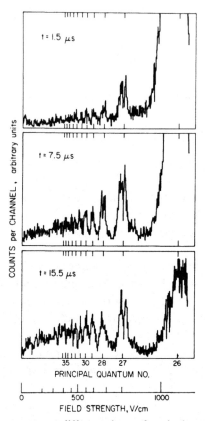

Fig. 5.5 Field ionization spectra at different times after the laser pulse excites Xe atoms to the 26f level showing the black body radiation induced transfer to higher levels (from ref. 1).

$n > 26$ peaks grow, as would be expected for progressively longer exposure times to the thermal radiation. In addition, due to the fact that the black body radiation induced transitions are dipole transitions, only d and g final states are expected. The fact that the field ionization spectra exhibit a sharp peak for each n is consistent with only one or two, not all, ℓ states being populated by the thermal radiation.[1] If all the ℓ states were populated there would be no sharp peaks in Fig. 5.5.

In an analogous experiment in which the initial population was placed in the Na 18s state, the transferred population was found to be in the higher lying p states, as expected for dipole transitions. Furthermore, the fraction of the atoms which had undergone transitions to each of the higher p states after 5 μs was in good agreement with values calculated on the basis of 300 K black body radiation induced transitions, as shown by Table 5.1.[2]

The other initial experiment was the observation of the increase in the decay rate due to 300 K black body radiation, shown by Eq. (5.13).[3] The decay rate of

Table 5.1. *Calculated and observed populations in higher lying* np *states 5 μs after the initial population of the Na* 18s *state expressed in terms of the initial* 18s *population.*[a]

Final state	Calculated yield	Observed yield
18p	4.2%	5.0%
19p, 20p	1.6%	1.6%
>20p,εp	1.0%	1.2%
Total	6.8%	7.8%

[a](see ref. 2)

Table 5.2. *Lifetimes of the Na* 17p *and* 18p *states.*[a]

State	τ (μs)	τ^{bb} (μs)	τ^T (calc) (μs)	τ^T(obs) (μs)
17p	48.4[b]	22.7	15.5	$11.4^{+5.6}_{-1.4}$
18p	58.4[b]	25.6	17.9	$13.9^{+8.8}_{-2.9}$

[a](from ref. 3)
[b](from ref. 17)

the Na 18s state at 300 K differs from its 0 K rate by about 20%, which is not a large enough difference that a measurement at 300 K alone is entirely convincing. The Na *np* states of $n \sim 20$, on the other hand, have long lifetimes[16,17] and offer a reasonable prospect for a convincing measurement, even if made only at 300 K. The Na 17p and 18p lifetimes were measured by exciting Na atoms in a beam to the 17p and 18p states and observing the populations in these states as a function of time after the laser pulse. The results of the experiment are given in Table 5.2, which shows clearly that the experimentally observed values are in good agreement with the lifetimes calculated with the black body radiation, but in stark disagreement with the 0 K lifetimes.

Temperature dependent measurements

Subsequent to the initial 300 K observations, measurements were done as a function of temperature. The first measurements, by Koch *et al.*,[18] were done at elevated temperatures, and later measurements were done at temperatures below room temperature. A critical issue in such measurements is knowing the temperature the atoms are experiencing. Typically an enclosure cooled to a temperature from 6 K to 300 K surrounds the atoms, shielding them from the 300 K radiation from the walls of the vacuum chamber. However, simply making an enclosure in which 90% of the solid angle seen by the atoms is cold is not adequate, because at far infrared wavelengths many materials, and virtually all metal surfaces, are excellent reflectors. Thus a 10% aperture to the warm world outside is adequate to raise substantially the temperature the atoms in the enclosure experience.

Two approaches have been used to ensure that the temperature experienced by the atoms is close to the temperature of the enclosure. Hildebrandt *et al.* lined the inside of the enclosure with graphite coated Cu wool, which they verified separately to be an effective way of absorbing far infrared radiation.[19] An alternative approach was followed by Spencer *et al.*[20] They covered the apertures through which laser beams propagated with glass, which is opaque at far infrared wavelengths, and those through which the atomic beam propagated with a fine, 70 μm, mesh, which blocks all radiation with $\lambda > 140\,\mu$m.

Experiments with Xe analogous to the previously described Xe experiment were carried out at both 90 K and 300 K. Specifically the time dependences of the populations in higher nd and ng states were measured subsequent to population of an nf state by a pulsed laser.[19] The time dependences of the observed populations were fit to a model which yielded the radiative transfer rates from the initially excited state to the final states. Not surprisingly, as the temperature was reduced from 300 K to 90 K, not only was the overall radiative transfer reduced, but the transfer to the highest lying states was most sharply reduced, as expected from the dependence of the photon occupation number on frequency and temperature. For example the transfer from the Xe 25f to 26d,g states was reduced by a factor of 3 while the 25f–27d,g transfer was reduced by a factor closer to 5.

Spencer *et al.* populated the Na 19s state and observed the sum of the populations in the 19s, 19p, and 18p states, at times from 2–32 μs after the laser excitation, to determine the sum of the radiative transition rates from the 19s to 18p and 19p states.[20] The measurements were made at temperatures from 6 K to 210 K with the results shown in Fig. 5.6. As shown by Fig. 5.6, the observed rates agree well with the calculated rates, shown by the line. At 0 K the rate is entirely due to the spontaneous 19s → 18p transition, and the increase above this rate is due to thermal radiative transfer to both the 18p and 19p states. It is interesting to note that above ~ 30 K the transfer rate increases linearly with T as expected from Eq. (5.16). The linear behavior only occurs above 30 K since only then is $kT > \omega$ for these two transitions.

While the Na 19s → 18p, 19p transfer rate is an excellent illustration of the validity of Eq. (5.16) for $\omega < kT$, the photoionization of Rydberg atoms of $n \approx 20$ by black body radiation is a test of the regime $\omega \sim kT$, and this regime has been explored by Spencer *et al.*[21] Specifically, they measured the relative rates for

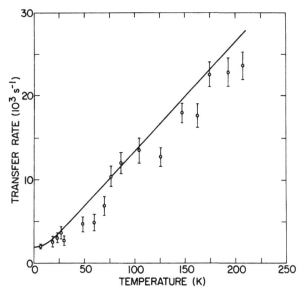

Fig. 5.6 Experimental and theoretical transfer rate of the Na 19s state to the 18p and 19p states vs temperature. The solid line is calculated with no adjustable parameters (from ref. 20).

photoionization of the Na 17d state by thermal radiation at temperatures of 90–300 K. The experiments were done by exciting the Na 17d state, allowing photoionization to occur for 500 ns, a time ten times shorter than the 17d lifetime, and then collecting the photoions by applying a small field pulse of 8 V/cm. A field this small was verified to produce negligible field ionization of bound states excited by black body radiation. When the relative rates are normalized to the calculated rate at 300 K, they are found to be in excellent agreement with the calculated rates.

Since the 17d state is bound by 380 cm^{-1}, which exceeds kT, the photoionization rate varies rapidly with temperature, roughly a factor of 100 between 90 K and 300 K, as shown by Fig. 5.7. The rapid, nearly exponential, dependence of the photoionization rate on temperature shown in Fig. 5.7 is to be contrasted with the linear temperature dependence of the 19s → 19p, 18p transfer rate shown in Fig. 5.6.

Suppression and enhancement of transition rates

We have described the effects of black body radiation in free space. In a closed cavity the radiation is confined to the allowed modes of the cavity. In essence all the thermal radiation is forced into the cavity modes, raising the intensity at the

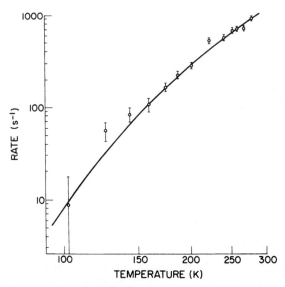

Fig. 5.7 Black body induced photoionization rate vs temperature for the 17d state in Na.
Scales are logarithmic. The solid line represents the calculated values. Experimental points
are normalized to the calculated value at 300 K (from ref. 21).

frequencies of the modes by a factor of Q relative to the free space value.[22] Here Q
is the quality factor of the cavity, the frequency of a mode divided by its linewidth
(FWHM), i.e. $Q = v/\Delta v$. Similarly, between the cavity modes the intensity is
substantially depressed. At a cavity mode resonance the radiation transfer rate is
increased, by a factor of Q, and away from a resonance, the transfer rate is
suppressed. As pointed out by Kleppner, Rydberg atoms, with their long
wavelength transitions, provide a nearly ideal system for the study of such
notions.[23]

    The first experimental realization of this notion was by Vaidyanathan *et al.* who
excited Na atoms to nd states between two parallel plates 0.337 cm apart.[24] The
region between these plates is cutoff for propagation of radiation of wavelengths
longer than 0.674 cm polarized parallel to the plate surfaces, and such radiation is
not present between the plates. Stated another way, frequencies less than the
cutoff frequency $v_c = 1.48$ cm^{-1} do not propagate. Radiation polarized perpendicular to the surfaces can propagate freely between the plates. The zero static
field 29d–30p interval is just under 1.48 cm^{-1}. However, the application of a small
static field increases the separation of the levels, and a spacing of 1.48 cm^{-1} occurs
at a static field of 2.4 V/cm. Thus as the static field is increased from zero to beyond
2.4 V/cm the 29d → 30p transition frequency passes from below to above cutoff,
and the black body radiation transition rate increases accordingly.

    In the experiment Na atoms are excited by two lasers from the ground state to
the nd state in an environment cooled to 180 K at which point $\bar{n} = 86$ for the
29d → 30p transition. The atoms are exposed to the thermal radiation for 20 μs,

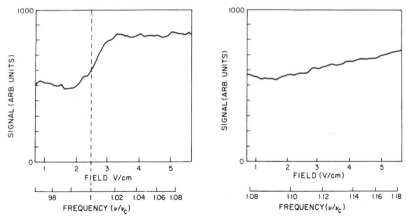

Fig. 5.8 Black body radiative transfer signals in Na located between parallel conducting plates for 29d → 30p (left-hand side) and 28d → 29p (right-hand side) as a function of the absorption frequency. The cutoff frequency is $\nu_c = 1/2d = 1.48$ cm^{-1}, where d is the plate separation. The increase in the transfer rate at $\nu = \nu_c$ (left-hand side) is due to the "switching on" of the radiation polarized parallel to the plates (from ref. 24).

after which they are exposed to a field ionization pulse to ionize atoms in the $(n + 1)$p state selectively. In Fig. 5.8 we show the results obtained when the 28d and 29d states are excited as functions of static field over the range 0–5 V/cm. As shown in Fig. 5.8, the 29d–30p interval is tuned through the cutoff frequency of the plates in this field range. Correspondingly, there is a sharp increase in the population found in the 30p state when the field tunes the 29d–30p interval through the cutoff frequency. In contrast, the 28d–29p interval always exceeds the cutoff frequency, and no sharp increase in the 29p population is observed as the field is tuned.

Just as the black body transitions can be suppressed, they can also be enhanced, by tuning to a cavity resonance. Raimond *et al.*[25] have passed a thermal beam of Na atoms excited to the $30s_{1/2}$ state through a 300 K Fabry–Perot cavity, which can be tuned to the $30s_{1/2}$–$30p_{1/2}$ resonance at 134 GHz, and detected the atoms downstream by field ionization. With the cavity tuned off resonance 5% of the atoms make the $30s_{1/2} \rightarrow 30p_{1/2}$ transition, but when it is tuned to resonance an additional 5% make the transition, in agreement with the calculated value.[25]

In any low angular momentum state the radiative decay rate is usually dominated by the high frequency transitions to low lying states, and as a result it is impossible to control completely the decay rate using a millimeter wave cavity. In a circular $\ell = m = n - 1$ state the only decay is the far infrared transition to the $n - 1$ level, and Hulet *et al.* have observed the suppression of the decay of this level.[26] They produced a beam of Cs atoms in the circular $n = 22, \ell = m = 21$ state by pulsed laser excitation and an adiabatic rapid passage technique.[27] The beam of circular state atoms then passed between a pair of plates 6.4 cm wide, 12.7 cm long, spaced by 230.1 μm, and held at 6 K. The 0 K radiative lifetime is 460 μs, and the transit time of the atoms through the plates to the detector is approximately one lifetime at the peak of the velocity distribution.

In zero static field the wavelength λ of the $n = 22$, $\ell = 21 \rightarrow n = 21$, $\ell = 20$ transition is $450\,\mu$m, and $\lambda = 0.98(2d)$ where d is the plate spacing. However, using the plates to apply a static field it is possible to reduce continuously the energy separation between the levels by up to 4%, increasing the wavelength of the transition to $1.02(2d)$. In other words it is possible to increase the wavelength of the transition from below cutoff to above cutoff by varying the tuning field. At this point it is worth recalling that in the static field the $\ell = 21$, $m = 21$ circular $n = 22$ state must interact with radiation polarized parallel to the field plates to decay to the $n = 21$, $\ell = m = 20$ state. The radiation polarized perpendicular to the plates cannot drive this transition. When the wavelength is longer than the cutoff wavelength , i.e. $\lambda > 2d$, radiative decay is suppressed, i.e. $\tau_{cav} = \infty$, but when the wavelength is less than the cutoff wavelength $\lambda < 2d$ the radiative decay rate is increased by 50%.[26]

They recorded time of flight spectra of atoms remaining in the circular state as a function of the spacing of the plates and the tuning field applied, and from these spectra determined the decay rates. Decay rates were measured and calculated for three conditions. First the plate spacing was increased by a factor of 30 to observe the free space decay. They observed a lifetime of $450(10)\,\mu$s, in good agreement with the calculated 6 K value of 451 μs. With the plate spacing returned to $230.1\,\mu$m the decay rates were measured with the electric field set so as to tune the wavelength both below and above the cutoff wavelength. With a low electric field, for which $\lambda < 2d$, the measured decay rate was 50% larger than the free space value, to within 5%. When the field was increased so that $\lambda > 2d$, roughly twice the number of atoms was detected and the decay rate was consistent with zero, to within 5% of the free space decay rate.

Level shifts

Rydberg states which are pure Rydberg states, not containing admixtures of valence states, all have the same positive energy shift in a thermal radiation field. To measure this shift requires careful measurement of a transition from a long lived low lying atomic state. Holberg and Hall have measured the thermal shift of Rb atoms exposed to a thermal source which ranged in temperature from 350 K to 1000 K using two photon, Doppler free spectroscopy between the ground state and the Rydberg states.[28] In their experiment a Rb beam passes through the two arms of a folded optical cavity, so the atoms interact with the optical radiation twice, providing a Ramsey interference pattern. An attraction of the Ramsey method is that the Ramsey interference pattern is located at the optical field free location of the 5s $\rightarrow$ ns transition. It is thus unaffected by laser light shifts. The central Ramsey interference fringe is 40 kHz wide, the inverse of the transit time of the atoms between the two arms of the optical cavity. The radiation from the heated black body source is focused on the atomic beam between its intersections with the two laser beams. Due to the small solid.angle subtended by the heated black body source at the intersection of the atomic and laser beams, it is only ~10% as effective as a complete enclosure at the same temperature. The radiation from the heated black body source is chopped and the position of the central fringe of the Ramsey pattern observed with and without the thermal radiation. In spite of the reduced efficiency of the source in producing a shift, blue shifts of up to 1.4 kHz were observed, at 876 K. The blue shift increases with temperature as predicted by Eq. (5.20), and while the results do not provide a stringent test, they are clearly consistent with Eq. (5.20).

The black body shifts are not confined to Rydberg atoms, but also alter the frequency of atomic clock transitions. Itano *et al.* have shown that the Cs 9 GHz ground state hyperfine interval, the definition of the second, is increased by one part in 10^{14} when the temperature is raised from 0 K to 300 K.[29].

Experimental manifestations and uses

As we have already pointed out, the presence of black body radiation invariably decreases the lifetime of a Rydberg state, and the decreased lifetimes have been reported in many systems.[3,30–32] What is most important, though, is the transfer of population to nearby states, and this effect must be taken into account in making measurements. To illustrate this point we consider the measurement of the 300 K lifetime of a Rydberg state using selective field ionization, an apparently straightforward measurement.[6] The 18s state of Na is calculated to have a 0 K radiative lifetime of 6.37 μs, which is reduced to 4.87 μs by the 300 K black body radiation. Na atoms were excited to the 18s state at time $t = 0$, and the population remaining at later times was determined by applying an electric field pulse high enough to ionize easily the atoms in the 18s state. This measurement yielded an apparent lifetime of 7.8 μs, a value longer than the 0 K radiative lifetime. The problem was that the field also ionized all the long lived, np states of $n \geq 17$ to which population was transferred by black body radiation. Only when the signal from higher lying states was detected separately was it possible to correct the observed 18s signal for the presence of the long lived np states. The final value was 4.78 μs, in excellent, and probably fortuitous, agreement with the calculated value of 4.87 μs.

As shown by Table 5.2, at 300 K the $n = 17$ and $n = 18$ black body decay rates are $\approx 5 \times 10^4$ s^{-1}, which might seem tolerable. However, in even slightly dense samples thermal radiation can trigger superradiance, which depopulates an initially populated state in 10 ns.[33,34] For superradiance to occur between two levels (crudely speaking, laser action without mirrors), a necessary condition is that the gain, G, be 1 along the length of the sample. This criterion may be expressed as[33]

$$G = NL\sigma \geq 1 , \tag{5.25}$$

where N is the difference in population densities of the upper and lower levels, L is the length of the sample, and σ, the optical cross section, is given by[35]

$$\sigma = \frac{g_\ell \lambda^2 A_{u\ell}}{8\pi g_u \Delta} , \tag{5.26}$$

where g_ℓ and g_u are the degeneracies of the upper and lower states. $A_{u\ell}$ is the Einstein coefficient for the transition, and Δ is the linewidth of the transition, to which the main contributions are usually radiative and Doppler broadening. Because of the long wavelength of the transitions between Rydberg states, the above gain criterion for superradiance can be met with a very low density of atoms, hundreds of atoms in a volume of $10^{-3}\,\mathrm{cm}^{-1}$, and is easily met even in a low density atomic beam. The mere presence of adequate gain does not, however, ensure superradiance. Some initial photons are required to trigger the superradiance, and due to the low spontaneous transition rates of the long wavelength transitions between Rydberg states, they are unlikely to come from spontaneous emission. On the other hand, the omnipresent black body radiation easily triggers superradiance.

Gounand *et al.* have used superradiant cascades initiated by black body radiation to populate efficiently states which would normally be inaccessible using optical excitation.[34] In vapor cell experiments they observed the population in Rb Rydberg states to undergo several very rapid superradiant cascades to populate lower lying levels. The population is transferred in times far shorter than would be possible by simple radiative decay.[34] Fig. 5.9 shows the population in various Rb Rydberg states after the initial population of the 12s state. Normal radiative decay would require $\sim 3\,\mu s$ to populate the 7f state, but as shown in Fig. 5.9, only 20 ns is required when superradiant cascades occur.

Stoneman and Gallagher have used black body radiation to make precise measurements of the avoided crossings between the K ns and $n-2$ Stark manifold states in electric fields,[36] and these measurements are described in Chapter 6.

Far infrared detection

An interesting potential application of Rydberg atoms is as a far infrared (or microwave) detector, a notion first suggested by Kleppner and Ducas.[37] The basic

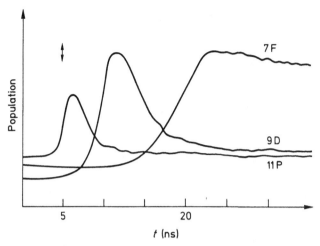

Fig. 5.9 Time dependence of the population in Rb Rydberg states after the initial population of the 12s state showing the rapid superradiant cascades (from ref. 35).

idea of the most straightforward type of device is to use a laser to make a target of Rydberg state atoms in one state, A. This target is exposed to the radiation to be detected at the infrared frequency ω_{AB}, which is equal to the frequency of the atomic transition from A to a higher energy state B. If the density of atoms in state A in the target is high enough that the target is optically thick to the radiation at frequency ω_{AB}, a photon at this frequency will be absorbed with nearly unit probability so that one atom will undergo the transition to state B. The fact that the atom has undergone the transition can be detected in either of two ways. The first is selective field ionization of atoms in state B but not those in state A, and the second is to detect the optical fluorescence from state B but not from state A.[38] In either case each incoming photon is converted into an ion and an electron or a visible photon, all of which are easily detected. Several variants of this approach have been examined.[2,39–41]

With optical detection it is unimportant whether state B lies above or below state A. If B lies below A, a second approach, triggered superradiance, is possible, and the technique has been carefully examined by Goy *et al.*[42] To detect the same radiation at ω_{AB} requires an optically thick target of atoms in the upper state A, which in a cooled environment would not become superradiant, since spontaneous emission is too slow to initiate the process. However, an incoming photon at frequency ω_{AB} would trigger a superradiant avalanche, converting one incoming photon into $\sim 10^4$ outgoing photons and leaving 10^4 atoms in state B, which would be trivial to detect. As shown by the sensitivity given in Table 5.3, this approach is equally as sensitive as the more straightforward approach described above.

While Rydberg atoms have been used to detect infrared radiation with wavelengths as short as 2.34 μm, they are most effective as detectors of far infrared

Table 5.3. *Far infrared detection sensitivity in Rydberg atom experiments.*

Method	Source	Wavenumber cm^{-1}	NEP W/cm^2 Hz$^{1/2}$
Absorption,[a] field ionization	300 K black body	22	10^{-14}
Absorption,[b] field ionization	Far infrared laser	20	5 × 10^{-15}
Absorption,[c] field ionization	Variable temperature black body	3.3	10^{-17}
Triggered[d] superradiance	Klystron harmonic	3.6	3 × 10^{-17}

[a] (see ref. 2)
[b] (see ref. 41)
[c] (see ref. 40)
[d] (see ref. 42)

radiation, and in Table 5.3 we present measured sensitivities for Rydberg atom detection of far infrared radiation.

References

1. E. J. Beiting, G. F. Hildebrandt, F. G. Kellert, G. W. Foltz, K. A. Smith, F. B. Dunning, and R. F. Stebbings, *J. Chem. Phys.* **70**, 3551 (1979).
2. T. F. Gallagher and W. E. Cooke, *Appl. Phys. Lett.* **34**, 369 (1979).
3. T. F. Gallagher and W. E. Cooke, *Phys. Rev. Lett.* **42**, 835 (1979).
4. R. Loudon, *The Quantum Theory of Light* (Oxford University Press, London, 1973).
5. T. F. Gallagher, in *Rydberg States of Atoms and Molecules*, eds. R. F. Stebbings and F. B. Dunning (Cambridge University Press, Cambridge, 1983).
6. W. E. Cooke and T. F. Gallagher, *Phys. Rev. A* **21**, 588 (1980).
7. H. A. Bethe and E. A. Salpeter, *Quantum Mechanics of One-and-Two-Electron Atoms* (Academic Press, New York, 1957).
8. U. Fano and A. R. P. Rau, *Atomic Collisions and Spectra* (Academic Press, New York, 1986).
9. A. Lindgard, and S. A. Nielsen, *Atomic Data and Nucl. Data Tables* **19**, 534 (1977).
10. C. H. Townes and A. L. Schawlow, *Microwave Spectroscopy* (McGraw-Hill, New York, 1955).
11. U. Fano and J. W. Cooper, *Rev. Mod. Phys.* **40**, 441 (1968).
12. J. W. Farley and W. H. Wing, *Phys. Rev. A* **23**, 2397 (1981).
13. H. G. Muller, A. Tip, and M. J. van der Wiel, *J. Phys. B* **16**, L679 (1983).
14. T. F. Gallagher, W. Sandner K. A. Safinya, and W. E. Cooke, *Phys. Rev. A* **23**, 2065 (1981).
15. M. Pimbert, *J. Phys. (Paris)* **33**, 331 (1972).
16. T. F. Gallagher, S. A. Edelstein, and R. M. Hill, *Phys. Rev. A* **14**, 2360 (1976).
17. F. Gounand, *J. Phys. (Paris)* **40**, 457 (1979).
18. P. R. Koch, H. Hieronymus, A. F. J. van Raan, and W. Raith, *Phys. Lett.* **75A**, 273 (1980).
19. G. F. Hildebrandt, E. J. Beiting, C. Higgs, G. J. Hatton, K. A. Smith, F. B. Dunning, and R. F. Stebbings, *Phys. Rev. A* **23**, 2978 (1981).
20. W. P. Spencer, A. G. Vaidyanathan, D. Kleppner, and T. W. Ducas, *Phys. Rev. A* **25**, 380 (1982).

21. W. P. Spencer, A. G. Vaidyanathan, D. Kleppner, and T. W. Ducas, *Phys. Rev. A* **26**, 1490 (1982).
22. E. M. Purcell, *Phys. Rev.* **69**, 681 (1946).
23. D. Kleppner, *Phys. Rev. Lett.* **47**, 233 (1981).
24. A. G. Vaidyanathan, W. P. Spencer, and D. Kleppner, *Phys. Rev. Lett.* **47**, 1592 (1981).
25. J. M. Raimond, P. Goy, M. Gross, C. Fabre, and S. Haroche, *Phys. Rev. Lett.* **49**, 117 (1982).
26. R. G. Hulet, E. S. Hilfer and D. Kleppner, *Phys. Rev. Lett.* **55**, 2137 (1985).
27. R. G. Hulet and D. Kleppner, *Phys. Rev. Lett.* **51**, 1430 (1983).
28. L. Holberg and J. L. Hall, *Phys. Rev. Lett.* **53**, 230 (1984).
29. W. M. Itano, L. L. Lewis, and D. J. Wineland, *Phys. Rev. A* **25**, 1233 (1982).
30. F. Gounand, P. R. Fournier, J. Cuvellier, and J. Berlande, *Phys. Lett.* **59A**, 23 (1976).
31. T. F. Gallagher and W. E. Cooke, *Phys. Rev. A* **20**, 670 (1979).
32. K. Bhatia, P. Grafstrom, C. Levinson, H. Lundberg, L. Nilsson, and S. Svanberg, *Z. Phys. A* **303**, 1 (1981).
33. M. Gross, P. Goy, C. Fabre, S. Haroche, and J. M. Raimond, *Phys. Rev. Lett.* **43** 343 (1979).
34. F. Gounand, M. Hugon, P. R. Fournier, and J. Berlande, *J. Phys. B* **12**, 547 (1979).
35. A. G. C. Mitchell and M. W. Zemansky, *Resonance Radiation and Excited Atoms* (Cambridge University Press, New York, 1971).
36. R. C. Stoneman and T. F. Gallagher, *Phys. Rev. Lett.* **55**, 2567 (1985).
37. D. Kleppner and T. W. Ducas, *Bull. Am. Phys. Soc.* **21**, 600 (1976).
38. R. M. Hill, and T. F. Gallagher, US Patent 4,024,396 (1977).
39. J. A. Gelbwachs, C. F. Klein, and J. E. Wessel, *IEEE J. Quant Electronics* **QE-14**, 77 (1978).
40. H. Figger, G. Leuchs, R. Straubinger, and H. Walther, *Opt. Comm.* **33**, 37 (1980).
41. T. W. Ducas, W. P. Spencer, A. G. Vaidyanathan, W. H. Hamilton, and D. Kleppner, *Appl. Phys. Lett.* **35**, 382 (1979).
42. P. Goy, L. Moi, M. Gross, J. M. Raimond, C. Fabre, and S. Haroche, *Phys. Rev. A* **27**, 2065 (1983).

6

Electric fields

The effect of an electric field on Rydberg atoms, the Stark effect, provides an interesting example of how the deviation from the coulomb potential at small r in, for example an alkali atom, significantly alters its behavior in a field from hydrogenic behavior. The difference is somewhat surprising since field effects are fundamentally long range effects, and this difference alone would make the Stark effect worthy of study. However, in addition to its intrinsic interest, the Stark effect is of great practical importance for the study of Rydberg atoms.

Hydrogen

We begin by considering the behavior of the H atom in a static field, ignoring the spin of the electron. We start with the familiar zero field $n\ell m$ angular momentum eigenstates with the immediate goal of showing that important qualitative features of the Stark effect are easily understood using familiar notions. If the applied field E is in the z direction, the potential seen by the electron is given by

$$V = -\frac{1}{r} + Ez. \tag{6.1}$$

Using the zero field $n\ell m$ states we calculate the matrix elements $\langle n\ell m|Ez|n'\ell'm'\rangle$ of the Stark perturbation to the zero field Hamiltonian. Writing the matrix element in spherical coordinates and choosing the z axis as the axis of quantization,

$$\langle n\ell m|Ez|n'\ell'm'\rangle = E\langle n\ell m|r\cos\theta|n'\ell'm'\rangle. \tag{6.2}$$

From the properties of the spherical harmonic angular functions we know that the Ez matrix elements are nonvanishing if $m' = m$ and $\ell' = \ell \pm 1$. The most obvious effect of the field is that it lifts the degeneracy of the ℓm states of a particular n. If we neglect the electric dipole coupling to other n states, the Hamiltonian matrix has the same entry, $-1/2n^2$, for all the diagonal elements and a set of off diagonal $\langle \ell|Ez|\ell\pm1\rangle$ matrix elements, all of which are proportional to E. If we subtract the common energy $-1/2n^2$ we find that all entries of the matrix are now proportional to E. Therefore, when the matrix is diagonalized to find its eigenvalues, they must all be proportional to E, i.e. the eigenvalues are given by $\lambda_i E$. Stated another way, the H atom exhibits a linear Stark shift. Since the eigenvalues are all proportional to E, the eigenvectors are independent of E, and the Stark states are thus field independent linear combinations of zero field ℓ states of the same n and m. The fact that the Stark states have linear Stark shifts means that they have permanent electric dipole moments. Stark states which have positive Stark shifts have the electron localized on the upfield side of the atom and Stark states which have negative Stark shifts have the electron localized on the downfield side of the atom.

Since m is a good quantum number, each set of m states is independent of the others. We consider first the circular $|m| = n - 1$ states. They have no first order Stark shift since there are no other states of the same m and n. There are, however, two $m = n - 2$ states, which exhibit small linear Stark shifts. The Stark shifts are not large since the radial matrix element is small,[1]

$$\langle n\ell|r|n\ell+1\rangle = \frac{-3n\sqrt{n^2 - \ell^2}}{2}. \tag{6.3}$$

The minus sign is for radial wavefunctions $R_{n\ell}(r) \sim +r^\ell$ as $r \to 0$. On the other hand, the $m = 0$ states also include the low ℓ states for which the radial matrix element is large, and the extreme $m = 0$ states have Stark shifts of approximately $\pm 3n^2 E/2$.

The most straightforward way to treat the Stark effect is to use parabolic coordinates, for in parabolic coordinates the problem remains separable even with the electric field.[1-3] The parabolic coordinates are defined in terms of the familiar Cartesian and spherical coordinates by

$$\xi = r + z = r(1 + \cos\theta)$$
$$\eta = r - z = r(1 - \cos\theta) \tag{6.4}$$
$$\phi = \tan^{-1} y/x,$$

and

$$x = \sqrt{\xi\eta}\cos\phi$$
$$y = \sqrt{\xi\eta}\sin\phi$$
$$z = (\xi - \eta)/2 \tag{6.5}$$
$$r = (\xi + \eta)/2.$$

Surfaces of constant ξ or η are paraboloids of revolution about the z axis. From Eqs. (6.4) it is apparent that $\xi = 0$ corresponds to the $-z$ axis, and $\xi = \infty$ corresponds to $r \to \infty$, $\theta \neq 0$. Correspondingly, $\eta = 0$ corresponds to the $+z$ axis and $\eta \to \infty$ corresponds to $r \to \infty$, $\theta \neq 0$. When the field is pointing in the z direction, corresponding to Eq. (6.1), the electron escapes to $\eta = \infty$. Equivalently, the motion in the ξ direction is bound, but the motion in the η direction is not.

In parabolic coordinates the Schroedinger equation for an electron orbiting a singly charged ion with an external field in the z direction, i.e. in the potential given by Eq. (6.1), is written using Eqs. (6.5) as[1-3]

$$\left[-\frac{\nabla^2}{2} - \frac{2}{\xi + \eta} + \frac{E(\xi - \eta)}{2} \right] \psi = W\psi, \tag{6.6}$$

where

$$\nabla^2 = \frac{4}{\xi + \eta} \frac{\partial}{\partial \xi} \left(\xi \frac{\partial}{\partial \xi} \right) + \frac{4}{\xi + \eta} \frac{\partial}{\partial \eta} \left(\eta \frac{\partial}{\partial \eta} \right) + \frac{1}{\xi \eta} \frac{\partial^2}{\partial \phi^2},$$

and W is the energy.

If we assume that solution can be written as the product,[1-3]

$$\Psi(\xi,\eta,\phi) = u_1(\xi) u_2(\eta) e^{im\phi}, \tag{6.7}$$

with m an integer or zero, and substitute Eq. (6.7) into Eq. (6.6), we find two independent equations for $u_1(\xi)$ and $u_2(\eta)$

$$\frac{d}{d\xi} \left(\xi \frac{du_1}{d\xi} \right) + \left(\frac{W\xi}{2} + Z_1 - \frac{m^2}{4\xi} - \frac{E\xi^2}{4} \right) u_1 = 0 \tag{6.8a}$$

and

$$\frac{\partial}{\partial \eta} \left(\eta \frac{du_2}{d\eta} \right) + \left(\frac{W\eta}{2} + Z_2 - \frac{m^2}{4\eta} + \frac{E\eta^2}{4} \right) u_2 = 0. \tag{6.8b}$$

From Eqs. (6.8a) and (6.8b) it is apparent that the sign of m is unimportant, and that the wavefunctions for $\pm m$ are degenerate, as might be expected for this cylindrically symmetric problem. In Eqs. (6.8a) and (6.8b) the separation constants Z_1 and Z_2 are related by[1,2]

$$Z_1 + Z_2 = 1. \tag{6.9}$$

Z_1 and Z_2 may be thought of as the positive charges binding the electron in the ξ and η coordinates. This point will become more apparent shortly. Note that the two differences between Eqs. (6.8a) and (6.8b) are that the field enters with different signs in the two equations and that Z_1 and Z_2 may have different values.

The classic way of determining the energies of hydrogenic levels in a field is to solve the zero field problem in parabolic coordinates and calculate the effect of the field using perturbation theory. The zero field parabolic wavefunctions obtained by solving Eqs. (6.8a) and (6.8b) have, in addition to the quantum numbers n and $|m|$, the parabolic quantum numbers n_1 and n_2, which are nonnegative integers.[1] n_1 and n_2 are the numbers of nodes in the u_1 and u_2 wavefunctions and are related to n and $|m|$ by

$$n = n_1 + n_2 + |m| + 1. \tag{6.10}$$

They are in addition related to the effective charges Z_1 and Z_2 by

$$Z_1 = \frac{1}{n} \left(n_1 + \frac{|m| + 1}{2} \right) \tag{6.11a}$$

and

$$Z_2 = \frac{1}{n} \left(n_2 + \frac{|m| + 1}{2} \right). \tag{6.11b}$$

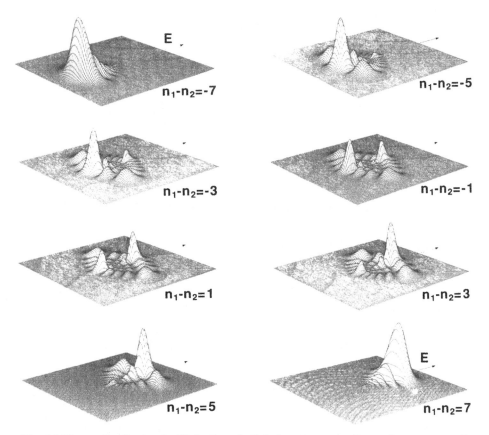

Fig. 6.1 Charge distribution for H, for "parabolic" eigenstates: $n = 8$, $m = 0$, $n_1 - n_2 = -7$ to 7. The dipole moments that give rise to the first order Stark effect are conspicuous (from ref. 4).

The wavefunctions are given in terms of associated Laguerre polynomials, and using the approximate form of the Laguerre polynomials for large arguments, we can write an approximate, unnormalized wavefunction as[1]

$$\psi_{nn_1n_2m} \propto e^{im\phi} \, \xi^{n_1 + |m|/2} \, \eta^{n_2 + |m|/2} \, e^{-(\xi+\eta)/2n}. \qquad (6.12)$$

Taking the squared absolute value of Eq. (6.12) and using the definition of the parabolic coordinates given in Eq. (6.4), we can write an expression for the electron probability distribution in spherical coordinates,[1]

$$|\psi_{nn_1n_2m}|^2 = r^{2n-2}(1 + \cos\theta)^{2n_1 + |m|}(1 - \cos\theta)^{2n_2 + |m|}e^{-2r/n}. \qquad (6.13)$$

Eq. (6.13) shows explicitly that the $n_1 - n_2 \approx n$ wavefunctions are localized along the $+z$ axis while the $n_2 - n_1 \approx n$ wavefunctions are localized along the $-z$ axis. Similarly, $n_1 - n_2 \approx 0$ wavefunctions are localized near the $z = 0$ plane. In Fig. 6.1 we show the $n = 8$, $m = 0$, $n_1 - n_2 = -7$ to 7 wavefunctions to illustrate the polarization along the field direction.[4] Careful inspection also reveals that the nodes lie on parabolas. Using the zero field wavefunctions the first order energies are[1]

$$W_{nn_1n_2m} = \frac{-1}{2n^2} + \frac{3}{2}E(n_1 - n_2)n.$$ (6.14)

While there is no explicit m dependence, there is an implicit m dependence due to the fact that $n_1 + n_2 + m + 1 = n$. For $m = 0$, allowed values of $n_1 - n_2$ are $n - 1$, $n - 3 \ldots, -n + 1$ while for $m = 1$ they are $n - 2, n - 4 \ldots -n + 2$. The even and odd m levels are interleaved. Furthermore, for the quantum numbers n and m there are $n - |m|$ Stark levels. Note that for the circular states, $|m| = n - 1, n_1 = n_2 = 0$, and the first order Stark shift vanishes, as we have already seen.

By taking into account the matrix elements off diagonal in n, the second and higher order contributions to the Stark effect can be calculated. If the calculation is carried through second order, the energies are given by[1]

$$W_{nn_1n_2m} = \frac{-1}{2n^2} + \frac{3\,En}{2}(n_1 - n_2) - \frac{E^2}{16}n^4\,[17n^2 - 3(n_1 - n_2)^2 - 9m^2 + 19].$$ (6.15)

The second order shift is always to lower energy, as expected from oscillator strength sum rules.[5] Equally important, the second order shift breaks the m degeneracy. The energies can be calculated to higher order,[6,7] but for many applications the first and second order shifts are adequate. This point is made explicitly by Fig. 6.2, a plot of the H $|m| = 1$ energy levels from $n = 8$ to $n=14$ in electric fields from 0 to 2×10^{-5} (0 to 10^5 V/cm). Fig. 6.2 makes several important points about the Stark effect in H. First, the levels exhibit apparently linear Stark shifts from zero field to the point at which field ionization occurs at significant rates, shown by the broken lines in Fig. 6.2. Only when looking along the diagram from the side of the page is the curvature of the levels evident. Second, the Stark levels of adjacent n cross; there is no coupling, at least at this level of resolution. Finally, the red, or down shifted, Stark states ionize near the classical ionization limit, but the higher energy blue, or upshifted, states ionize only at higher fields.

As shown by Fig. 6.2, the Stark shifts are quite linear, except at the highest fields shown, and the first order energies of Eq. (6.14) are adequate for many purposes. On the other hand, even the second order energies of Eq. (6.15) are not adequate for comparisons to precise measurements of the energy levels in field. For example, using a fast H beam Koch[8] observed the $(10,8,0,1) \rightarrow (25,21,2,1)$ and $(10,0,9,0) \rightarrow (30,0,29,0)$ transitions in fields of 2.514 and 0.689 kV/cm respectively using the 10 μm R24 and P24 CO_2 laser lines. Here the states are denoted $(n,n_1,n_2,|m|)$. From his results it is apparent that the energy of the red shifted $(30,0,29,0)$ state is given accurately by perturbation theory results at about tenth order. In contrast, the blue shifted $(25,21,2,1)$ state is never given to the same accuracy, and, in fact, the perturbation theory result does not converge but oscillates about the experimental result with the minimum amplitude of the oscillations at about tenth order.[8]

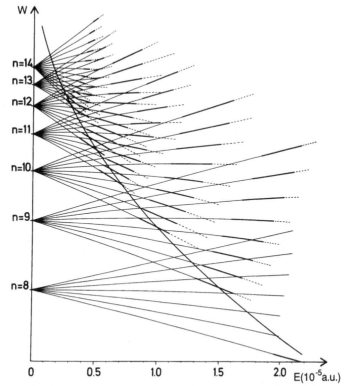

Fig. 6.2 Stark structure and field ionization properties of the $|m| = 1$ states of the H atom. The zero field manifolds are characterized by the principal quantum number n. Quasi-discrete states with lifetime $\tau > 10^{-6}$ s (solid line), field broadened states 5×10^{-10} s $< \tau < 5 \times 10^{-6}$ s (bold line), and field ionized states $\tau < 5 \times 10^{-10}$ s (broken line). Field broadened Stark states appear approximately only for $W > W_c$. The saddle point limit $W_c = -2\sqrt{E}$ is shown by a heavy curve (from ref. 3).

The second point made by Fig. 6.2, although not with very high resolution, is that the levels of n and $n + 1$ cross. The extreme $m = 0$ red and blue states have Stark shifts of approximately $\pm 3n^2 E/2$, which combined with the n − (n + 1) energy spacing of $1/n^3$, yields a crossing field of

$$E = 1/3n^5. \tag{6.16}$$

This field is associated with the Inglis–Teller limit, where Stark broadening in a plasma causes levels of adjacent n to become unresolvable.[9] The fact that the red $n + 1$ and blue n levels actually cross, in spite of the fact that they have the same m, is a result of the fact that the use of the parabolic coordinates diagonalizes the Runge–Lenz vector, which has a different eigenvalue in the two crossing states. The fact that both the Runge–Lenz vector and angular momentum are conserved in H in zero field was exploited by Park[10] to show in an elegant way that the transformation coefficient between the parabolic nn_1n_2m states and the spherical

$n\ell m$ states was simply a Clebsch–Gordon coefficient. If we expand the Stark state $|nn_1n_2m\rangle$ as

$$|nn_1n_2m\rangle = \sum_\ell |n\ell m\rangle\langle n\ell m|nn_1n_2m\rangle, \qquad (6.17)$$

then the transformation coefficient is given in terms of Wigner 3J symbols as

$$\langle nn_1n_2m|n\ell m\rangle = (-1)^{(1-n+m+n_1-n_2)/2 + \ell}$$

$$\times \sqrt{2\ell + 1}\begin{pmatrix} \dfrac{n-1}{2} & \dfrac{n-1}{2} & \ell \\[2mm] \dfrac{m+n_1-n_2}{2} & \dfrac{m-n_1+n_2}{2} & -m \end{pmatrix}. \qquad (6.18)$$

An equivalent form is given by Englefield.[11] It is possible to find quite a variety of phases for the transformation coefficients of Eq. (6.18).[10–13] The phase depends on the phase conventions established for the spherical and parabolic states. The choice of phase in Eq. (6.18) is for spherical functions with an r^ℓ, as opposed to $(-r)^\ell$, dependence at the origin and the spherical harmonic functions of Bethe and Salpeter. A few examples of the spherical harmonics are given in Table 2.2. The parabolic functions are assumed to have an $(\xi n)^{|m|/2}$ behavior at the origin and an $e^{im\phi}$ angular dependence. This convention means, for example, that for all Stark states with the quantum number m, the transformation coefficient $\langle nn_1n_2m|nmm\rangle$ is positive. To the extent that the Stark effect is linear, i.e. to the extent that the wavefunctions are the zero field parabolic wavefunctions, the transformation of Eqs. (6.17) and (6.18) allows us to decompose a parabolic Stark state in a field into its zero field components, or vice versa.

Harmin[14] derived an approximate form of the transformation of Eq. (6.18) which is particularly useful. The information contained in the quantum numbers n_1 and n_2 always appears as $\pm(n_1 - n_2)$ and can also be represented by Z_1 and Z_2, the separation constants, a notion which easily passes into the regime in which n_2 is not a well defined quantum number. (In a field n_1 is always a good quantum number.) Explicitly, Harmin showed that Eq. (6.18) can also be written as[14]

$$\langle nn_1n_2m|n\ell m\rangle = \sqrt{\frac{2}{n}}(-1)^\ell P_{\ell m}(Z_1 - Z_2), \qquad (6.19)$$

where $P_{\ell m}$ is a normalized associated Legendre polynomial of Bethe and Salpeter.[1] It is defined for $m < 0$, and $\int_{-1}^1 P_{\ell m}^2(x)dx = 1$.[1] Eq. (6.19) is only defined over the region $-1 \le Z_1 - Z_2 \le 1$, which is, however, a less restrictive requirement than that both n_1 and n_2 quantum numbers have integral values. We may also write Eq. (6.19) in terms of the parabolic n_1 and n_2 quantum numbers using $Z_1 - Z_2 = (n_1 - n_2)/n$ from Eq. (6.11). From Eq. (6.19) it is easy to see how the parabolic states are composed of the $n\ell m$ states. From either the properties of the 3J symbol or the fact that $P_{00}(x) = 1/\sqrt{2}$ it is apparent that the *ns* state is evenly

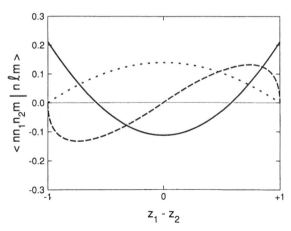

Fig. 6.3 Plots of the projection of the $n = 100$ Stark states of $m = 0, 1$, and 2 on the $n = 100$ zero field d states, $<100\, n_1 n_2 m|100\, 2m>$ for $m = 0, 1$, and 2: $m = 0$ (——), $m = 1$ (– – –) $m = 2$ (. . .). The values are obtained from Eqs. (6.11) and (6.19).

spread over the nn_1n_20 states. The $\ell \geq 1$ states are more interesting. In Fig. 6.3 we plot Eq. (6.19) for $\ell = 2\, |m| = 0, 1$, and 2. From Fig. 6.3 it is clear that the $m = 0$ states at the edge of the Stark manifold, i.e. those for which $|n_1 - n_2| \sim n$, have much more d character than those in the center. On the other hand, the $|m| = 2$ states at the center of the Stark manifold have substantial d character while the states at the edges do not. For $n = 10$ and $|m| = 2$ the largest value of $|Z_1 - Z_2| = |n_1 - n_2|/n$ is 0.7, but at higher n it approaches 1, at which point the $|m|=2$ Stark states at the edge of a Stark manifold have vanishingly small d character.

Working out the parabolic wavefunctions in terms of the Laguerre polynomials is useful in the analytic calculation of the Stark effect using perturbation theory. However, it is not useful in very strong electric fields. Here we outline a more general procedure which is valid in strong fields and lends itself to numerical computations. If we replace $u_1(\xi)$ and $u_2(\eta)$ in Eqs. (6.8a) and (6.8b) by[1–3,15]

$$u_1(\xi) = \frac{\chi_1(\xi)}{\sqrt{\xi}} \tag{6.20a}$$

$$u_2(\eta) = \frac{\chi_2(\eta)}{\sqrt{\eta}} \tag{6.20b}$$

and introduce a normalizing factor of $\sqrt{2\pi}$ to conform to common usage, the wavefunction is given by

$$\psi(\xi, \eta, m) = \frac{\chi_1(\xi)\, \chi_2(\eta) e^{im\phi}}{\sqrt{2\pi\varepsilon\eta}}, \tag{6.21}$$

which leads to the separated equations written as

$$\frac{d^2\chi_1}{d\xi^2} + \frac{\chi_1}{2}[W - V(\xi)] = 0 \tag{6.22a}$$

and

$$\frac{d^2\chi_2}{d\eta^2} + \frac{\chi_2}{2}[W - V(\eta)] = 0, \tag{6.22b}$$

where

$$V(\xi) = 2\left(-\frac{Z_1}{\xi} + \frac{m^2 - 1}{4\xi^2} + \frac{E\xi}{4}\right), \tag{6.23a}$$

and

$$V(\eta) = 2\left(-\frac{Z_2}{\eta} + \frac{m^2 - 1}{4\eta^2} - \frac{E\eta}{4}\right) \tag{6.23b}$$

The χ_1 and χ_2 wavefunctions describe the motion of particles of total energy $W/4$ in the potentials $V(\xi)/4$ or $V(\eta)/4$. It is also interesting to note that for the case $E=0$ the wave equations of Eqs. (6.22a) and (6.22b) are similar to the radial equation for the coulomb potential in spherical coordinates. Explicitly, making the substitutions

$$2W \to W/2$$
$$\ell(\ell + 1) \to m^2 - 1 \tag{6.24}$$
$$Z \to Z_1, Z_2$$
$$r \to \xi, \eta$$

in Eq. (2.10) for $\rho(r) = rR(r)$ yields Eqs. (6.22a) and (6.22b).

When $E = 0$, the potential $V(\eta)$ with Z_2 substituted for Z_1 is the same as $V(\xi)$. It is apparent that for small values of Z_1 and Z_2 the potentials $V(\xi)$ and $V(\eta)$ are not as deep as for the higher values of Z_1 and Z_2. Thus at the same energy W the number of nodes which occur in the wavefunction increases with Z_1 or Z_2. Inspecting Eq. (6.24), we can see that when $E \neq 0$ that $V(\xi)$ and $V(\eta)$ are no longer identical for the same values of Z_1 and Z_2. The motion in the ξ direction is bounded for all energies since $V(\xi) \to E\xi$ as $\xi \to \infty$. In contrast, the motion in the η direction is unbounded, for any energy, since $V(\eta) \to -E\eta$ as $\eta \to \infty$. The different behaviors of $V(\xi)$ and $V(\eta)$ as ξ and $\eta \to \infty$ lead to qualitatively different wavefunctions. The motion in the ξ direction has a well defined integer quantum number, n_1, and the motion in the η direction is, in principle, a continuum, containing resonances. In practice, at energies far below the saddle point in $V(\eta)$ the motion in the η direction also has a good quantum number, n_2. In Fig. 6.4 we show the potentials $V(\xi)$ and $V(\eta)$ for $E=10^{-6}$, and $|m| = 1$. The potentials are shown for Z_1 and $Z_2 = 0.1, 0.5$, and 0.9, corresponding to a blue shifted Stark state, a middle state, and a red shifted state. As shown by Fig. 6.4(a), the ξ motion is always bounded. In Fig. 6.4(b) it is apparent that the saddle point in $V(\eta)$ occurs at lower energy for the red state ($Z_2 = 0.9$) than for the blue state. Had we drawn the potential $V(\eta)$ for $m=0$, it would be deeper at $\eta = 0$ and would lie generally lower than the $|m| = 1$ potential. For $|m| > 1$ the potential has a $1/\eta^2$ centrifugal barrier at small η and lies generally above the $m = 1$ potentials of Fig. 6.4.

To solve Eqs. (6.22a) and (6.22b) for $E \neq 0$ we take advantage of the fact that the motion in ξ is bounded. This motion always has a bound wavefunction and a

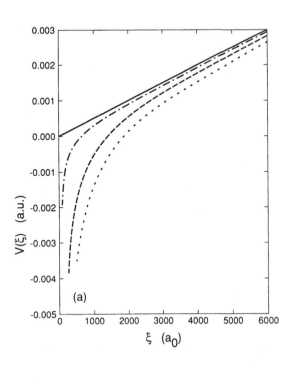

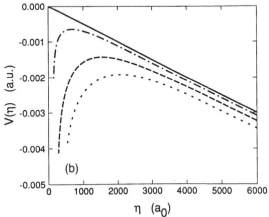

Fig. 6.4 Plots of the potentials $V(\xi)$ and $V(\eta)$ in a field $E = 10^{-6}$ for $|m| = 1$. (a) $V(\xi)$ for $Z_1 = 0$ (——), $Z_1 = 0.1$ (– · · –), $Z_1 = 0.5$ (– – – –), and $Z_1 = 0.9$ (· · ·) (b) $V(\eta)$ for $Z_2 = 0$ (——), $Z_2 = 0.1$ (– · · –), $Z_2 = 0.5$(– – – –), and $Z_2 = 0.9$ (· · ·). Note that the saddle point in $V(\eta)$ occurs at lower energy for larger values of Z_2. As $\eta \to \infty$ the Stark effect dominates both potentials, so that the ξ motion is always bounded.

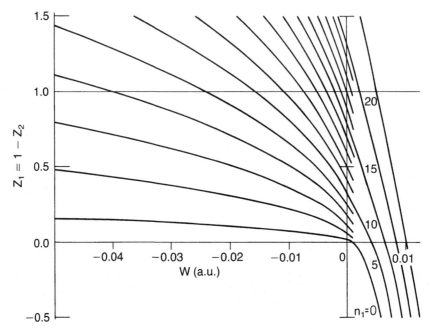

Fig. 6.5 Eigenvalues $Z_1 = 1 - Z_2$ as a function of energy for $m = 0$, $E = 1.5 \times 10^{-5}$ au $= 77$ kV/cm, and $n_1 = 0–20$ (from ref. 15).

well defined, integral quantum number, n_1, which is simply the number of nodes in the $\chi_1(\xi)$ wavefunction. For any energy W and any value of n_1 we can find the separation constant, or eigenvalue, Z_1. A straightforward way of finding the eigenvalues of Z_1 is to use the unnormalized WKB approximation i.e.[1,3,15]

$$\chi_1(\xi) = \frac{1}{[T(\xi)]^{1/4}} \cos\left[\int_{\xi_1}^{\xi} \sqrt{T(x)}\, dx - \frac{\pi}{4}\right], \tag{6.25}$$

where $T(\xi) = [W - V(\xi)]/2$, and ξ_1 is the inner turning point. The WKB wavefunction of Eq. (6.25) satisfies the quantization condition

$$\int_{\xi_1}^{\xi_2} \sqrt{T(\xi)}\, d\xi = \left(n_1 + \frac{1}{2}\right)\pi, \tag{6.26}$$

where n_1 is a positive integer and ξ_1 and ξ_2 are the inner and outer classical turning points, where $W = V(\xi)$. An alternative approach is use a Numerov algorithm to find the allowed wavefunctions, as discussed by Luc-Koenig and Bachelier.[3] In either case it is important to remember that for the same values of W, m, and E there exist many solutions to Eq. (6.26) corresponding to different values of the quantum number n_1 and the eigenvalue Z_1. These solutions are orthogonal, which is why the H Stark energy levels of the same m cross, as shown by Fig. 6.2.

In Fig. 6.5 we show the eigenvalues Z_1 associated with several values of the quantum number n_1 as a function of the energy W. These values of Z_1 are obtained by means of the WKB quantization condition of Eq. (6.26). For a constant value of n_1, the value of Z_1 decreases with increasing energy. Keeping n_1 constant means the WKB integral of Eq. (6.26) must remain a constant. The WKB integral is an integral of the momentum, or the square root of twice the kinetic energy, over the classically allowed spatial region. Increasing the energy increases both the spatial range covered by the integral and the kinetic energy. To keep the same value of n_1 as W is increased, the depth of the small ξ part of the potential is reduced, by reducing Z_1.

For fixed W, m, and E there exists a series of values of n_1 and Z_1 implying a series of allowed values of $Z_2 = 1 - Z_1$. The known value of Z_2 can now be used to solve Eq. (6.22b). Since $V(\eta) \rightarrow -E\eta/2$ as $\eta \rightarrow \infty$, in principle, the solution to Eq. (6.22b) is a continuum wave for all values of the energy. Accordingly we normalize the $\chi_2(\eta)$ wavefunction per unit energy as $\eta \rightarrow \infty$. The asymptotic form of the χ_2 wavefunction is easily obtained using the WKB approximation as[3]

$$\chi_2(\eta) = \frac{N}{(E\eta + 2W)^{1/4}} \sin\left[\frac{1}{3E}(E\eta + 2W)^{3/2} + \alpha\right], \tag{6.27}$$

where α is a phase. In Eq. (6.27) we see the expected oscillatory behavior and an amplitude decreasing as the square root of the momentum. Normalization per unit energy is often obtained by requiring that the inward or outward flux $I = 1/2\pi$. An approach consistent with the one given in Chapter 2 is to require

$$N^2 \int \psi^* d\tau \int_{W-\Delta W}^{W+\Delta W} \psi dW' = 1. \tag{6.28}$$

In parabolic coordinates the volume element $d\tau = (\xi + \eta) d\xi d\eta d\phi/4$. Using the general form of the wavefunction given in Eq. (6.21) and carrying out the angular integration, leads to

$$N^2 \int_0^\infty \int_0^\infty \chi_1^*(\xi) \chi_2^*(\eta) \left(\frac{1}{\eta} + \frac{1}{\xi}\right) \frac{d\xi d\eta}{4} \int_{W-\Delta W}^{W+\Delta W} \chi_1(\xi) \chi_2(\eta) dW'. \tag{6.29}$$

The major contribution to the integral of Eq. (6.29) comes from $\eta \rightarrow \infty$, in which case the $1/\eta$ term may be ignored, leaving

$$N^2 \int_0^\infty \frac{(\chi_1^*(\xi)\chi_1(\xi))}{\xi} d\xi \int_0^\infty \chi_2^*(\eta) d\eta \int_{W-\Delta W}^{W+\Delta W} \chi_2(\eta) dW' = 1. \tag{6.30}$$

Requiring that the bound $\chi_1(\xi)$ wavefunction be normalized according to

$$\int_0^\infty \frac{\chi_1^2(\xi)}{\xi} d\xi = 1 \tag{6.31}$$

yields

$$N = \frac{1}{\sqrt{2\pi}}. \tag{6.32}$$

Thus, the amplitude of the continuum $\chi_2(\eta)$ wave varies insignificantly with energy as $r \rightarrow \infty$.

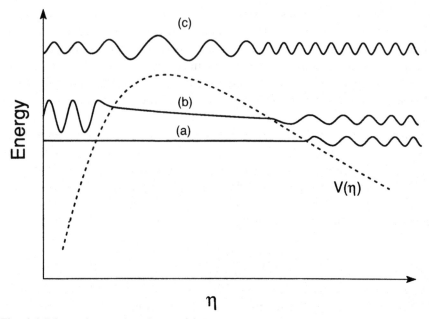

Fig. 6.6 Schematic wavefunctions $\chi_2(\eta)$ for several choices of energy W. (a) An energy below V_b the peak of the potential $V(\eta)$ but away from a resonance of the inner well of the potential. The wavefunction in vanishingly small in the inner well of the potential. (b) An energy below V_b and equal to one of the resonances of the inner well. (c) An energy above V_b.

We are not so much interested in the wavefunction at $r = \infty$ as we are in the wavefunction in the inner potential well of $V(\eta)$. Examining plots of the potential $V(\eta)$, it is apparent that there is, in general, a peak V_b in the potential for $\eta > 0$. If the energy W is above the peak of the potential, $W > V_b$, the entire region is classically accessible. If the energy is less than the peak of the potential barrier, V_b, there are two classically allowed regions, which are connected by a classically forbidden region through which tunnelling occurs. The ratio of the amplitude of the wavefunction in the inner well to that in the outer well is the quantity of interest.

Consider for a moment the energy dependence of $\chi_2(\eta)$ shown qualitatively in Fig. 6.6. Outside the barrier it has the standing wave form of Eq. (6.27) with virtually the same amplitude irrespective of energy. At an energy below the peak of the barrier the small transmission through the barrier allows some of the exterior wavefunction to penetrate into the inner well. Due to the small transmission, in general the wavefunction in the inner well of $V(\eta)$ is negligible, as shown by Fig. 6.6(a). However, at a resonance of the inner well, when an integral number, n_2, of nodes fit, the amplitude of the wavefunction in the inner well becomes very large, as shown in Fig. 6.6(b). This situation is no different from a microwave cavity tenuously coupled to a transmission line.

Far below the barrier in $V(\eta)$ the transmission through the barrier is vanishingly small and the resonances are so sharp as to be bound states for all practical purposes, i.e. n_2 is, for all intents and purposes, a good quantum number. As the energy is increased the transmission through the barrier increases, increasing the width of the resonances and decreasing the amplitude of $\chi_2(\eta)$ in the inner well. Finally, when the energy exceeds the top of the barrier, the wavefunction has an amplitude which is spatially smoothly varying, displaying minimal energy dependence as shown by Fig. 6.6(c). While Fig. 6.6 shows the variation of the amplitude of the wavefunction in the inner well of $V(\eta)$ in a qualitative way, we need a quantitative measure. A useful one takes advantage of the fact that for small values of ξ and η the regular solutions of $\chi_1(\xi)/\sqrt{\xi}$ and $\chi_2(\eta)/\sqrt{\eta}$ behave as $\xi^{|m|/2}$ and $\eta^{|m|/2}$. A useful approach is to define a small ξ, η wavefunction as[3]

$$\psi = \frac{\sqrt{C_{n_1}^m}}{\sqrt{2\pi}} (\xi\eta)^{|m|/2} e^{im\phi},\tag{6.33}$$

where $C_{n_1}^m$, the density of states, is a measure of the probability of finding the electron near the origin, which is of practical interest for photoexcitation from the ground state and for field ionization of nonhydrogenic atoms. $C_{n_1}^m$ is easily obtained from the normalized $\chi_1(\xi)$ and $\chi_2(\eta)$ wavefunctions.

Figure 6.7 is a plot of $\sqrt{C_{n_1}^m}$, the square root of the density of states for $m = 0$, $n_1 = 7$, and $E = 1.5 \times 10^{-5}$ au. In this case the maximum in the potential barrier of $V(\eta)$ is at $V_b = -0.00372$. Above V_b there is, in essence, a continuum. Below V_b there is a series of sharp states which each have the approximate quantum numbers n_2 shown, corresponding to n_2 nodes of the $\chi_2(\eta)$ wavefunction in the inner potential well of $V(\eta)$. For any m, each value if the quantum number n_1 has a spectrum of $\sqrt{C_{n_1}^m}$ analogous to the one shown in Fig. 6.7.

Field ionization

Field ionization, which is of great practical importance in the study of Rydberg atoms, has been introduced implicitly in Figs. 6.2, 6.6, and 6.7. We now consider it explicitly. The order of magnitude of the field required for ionization can be estimated using the method outlined in Chapter 1. The combined coulomb–Stark potential,

$$V = -1/r + Ez,\tag{6.34}$$

has its saddle point on the z axis at $z = -1/\sqrt{E}$ where the potential has the value $V = -2\sqrt{E}$. Thus if an atom is in a state of $m = 0$, so that there is no additional centrifugal potential, and if the electron is bound by an energy W, a field given by

$$E = \frac{W^2}{4}\tag{6.35}$$

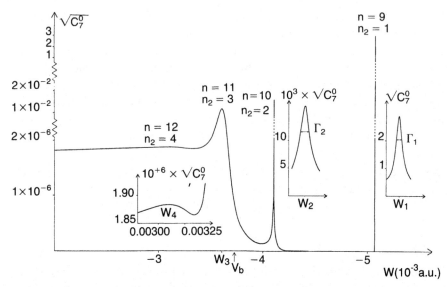

Fig. 6.7 Typical energy dependence of the square root of the density of states $\sqrt{C_{n_1}^m}$ for the states $n_1 = 7$, $|m| = 0$ in the field $E = 1.5 \times 10^{-5}$ au. V_b is the energy of the peak in the potential $V(\eta)$. Each resonance is characterized by the parabolic quantum number n_2 and the width Γ_{n_2} which varies rapidly with n_2. For $W < V_b$ the spectrum exhibits quasi-discrete resonances; for $W > V_b$ the continuum exhibits oscillations which damp out with increasing n_2. The energies and widths of the prominent resonances are.

$$W_1 = -0.005066 \qquad \Gamma_1 = 3.7 \times 10^{-16} \text{ au}$$
$$W_2 = -0.004097 \qquad \Gamma_2 = 6.368 \times 10^{-8} \text{ au}$$
$$W_3 = -0.003601 \qquad \Gamma_3 = 1.353 \times 10^{-4} \text{ au}$$
$$W_4 = -0.003100 \qquad \Gamma_4 = 1.637 \times 10^{-3} \text{ au}$$

(from ref. 3).

is adequate for ionization to occur classically. This field is usually termed the classical field for ionization. If we ignore the Stark shift of a Rydberg state of principal quantum number n and write the binding energy in terms of n we obtain the familiar result for the classical field for ionization in terms of the principal quantum number,

$$E = \frac{1}{16n^4}. \tag{6.36}$$

These results, Eqs. (6.35) and (6.36), are only valid for $m = 0$ states. In higher $|m|$ states there is a $1/(x^2 + y^2)$ centrifugal potential keeping the electron away from the z axis, and the centrifugal barrier raises the threshold field of $m \neq 0$ states.[16] Specifically, for $m \neq 0$ states the fractional increase in the field required for ionization, compared to an $m = 0$ state of the same energy is[16]

$$\frac{\Delta E}{E} = \frac{|m|\sqrt{W}}{\sqrt{2}} = \frac{|m|}{2n}. \tag{6.37}$$

Actually, the classical picture outlined above has two serious defects. First, Eq. (6.36) ignores the Stark shifts, as does Eq.(6.37) if we choose to use the $|m|/2n$ form. Eq. (6.35) and the $|m|\sqrt{W}/\sqrt{2}$ form of Eq. (6.37) do not suffer from this defect. Second, the classical approach ignores the spatial distribution of the wavefunctions. As shown by Fig. 6.2 for H states of the same n and $|m|$, the higher energy, or blue, $n_1 - n_2 \sim n$, Stark states require higher fields to ionize than the lower energy, or red, $n_2 - n_1 \sim n$ states. This point is made quite graphically by Bethe and Salpeter[1] using the data of Rausch von Traubenberg.[17] A simple physical argument for why blue states are harder to ionize than red states is that the electron is located on the side of the atom away from the saddle point in the potential in the blue states, whereas the electron is adjacent to the saddle point in the red states.

In parabolic coordinates the motion in the ξ direction is bounded. Thus for ionization to occur the electron must escape to $\eta = \infty$. Classically, ionization only occurs for energies above the peak V_b of the potential barrier in $V(\eta)$. If we ignore the short range $1/\eta^2$ term in Eq. (6.23b), an approximation good for low m, and set $W = V_b$ we find the required field for ionization to be

$$E = \frac{W^2}{4Z_2}. \tag{6.38}$$

Eq. (6.38) differs from Eq. (6.35) by the factor of Z_2, the effective charge. Consider three low $|m|$ states of the same energy. A blue state of $n_1 - n_2 \approx n$ has $Z_2 \approx 1/n$, a central state has $n_1 = n_2 \sim n/2$ and $Z_2 \approx 1/2$, and a red state has $n_2 - n_1 \sim n$ and $Z_2 \sim 1$. If the three states have the same energy the red one will be most easily ionized.

For the extreme red Stark state of high n Eq. (6.38) reduces to Eq. (6.35) since $Z_2 \approx 1$. For this Stark state the Stark shift increases the binding energy, and for an $m = 0$ state the energy is adequately given using the linear Stark effect as

$$W = -\frac{1}{2n^2} - \frac{3n^2 E}{2}. \tag{6.39}$$

Using this energy we find a threshold field

$$E = \frac{1}{9n^4}. \tag{6.40}$$

The numerical factor of 1/9 instead of 1/16 is due to the Stark shift of the level.

For the blue states it is not possible to estimate simply the threshold field. However, blue and red states of the same n and $m=0$ often have threshold fields differing by a factor of 2.

We have thus far discussed field ionization as occurring only when it is classically allowed, when $E > W^2/4Z_2$, i.e. when the energy is above V_b, the peak of the barrier in $V(\eta)$, as shown in Fig. 6.6. At lower energies tunnelling occurs, and accurate calculations of ionization rates have been made.[2,18–20] As an example, we show in Fig. 6.8 the ionization rates calculated by Bailey *et al.*[20] using

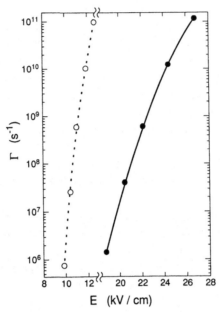

Fig. 6.8 Calculated ionization rates of the (n, n_1, n_2, m) states as a function of field using the values of ref. 20: the $(15, 0, 14, 0)$ red state (○) and the $(15, 14, 0, 0)$ blue state (●).

the WKB method of Rice and Good[18] for the H $n=15$ extreme red and blue $m = 0$ Stark states over the range of ionization rates from $10^6 \, \text{s}^{-1}$ to $10^{11} \, \text{s}^{-1}$. From Fig. 6.8 it is apparent that the blue states ionize at higher fields than red states of the same n. It is also useful to note that the ionization rates increase rapidly with field, implying that simply computing when ionization is classically allowed is often quite adequate for practical applications.

Using a fast (8 keV) beam of H Koch and Mariani[21] have made precise measurements of the ionization rates of state selected $|m| = 0$ and 1, $n = 30\text{--}40$ blue Stark states of H. As can be seen in Fig. 6.9, the measured rates are in quite good agreement with, but typically offset from, the analytic results obtained by Damburg and Kolosov,[22] using the approximation that the field is weak enough that n_2 is very nearly a good quantum number. However, as shown, the measured rates are in excellent agreement with the exact numerical calculations of Damburg and Kolosov,[23] which are in good agreement with the ionization rates calculated by Bailey et al.[20] The excellent agreement between theory and experiment suggests that a nonrelativistic treatment is quite adequate. However, Bergeman has calculated that it should be possible to see the effects of spin orbit coupling in field ionization.[24]

The ionization measurements of Koch and Mariani[20] are for blue states of $n_1 \approx n$. Rottke and Welge[25] carried out measurements on the relatively red $|m| = 0$ and 1 states of $n = 18$ and blue states of $n = 19$ verifying the rapid decrease in ionization rate with n_1 shown theoretically in Fig. 6.9. In their measurements

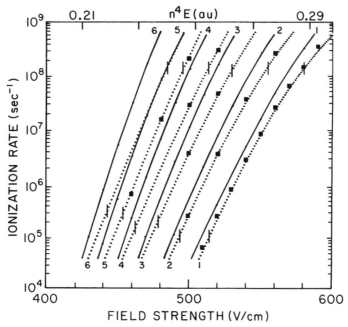

Fig. 6.9 Dotted lines, experimental ionization rate curves for the $(n,n_1,n_2,|m|)$ states 1: (40,39,0,0); 2: (40,38,0,1); 3: (40,38,1,0); 4: (40,37,1,1); 5: (40,37,2,0); 6: (40,36,2,1). The tick marks represent the range of validity of the experiment. Solid line, theoretical curves calculated with Eq. (6) of ref. 22. Squares, numerical theoretical results from ref. 23 (from ref. 21).

they used pulsed laser excitation of a beam of H atoms in a field of 5714 V/cm followed by time resolved detection of the electrons from the ionizing H atoms.

Most of the discussion of field ionization thus far has been focused on states of low $|m|$. For states of high $|m|$ we may no longer neglect the centrifugal potential, and it is not possible to derive a simple expression such as Eq. (6.38). Nonetheless, since the centrifugal term raises the barrier in the potential $V(\eta)$ it is clear that the ionizing field must be higher for high m states. A graphic illustration of this point is provided by Fig. 6.10, a simulation of the ionizing fields of all the $|m| \geq 3$ states of $n = 31$.[26] Measurements of the field ionization profiles of collisionally induced mixtures of the degenerate high $\ell,|m|$ states of the same n are in substantial agreement with the profile predicted in Fig. 6.10.[26]

Nonhydrogenic atoms

Atoms other than H have characteristics essentially similar to those of the H atom in an electric field, but there are important differences due to the presence of the finite sized ionic core. In zero field the presence of the core simply depresses

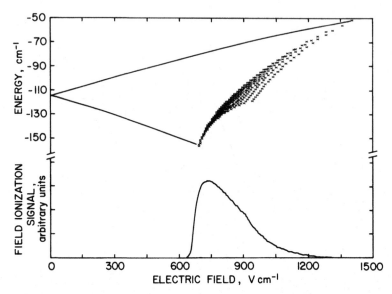

Fig. 6.10 Calculated SFI profile for diabatic ionization of the H like $|m| \geq 3$ states. Top, extreme members of the $n = 31$, $|m| = 3$ Stark manifold. The crosses represent the points at which each $|m| \geq 3$ Stark state achieves an ionization rate of 10^9 s^{-1}. Bottom, calculated SFI profile for diabatic ionization of a mixture containing equal numbers of atoms in each $|m| \geq 3$ Stark level for $n = 31$ at a slew rate of 10^9 V/cm s. (from ref. 26).

the energies, particularly those of the lowest ℓ states. As long as the core is spherically symmetric, it does not alter the spherical symmetry of the problem, and the effect is relatively minor. However, with a finite sized ionic core the wavefunction is no longer separable in parabolic coordinates. As a result the parabolic quantum number n_1, which is a good quantum number in H, is no longer good.

The most important implication of n_1 not being a good quantum number is that blue and red states are coupled by their slight overlap at the core. In the region below the classical ionization limit blue and red states of adjacent n do not cross as they do in H, but exhibit avoided crossings as a result of their being coupled. Above the classical ionization limit blue states, which would be perfectly stable in H, are coupled to degenerate red states, which are unbound, and ionization occurs rapidly compared to radiative decay. It is really an autoionization process in which the blue state is coupled to the red continuum state at the ionic core.

In this chapter we shall treat the Stark effect in nonhydrogenic atoms by methods which, while not the most powerful, provide physical insight. We begin by calculating the energy levels of Na, as an example, in the regime where the field ionization rates are negligible. Ignoring the spin of the Rydberg electron, the Hamiltonian is given by

$$H = -\frac{\nabla^2}{2} + \frac{1}{r} + V_d(r) + Ez, \qquad (6.41)$$

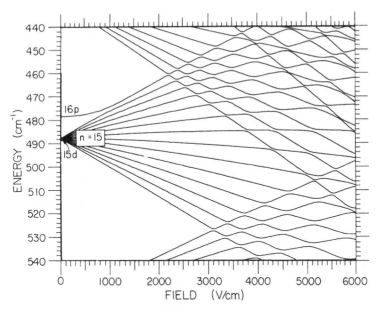

Fig. 6.11 Calculated Na $|m| = 1$ energy levels (from ref. 27).

where $V_d(r)$ is the difference between the Na potential and the coulomb potential, $-1/r$. Consequently, $V_d(r)$ is only nonzero near $r = 0$. A straightforward and effective method of calculating the energies and wavefunctions is direct diagonalization of the Hamiltonian matrix. If we use the Na $n\ell m$ spherical states as basis functions, the Hamiltonian has diagonal elements given by the known energies of the zero field $n\ell m$ states, i.e.

$$\langle n\ell m|H|n\ell m\rangle = \frac{1}{2(n - \delta_\ell)^2}, \tag{6.42}$$

and off diagonal matrix elements $\langle n\ell m|Ez|n'\ell \pm 1m\rangle$ which may be calculated using the Numerov method outlined in Chapter 2. If several n manifolds of ℓ states are included, the eigenvalues of the matrix give the energy levels of the Na atom in the field, and the eigenvectors give the Stark states in terms of the zero field $n\ell m$ states.

In Figs. 6.11 and 6.12 we show the energy levels for the Na states of $n \sim 15$ for $|m| = 1$ and 0, respectively, obtained by direct diagonalization of the Hamiltonian matrix.[27] In zero field the quantum defects 1.35 and 0.85, of the Na s and p states displace them from the high ℓ states and they only exhibit large Stark shifts when they intersect the manifold of Stark states. A second obvious difference between Figs. 6.11 and 6.12 and Fig. 6.2 is that the energy levels of different n do not cross as they do in H. There are avoided crossings, which are clearly visible for the $|m| = 1$ states of Fig. 6.11 and overwhelming for the $m = 0$ states of Fig. 6.12. The energy level diagram for the Na $|m| = 2$ states would not be observably different from the analogous diagram for H on the scale of Figs. 6.11 and 6.12. At zero field

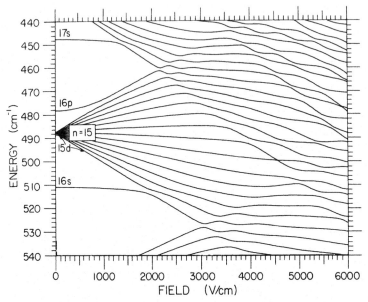

Fig. 6.12 Calculated Na $m = 0$ energy levels (from ref. 27).

the levels would be degenerate, and for $E > 1/3n^5$ the blue and red levels of adjacent n would cross.

A subtle point which is not immediately apparent when inspecting Figs. 6.11 and 6.12 is that the $|m| = 0$ and 1 levels are not interleaved as they are in H. Note that in Fig. 6.12 the 16s state joins the manifold at a field of 2000 V/cm. At fields below 1500 V/cm the $|m| = 0$ and $|m| = 1$ levels of the lower half of the $n = 15$ Stark manifold lie practically on top of one another. This point was demonstrated explicitly by Fabre *et al*[28] who used a diode laser to excite Na atoms from the 3d state to states of the $n = 29$ Stark manifold with high resolution. In Fig. 6.13 we show the result of scanning the wavelength of the diode laser across three high lying pairs of $|m| = 0$ and 1 levels of the $n = 29$ Stark manifold in a field of 20.5 V/cm. The $n = 29$ and $n = 30$ Stark manifolds intersect at a field of 83 V/cm, so the field of 20.5 V/cm corresponds roughly to a field of 550 V/cm in Figs. 6.11 and 6.12, i.e. the 30p state has not yet joined the $n = 29$ Stark manifold. The $|m| = 0$ and 1 states are <200 MHz apart while the adjacent pairs of Stark levels are ~2 GHz apart. In H the $|m| = 0$ and 1 levels would be 1 GHz apart. The $|m| = 0$ levels lie above the $|m| = 1$ levels, i.e. they are further displaced from the center of the manifold, or the zero field $n = 29$ high ℓ levels.

Fabre *et al.*[28] used a projection operator technique to describe the Stark shifts at fields below where low ℓ states of large quantum defects join the manifold. A less formal explanation is as follows. If, for example, the s and p states are excluded, as in Fig. 6.13 below 800 V/cm, effectively only the nearly degenerate $\ell \geq 2$ states are coupled by the electric field. The only differences among the $|m| = 0,1$, and 2 manifolds occur in the angular parts of the matrix element, i.e.[1]

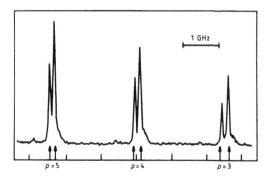

Fig. 6.13 Part of the excitation spectrum of the Na $n = 29$ Stark levels from the 3d state in an electrostatic field of 20.5 V/cm corresponding to n_1 values of $n - 3$, $n - 4$, and $n - 5$. The energy splitting between $m = 0$ (highest energy line in the doublets) and $|m| = 1$ states is of the order of 180 MHz. The arrows indicate theoretical positions of energy levels obtained by a numerical diagonalization of the Stark Hamiltonian (from ref. 28).

$$\langle n\ell m | \cos \theta | n\ell + 1m \rangle = \sqrt{\frac{(\ell + 1)^2 - m^2}{(2\ell + 3)(2\ell + 1)}}. \tag{6.43}$$

Only in a few low ℓ states is the m dependence significant, and as a result the $|m| = 0,1$, and 2 energy levels are almost degenerate. However, due to the low ℓ states, the $|m| = 0$ levels are displaced more than the $|m| = 1$ levels, which are displaced insignificantly more than the $|m| = 2$ levels.

Reexamining Figs. 6.11 and 6.12, the avoided crossings of the blue and red states of adjacent n are evident. When the avoided crossings are large they can be measured optically as was done by Zimmerman *et al.*[27] They excited Li atoms in a beam from the 2s state to the 2p state and then to the 3s state by two pulsed dye lasers at fixed wavelengths. The wavelength of a third laser was scanned to drive atoms from the 3s state to Rydberg Stark states in the presence of a static field. The transition to the Rydberg state was detected by applying a high field pulse to the atoms subsequent to the laser pulses. The ions were detected with a particle multiplier as the wavelength of the third laser was scanned. In Fig. 6.14 we show their observation of the avoided crossing of the (18,16,0,1) and (19,1,16,1) states. The experimental picture of Fig. 6.14 is built up by scanning the third laser over the energy of the avoided crossing for several fields near the avoided crossing field of 943 V/cm. As shown by Fig. 6.14, the observed energies of the levels match those calculated by matrix diagonalization. It is also interesting to note that the oscillator strength to the upper state vanishes at the anticrossing. At the anticrossing the eigenstates have the form $(1/\sqrt{2})$ $(\Psi_A \pm \Psi_B)$ where Ψ_A and Ψ_B are the eigenstates away from the crossing. As shown by Fig. 6.14, away from the avoided crossing both states apparently have similar amounts of p character, and at the crossing the upper state loses it all to the lower state.

The method used by Zimmerman *et al.*[27] to measure the Li avoided crossing shown in Fig. 6.14 requires that the resolution of the laser be finer than the size of

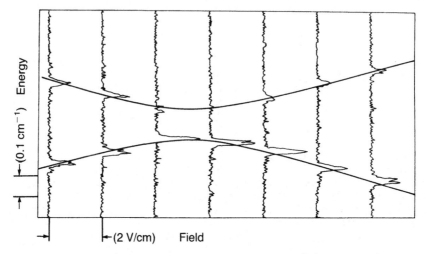

Fig. 6.14 Level anticrossing in Li. Intersection of the $(n,n_1,n_2,|m|)$ states $(18,16,0,1)$ and $(19,1,16,1)$ at 321.5 cm^{-1} and 943 V/cm. The calculated level structure is superimposed on the data. Note the vanishing oscillator strength to the upper level at the avoided crossing (from ref. 27).

the avoided crossing. An alternative approach used by Stoneman *et al.*[29] bypasses this requirement. They observed the narrow anticrossings between the K $(n + 2)$s states, which have a quantum defect of 2.18, and the low lying members of the n Stark manifold, which is composed of $\ell \geq 3$ states. They excited atoms in a beam from the 4s state to the 4p state and then to the ns state with two pulsed dye lasers with linewidths of 1 cm^{-1}. Two to three microseconds after the laser excitation the atoms were selectively field ionized, and only atoms which had undergone black body radiation induced transitions to the higher lying $(n + 2)$p states were detected. The $(n + 2)$p signal was recorded as the static field was slowly scanned through the avoided crossing field over many shots of the laser. Away from the avoided crossing only the $(n + 2)$s state was excited, and it could undergo black body radiation induced transitions to the $(n + 2)$p state. The n Stark state, composed of $\ell \geq 3$ states, cannot be excited from the 4p state or to the $(n + 2)$p state. At the avoided crossing the eigenstates are 50–50 mixtures of the two states, and both are excited from the 4p state, with the result that the same number of Rydberg atoms is always excited. However, each of these two states has half the $(n + 2)$s–$(n + 2)$p black body radiation induced transition rate, and the observed $(n + 2)$p signal decreases. An example of their signals, the avoided crossing signals of the K 20s state with the lowest lying $|m| = 0$ and 1 states of the $n = 18$ Stark manifold, is shown in Fig. 6.15.[29] First we note the decrease in the 20p ion signal at each of the avoided crossings, as described above. Second we note that the $|m| = 0$ and 1 Stark states are nearly degenerate. If they were interleaved as in H, the two avoided crossings would be separated by 50 V/cm not 5 V/cm as they are in Fig. 6.15. We have again encountered the noninterleaving of the energy

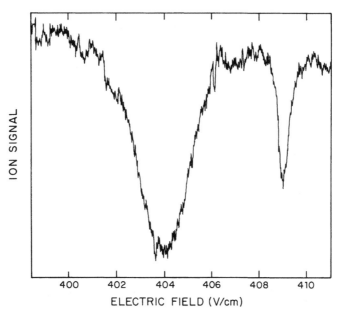

Fig. 6.15 Anticrossing signal from the avoided crossing of the K 20s state with the lowest energy $n = 18$ Stark manifold state. The left-hand peak corresponds to the $m = 0$ anticrossing, and the right-hand peak to the $|m| = 1$ anticrossing (from ref. 29).

levels of different m, as shown in Fig. 6.13. The spin orbit interaction, which we have ignored, is responsible for the avoided crossing between the 20s state and the $|m| = 1$ Stark state. Diagonalizing the Hamiltonian matrix, including the spin orbit interaction, yields 450 and 47 MHz for the widths of the $|m| = 0$ and 1 anticrossings of Fig. 6.16, in reasonably good agreement with the measured values of 510 MHz and 80 MHz.[29] In the heavier alkali atoms the spin orbit interaction plays an even more dominant role.[27]

When the quantum defects are large and $|m|$ is low enough that several states have large quantum defects, matrix diagonalization using the $n\ell m$ states as basis functions is probably the most efficient approach. However, if the contributing quantum defects are small, a different approach to Eq. (6.41), suggested by Komarov *et al.*[30] is useful. They suggested using the hydrogenic parabolic states as a basis instead of the zero field $n\ell m$ states, in which case the Hamiltonian of Eq. (6.41) has diagonal matrix elements

$$\langle nn_1n_2m|H|nn_1n_2m\rangle =$$
$$-\frac{1}{2n^2} + \frac{3}{2}n(n_1 - n_2)E + O(E^2) + \langle nn_1n_2m|V_d(r)|nn_1n_2m\rangle, \quad (6.44)$$

and nonzero off diagonal elements,

$$\langle nn_1n_2m|H|n'n_1'n_2'm\rangle = \langle nn_1n_2m|V_d(r)|n'n_1'n_2'm\rangle. \quad (6.45)$$

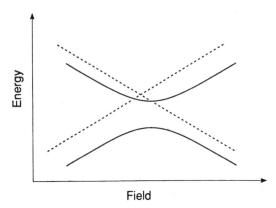

Fig. 6.16   Hydrogenic Stark energy levels (- -) before considering the effect of the core perturbation $V_d(r)$. The alkali levels after incorporating the diagonal and off diagonal matrix elements of $V_d(r)$ (——). The diagonal elements shift the levels and the off diagonal elements lift the degeneracy at the level crossing. Note that the different shifts of the levels move the avoided crossing from the field of the hydrogenic crossing.

The diagonal elements are the hydrogenic energies plus a small energy shift, and the off diagonal elements are the couplings between levels. In Fig. 6.16 we show two hydrogenic levels as dotted lines and the non hydrogenic levels as solid lines. The shifts of the nonhydrogenic levels from their hydrogenic counterparts are due to the diagonal matrix elements of $V_d(r)$ and the avoided crossings to the off diagonal matrix elements of $V_d(r)$.

It is straightforward to evaluate the matrix elements of $V_d(r)$ using the known transformation between parabolic and spherical states given in Eqs. (6.18) and (6.19). If we expand the parabolic states in terms of hydrogenic spherical states, the matrix elements of $V_d(r)$ may be written as[30]

$$\langle nn_1n_2m|V_d(r)|n'n_1'n_2'm\rangle = \sum_\ell \langle nn_1n_2m|n\ell m\rangle \langle n\ell m|V(r)|\acute{n}\ell m\rangle \langle n'\ell m|n'n'm\rangle,$$

(6.46)

where we have taken advantage of the spherical symmetry of $V_d(r)$ to require the same ℓ on both sides of the matrix element. We recall that the core interaction causes the depression of the energy levels below their H analogues. Thus,

$$\langle n\ell m|V_d(r)|n\ell m\rangle = \frac{-\delta_\ell}{n^3}.$$

(6.47)

This expression may be generalized by realizing that $V_d(r)$ is only nonzero for small r, where the dependence on n is through the normalization factor $n^{-3/2}$ Accordingly Eq. (6.47) can be generalized to[30]

$$\langle n\ell m|V_d(r)|n'\ell m\rangle = \frac{-\delta_\ell}{\sqrt{n^3n'^3}}.$$

(6.48)

Thus,

$$\langle nn_1n_2m|V_d|n'n_1'n_2'm\rangle = \sum_\ell \langle nn_1n_2m|n\ell m\rangle \frac{-\delta_\ell}{\sqrt{n^3n'^3}} \langle n'\ell m|n'n_1'n_2'm\rangle. \qquad (6.49)$$

Since $|\langle nn_1n_2m|n\ell m\rangle|^2$ is the $n\ell m$ fraction of the nn_1n_2m state, it is evident that the depression of the nonhydrogenic Stark levels below the hydrogenic value, given by the diagonal matrix elements of $V_d(r)$, is simply the amount of each low ℓ state it possesses multiplied by its quantum defect. The nonhydrogenic energy must be lower than the hydrogenic energy since $\delta_\ell > 0$ for all ℓ.

At the field where the diagonal energies of two levels, given by Eq. (6.43), are equal, the off diagonal matrix element lifts the degeneracy. At the avoided crossing the energy levels are split by

$$\Delta W = 2|\langle nn_1n_2m|V_d(r)|n'n_1'n_2'm\rangle|, \qquad (6.50)$$

and the eigenstates are symmetric and antisymmetric combinations of the two Stark states. Eq. (6.49) shows how the avoided crossings scale with n. Averaged over ℓ, $\langle nn_1n_2m|n\ell m\rangle^2$ must be equal to $1/(n-m)$. Thus, for low $|m| \ll n$ and $n \sim n'$, the matrix element of Eq. (6.49) is proportional to $1/n^4$, and the avoided crossings exhibit a $1/n^4$ scaling.

Ionization of nonhydrogenic atoms

A nonhydrogenic atom, Na for example, differs from H not only in its energy level spectrum but also in how it is ionized by a field. In particular, in a nonhydrogenic atom there are two forms of field ionization. The first is exactly like H. A state of approximate quantum number n_1 can itself ionize, keeping the same value of n_1, in a field E if its energy W is enough for the electron to surmount the potential barrier V_b in $V(\eta)$. Rates for this form of ionization increase rapidly in the vicinity of the threshold for classical ionization, just as in H. In this form of ionization the blue states require higher fields for ionization than red states of the same energy.

The second form of ionization is similar to autoionization.[31] In a nonhydrogenic atom n_1 is not a good quantum number and bound states of high n_1 are coupled to Stark continua of low n_1. This form of ionization applies to those states other than the reddest Stark states. The extreme red Stark states have $n_1 = 0$ and ionize, as do the red H states, at the classical ionization limit given by Eq. (6.35), modified by Eq. (6.37) for $m \neq 0$ states. This point has been demonstrated explicitly by Littman et al,[32] who measured the time resolved ionization of Na $|m| = 2$ states subsequent to pulsed laser excitation. Their measured rates are in good agreement with the rates obtained by extrapolation of the rates of Bailey et al.[20] shown in Fig. 6.8.

In H it is possible to have $n_1 \geq 1$ states which are quite stable against ionization even though energetically degenerate states of lower n_1 have been converted to continua. In other atoms these blue $n_1 \geq 1$ states are coupled by the finite core to the red continua and autoionize. A simple physical picture is that an electron in a blue orbit comes near the core once on each orbit and has a finite probability of being scattered from the stable blue orbit into the degenerate red continuum. A significant difference between this form of ionization and hydrogenic ionization is that its rate is not exponentially dependent on the field.

The existence of the second form of ionization means that there is a large regime of field and energy in which nonhydrogenic atoms have states which decay more rapidly by ionization than by radiative decay, but are still spectroscopically narrow. In H all states have ionization rates which are exponentially dependent on the field so the analogous range is much smaller. The fact that there exist sharp but ionizing levels above the classical ionization limit but below the hydrogenic ionization limit was first observed by Jacquinot *et al.* in Rb.[33] It was most clearly shown by Littman *et al.* in Li.[31] They excited a beam of Li atoms in static fields by two fixed frequency lasers from the ground state 2s via the 2p state to the 3s state. They then scanned the wavelength of a third laser to drive transitions from the 3s state to Rydberg states having np character. The excitation to a Rydberg state was detected by field ionization of the atom and subsequent detection of the ion with a particle multiplier. Spectra were recorded at many static fields, as shown by Fig. 6.17, which is composed of spectra recorded with the third laser polarized perpendicular to the static field so as to excite only $|m| = 1$ states. The spectra of Fig. 6.17 were recorded in two different ways. The spectra of Fig. 6.17(a) were recorded by applying a field ionization pulse $3 \, \mu s$ after laser excitation and only detecting the ions which result from ionization by the pulse, not spontaneous ionization. The resulting spectra are then composed of states which are stable for at least $3 \, \mu s$ against field ionization, or radiative decay, which is longer than $3 \, \mu s$ for the states of Fig. 6.17. It is quite apparent that there are no stable states above the classical ionization limit.

On the other hand, if no field ionization pulse is applied and the ions from atoms which ionize in the first $3 \, \mu s$ after the laser excitation are detected, the result is Fig. 6.17(c). The spectra begin at the classical limit and extend to much higher static fields. What is most important to note is that levels appear over a field range of typically a factor of 2 with no apparent change in width, a phenomenon which does not occur in H. Fig. 6.17(b) shows the calculated H $|m| = 1$ energy levels. Inspecting Fig. 6.17(c) it is easy to see that the levels disappear from the experimental spectrum when their decay rates exceed the laser linewidth by a large margin.

We may develop a more quantitative understanding of nonhydrogenic ionization by starting with the hydrogenic Stark states and adding the core perturbation $V_d(r)$, just as we calculated the sizes of the avoided level crossings. The level crossings described earlier are the result of core couplings between the bound

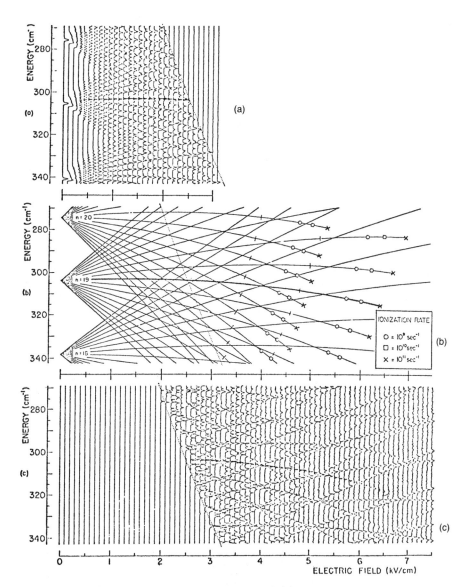

Fig. 6.17 Tunnelling and saddle point ionization in Li. (a) Experimental map of the energy levels of Li $|m| = 1$ states in a static field. The horizontal peaks arise from ions collected after laser excitation. Energy is measured relative to the one-electron ionization limit. Disappearance of a level with increasing field indicates that the ionization rates exceed $3 \times 10^5 \, \mathrm{s}^{-1}$. The dotted line is the classical ionization limit given by Eqs. (6.35) and (6.36). One state has been emphasized by shading. (b) Energy levels for H ($n = 18$–20, $|m| = 1$) according to fourth order perturbation theory. Levels from nearby terms are omitted for clarity. Symbols used to denote the ionization rate are defined in the key. The tick mark indicates the field where the ionization rate equals the spontaneous radiative rate. (c) Experimental map as in (a) except that the collection method is sensitive only to states whose ionization rate exceeds $3 \times 10^5 \, \mathrm{s}^{-1}$. At high fields, the levels broaden into the continuum in agreement with tunnelling theory for H (from ref. 32).

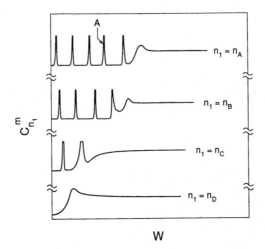

$C^m_{n_1}$

W

Fig. 6.18 The density of states $C^m_{n_1}$ for four different values of the quantum number n_1: $n_A > n_B > n_C > n_D$. For each value of n_1 $C^m_{n_1}$ is composed of narrow resonances up the $W = V_b$, the energy of the saddle point in $V(\eta)$. Above $W = V_b$ $C^m_{n_1}$ is continuous as shown. The bound state A in $n_1 = n_A$ can ionize into the $n_1 = n_C$ and n_D continua in a nonhydrogenic atom. There is often a peak in $C^m_{n_1}$ at the onset of the continuous spectrum, as shown.

nn_1n_2nm and $n'n'_1n'_2m$ states, couplings between bound levels of different n_1. The same coupling between a hydrogenically bound n_1m state and a continuum state of lower n_1 results in ionization.

Ignoring for the moment coupling between nominally bound states of different n_1, the ionization rate of the state nn_1m to the $\varepsilon n'_1m$ continua is given by Fermi's Golden Rule,

$$\Gamma = 2\pi \sum_{n_1} |\langle nn_1m|V_d(r)|\varepsilon n'_1m\rangle|^2. \tag{6.51}$$

The summation extends over all values of n'_1 which are continua at the energy and field in question.

Since $V_d(r)$ is only nonzero near $r = 0$ the matrix element of Eq. (6.51) reflects the amplitude of the wavefunction of the continuum wave at $r \sim 0$. Specifically, the squared matrix element is proportional to $C^m_{n_1}$, the density of states defined earlier and plotted in Fig. 6.18. From the plots of Fig. 6.18 it is apparent that the ionization rate into a continuum substantially above threshold is energy independent. However, as shown in Fig. 6.18, there is often a peak in the density of continuum states just at the threshold for ionization, substantially increasing the ionization rate for a degenerate blue state of larger n_1. This phenomenon has been observed experimentally by Littman *et al.*[32] who observed a local increase in the ionization rate of the Na (12,6,3,2) Stark state where it crosses the 14,0,11,2 state, at a field of 15.6 kV/cm, as shown by Fig. 6.19. In this field the energy of the

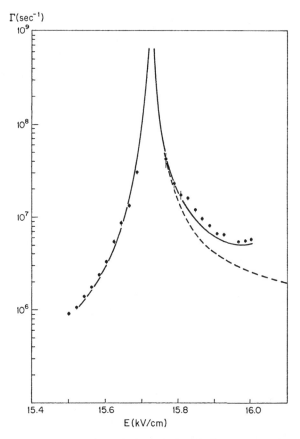

Fig. 6.19 Ionization rates for the (12,6,3,2) level of Na in the region of the crossing with the rapidly ionizing (14,0,11,2) level. The levels are specified as $(n, n_1, n_2, |m|)$. The solid line is a theoretical curve. The dashed line is calculated neglecting another avoided level crossing at 17.3 kV/cm (from ref. 31).

(14,0,11,2) state is just below the top of the barrier in $V(\eta)$ and its ionizing rate is 10^{10} s^{-1}. With this low an ionization rate the (14,0,11,2) wavefunction still has a large amplitude at the core, and the autoionization rate of blue states into it is at its highest.

Finally, it is useful to obtain an estimate of the typical ionization rates in the region between the classical ionization limit and the hydrogenic limit using Eq. (6.51). Each term in the sum of Eq. (6.51) can be written as

$$|\langle nn_1m|V_d(r)|\varepsilon n_1'm\rangle|^2 = |\langle nn_1m|n\ell m\rangle \langle n\ell m|V_d|\varepsilon\ell m\rangle \langle \varepsilon\ell m|\varepsilon n_1'm\rangle|^2 \qquad (6.52)$$

The maximum value of n_1' in Eq. (6.51) is the value of n_1 for which $Z_2 = W^2/4E$, or $Z_1 = 1 - Z_2 = 1 - W^2/4E$. As shown by Harmin,[14] a continuum form of Eq. (6.19), valid for small radii, is

$$\langle \varepsilon\ell m|\varepsilon n_1'm\rangle = (-1)^\ell P_{\ell m}(Z_1 - Z_2). \qquad (6.53)$$

We are now in a position to evaluate Eq. (6.51) using the explicit form of Eq. (6.52) and replacing the sum over n_1 by an integral over $Z_1 - Z_2$. The first term of the matrix element of Eq. (6.52) is given by Eqs. (6.11) and (6.19), the second one by the continuum modification of Eq. (6.48), and the third by Eq. (6.53). In addition to these substitutions we convert the summation of Eq. (6.51) to an integral over $Z_1 - Z_2$. According to Eq. (6.38), the classical ionization limit occurs at $E = W^2/4Z_2$. Consequently, the contributing range of Z_2 is from $W^2/4E$ to 1. Written in terms of $Z_1 - Z_2$ the integral corresponding to the ionization rate of Eq. (6.51) is given by

$$\Gamma = 2\pi P_{\ell m}^2 \left[\left(\frac{2n_1}{n} \right) - 1 \right] \frac{\sin^2 \delta_\ell'}{n^3} \int_{-1}^{1-W^2/2E} P_{\ell m}^2 (Z_1 - Z_2) \mathrm{d}(Z_1 - Z_2), \tag{6.54}$$

where δ_ℓ' is the magnitude of the difference between δ_ℓ and the nearest integer. For example, $\delta_\ell = 0.85$ implies $\delta_\ell' = 0.15$. The first term in Eq. (6.54) reflects the amount of ℓ character in the initial nn_1m state, the second term is the coupling to the ℓ part of the continuum, and the third, the integral, reflects how much of the continuum is available having ℓ character. The value of the integral ranges from zero at energies just above the classical ionization limit to one when $W = -\sqrt{2E}$.

Approximating $\langle nn_1n_2m|n\ell m \rangle^2$ by $1/n$ and the integral of Eq. (6.54) by 1/2 we find the approximate result

$$\Gamma \sim \pi \frac{\sin^2 \delta_\ell'}{n^4} \tag{6.55}$$

Observations to date indicate that Na $|m| = 0, 1$, and 2 Stark states of $n \approx 20$ above the classical ionization limit have widths of $\sim$30–90 GHz, 1–3 GHz, and 100 MHz,[34] in rough accord with Eq. (6.55).

In estimating ionization rates we have ignored all interactions between bound states of high n_1. These interactions also lead to local variations in the autoionization rates which can be quite striking and can be understood in a straightforward way. For a moment let us disregard the coupling to the continuum and focus on two nominally bound Stark states which have an avoided crossing. As we have already discussed, at the avoided crossing the eigenstates are linear combinations of the two Stark states. If away from the avoided crossing one of the two Stark levels is rapidly ionizing and the other slowly ionizing, at the crossing the ionization rates of the two eigenstates will most likely be about the same and half the rate of the rapidly ionizing state.

A more interesting case is one in which the two Stark states have equal autoionization rates. If the coupling is to the same continuum in both cases, then at the avoided crossing the coupling will vanish in one of the eigenstates and double in the other one. The phenomenon is basically the same as the vanishing of the oscillator strength to the upper level at the avoided crossing of Fig. 6.14. If the two Stark states do not have precisely the same ionization rates away from the avoided crossing, one of the ionization rates cannot vanish at the avoided

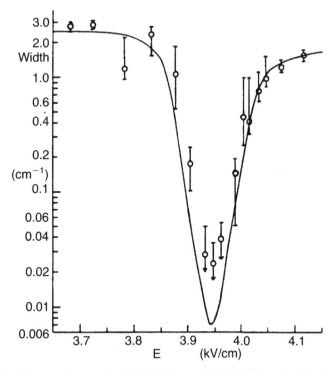

Fig. 6.20 Ionization width vs electric field for the Na (20,19,0,0) level near its crossing with the (21,17,3,0) level from experiment (data points) and from WKB–quantum defect theory (solid line). The levels are specified as $(n,n_1,n_2,|m|)$ Because the lineshapes are quite asymmetric (except for very narrow lines), the width in this figure is taken to be the FWHM of the dominant feature corresponding to the (20,19,0,0) level in the photoionization cross section. For the narrowest line, experimental widths are limited by the 0.7 GHz laser linewidth. Error limits are asymmetric because of the peculiar line shapes and because of uncertainties due to the overlapping $|m| = 1$ resonance (from ref. 37).

crossing. This phenomenon was first observed by Feneuille *et al.*[35] and discussed further by Luc–Koenig and LeCompte.[36] However, it is shown most clearly in the data of Liu *et al.*[37] who measured the width of the Na (20,19,0,0) state in the vicinity of its intersection with the (21,17,3,0) state at 3.95 kV/cm. As shown by Fig. 6.20, the width of the state decreases from 2 cm^{-1} away from the avoided crossing to 0.02 cm^{-1} at the avoided crossing. Correspondingly, the width of the (21,17,3,0) state approximately doubles at the avoided crossing.

References

1. H. A. Bethe and E. A. Salpeter, *Quantum Mechanics of One and Two Electron Atoms* (Academic Press, New York, 1957).
2. L. D. Landau and E. M. Lifshitz, *Quantum Mechanics* (Pergamon Press, 1977).
3. E. Luc-Koenig and A. Bachelier, *J. Phys. B* **13**, 1743 (1980).

4. D. Kleppner, M. G. Littman, and M. L. Zimmerman, in *Rydberg States of Atoms and Molecules*, eds. R. F. Stebbings and F. B. Dunning (Cambridge University Press, Cambridge,1983).

5. U. Fano and J. W. Cooper, *Rev. Mod. Phys.* **40**, 441 (1965).

6. H. J. Silverstone, *Phys. Rev. A* **18**, 1853 (1978).

7. R. J. Damburg and V. V. Kolosov, in *Rydberg States of Atoms and Molecules*, eds. R. F. Stebbings and F. B. Dunning (Cambridge University Press, Cambridge, 1983).

8. P. M. Koch, *Phys. Rev. Lett.* **41**, 99 (1978).

9. D. R. Inglis and E. Teller, *Astrophys. J.* **90**, 439 (1939).

10. D. A. Park, *Z. Phys.* **159**, 155 (1960).

11. M. J. Englefield, *Group Theory and the Coulomb Problem* (Wiley Interscience, New York, 1972).

12. D. R. Herrick, *Phys. Rev. A* **12**, 1949 (1975).

13. V. L. Jacobs and J. Davis, *Phys. Rev. A* **19**, 776 (1979).

14. D. A. Harmin in *Atomic Excitation and Recombination in External Fields*, eds. M. H. Nayfeh and C. W. Clark (Gordon and Breach, New York, 1985).

15. D. A. Harmin, *Phys. Rev. A* **24**, 2491 (1981).

16. W. E. Cooke and T. F. Gallagher, *Phys. Rev. A* **17**, 1226 (1978).

17. H. Rausch v. Traubenberg, *Z. Phys.* **54**, 307 (1929).

18. M. H. Rice and R. H. Good, Jr., *J. Opt. Soc. Am.* **52**, 239 (1962).

19. J. O. Hirschfelder and L. A. Curtis, *J. Chem. Phys.* **55**, 1395 (1971).

20. D. S. Bailey, J. R. Hiskes, and A. C. Riviere, *Nucl. Fusion* **5**, 41 (1965).

21. P. M. Koch and D. R. Mariani, *Phys. Rev. Lett.* **46**, 1275 (1981).

22. R. J. Damburg and V. V. Kolosov, *J. Phys. B* **12**, 2637 (1979).

23. R. J. Damburg and V. V. Kolosov, *J. Phys. B* **9**, 3149 (1976).

24. T. Bergeman, *Phys. Rev. Lett.* **52**, 1685 (1984).

25. H. Rottke and K. H. Welge, *Phys. Rev. A* **33**, 301 (1986).

26. F. H. Kellert, T. H. Jeys, G. B. MacMillan, K. A. Smith, F. B. Dunning, and R. F. Stebbings, *Phys. Rev. A* **23**, 1127 (1981).

27. M. L. Zimmerman, M. G. Littman, M. M. Kash, and D. Kleppner, *Phys. Rev. A* **20**, 2251 (1979).

28. C. Fabre, Y. Kaluzny, R. Calabrese, L. Jun, P. Goy, and S. Haroche, *J. Phys. B.* **17**, 3217 (1984).

29. R. C. Stoneman, G. Janik, and T. F. Gallagher, *Phys. Rev. A* **34**, 2952 (1986).

30. I. V. Komarov, T. P. Grozdanov, and R. K. Janev, *J. Phys. B.* **13**, L573 (1980).

31. M. G. Littman, M. M. Kash, and D. Kleppner, *Phys. Rev. Lett.* **41**, 103 (1978).

32. M. G. Littman, M. L. Zimmerman, and D. Kleppner, *Phys. Rev. Lett.* **37**, 486 (1976).

33. P. Jacquinot, S. Liberman, and J. Pinard, in *Etats Atomiques et Moleculaires couples a un continuum. Atomes et molecules hautement excites*, eds. S. Fenuille and J. C. Lehman (CNRS, Paris, 1978).

34. J. Y. Liu, P. McNicholl, D. A. Harmin, T. Bergeman, and H. J. Metcalf, in *Atomic Excitation and Recombination in External Fields*, eds. M. H. Nayfeh and C. W. Clark (Gordon and Brench, New York, 1985).

35. S. Feneuille, S. Liberman, E. Luc-Koenig, J. Pinard, and A. Taleb, *J. Phys. B* **15**, 1205 (1982).

36. E. Luc-Koenig and J. M. LeCompte, in *Atomic Excitation and Recombination in External Fields*, eds. M. H. Nayfeh and C. W. Clark (Gordon and Breach, New York, 1985).

37. J. Y. Liu, P. McNicholl, D. A. Harmin, I. Ivri, T. Bergeman, and H. J. Metcalf, *Phys. Rev. Lett.* **55**, 189 (1985).

7

Pulsed field ionization

Because it can be efficient and selective, field ionization of Rydberg atoms has become a widely used tool.[1] Often the field is applied as a pulse, with rise times of nanoseconds to microseconds,[2-4] and to realize the potential of field ionization we need to understand what happens to the atoms as the pulsed field rises from zero to the ionizing field. In the previous chapter we discussed the ionization rates of Stark states in static fields. In this chapter we consider how atoms evolve from zero field states to the high field Stark states during the pulse. Since the evolution depends on the risetime of the pulse, it is impossible to describe all possible outcomes. Instead, we describe a few practically important limiting cases.

Although we are not concerned here with the details of how to produce the pulses, it is worth noting that several different types of pulse, having the time dependences shown in Fig. 7.1, have been used. Fig. 7.1(a) depicts a pulse which rises rapidly to a plateau. Atoms in a fast beam experience this sort of pulse when passing into a region of high homogenous field. Fig. 7.1(b) shows a rapidly rising pulse which decays rapidly after reaching its peak. While not elegant, such pulses are easily produced. For pulse shapes such as those of Figs. 7.1(a) and (b) the ability to discriminate between different states comes mostly from adjustment of the amplitude of the pulse. Fig. 7.1(c) depicts a linearly rising field ramp in which the discrimination between different states comes from the time in the rising field at which they are ionized. The definition of the ionization threshold field depends on the experiment. If the field pulse reaches its peak for 200 ns, then, if the field produces an ionization rate of $3.5 \times 10^6 \, \text{s}^{-1}$, 50% ionization results, and this is a possible definition of the threshold field.

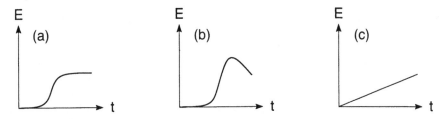

Fig. 7.1 Typical field pulses: (a) a pulse which rises in time to a flat topped pulse; (b) a pulse which rises to a peak then falls quickly ; (c) a linearly rising field.

103

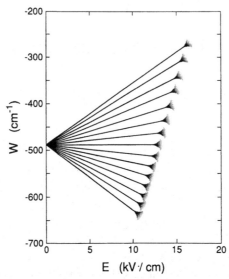

Fig. 7.2 Energy levels of the H $n = 15$, $m = 0$ Stark levels. The broadening of the levels corresponds to an ionization rate of $10^6\,\mathrm{s}^{-1}$. The extreme red and blue state ionization rates are taken from the calculations of Bailey *et al.* (ref. 5), and those of the intermediate states are simply interpolated.

Hydrogen

Let us consider first the H atom. The essential question is, how does a H atom pass from near zero field to the high field in which ionization occurs? If the atom is initially in a zero field parabolic state, as might be formed by passage through a foil, it keeps the same nn_1n_2m quantum numbers as the field is increased to the ionization field, which depends on n, n_1, n_2, and m of the state in question. To a reasonable approximation the ionizing field is near the classical field defined by Eq. (3.45). For the red, $n_1 \sim 0$ states $E \simeq 1/9n^4$. However, in the blue $n_1 \gg n_2$ states the required fields are higher. Fig. 7.2 is an energy level diagram of the hydrogenic $n = 15$, $m = 0$ Stark states. Each Stark state simply follows its energy level as the field is increased from zero to the high ionizing field. As shown by Fig. 7.2 the blue state requires a higher ionization field than the red state.

In Fig. 7.2 the ionization fields shown correspond to an ionization rate of 10^6 s^{-1}. The fields for the extreme blue and red states are taken from the calculations of Bailey *et al*,[5] and the fields for intermediate states are simply interpolated. We have also not shown any other n states since the Stark states of different n, which cross each other, have different values of n_1, and do not interact, as described in Chapter 6.

Usually the H Rydberg states are not prepared in the zero field parabolic states, but in the spherical states by, for example, optical excitation. In this case, when the field is applied we would expect it to project the single $n\ell m$ state onto the degenerate nn_1n_2m states, each of which would then follow its own path to

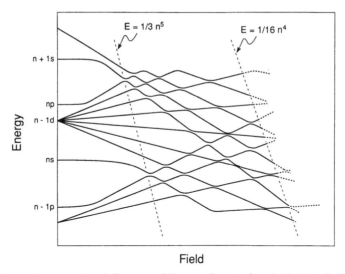

Fig. 7.3 Schematic energy level diagram of Na $m = 0$ states in a field. The displacement of the zero field ns and np states from the approximately hydrogenic $\ell \geq 2$ states is evident. In the region $1/3n^5 < E < 1/16n^4$ the levels of different n have avoided crossings, and above $E = 1/16n^4$ the states ionize by coupling to degenerate red continua of low n_1. In this region the levels are shown by dotted lines. In a slowly rising pulse atoms follow the adiabatic levels shown, and each zero field state is correlated to one state, and thus one ionizing field, at $E = 1/16n^4$.

ionization as shown in Fig. 7.2. The precise projections onto the nn_1n_2m states are given by Eq. (6.19) and Fig. 6.3.[6] For example, the 15s state would be projected equally onto all the $n = 15$ Stark states of Fig. 7.2, while the 15p $m = 0$ state would be preferentially projected onto states at the red and blue edges of the manifold.

The H atom is a special case. Since the states of the same n and m are all degenerate in zero field, no matter how slowly we apply a field we project the $n\ell m$ states onto the nn_1n_2m states, i.e. the transition is always diabatic. On the other hand, as long as the field rises slowly compared to the Δn interval, a H atom in an n_1 state remains in the same n_1 state and ionization always occurs at the same field, irrespective of the risetime of the pulse.

Nonhydrogenic atoms

Pulsed field ionization of an alkali atom differs from the description just given for H because of the finite sized ionic core, or equivalently, the nonzero quantum defects. There are three important effects. First, the zero field levels can only be spherical $n\ell m$ levels, not parabolic levels. Second, in the $E > 1/3n^5$ regime there are avoided crossings of states of different n. Third, ionization can occur at lower fields than in H. Specifically, in H blue states have higher ionization fields than red states, but in an alkali atom this is not the case due to n_1 changing ionization.

The easiest way of thinking about field ionization of, for example, Na is to start with a static field energy level diagram such as Fig. 7.3, which shows schematically the Na $m = 0$ states. In low fields ℓ is a good quantum number. What is a low field obviously depends on ℓ. Intermediate fields are those in which ℓ is not good, but $E < 1/3n^5$. High fields are those in excess of $1/3n^5$. At fields $E > W^2/4$ ionization is classically allowed and occurs by coupling to the degenerate red continua. In this region of Fig. 7.3 the levels are shown as broken lines. Finally, at very high fields, not shown in Fig. 7.3, the states themselves ionize by tunneling, as in H.

How ionization occurs depends upon how quickly the field rises from zero to the high field required for ionization. Consider a slowly rising pulse. In this case the passage from the zero field $n\ell m$ states to the intermediate field Stark states is adiabatic. The zero field $n\ell m$ state slowly evolves into a single Stark state. When the field reaches $1/3n^5$, the avoided crossings with Stark states of the same m but adjacent n are encountered. If the field is rising slowly, the level crossings are traversed adiabatically and the atom remains in the same adiabatic energy level. At each avoided crossing the atom passes smoothly from one Stark state into another, as shown by the differing slopes of the adiabatic energy levels on either side of an avoided crossing of Fig. 7.3. Finally, when the field reaches $E = W^2/4$ ionization occurs by coupling to the underlying red Stark continua composed of red $n_1 \simeq 0$, states of higher n.

Fig. 7.3 illustrates the most important features of adiabatic ionization. First, the energy ordering of the zero field states is preserved as they pass to the high ionizing field. Second, the adiabatic states which connect to the zero field states of one n are trapped in the energy range $-1/2(n - 1/2)^2 < W < -1/2(n + 1/2)^2$. A way of representing the energy in the field relative to the zero field limit is to use the effective quantum number in the field,[3] n_s, defined by $W = -1/2n_s^2$. From Fig. 3 it is apparent that the Na $(n + 1)$s and nd states pass to states of $n_s \simeq n - 1/2$ while the $(n + 1)$p states pass to states of $n_s \simeq n + 1/2$, which are much more easily ionized. There is only a small difference in the ionization fields of the nd and $(n + 1)$ s states in spite of their large zero field separation. In contrast, there is a large separation between the fields required to ionize the nd and $(n + 1)$p states in spite of their zero field separation's being less than half the $(n + 1)$s–nd separation.

If we assume that the energies in the high field are evenly spaced and group the $(n+1)$s and $n\ell$ states together, we may estimate the n_s values of the series of $m = 0$ states as follows,[3]

$$(n + 1)\text{s} \to n_s = n - 1/2$$
$$n\text{d} \to n_s = n - 1/2 + 1/n$$
$$n\text{f} \to n_s = 1/2 + 2/n \qquad\qquad (7.1)$$
$$n\ell \to n_s = n - 1/2 + (\ell - 1)/n$$
$$(n + 1)\text{p} \to n_s = n + 1/2 - 1/n.$$

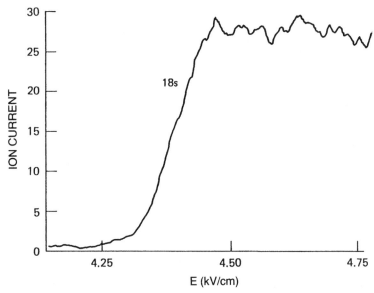

Fig. 7.4 Na 18s ionization threshold. The field ionization current is plotted vs the peak ionizing field (from ref. 3).

Using these values of n_s we can immediately compute the field required to ionize any $m = 0$ state,

$$E = \frac{W^2}{4} = \frac{1}{16n_s^4}. \tag{7.2}$$

In adiabatic ionization what is important is the ordering of the zero field energy levels, not their precise zero field energies.

In the Na atom the ionization of optically accessible states of $n < 18$ by 1 μs risetime pulses is often adiabatic, and is described by Eqs. (7.1) and (7.2). In Fig. 7.4 we show the ionization signal from the Na 18s state exposed to a 0.5 μs risetime pulse which is within 5% of its peak field for 200 ns. As expected, there is a single threshold, at 4.38 kV/cm, close to the 4.33 kV/cm predicted by Eqs. (7.1) and (7.2). The width of the threshold, from 10 to 90% ionization, is $\approx$3%, as might be expected for a pulse of this shape. There is a small, 5%, signal below the threshold due to black body radiation driving the atoms to higher states between the laser excitation and the field ionization pulse.[7]

When field ionization threshold curves such as the one shown in Fig. 7.4 are measured for many states, they can be plotted together to exhibit the n dependence of the ionization threshold field. In Fig. 7.5 we show a plot of the threshold fields (50% ionization) for the Na $|m| = 0$, 1, and 2 states obtained with a 0.5 μs risetime field pulse similar to the one shown in Fig. 7.1(b).[8] In Fig. 7.5 it is apparent that, while the threshold fields of the $m = 0$ states are described by Eq. (7.2), $|m| = 1$ and 2 states ionize at slightly higher fields, which can be easily understood. Due to the centrifugal barrier, $m \neq 0$ states are excluded from the z

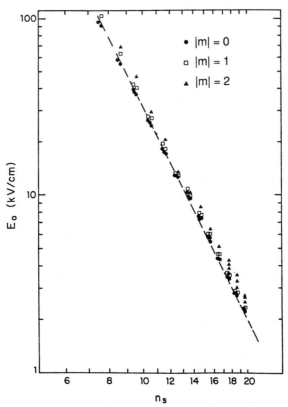

Fig. 7.5 Plot of observed ionizing fields for Na $|m| = 0$, 1, and 2 states for $n = 8$–20 plotted vs n_s, the effective quantum number in the strong field. The line indicates the classical ionization threshold (from ref. 8).

axis, the location of the saddle point in the potential. Thus the fields required to ionize $m \neq 0$ states are higher than the value given in Eq. (7.2). An approximate expression for the increase in field required for ionization of a state with $m \neq 0$ above the field required to ionize an $m = 0$ state of the same n_s is[9]

$$\Delta E/E = |m|/2n. \tag{7.3}$$

For $n = 15$, $|m| = 1$ and 2 states are predicted to ionize at fields 3% and 5% higher than their $m = 0$ counterparts, in reasonable agreement with the data shown in Fig. 7.5.

It is useful to compare the difference between the thresholds of adjacent $|m|$ states of the same ℓ to the difference between adjacent ℓ states of the same m. From the assignments of Eq. (7.1) the fractional change in ionizing field for an $m = 0$ state when ℓ is increased by one, corresponding to $\Delta n_s = 1/n_s$, is

$$\Delta E/E = 4/n_s^2. \tag{7.4}$$

At $n = 20$ changing ℓ by one produces a change in ionizing field of 1%, while changing $|m|$ from 0 to 1 produces a change of 2.5%. Comparing these two fractional changes to the width of the threshold field shown in Fig. 7.4, it is apparent that it should be straightforward to resolve adjacent $|m|$ states, but more difficult to resolve adjacent ℓ states of the same $|m|$.

What conditions must be fulfilled for the ionization process to occur in the adiabatic fashion we have just described? First, the transition from the zero field $n\ell m$ states to the intermediate field Stark states must be adiabatic. Second, the traversal of the avoided crossings in the strong field regime, $E > 1/3n^5$ must be adiabatic as well. Finally, ionization only occurs at $E > W^2/4$ if the ionization rate exceeds the inverse of the time the pulse spends with $E > W^2/4$.

Let us consider the first issue, the passage from zero field to the intermediate field. The zero field spacing between ℓ levels is $(\delta_\ell - \delta_{\ell+1})/n^3$. When the field reaches a value such that the separation between Stark states, $3nE$, is equal to the zero field spacing of the ℓ states, the ℓ states are no longer good eigenstates. If the field reaches this value in a time which is long compared to the inverse of the zero field splitting the passage is adiabatic. On the other hand if the field reaches this value in a time short compared to the inverse of the zero field splitting, the passage is diabatic, and the zero field $n\ell m$ state is projected onto several Stark states. Assuming the field rises linearly in time, we can write a useful criterion for deciding if the passage from low to intermediate field is adiabatic or not in terms of the slew rate $S = dE/dt$. We define a critical slew rate S_ℓ for the passage from a zero field ℓ state by

$$S_\ell = \frac{(\delta_\ell - \delta_{\ell+1})^2}{3n^7}. \tag{7.5}$$

In practice Eq. (7.5) means that $n \leq 20$ states with quantum defect differences of 10^{-3} satisfy the adiabaticity requirements for pulses with risetimes of 1 μs.

On the other hand the high ℓ states have quantum defect differences which are much smaller than 10^{-3} and they do not satisfy the adiabatic criterion of Eq. (7.5) for the same risetime. As a result, when the field is turned on they are projected diabatically onto the intermediate field states. From Eq. (7.5) it is clear that if the risetime is kept constant, the ℓ at which diabatic passage from zero field occurs becomes lower as n is increased. Since it is impossible to excite optically high ℓ states, the statement that they pass diabatically from low field to the intermediate regime has not been tested, but it has been experimentally established that the optically accessible low ℓ states of $n \approx 20$ Na atoms do in fact pass adiabatically to the intermediate field regime for pulses with 1 μs risetimes.

We now consider how the avoided crossings in the high field regime, $E > 1/3n^5$, are traversed. Consider an isolated avoided crossing of magnitude ω_0 between two levels 1, and 2 with different Stark shifts dW_1/dE and dW_2/dE, as shown in Fig. 7.6. If the atom is initially in state 1 and the crossing is traversed slowly compared to $1/\omega_0$, the inverse of the magnitude of the crossing, the traversal is adiabatic, as shown by the solid arrow of Fig. 7.6. If the crossing is traversed rapidly compared to $1/\omega_0$ the passage is diabatic, as shown by the broken arrow of

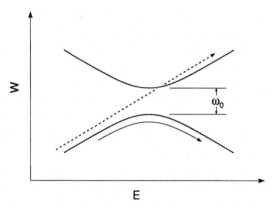

Fig. 7.6 Two Stark levels with an avoided crossing ω_0. If the field is slewed through the avoided crossing in a time long compared to $1/\omega_0$ the passage is adiabatic (solid arrow), while if it is slewed rapidly through the crossing the passage is diabatic (broken arrow).

Fig. 7.6. The criterion for the critical slew rate for the crossings, S_x, may be written as

$$S_x = \frac{\omega_0^2}{\left(\dfrac{dW_1}{dE} - \dfrac{dW_2}{dE}\right)}. \tag{7.6}$$

If $S \gg S_x$ the avoided crossing is traversed diabatically and if $S \ll S_x$ it is traversed adiabatically. Between the two extremes of purely adiabatic and purely diabatic traversals, the probability of making the diabatic transition is given by the Landau–Zener transition probability, as has been demonstrated by Rubbmark *et al.*[10]

To apply Eq. (7.6) we approximate the size of an avoided crossing by assuming that $\langle nn_1n_2m|n\ell m\rangle \sim 1/\sqrt{n}$ and only use the contribution from the lowest contributing ℓ state, $\ell = m$. In this approximation we obtain

$$\omega_0 \simeq \delta'_m \, 1/n^4, \tag{7.7}$$

where δ'_m is the absolute value of the quantum defect δ_ℓ (modulo 1). Eq. (7.7) makes it apparent that the avoided crossings are traversed more diabatically as n or $|m|$ increases. For Na $\delta'_0 = 0.35$ $\delta'_1 = 0.15$ $\delta'_2 = 0.015$. If we use these values for ω_0 in Eq. (7.6) with the largest possible value of $(dW_1/dE - dW_2/dE) = 3n^2$, we conclude that the level crossings should be transversed adiabatically for pulses of risetime $\sim 1\,\mu s$ even for $n \sim 100$ for $|m| = 0, 1$, and 2 states. Above $E = W^2/4$ the core coupling is between a discrete state and a red Stark continuum, and the ionization rate Γ to the available continua is given by Eq. (6.54). Ionization occurs at a field dependent rate, and the time required for nearly complete, 90%, ionization is $2/\Gamma$. One can estimate the ionization rates just above $E = W^2/4$ to be $\Gamma \sim \delta'_m(2/n^4) \cdot (1/n)$. The ionization rate is estimated assuming that one n_1 channel is available, so that the integral of Eq. (6.54) is $\sim 1/n$. Using these rates one expects to observe ionization in ~ 1 ns for $|m| = 0$ and 1 states and 30 ns for $|m| = 2$ states at $n = 20$. For most practical purposes these times are fast enough that ionization appears to occur as soon as $E \geq W^2/4$.

Contrary to the estimate above, for Na nd states of $m = 2$ and $n \geq 18$ non adiabatic ionization is observed. Specifically, when Na nd states of $m = 2$ are exposed to 0.5 μs risetime pulses similar to Fig. 7.1(b), they exhibit multiple ionization thresholds.[3] There are, in principle, two possible explanations. The first is that above the classical limit the atom passes through regions in which the ionization rate varies enormously. If this were the case we would expect to see occasional decreases in the ionization probability as a function of pulse amplitude, not a monotonic increase with ionizing field. A more likely explanation for the multiple $m = 2$ ionization thresholds is that some of the avoided crossings in the $E > 1/3n^5$ region are traversed partially diabatically, resulting in the atoms following several paths to the classical ionization limit, resulting in several ionizing fields. Support for this explanation comes from the work of Vialle and Duong[11] who used a stepped pulse to study field ionization in Na, and of Jeys *et al.*[4] who traced the evolution from adiabatic to diabatic ionization in Na.

Jeys *et al.*[4] observed the transition from adiabatic to diabatic passage with increasing n. They studied the ionization of the Na nd states of $n \geq 30$ with a linearly rising pulse as shown in Fig. 7.1(c), using the time at which electrons were detected to determine the field in which the atoms were ionized. Their results for nd states $30 \leq n \leq 36$ are shown in Fig. 7.7. Let us focus on the 32d state. Most of the atoms ionize when the field reaches ~ 400 V/cm. A small number of atoms ionizes between 450 and 600 V/cm, and a significant number of atoms ionize at 600 V/cm. Ionization at ~ 400 V/cm corresponds to following the adiabatic path shown by the broken line of Fig. 7.7(b). The $|m| = 0$ and 1 states excited certainly follow this path. Ionization at 600 V/cm is due to $|m| = 2$ atoms and corresponds to following the completely diabatic path shown by the bold line in Fig. 7.7(b), the diabatic hydrogen-like path of the reddest $|m| = 2$ Stark state. Ionization at fields between 450 and 600 V/cm is presumably due to $|m| = 2$ atoms which follow paths between the adiabatic and diabatic paths of Fig. 7.7(b) and ionize at the classical limit. It is also clear that the high field $|m| = 2$ feature grows in size as n is increased, suggesting that, as n is increased, progressively larger fractions of the atoms make the completely diabatic passage as shown by the bold lines of Fig. 7.7(b). The observation that diabatic passage is more likely with increasing n is consistent with the onset of multiple $|m| = 2$ thresholds at $n = 18$. It is also important to notice that no ionization occurs at fields above the $|m| = 2$ feature. Thus, no H-like states other than the extreme red ones are being populated. The ionization of the Na nd states at $n = 36$ corresponds exactly to the ionization of the red H $n = 36$, $n_1 = 0$, $|m| = 2$ state, which ionizes when classically allowed at a field $E \simeq 1/9n^4$. The form of ionization exhibited by the $|m| = 2$ states is often termed diabatic ionization. However, it is really only when $E > 1/3n^5$ that it is diabatic.

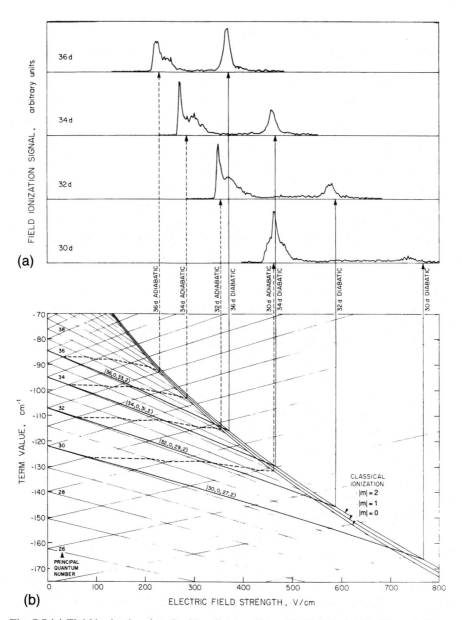

(a)

(b)

Fig. 7.7 (a) Field ionization data for Na *nd* states of *n* = 30, 32, 34, and 36. (b) Light lines: extreme members of |*m*| = 0 Stark manifolds (fourth order perturbation theory); dotted lines: adiabatic paths to ionization for *n* = 30, 32, 34, and 36; dark lines: diabatic paths to ionization for lowest members of |*m*| = 2 manifolds for *n* = 30, 32, 34, and 36. The lines indicating the classical ionization fields are calculated on the basis of Ref. 5 (from ref. 4).

The passage from the zero field nd state to a single Stark state is evidently adiabatic. A hydrogen nd $|m|=2$ state would pass diabatically from zero field to many $nn_1n_2 2$ Stark states and exhibit multiple ionization fields.

Exposure of blue Stark states to rapidly rising fields also results in diabatic, or H like, ionization, at fields far above the classical field for ionization. This point has been demonstrated by Neijzen and Donszelmann[12] using high lying, $n = 66$, states of In and by Rolfes *et al.*[13] using Na atoms of $n = 34$ and $|m| = 2$.

Apparently the Na $|m| = 2$ states of $n \approx 35$ traverse the avoided crossings diabatically and do not ionize rapidly if they are above the classical ionization limit, in clear contradiction to our earlier estimates that all the avoided crossings for $E > 1/3n^5$ would be traversed adiabatically, and ionization would occur at the classical limit $E = W^2/4$. What is wrong with the estimates? In our estimates of $\Delta\omega_0$ and Γ we assumed that the d state was evenly spread over the $|m| = 2$ Stark states, that is $\langle nn_1n_2 2|n22\rangle \sim 1/\sqrt{n}$, irrespective of n_1 and n_2. In fact, for the extreme Stark states $|n_1 \mp n_2| = n - 3$, $|\langle n(n - 3)02|n22\rangle| = |\langle n\,0\,n - 3\,2|\,n22\rangle| = 6\sqrt{5}/n^{3/2}$ a factor $\sim n$ smaller than our estimate. The d states are concentrated in the central Stark states where $n_1 \sim n_2 \sim n/2$, as shown by Fig. 6.3. As a result, the avoided crossings between the extreme states, which are traversed most rapidly, are a factor of n^2 smaller than our estimate based on the average value of the transformation coefficients. Therefore the risetime must be a factor of n^2 slower than our estimates for the passage to be adiabatic. At the slew rates actually used it is not surprising that the crossings are traversed diabatically. Similarly, at the classical limit, the coupling of an extreme blue Stark state to the underlying red continuum is reduced by a factor of n^2. For this reason the blue state does not ionize when it crosses the classical ionization limit, rather, only when it reaches its own hydrogenic ionization limit, as shown by Fig. 7.2.

In the discussion of adiabatic ionization we have implicitly assumed that this ionization occurs if the field exceeds the classical ionization limit. As shown in Chapter 6, there are substantial variations in the ionization rates due to variations in the density of states of the red continua and avoided crossings with more rapidly ionizing levels. Although the variations in the ionization rate above the classical limit are not the cause of the multiple ionization thresholds observed in Na, they can lead to nonmonotonic ionization curves, which have been observed in several cases.[8,14] A particularly clear example is shown in Fig. 7.8.[14] Van de Water *et al.* excited fast, 11 keV, He atoms to $n = 19$ triplet states in fields of 2.1–2.3 kV/cm, just below the classical ionization limit. The atoms then passed adiabatically through a buffer field and into a field ionization region 8.24 cm long. Subsequent to the field ionization region was an ionizing region in which all Rydberg atoms not ionized in the ionization region were ionized. They observed the number of atoms surviving the ionizing field as a function of its magnitude near the classical ionization limit. As shown by Fig. 7.8, the ionization curves are anything but monotonic, indicating the variation of the ionization rate with fields in the continuum of the red Stark states. The atoms are exposed to the ionizing field for

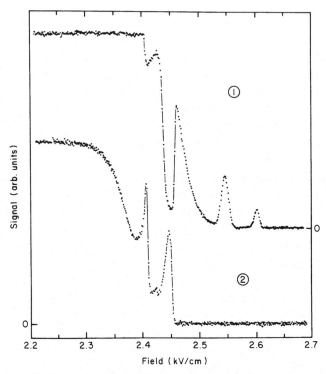

Fig. 7.8 Signal of highly excited triplet He atoms of the $m = 0$ manifold which survived exposure to the ionizing electric field E as a function of its strength. The curves show the ionization behavior of two different adiabatic states of $n_s \cong 19$ (from ref. 14).

200 ns, roughly the same time as for the data shown in Fig 7.4, yet show markedly more structure. The clarity of the structure in the threshold of Fig. 7.8 is probably due primarily to the difference in the pulse shapes. The data of Fig. 7.8 were taken with the pulse shape of Fig. 7.1(a), while the data of Fig. 7.4 were taken with the pulse shape of Fig. 7.1(b). Multiple thresholds occur because of partially adiabatic traversals of avoided crossings of bound levels or incomplete ionization upon reaching the classical ionization limit. These two causes need not occur separately, but can occur together, and observations consistent with their occurring together have been made by McMillian *et al.*[15]

It is useful to present visually the difference between adiabatic and diabatic field ionization. In Fig. 7.9 we show schematically how adiabatic and diabatic ionizations occur for three $n = 15$ states. Diabatic ionization, shown by the solid bold lines, is exactly like hydrogen. Only the red state ionizes at the classical ionization limit; the fields for other states are higher. In adiabatic ionization, shown by the bold broken lines, the $n = 15$ levels are trapped between the $n = 14$ and $n = 16$ levels and ionize at the classical ionization limit. In reality the true adiabatic levels, are not field independent, as they are shown in Fig. 7.9, but exhibit the avoided crossings shown in Fig. 7.3. However this simplification in the drawing

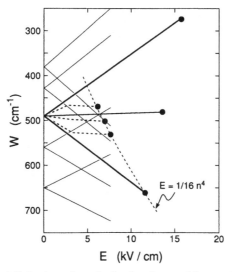

Fig. 7.9 Adiabatic and diabatic paths to ionization for n = 15 states in the center and on the edges of the Stark manifold. The diabatic paths are shown by solid bold lines and the adiabatic paths by broken bold lines. In both cases ionization occurs at the large black dots. The diabatic paths are identical to hydrogenic behavior. The adiabatic n = 15 paths are trapped between the adiabatic $n = 14$ and $n = 16$ levels. Adiabatic ionization always occurs at lower fields than diabatic ionization.

does not affect the reasoning. From Fig. 7.9 it is clear that adiabatic ionization occurs at lower fields than diabatic ionization in all cases.

Spin orbit effects

We have until now ignored the spin of the electron. It has two effects; it splits the zero field $\ell > 0$ states, and it alters the avoided crossings in high fields. In H, Li, and Na the second, high field, effect is negligible so we shall for the moment ignore it. In H the fine structure lifts the zero field ℓ degeneracy and has, in principle, the effect of allowing adiabatic passage from zero field spherical states to Stark states in the field. However, since the fine structure splittings are not large compared to the radiative decay rates, it does not appear that this possibility is of much practical importance.

In light alkali atoms, Li and Na, the fine structure splitting of a low ℓ state is typically much larger than the radiative decay rate but smaller than the interval between adjacent ℓ states. In zero field the eigenstates are the spin orbit coupled $\ell s j m_j$ states in which ℓ and s are coupled. However, in very small fields ℓ and s are decoupled, and the spin may be ignored. From this point on all our previous analysis of spinless atoms applies. How the passage from the coupled to the uncoupled states occurs depends on how rapidly the field is applied. It is typically a simple variant of the question of how the $|m|$ states evolve into Stark states. When

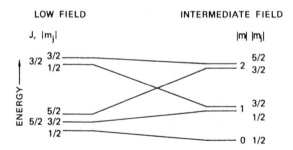

Fig. 7.10 Adiabatic correlation diagram for the Na nd states obtained from the known d state fine structure splitting, the intermediate field energy ordering, and applying the no-crossing rule for states of the same m_j (from ref. 3).

the field reaches the point that the Stark splitting of the $|m|$ levels is equal to the zero field fine structure interval the states are uncoupled mm_s states. This occurs when

$$a_2 E^2 = W_{FS}, \qquad (7.8)$$

where W_{FS} is the fine structure interval and a_2 is the tensor polarizability. If this field is reached rapidly compared to the inverse of the fine structure interval, the passage is diabatic, and the ℓsjm_j states are simply projected onto the uncoupled ℓmsm_s states, with the projections given by Wigner 3J symbols or Clebsch–Gordon coefficients. If this field is reached slowly the passage is adiabatic, each ℓsjm_j state passing into one ℓmsm_s state. In an adiabatic passage the ordering of the states is critical as in all adiabatic processes. Consider Na, for example. The zero field nd states are inverted, and in the field the uncoupled $|m| = 0$, 1, and 2 states are ordered as shown in Fig. 7.10. The m ordering is derived from the fact that the d–f dipole matrix elements have magnitudes which decrease as $|m|$ is increased from 0 to 2. In Fig. 7.10 we have also shown the adiabatic correlation in which $m_j = m + m_s$ is conserved.[3] The $d_{3/2}$ state leads to the $|m| = 1$ and 2 states and the $d_{5/2}$ state leads to $|m| = 0$, 1 and 2 states. Experimentally the field ionization threshold fields of Fig. 7.11 are observed, when Na 17d atoms are exposed to a pulse similar to Fig. 7.1(b) which rises in ~0.5 μs to the peak field shown on the horizontal axis. As expected from Fig. 7.10 there are only $|m| = 1$ and 2 thresholds from the $17d_{3/2}$ state and $|m| = 0$, 1, and 2 thresholds from the $17d_{5/2}$ state. Had the Na nd fine structure intervals been normal, the $nd_{3/2}$ states would only correlate to the $|m| = 0$ and 1 states and the $nd_{5/2}$ states to the $|m| = 1$ and 2 states.

The fine structure intervals of the alkali atoms often fall in the 1–10 MHz range, in which case the transition between spin orbit and uncoupled states can be made either diabatically or adiabatically. Jeys *et al.*[16] have observed the transition from an adiabatic to a diabatic passage from the coupled fine structure states to the uncoupled states. With a pulsed laser, they excited Na atoms from the $3p_{1/2}$ state to the $34d_{3/2}$ state with σ polarized light, which leads to 25% $|m_j| = 1/2$ atoms and

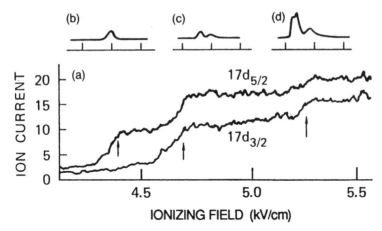

Fig. 7.11 (a) Experimental traces of the ion current vs peak ionization voltage for the $17d_{3/2}$ and $17d_{5/2}$ states. The approximate locations of the $|m| = 0$, 1, and 2 thresholds are indicated by arrows. (b), (c), (d) Oscilloscope traces of ion signals at different peak ionizing fields. In each case the center time marker corresponds to the peak of the ionizing, high voltage pulse. The horizontal scale is 200 ns/division. (b) $m = 0$ ion pulse, peak field = 4.58 kV/cm. (c) $m = 0$ followed by $|m| = 1$ ion pulse, peak field = 4.98 kV/cm. (d) Overlapping $m = 0$ and $|m| = 1$ ion pulses followed by $|m| = 2$ ion pulse, peak field = 5.27 kV/cm (from ref. 3).

75% $|m_j| = 3/2$ atoms. After 100ns they applied a linearly rising field ramp which rose to 0.4 V/cm at slew rates from 0.1 to 50 V/cm μs, a range sufficient to encompass both adiabatic and diabatic passage. Afterwards a fixed ramp of up to 800 V/cm was applied which allowed the field ionization signals of $|m| = 0, 1$, and 2 states to be resolved. As shown by Fig. 7.10 a purely adiabatic passage yields 25% $|m| = 1$ and 75% $|m| = 2$ atoms. A purely diabatic passage yields 10% $m = 0$, 30% $|m| = 1$ and 60% $|m| = 2$ atoms. When the slew rate was varied, the results shown in Fig. 7.12 were obtained. At slew rates less than 0.5 V/cm μs the passage is purely adiabatic and at rates in excess of 10 V/cm μs it is purely diabatic. These observations are in reasonable agreement with estimates based on Eq. (7.6). The Na 34d fine structure interval is 2.5 MHz[16] and its tensor polarizability is 350 MHz/(V/cm)^2.[17] Eq. (7.6) is satisfied for $E = 0.1$ V/cm, and reaching this field in a time of $(400/2\pi)$ ns, a slew rate of ~1.5 V/cm μs should be the approximate borderline between adiabatic and diabatic passage, and it is. Making the transition from the coupled fine structure states diabatically has been used in quantum beat experiments by Leuchs and Walther[2] and Jeys et al.[18]

We now consider the effect of the spin orbit coupling on the avoided level crossings in high fields. With spin it is the projection on the z axis of the total angular momentum, $m_j = m + m_s$, which is conserved, not m_s and m separately. The spin orbit couplings between Stark states may be computed just as we computed the quantum defect couplings. In essence the spin orbit splittings are important if the ℓ states involved have large enough spin orbit splittings to produce noticeable differences in the quantum defects of different j states of the

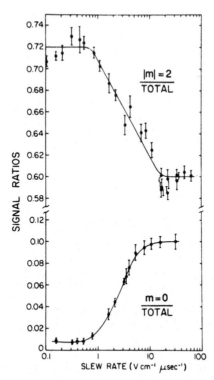

Fig. 7.12 Ratio of the signal resulting from ionization of $|m| = 2$ states (upper curve), and ionization of $m = 0$ states (lower curve) to the total ionization signal as a function of the slew rate from low to intermediate fields following excitation of the $34d_{3/2}$ state via the $3p_{1/2}$ state with σ polarization (from ref. 16).

same ℓ. For example in Na, the p fine structure interval produces a difference of 10^{-3} in the quantum defects of the np states. This is a factor of 15 smaller than the nd quantum defect. As a result the avoided crossing of an $m = 0$, $m_s = 1/2$ state with an $m = 1$, $m_s = -1/2$ state is negligibly small. Avoided crossings in which $\Delta m \neq 0$ are traversed diabatically, and the spin may be safely ignored, as it was in the earlier discussion of spinless atoms. In Li and H, which have smaller fine structure intervals than Na, the effect of the fine structure in high field can also be ignored.

On the other hand, in K the spin orbit splitting of the p states is large enough that the $|m| = 0$ and 1 levels are coupled strongly enough by the spin orbit interaction that avoided level crossings are no longer traversed purely adiabatically. Traversals of level crossings which are neither purely adiabatic nor purely diabatic lead to multiple threshold fields. As a result, states of $|m| = 0$ and 1 of the same energy exhibit similar multiple threshold fields.[19] However, the optically accessible $|m| = 2$ states exhibit single adiabatic thresholds as expected from the small spin orbit splitting and large quantum defect of the nd states. In Rb and Cs the np fine structure is also significant, and multiple thresholds are observed for all the optically accessible states, making field ionization a much less selective detection technique.[20]

References

1. F. B. Dunning and R. F. Stebbings, in *Rydberg States of Atoms and Molecules*, eds. R. F. Stebbings and F. B. Dunning (Cambridge University Press, Cambridge, 1983).
2. G. Leuchs and H. Walther, *Z. Phys. A* **293**, 93 (1979).
3. T. F. Gallagher, L. M. Humphrey, W. E. Cooke, R. M. Hill, and S. A. Edelstein, *Phys. Rev. A* **16**, 1098 (1977).
4. T. H. Jeys, G. W. Foltz, K. A. Smith, E. J. Beiting, F. G. Kellert, F. B. Dunning, and R.F. Stebbings, *Phys. Rev. Lett.* **44**, 390 (1980).
5. D. S. Bailey, J. R. Hiskes, and A. C. Riviere, *Nucl. Fusion* **5**, 41 (1965).
6. D. A. Park, *Z.Phys.* **159**, 155 (1960).
7. W. E. Cooke and T. F. Gallagher, *Phys. Rev. A* **21**, 580 (1980).
8. J. L. Dexter and T. F. Gallagher, *Phys. Rev. A* **35**, 1934 (1987).
9. W. E. Cooke and T. F. Gallagher, *Phys. Rev. A* **17**, 1226 (1978).
10. J. R. Rubbmark, M. M. Kash, M. G. Littman, and D. Kleppner, *Phys. Rev. A* **23**, 3107 (1981).
11. J. L. Vialle and H. T. Duong, *J. Phys. B* **12**, 1407 (1979).
12. J. H. M. Neijzen and A. Donszelmann, *J. Phys. B* **15**, L87 (1982).
13. R. G. Rolfes, D. B. Smith, and K. B. MacAdam, *J. Phys. B* **16**, L533 (1983).
14. W. van de Water, D. R. Mariani, and P. M. Koch, *Phys. Rev. A* **30**, 2399 (1984).
15. G. B. McMillian, T. H. Jeys, K. A. Smith, F. B. Dunning, and R. F. Stebbings, *J. Phys. B* **15**, 2131 (1982).
16. T. H. Jeys, G. B. McMillian, K. A. Smith, F. B. Dunning, and R. F. Stebbings, *Phys. Rev. A* **26**, 335 (1982).
17. T. F. Gallagher, L.M. Humphrey, R. M. Hill, W. E. Cooke, and S. A. Edelstein, *Phys. Rev. A* **15**, 1937 (1977).
18. T. H. Jeys, K. A. Smith, F. B. Dunning, and R. F. Stebbings, *Phys. Rev. A* **23**, 3065 (1981).
19. T. F. Gallagher and W. E. Cooke, *Phys. Rev. A* **19**, 694 (1979).
20. T. F. Gallagher, B. E. Perry, K. A. Safinya, and W. Sandner, *Phys. Rev. A* **24**, 3249 (1981).

8

Photoexcitation in electric fields

Hydrogenic spectra

A good starting point is photoexcitation from the ground state of H. The problem naturally divides itself into two regimes: below the energy of classical ionization limit, where the states are for all practical purposes stable against ionization, and above it where the spectrum is continuous.

As an example, we consider first the excitation of the $n = 15$ Stark states from the ground state in a field too low to cause significant ionization of $n = 15$ states. From Chapter 6 we know the energies of the Stark states, and we now wish to calculate the relative intensities of the transitions to these levels. One approach is to calculate them in parabolic coordinates. This approach is an efficient way to proceed for the excitation of H; however, it is not easily generalized to other atoms. Another, which we adopt here, is to express the $n = 15$ nn_1n_2m Stark states in terms of their $n\ell m$ components using Eqs. (6.18) or (6.19) and express the transition dipole moments in terms of the more familiar spherical $n\ell m$ states.[1,2]

In the excitation of the Stark states of principal quantum number n from the ground state only p state components are accessible via dipole transitions, so the relative intensities for light polarized parallel and perpendicular to the static field, π and σ polarizations, are proportional to the squared transformation coefficients $|\langle nn_1n_2m|n\ell m\rangle|^2$ from the nn_1n_2m parabolic states to the $n\ell m$ states for $\ell = 1$ and $m = 0$ and 1. In Fig. 8.1 we show the relative intensities by means of the squared transformation coefficients $|\langle 15n_1n_2m|15pm\rangle|^2$ for $m = 0$ and 1. When the exciting light is σ polarized, the distribution rises smoothly from both sides of the Stark manifold. On the other hand, for π polarization, the distribution drops abruptly at both edges of the manifold. When the manifolds of adjacent n overlap, the beginning of a new n manifold is much more apparent with π polarized light.

We can explicitly write out the statements above in the following way. We denote the oscillator strength from the $n'l'm'$ state to the nn_1n_2m state by $f_{nn_1m,n'l'm'}$. From the 1s state to the nn_1n_2m state it is given by

$$f_{nn_1m,1s0} = |\langle nn_1n_2m \,|npm\rangle|^2 f_{npm,1s0} \tag{8.1}$$

where $f_{npm,1s0} = 2\omega|\langle npm|r_m|1s0\rangle|^2$, $r_m = z$ for π polarization and $(x \pm iy)/2$ for $\sigma \pm$ polarization, and ω is the photon energy. Since n_2 is redundant, we do not use it in the oscillator strength. Of the two factors in Eq. (8.1) the first varies within an n manifold as shown by Eq. (6.19) and Fig. 8.1, and the second varies as n^{-3}.

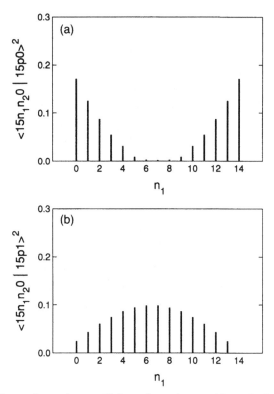

Fig. 8.1 Squared transformation coefficients from the $n = 15$ parabolic 15 $n_1 n_2 m$ states to spherical 15p states for (a) $m = 0$ and (b) $m = 1$. Note that the 15p state is concentrated in the edges of the $m = 0$ Stark manifold but at the center of the $m = 1$ manifold.

The procedure just outlined is perfectly adequate in the regime in which the Stark states are all discrete. However, it breaks down above the classical limit for ionization, when some of the Stark states become continuous and n and n_2 are ill defined. In this region the simplest approach is to treat the problem as excitation to continua. We take advantage of the fact that n_1 is a good quantum number, compute the oscillator strengths to the n_1 continua at energy W, and add these oscillator strengths to find the total oscillator strength. Explicitly, the analogue to Eq. (8.1) is[3]

$$\frac{df_{Wn_1m,1s0}}{dW} = 2\omega|\langle Wn_1m|r_m|1s0\rangle|^2. \tag{8.2}$$

Summing over n_1 gives the total oscillator strength

$$\frac{df_{Wm,1s0}}{dW} = \sum_{n_1} \frac{df_{Wn_1m,1s0}}{dW}. \tag{8.3}$$

To calculate the oscillator strengths of Eq. (8.2) we transform the Wn_1m continua to $W\ell m$ continua and calculate the excitation in spherical coordinates.

First, we extend the transformation of Eq. (8.1) between bound spherical and parabolic states into the continuum. How to go about this was pointed out by Fano[4] and Harmin.[5] We are interested in the $E \neq 0$ Stark state wavefunctions near the origin, where the applied field is negligible compared to the coulomb field, i.e. for $r \ll 1/\sqrt{E}$. In this region the wavefunctions for an $E \neq 0$ and an $E = 0$ parabolic state with the same energy and separation parameters are functionally identical, and we may replace the wavefunction for the parabolic $E \neq 0$ state by its zero field analogue with the same energy and separation parameters Z_1 and Z_2. Harmin[2] has shown that for continuum waves normalized per unit energy, in zero field

$$\langle Wn_1m|W\ell m\rangle = (-1)^\ell P_{\ell m}(Z_1 - Z_2), \tag{8.4}$$

where $P_{\ell m}(Z_1 - Z_2)$ is a normalized associated Legendre polynomial. Recall from chapter 6 that for any W, E, and m, fixing the value of n_1 fixes also the value of Z_1.

In the small r region, the only difference between the $E = 0$ and $E \neq 0$ wavefunctions with the same separation parameters is in the normalization. That is for small r

$$|Wn_1m\rangle_E = \frac{\sqrt{\left(C_{n_1}^m\right)}|Wn_1m\rangle_0}{\sqrt{C_{n_10}^m}}. \tag{8.5}$$

Here $C_{n_1}^m$ is the density of states at the origin defined by Eq. (6.33), and $C_{n_10}^m$ is its zero field analogue. Using Eqs. (8.4) and (8.5) as well as the dipole selection rule, implying that $\ell = 1$ in the final state, we can write the oscillator strength of Eq. (8.2) as

$$\frac{df_{Wn_1m,1s0}}{dW} = 2\omega|\langle Wpm|r_m|1s0\rangle|^2 P_{1m}^2(Z_1 - Z_2)\frac{C_{n_1}^m}{C_{n_10}^m}$$

$$= \frac{df_{W1m,1s0}}{dW} P_{1m}^2(Z_1 - Z_2)\frac{C_{n_1}^m}{C_{n_10}^m}. \tag{8.6}$$

In this form it is evident that the oscillator strength is simply the zero field oscillator strength multiplied by $P_{1m}^2(Z_1 - Z_2)$, which gives the amount of p character as n_1 or Z_1 is changed, and $C_{n_1}^m/C_{n_10}^m$ which gives the density of the n_1 continuum at the origin. The zero field cross section and $C_{n_10}^m$ are very slowly varying functions of energy, so any structure is due to either $C_{n_1}^m$ or $P_{1m}^2(Z_1 - Z_2)$. As shown by Fig. 6.7, far below the classical limit for ionization, $C_{n_1}^m$ is composed of delta functions corresponding to Stark states of good n_2, and the spectrum can be computed using the approach of Eq. (8.1). Near the classical limit $C_{n_1}^m$ is rapidly varying, reflecting the presence of broader resonances, the Stark states which ionize by tunneling. Above the classical limit $C_{n_1}^m$ is relatively smooth. In Fig. 8.2 we show the computed oscillator strength distributions from the ground

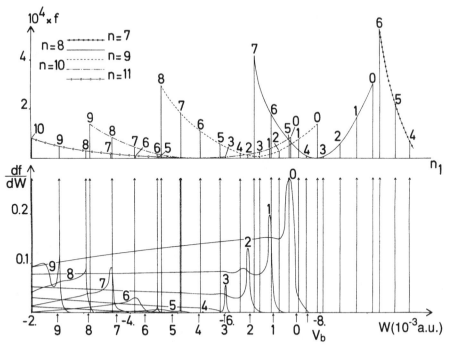

Fig. 8.2 n_1 continuum oscillator strengths $\mathrm{d}f_{Wn_1m,1s0}/\mathrm{d}W$ (shown as df/dW) and bound oscillator strengths $f_{nn_1m,1s0}$ (shown as f) for transitions from the ground state towards nn_1m states in the H atom in the presence of the field of strength $E = 1.5 \times 10^{-5}$ au with π polarization ($m = 0$) and for energies smaller than the ionization potential of the unperturbed atom. V_b is the critical energy in the saddle point model; the parabolic critical energies V_b for the different n_1 values are indicated by numbered arrows, the numbers giving the values of n_1. Continuum oscillator strengths $\mathrm{d}f_{Wn_1m,1s0}/\mathrm{d}W$ (lower figure) exhibit resonances of very different widths. Structures associated with quasi-discrete upper states, having a negligible width, but corresponding to a large value of df/dW; are marked by a line. The resonances can be labelled by the quantum numbers n. n_1, and m. By integration of df/dW over the Lorentzian profile of a resonance, a total value of the oscillator strength f can be defined. The total oscillator strengths are presented in the upper figure; a line connects the different total oscillator strengths corresponding to different values of n_1, but the same value of the principal quantum number $n = 7$ (–●–●–), $n = 8$ (——), $n = 9$ (– – –), $n = 10$ (– · – · –), $n = 11$ (–O–O–). Each maximum in a curve $\mathrm{d}f_{Wn_1m,1s0}/\mathrm{d}W$ is related to a well defined $f_{nn_1m,1s0}$ with the same n_1 and m values (from ref. 3).

state with π polarization for $n = 7$ to $n = 11$ states in a field of 1.5×10^{-5} au (7.6×10^4 V/cm). Fig. 8.2 is analogous to Fig. 8.1(a).

For $n = 7$ to $n = 9$ the spectra are similar to the spectra of Fig. 8.1(a). The lower $0 \leq n_1 \leq 3$ Stark states of $n = 10$ lie near the classical limits for their n_1 channels and mark the onset of continuous absorption in these channels. However, the higher $n = 10$ Stark levels of $n_1 > 5$ still appear as sharp lines. Of the $n = 11$ states, only the highest energy, blue, Stark states are recognizable as resonances; the lower Stark states have become part of the continuum absorption. A point which is important to note in Fig. 8.2 is that absorption corresponding to the center of the

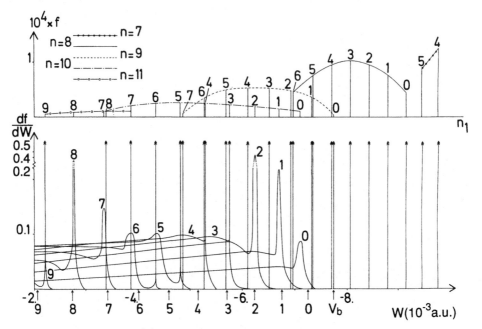

Fig. 8.3 The same as in Fig. 8.2 but with σ polarization, $m = 1$ (from ref. 3).

manifold, where $n_1 = n_2 = n/2$ and $Z_1 = Z_2 = 1/2$, is always weak whether the final states are discrete or a continuum.

In Fig. 8.3 we show the oscillator strengths for excitation from the H ground state with σ polarization, to excite final $m = \pm 1$ states. The oscillator strength distribution for $7 \leq n \leq 9$ resembles the oscillator strength distribution of Fig. 8.1(b). At $n = 10$ the onset of the continuous spectrum is observed in the lower Stark states but not the upper Stark states, as in Fig. 8.2. The fundamental difference between Figs. 8.2 and 8.3 is that in Fig. 8.3 there is appreciable oscillator strength from the ground state to the center of the Stark manifold, not to the edges.

Above the zero field limit, where one would not expect to see much, if any, structure in the photoionization cross section, there exist regular oscillations, as first observed by Freeman *et al.* in the photoexcitation of Rb from its ground state.[6,7] They excited a beam of Rb atoms with a frequency doubled pulsed dye laser to obtain the spectra shown in Fig. 8.4. They observed oscillations with π polarized light but no oscillations with σ polarized light. These oscillations are not peculiar to Rb but exist in H as well and can be described by computing the oscillator strength using Eq. (8.6). Since the εp oscillator strength is so slowly varying as to be effectively constant, the modulation in the signal apparently originates in the density of states, $C_{n_1}^m$ and $P_{1m}^2(Z_1 - Z_2)$.

The density of states near the origin, $C_{n_1}^m$, for a specific n_1, is only nonvanishing over an energy range corresponding to a range of values of Z_1 and Z_2. While the

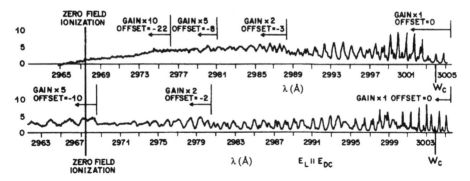

Fig. 8.4 Relative ground state photoionization cross section as a function of laser wavelength in Rb in the presence of a 4335 V/cm field. Note the relative gain and offset settings. For the light polarization parallel to the electric field (lower trace), field dependent resonance structure extends beyond the zero field limit. No structure is observed for the case of light polarized perpendicular to the field (upper trace) (from ref. 6).

statement is true for all m, it is particularly easy to compute the energy range for m = 1, in which case the potentials $V(\xi)$ and $V(\eta)$, from Eq. (6.23), have the forms

$$V(\xi) = 2\left(-\frac{Z_1}{\xi} + \frac{E\xi}{4}\right) \tag{8.7a}$$

and

$$V(\eta) = 2\left(-\frac{Z_2}{\eta} - \frac{E\eta}{4}\right). \tag{8.7b}$$

From Eq. (8.7a) it is apparent that for energies near the ionization limit the electron is excluded from the origin in the ξ motion if $Z_1 < 0$. Similarly if $Z_2 < 0$ the electron is excluded from the origin in the η motion. Therefore $Z_1 = 0$ and $Z_2 = 0$ represent the limits of the significant values of $C_{n_1}^1$. Stated in terms of Z_1 only, $C_{n_1}^1$ is nonzero over the range $0 \leq Z_1 \leq 1$. For a fixed n_1, Z_1 decreases with increasing energy. We may estimate the energies at which $Z_1 = 0$ and 1 using the WKB approximation. Examining Eqs. (6.22a) and (6.23a) we can see that if we set

$$\int_0^{\xi_m} \left(\frac{W}{2} + \frac{Z_1}{\xi} - \frac{E\xi}{4}\right)^{1/2} d\xi = \left(n_1 + \frac{1}{2}\right)\pi, \tag{8.8}$$

where ξ_m is the classical outer turning point, and evaluate it for $Z_1 = 0$ and 1 we can determine the upper and lower energy bound of $C_{n_1}^1$ for each value of n_1.[3,8]

In Fig. 8.5 we show the computed values of $C_{n_1}^m$ for $n_1 = 13\ m = 0$ and $n_1 = 12\ m$ = 1.[3] Both are nonzero over similar energy ranges, as expected. However, C_{13}^0 has sharper edges than C_{12}^1. Also shown in Fig. 8.5 are the oscillator strengths $df_{Wn_1m,1s0}/dW$ which are proportional to $C_{n_1}^m P_{1m}^2(Z_1 - Z_2)$. Due to the fact that $P_{10}^2(Z_1 - Z_2)$ vanishes for $Z_1 - Z_2 = 0$ and is peaked at $Z_1 - Z_2 = \pm 1$, the edges of $C_{n_1}^0$ contribute peaks to the $m = 0$ cross section while the center contributes

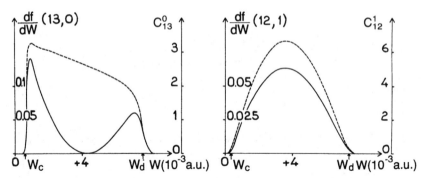

Fig. 8.5 Comparison between the energy dependences of the partial density of states $C_{n_1}^m$ (– –) and the continuum oscillator strengths $df_{Wn_1m,1s0}/dW$ (shown as $df(n_1,m)/dW$) (——) in the photoabsorption spectrum from the ground state of H in the presence of an external electric field $E = 1.5 \times 10^{-5}$ au at energy greater than 0, the ionization potential of the unperturbed atom, for $n_1 = 13, m = 0$ and $n_1 = 12, |m| = 1$; W_c is the energy at which the n_1 channel becomes continuous, corresponding to $Z_1 \approx 1$ and $Z_2 \approx 0$. W_d is the energy at which $Z_1 \approx 0$ and $Z_2 \approx 1$ (from ref. 3).

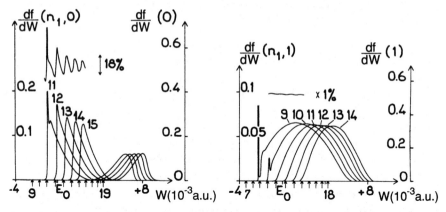

Fig. 8.6 Energy dependences of the continuum oscillator strengths $df_{Wn_1m,1s0}/dW$ (shown as $df(n_1,m)/dW$) for $|m| = 0$ and 1 in the photoabsorption spectrum from the ground state of H in the presence of an external electric field $E = 1.5 \times 10^{-5}$ au. The total oscillator strengths $df_{Wm,1s0}/dW$ (shown as $df(0)/dW$) are presented at the top of the figure; the depth of the modulation at the zero field ionization potential is equal to 18% and 1% for $m = 0$ and 1 respectively (from ref. 3).

nothing. In contrast, when multiplied by $P_{11}^2(Z_1 - Z_2)$ the center of $C_{n_1}^1$ contributes a single broad peak. When the oscillator strengths for all values of n_1 are summed, the sharp peaks on the low energy sides of the $m = 0$ n_1 channels are still evident, while the broad peaks of the $m = 1$ channels add to give a nearly structureless oscillator strength, as shown by Fig. 8.6. Examining C_{13}^0 and C_{12}^1 in Fig. 8.5 it is tempting to conclude that the relatively sharp low energy edge of $C_{n_1}^0$ would, by itself, lead to observable structure in the cross section. However, as pointed out by Luc-Koenig and Bachelier, the modulation in the cross section would then be ~3%, far less than observed experimentally.[3,9] The variations of both $C_{n_1}^0$ and $P_{10}^2(Z_1 - Z_2)$ are needed for observable modulations in the oscillator strengths.

If we are only interested in the frequency of the modulations in the vicinity of the zero field limit we may employ a different approach, used by Freeman *et al.*[6,7] and Rau[10]. They used the fact that the motion in the ξ direction is bound and found the energy separation between successive eigenvalues. Specifically, they used Eq. (8.8), the WKB quantization condition for the bound motion in the ξ direction, and differentiated it to find the energy spacing between states of adjacent n_1 or, equivalently, between the oscillations observed in the cross sections. Differentiating Eq. (8.8) with respect to energy yields

$$\int_0^{\xi_m} \left(\frac{W}{2} + \frac{Z_1}{\xi} - \frac{E\xi}{4} \right)^{-1/2} \frac{d\xi}{4} = \frac{dn_1}{dW} \pi. \tag{8.9}$$

Since we are interested in the result for $Z_1 = 1$ near $W = 0$, we set $W = 0$ and $Z_1 = 1$ and choose $\xi_m = 2/\sqrt{E}$. Evaluating the integral by making the substitution $\sin \theta = \sqrt{E}\xi/2$ we find

$$\frac{dW}{dn_1} = 3.70E^{3/4}. \tag{8.10}$$

In laboratory units, $dW/dn_1 = 22.5$ cm^{-1} when $E = 4335$ V/cm. In other words, the spacing between the high energy edge of the continuous spectrum of $C_{n_1}^1$ for adjacent values of n_1 scales as $E^{3/4}$ in the vicinity of the ionization limit, $W = 0$. Eq. (8.10) correctly describes the frequency of the oscillations in the cross section, as shown in Fig. 8.4.

Another approach to calculating the spectra such as the one of Fig. 8.4 was suggested by Reinhardt,[11] who proposed using a wave packet approach. The essential idea is that the laser light creates a wave packet at the origin (more precisely in the volume of the ground state) which propagates radially outward from the origin. If the wave packet encounters a potential barrier, such as the Stark potential on the upfield side, it is reflected. If the wave packet returns to the origin during the laser pulse a standing wave is created, leading to the modulation in the cross section. How strong the modulation is depends upon how large a fraction of the outgoing wave packet returns to the origin. This picture clarifies why there is only modulation with π polarization. With π polarization half the electrons are preferentially ejected in the upfield, $+z$, direction and are reflected back to the origin. With σ polarization no electrons are ejected in the upfield direction and none are reflected back to the origin.

A related approach to the calculation of the strong field mixing spectrum has been pursued by Gao *et al.*[12] They start the calculation in the same way Reinhardt does, by computing the outgoing quantum mechanical wavefunction of the photoelectron produced at the origin by photoabsorption. At a distance ~50 a_0

from the nucleus the outgoing wave fronts are converted to classical trajectories, perpendicular to the wave fronts. The classical trajectories are then propagated. Some of the trajectories in the upfield direction are reflected back to the origin, and, before they approach the origin, they are converted back to quantum mechanical wave fronts to recreate the quantum mechanical wavefunction. The constructive and destructive interference of the returning wavefunction with the outgoing wavefunction leads to the modulations in the strong field mixing spectrum. This approach leads to an excellent representation of the depth of the modulation as well as the locations of the maxima and minima of strong field mixing resonances in Na.[12] These are two quantities which are not specified by the WKB approach of Eqs. (8.9) and (8.10).

The oscillations in the photoionization cross section are not restricted to starting from an s state. They can be observed from other ℓ initial states as well, as demonstrated by Sandner *et al.*,[13] who observed the photoexcitation from the excited Ba 6s6p states in electric fields using the two laser excitation scheme Ba $6s^2$ $\rightarrow$ 6s6p $\rightarrow$ 6sεs, 6sεd. They used the four choices of linear polarization π–π, π–σ, σ–π, and σ–σ., and the σ–π and π–π spectra are shown in Fig. 8.7. The features at $W = 42,117$ and $41,841$ cm^{-1} are doubly excited states, and the ionization limit is at $42,035$ cm^{-1}. For π–π polarization strong oscillations are observed, for σ–σ polarization weak oscillations are observed, and for σ–π and π–σ polarizations no oscillations are observed. The oscillations in the π–π and σ–σ spectra come from the excitation to the εd $m = 0$ continuum (The oscillator strength to the s continuum is much weaker and can be ignored.). For the π–π spectrum of Fig. 8.7(b) the analogue to Eq. (8.6) is given by

$$\frac{\mathrm{d}f_{Wn_10,6p0}}{\mathrm{d}W} = \frac{\mathrm{d}f_{Wd0,6p0}}{\mathrm{d}W} P_{20}^2 (Z_1 - Z_2) \frac{C_{n_1}^0}{C_{n_10}^0}. \tag{8.11}$$

The only functional difference between Eqs. (8.6) and (8.11) is in the Legendre polynomial. $P_{20}^2(Z_1 - Z_2)$ is even more sharply peaked at $Z_1 - Z_2 = \pm 1$ than is $P_{10}^2(Z_1 - Z_2)$. Consequently, in the oscillator strength the low energy side of C_{n1}^0 is more strongly emphasized in Eq. (8.11) than in Eq. (8.6). Accordingly, the π–π oscillations in Fig. 8.7(b) are more apparent than those in the π spectrum of Fig. 8.4. The expressions analogous to Eq. (8.11) for the σ–π and π–σ polarizations contain $P_{21}^2(Z_1 - Z_2)$, while for σ–σ polarization it contains both $P_{20}^2(Z_1 - Z_2)$ and $P_{22}^2(Z_1 - Z_2)$, the latter being dominant. Neither $P_{21}^2(Z_1 - Z_2)$ nor $P_{22}^2(Z_1 - Z_2)$ is strongly peaked at $Z_1 - Z_2 = 1$, and there are no visible oscillations in the π–σ or σ–π spectra and only a small oscillatory component in the σ–σ spectrum.

In the H atom it is not possible to excite Rydberg states in a strong electric field from an angular momentum state other than the 1s state. All other states are converted to Stark states by the field, leading to a pronounced asymmetry in the excitation of red and blue Rydberg Stark states from them. The asymmetry is easily understood by considering excitation from the $n = 2, m = 0$ Stark states. In

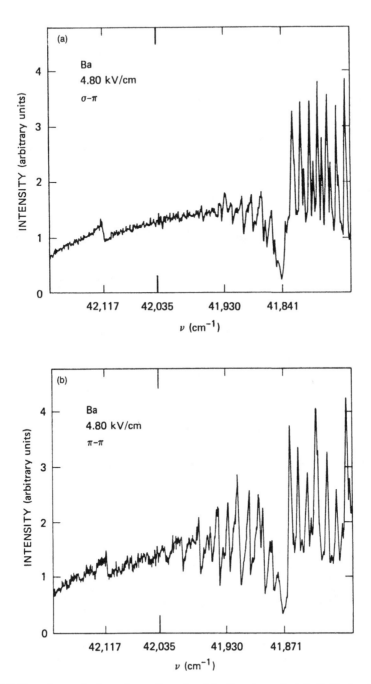

Fig. 8.7 Experimental photoabsorption signal in Ba at an electric field strenth of 4.80 kV/cm, obtained by resonant two photon absorption via the intermediate 6s6p $^1P_1$ state with (a) σ-π, and (b) π-π, polarization (from ref. 13).

the red state the electron is on the $-z$ side of the proton and the blue state it is on the $+z$ side of the proton. It is therefore hardly surprising that photoexcitation of red states is to red states and blue states is to blue states. A straightforward way to make this notion quantitative is to express the H $m = 0, n = 2 |\psi_{nn_1m}\rangle$ Stark states in terms of the $n = 2 |nlm\rangle$ states. Explicitly,

$$|\psi_{200}\rangle = |\psi_{\text{red}}\rangle = \frac{1}{\sqrt{2}} (|2s0\rangle + |2p0\rangle), \tag{8.12a}$$

and

$$|\psi_{210}\rangle = |\psi_{\text{blue}}\rangle = \frac{1}{\sqrt{2}} (|2s0\rangle - |2p0\rangle). \tag{8.12b}$$

In calculating photoexcitation cross sections from either of these two parabolic states we must coherently add the transition amplitudes from the s and p parts of the $n = 2$ state. Consider, for example, the excitation of the $m = 0$ Stark states of principal quantum number n at fields well below the fields required for ionization. The relative strengths of the transitions are given by expressions analogous to Eqs. (8.1) and (8.2). Explicitly, the oscillator strength $f_{nn_10,2n'_10}$ for excitation of the nn_1n_20 state from the $n = 2, m = 0$ Stark states is given by

$$
\begin{aligned}
f_{nn_10,2n'0} = 2\omega \frac{1}{2} \Big[&\langle nn_1n_20|np0\rangle \langle np0|z|2s0\rangle \\
&\pm \{\langle nn_1n_20|ns0\rangle \langle ns0|z|2p0\rangle \\
&+ \langle nn_1n_20|nd0\rangle \langle nd0|z|2p0\rangle\} \Big]^2,
\end{aligned}
\tag{8.13}
$$

where the $+$ and $-$ signs refer to the red ($n'_1 = 0$) and blue ($n'_1 = 1$) $n = 2, m = 0$ states respectively. Using the associated Legendre polynomial form of Eq. (6.19) for the transformation coefficients, Eq. (8.13) may be written as

$$
\begin{aligned}
f_{nn_10,2n'_10} = \frac{2\omega}{n} \Big\{ &(-1)P_{10}\left(\frac{n_1 - n_2}{n}\right) \langle np0|z|2s0\rangle \pm \\
&\left[P_{00}\left(\frac{n_1 - n_2}{n}\right) \langle ns0|z|2p0\rangle + P_{20}\left(\frac{n_1 - n_2}{n}\right) \langle nd0|z|2p0\rangle \right] \Big\}^2.
\end{aligned}
\tag{8.14}
$$

As $n \to \infty$, $\langle np0|z|2s0\rangle$, $\langle ns0|z|2p0\rangle$, and $\langle nd0|z|2p0\rangle$, are given by $3.83 \times n^{-3/2}$, $1.11 \times n^{-3/2}$, and $3.95 \times n^{-3/2}$, respectively.[14] Thus, the dominant interference among the three terms of Eq. (8.14) is between the excitations to the np and nd components. Both $P_{10}\left((n_1 - n_2)/n\right)$ and $P_{20}\left((n_1 - n_2)/n\right)$ have their maximum magnitudes at $(n_1 - n_2)/n \cong \pm 1$, and consequently their interference is most pronounced in the extreme blue, $n_1 \approx n$, or extreme red, $n_1 \approx 0$, states of the Stark manifold. Since $P_{20}\left((n_1 - n_2)/n\right)$ is positive for $(n_1 - n_2)/n \cong \pm 1$ but $P_{10}\left((n_1 - n_2)/n\right)$ changes sign with $(n_1 - n_2)/n$, the excitation amplitudes to the blue or red final state add constructively or destructively depending on whether the initial state is a blue or red $n = 2$ state.

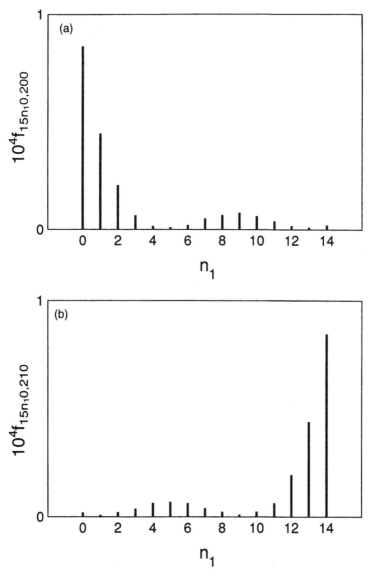

Fig. 8.8 Calculated oscillator strengths to the $n = 15$, $m = 0$ Stark states from the (a) red $n = 2$, $n_1 = 0$, $m = 0$ Stark state (b) blue $n = 2$, $n_1 = 1$, $m = 0$ Stark state.

In Fig. 8.8 we show a graph of the oscillator strengths from the red and blue $n = 2$ Stark states to the $n = 15$ Stark states, i.e. $f_{15n_10,2\pm0}$, assuming that ω has its zero field value of 0.123. As shown by Fig. 8.8, from the red $n = 2$ state predominantly the red $n = 15$ states are excited, while from the blue $n = 2$ state predominantly the blue $n = 15$ states are excited. The result of exciting from the red instead of the blue $n = 2$ Stark state is simply to reverse the asymmetry of the excitation of the Stark manifold.

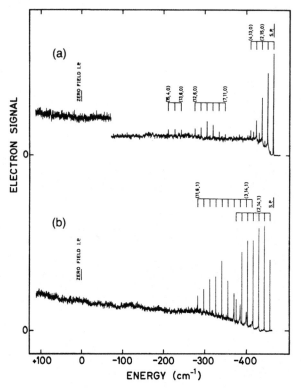

Fig. 8.9 Observed spectrum from the red $n = 2$, $n_1 = 0$, $m = 0$ Stark state in a field of 5714 V/cm with (a) π and (b) σ polarization of the second laser. Note the sharp onset of the continuous spectrum (from ref. 15).

In stronger fields, the difference between starting from the red and blue states becomes more pronounced. Since the red states are more easily ionized than the blue states, we expect to observe the onset of a continuous spectrum sooner and more abruptly when exciting from the red $n = 2$ state than from the blue $n = 2$ state. This point has been clearly shown by Rottke and Welge[15] and Glab et al.[16] Rottke and Welge excited H atoms in a beam first to one of the four $n = 2$ Stark states with a vacuum ultraviolet laser at 1216 Å which was fixed in frequency, and then used a second tunable laser to excite the H atoms from the selected $n = 2$ Stark state to the region of the zero field limit. The experiment was done in the presence of a static electric field of 5–10 kV/cm and the electrons resulting from the excitation and subsequent ionization of Rydberg states were detected as the wavelength of the second laser was scanned. They detected atoms which ionized with rates of 5×10^5 s^{-1} or faster. In Figs. 8.9 and 8.10 we show the spectra obtained using the red and blue $n = 2$, $m = 0$ Stark states as intermediate states. Fig. 8.9(a) shows the spectrum obtained by sweeping the second, π polarized, laser using the red $n = 2$ Stark state as the intermediate state. As shown, the spectrum is dominated by the continuous excitation of the red Stark states. The

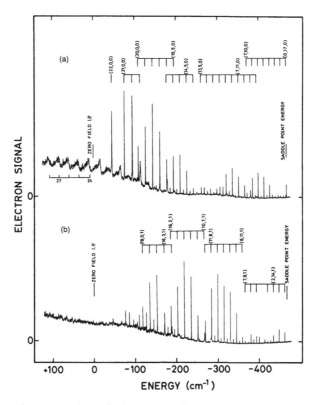

Fig. 8.10 Observed spectrum from the blue H $n = 2$, $n_1 = 1$, $m = 0$ Stark state in a field of 5714 V/cm with (a) π polarization and (b) σ polarization of the second laser. Note the progression from the extreme blue Stark states to the strong field mixing resonances (from ref. 15).

most prominent features are the relatively low n_1 Stark states of $n = 18$ which lie just above their classical ionization limits. Fig. 8.10(a) shows the analogous excitation spectrum obtained with π polarization using the blue $n = 2$ Stark state as the intermediate state. There are several marked differences between Figs. 8.9(a) and 8.10(a). First, there is no sharp onset of continuous excitation in Fig. 8.10(a), rather a gradually growing continuum excitation. Second, there are many more sharp states excited, and they are mostly to the blue sides of the Stark manifolds. From Fig. 8.8 we could reasonably expect predominantly blue states to be excited. The fact that there are far more sharp blue states than red states is due to the fact that their ionization rates do not increase as rapidly with field as do those of the red states, as shown by Fig. 6.8. Finally, we note that evolution of the extreme blue Stark states into the strong field mixing resonances is clearly evident in Fig. 8.10(a). The resonances are spaced as $E^{3/4}$, as described by Eq. (8.10), but the modulation is in this case deeper than it is when starting from an initial state which is spherical. The modulation is deeper for the same reason that Fig. 8.8 is

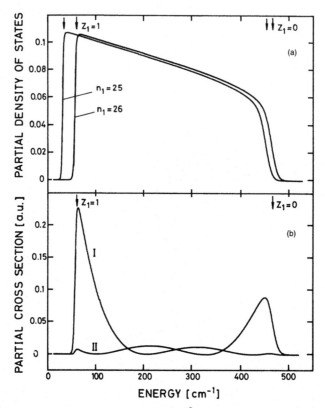

Fig. 8.11 (a) Calculated density of $m = 0$ states $C^0_{n_1}$ above the zero field ionization threshold for final states with quantum numbers $n_1 = 25,26$ at $E = 5714$ V/cm. Positions $Z_1 = 0$ and 1 are indicated by the arrows. (b) Calculated oscillator strengths for excitation from the $n = 2$ parabolic states 210 (blue) and 200 (red) into the channel with quantum number $n_1 = 26$ at 5714 V/cm in the energy region $W \geq 0$. Curve I, $df_{Wn_10,210}/dW$; curve II, $df_{Wn_10,200}/dW$ (from ref. 15).

not symmetric in the red and blue final states. The oscillator stength to the n_1 continuum from the $n = 2$, $m = 0$ red and blue Stark states is given by

$$\frac{df_{Wn_10,2n_1'0}}{dW} = 2\omega \frac{C^0_{n_1}}{C^0_{n_10}}[(-1)P_{10}(Z_1 - Z_2)\langle \varepsilon p0|z|2s0\rangle$$

$$\pm (P_{00}(Z_1 - Z_2)\langle \varepsilon s0|z|2p0\rangle + P_{20}(Z_1 - Z_2)\langle \varepsilon d0|z|2p0\rangle)]^2, \quad (8.15)$$

where the $+$ and $-$ sign refer to excitation from the red ($n_1' = 0$) and blue ($n_1' = 1$) $n = 2$ state respectively.

Inspecting Eq. (8.15) we can see that, apart from constant factors, it is simply a continuous version of Eq. (8.14), or Fig. 8.8, multiplied by $C^0_{n_1}$. This point is made explicitly in Fig. 8.11 in which $C^0_{n_1}$ and the oscillator strength are plotted. For an initial blue $n = 2$ state Fig. 8.8 peaks at $n_1 = n$, or $Z_1 = 1$, corresponding to the sharp low energy edge of $C^0_{n_1}$ (curve I of Fig. 8.11(b)). In contrast, for the red initial state Fig. 8.8 peaks at $n_1 = 0$, or $Z_1 = 0$, corresponding to the more gradually sloped high energy side of C_{n1}^0 (curve II of Fig. 8.11(b)). When the oscillators strengths of many n_1 channels are added, the oscillations in the spectrum from the $n = 2$ blue state are stronger than those in the spectrum from the 1s state, while those in the spectrum from the red $n = 2$ state almost vanish.

Realizing that the strong field mixing resonances evolve from the extreme blue Stark states we can write their energies to first order in the electric field as

$$W = -\frac{1}{2n^2} + \frac{3n^2E}{2},$$ (8.16)

and the $E^{3/4}$ spacing of the resonances at W=0 follows immediately from Eq. (8.16).

Nonhydrogenic spectra

The photoexcitation of nonhydrogenic atoms is in many ways similar to the photoexcitation of H. For example, the strong field mixing resonances observed in Rb, Ba, and Na are well described by a hydrogenic theory.[6,7,13] However, all features of the photoexcitation spectra of nonhydrogenic atoms are not equally well described by a hydrogenic theory, and we now describe the deviations. It is convenient to consider three spectral regions, below the classical ionization limit, above the classical ionization limit but below the zero field limit, and above the zero field limit.

Below the classical ionization limit all the states are effectively bound, and the probability of exciting a Stark state from a low lying $n\ell m$ state may be computed in the same way we computed it for H. Specifically, we project the Stark state of interest onto the zero field ℓ states and add the excitation amplitudes from the low lying state. Unfortunately, we cannot use the transformation from the Stark states to the $n\ell m$ states given by Eq. (6.19).

An excellent illustration of the deviation of the transformation coefficients from the hydrogenic values is afforded by the excitation of the Li $m = 0$ states of $n = 15$ from the 3s state, shown in Fig. 8.12. The spectra shown in Fig. 8.12 were obtained by Zimmerman *et al.*[17] by exciting Li atoms in a beam from the ground state via the 2p state to the 3s state using two pulsed dye lasers. A third dye laser, polarized parallel to the field, drove the 3s → Rydberg transition and was scanned in frequency to produce each of the spectra of Fig. 8.12. The Rydberg atoms were detected by applying a field pulse subsequent to excitation to field ionize them. Since only the p $m = 0$ components of the Stark states are accessible, we might expect a field independent distribution of oscillator strength among the Stark states, as implied by Fig. 8.1. As shown by Fig. 8.12, in fields up to ~1500 V/cm all the Stark states are excited, but not with the field independent hydrogenic $m = 0$

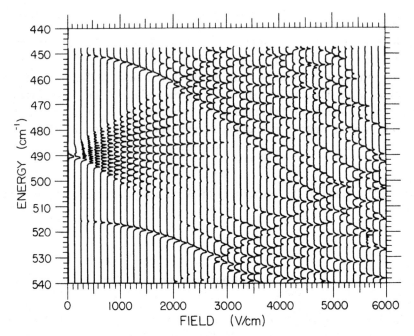

Fig. 8.12 Spectrum of the Li $m = 0$ Stark states from an initially excited 3s state. The line intensities are proportional to the amount of 15p $m = 0$ character in each Stark state (from ref. 17).

pattern shown in Fig. 8.1(a), rather with a pattern more similar to the $m = 1$ pattern of Fig. 8.1(b). Furthermore, the pattern is not field independent; for $E > 1500$ V/cm the excitation of the members near the edges of the Stark manifold disappears, only to reappear at higher fields.

In this regime, where the levels are discrete, it is possible to calculate the intensities of the transitions by matrix diagonalization, just as the energies are calculated. It is simply a matter of computing the eigenvectors of the Hamiltonian as well as its eigenvalues. For example, to calculate the intensities in the spectra shown in Fig. 8.12 we calculate the np amplitude in each of the Stark states and multiply it by the matrix element connecting the 3s state to the n'p state,

$$f_{nn_10,3s0} = 2\omega \left| \sum_{n'} \langle nn_1n_{2}0|n'\text{p}0\rangle \langle n'\text{p}0|z|3\text{s}0\rangle \right|^2 . \tag{8.17}$$

Often only one value of n' is really significant. For example for the $n = 15$ states of Fig. 8.12 only the $n' = 15$ term of Eq. (8.17) makes a large contribution. As shown by Zimmerman *et al.*, this procedure allows the intensities of Fig. 8.12 to be predicted accurately.[17]

As shown in Fig. 8.12, even the small Li$^+$ ionic core, producing an s state quantum defect of 0.3, radically alters the spectrum of Li $m = 0$ atoms from the

H $m = 0$ spectrum. With larger quantum defects the departures can be even more extreme, as shown by the energy level diagrams of the Na $m = 0$ and 1 states of Figs. 6.11 and 6.12. The energies of the Na $m = 0$ levels bear no obvious relation to hydrogenic levels for $E > 1/3n^5$. Furthermore, it is often difficult to pick out any pattern. Nonetheless, there does exist a similarity to the hydrogenic spectrum, which is brought out by the technique of scaled energy spectroscopy. The basic notion is as follows. Classically, the Hamiltonian for the H atom in the field $E\hat{z}$ is

$$H = p^2/2 - 1/r + Ez, \tag{8.18}$$

where p^2 is the momentum of the electron. It obeys classical scaling laws, and if we make the replacements

$$\tilde{r} = r E^{1/2}$$
$$\tilde{p} = pE^{-1/4} \tag{8.19}$$
$$\tilde{E} = 1.$$

then

$$\tilde{H}(\tilde{p},\tilde{r},1) = H(p,r,E)/\sqrt{E}, \tag{8.20}$$

and the scaled energy $\tilde{W} = W/\sqrt{E}$. Eichmann et al.[18] obtained spectra of Na at constant scaled energy $\tilde{W} = -2.5$ by simultaneously scanning both the electric field and the wavelength of the exciting laser so as to keep $\tilde{W} = -2.5$. This scaled energy lies slightly below the classical ionization limit, where $\tilde{W} = -2.0$. They excited Na atoms in a beam to either the $3p_{1/2}$ or $3p_{3/2}$ state using a laser linearly polarized in the field direction. With this polarization $m = 0$ and 1 3p atoms are produced. They then excited atoms from the $3p_j$ state to states of $\tilde{W} = -2.5$ by scanning the frequency of a second laser while the field was scanned to keep $\tilde{W} -2.5$. Subsequent to the laser excitation a field ionization pulse was applied to the atoms to ionize them and accelerate the resulting ions toward the detector. The resulting spectrum, plotted vs laser frequency, is a bewildering mass of lines suggesting no obvious interpretation. However, when it is Fourier transformed to obtain the spectrum vs the classical action S, it exhibits surprising regularity and is very similar to the Fourier transform spectra computed for H. Fig. 8.13 illustrates this point. In Fig. 8.13(a) we show the measured Fourier transform spectrum obtained with σ polarization of the second laser, so as to produce predominantly $m = 2$ final states. The experimentally derived spectrum of Fig. 8.13(a) is very similar to the theoretical spectrum of Fig. 8.13(b) computed for the excitation H $m = 2$ states from the 2p $m = 1$ states (to observe such transitions is, of course, impossible). Since the Na $m = 2$ states are very nearly hydrogenic, the similarity of Figs. 8.13(a) and 8.13(b) demonstrates primarily the agreement of theory and experiment in a known test case. In Fig. 8.13(c) we show the observed spectrum when the second laser is π polarized, so as to produce $m = 0$ and 1 final states. This experimental spectrum is similar to Fig. 8.13(d), the calculated spectrum for $\Delta m = 0$ transitions from the H 2p $m = 1$ state, with the largest discrepancies occurring at the highest values of the action S.

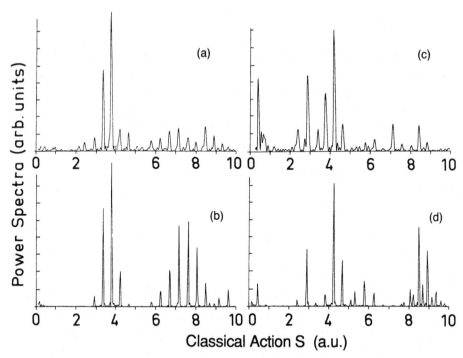

Fig. 8.13 Power spectra of measured Na photoexcitation cross sections from the 3p states vs the classical action S in atomic units: (a) σ polarization and (c) π polarization; and calculated power spectra of H (b) from the 2p $m = 1$ state with σ polarization and (d) from the 2p $m = 1$ state with π polarization. All are for fixed scaled energy $\bar{W} = W\sqrt{E} = -2.5$ (from ref. 18).

Above the classical ionization limit the blue states, which are stable in H, are degenerate with the red continua. In H they are not coupled, but in any other atom they are. We have already described how this coupling leads to ionization, which is a form of autoionization.[19] Similarly, the coupling leads to the same sort of interference in the excitation amplitudes that produces the asymmetric Beutler–Fano profiles common in the excitation spectra of autoionizing states.[20] Ultimately the electron finds its way into the red continuum, but it may reach the continuum by either direct excitation or by passing through the blue state, and the amplitudes for these two paths interfere. Since the strength of the blue state excitation varies across the state and the phase of the continuum changes by π across the state, at some point the two amplitudes are equal and of opposite sign, resulting in the cross section vanishing. Of course if there is a second, noninteracting, continuum, it contributes a constant cross section, and the total cross section does not vanish at any energy.[20]

In Fig. 8.14 we show an example of an asymmetric Beutler–Fano profile observed in a $m = 0$ state of Rb in a field of 158 V/cm.[21] The spectrum of Fig. 8.14 was obtained by Feneuille *et al.*[21] by exciting Rb atoms in a well collimated beam

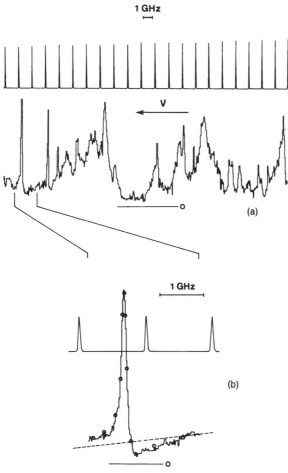

Fig. 8.14 (a) Photoionization spectrum of the ground state of Rb in the presence of a 158 V/cm static field for an excitation energy in the vicinity of 33,614 cm^{-1} and a light polarization parallel to the field. (b) Example of a charcteristically asymmetric profile. Dots represent the best fit to a Beutler–Fano profile. This fit has been obtained by assuming a linear variation of the ionziation background vs the excitation energy (dotted line). Fabry-Perot fringes 1.3 GHz apart provide the frequency scale (from ref. 21).

with a single mode pulsed dye laser, frequency doubled to provide uv pulses of 40 MHz linewidth. The atoms are excited in the presence of a static field which accelerates the Rb$^+$ ions formed to a particle detector.

As shown by Fig. 8.14, in most Stark spectra above the classical ionization limit there is never one isolated resonance but, more often, an irregular jumble of them. For example, in Fig. 8.15 we show the observed[22] and calculated[23] Na spectra near the ionization limit in a field $E = 3.59$ kV/cm.[22] The experimental spectrum of Fig. 8.15(a) was obtained by Luk *et al.*[22] by exciting a Na beam with two simultaneous dye laser pulses from the $3s_{1/2}$ to $3p_{3/2}$ state and then to the ionization limit. Both lasers were polarized parallel to the field, and the ions

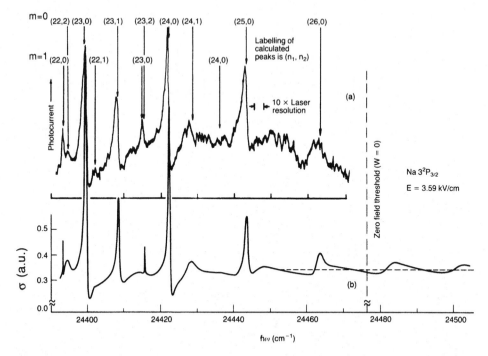

Fig. 8.15 (a) Experimental photionization spectrum from the Na $3^2P_{3/2}$ state in a field $E = 3.59$ kV/cm, vs photon energy $\hbar\omega$, within 0.01 eV of threshold (from ref. 21). Both the laser populating the $3p_{3/2}$ state and the second scanned laser are polarized parallel to the static field. Note labeling of Stark resonances (n_1, n_2) for $m = 0$ and 1. (b) Theoretical cross section, from the density of states theory. Both $m = 0$ and 1 final states are present (from ref. 23).

resulting from photoexcitation were collected as the wavelength of the second laser was scanned. The resulting spectrum of Fig. 8.15(a) is a composite of $m = 0$ and $m = 1$ final states since the spin orbit splitting of the 3p states, 17 cm^{-1}, is far greater than the inverse of the pulse duration of the lasers. Reproducing an entire experimental spectrum such as Fig. 8.15(a) poses a formidable challenge to the theorist, for the entire spectrum can not be interpreted using an isolated resonance approach used to interpret Fig. 8.14. However, Harmin's density of states approach[23] has been very successful in reproducing experimental spectra, as shown by Fig. 8.15(b). The density of states theory, a form of quantum defect theory applied to the Stark effect, also accounts for, quite naturally, the variation in the ionization rates, which is a quite general effect occurring in all three or more channel systems.[24]

In the density of states theory space is divided into three regions. For $r < r_0$ the effect of the external field may be neglected, and for $r > r_c$ the potential from the ionic core is negligible. Thus only in the region $r_c < r < r_0$ is the potential a pure coulomb potential, and in this and only this region may we transform freely between the spherical and parabolic wavefunctions. The noncoulomb potential at $r < r_c$ introduces a radial phase shift of δ_ℓ into the spherical wavefunctions with the result that they are given by a linear combination of the regular and irregular coulomb functions in the region $r_c < r < r_0$. In this region the wavefunction in spherical coordinates is given by

$$\psi = \frac{P_{\ell m}(\theta)e^{im\theta}}{\sqrt{2\pi}} (\cos \delta_\ell rf(r) + \sin \delta_\ell rg(r)), \tag{8.21}$$

where $f(r)$ and $g(r)$ are the regular and irregular coulomb functions. Since the g function is irregular at the origin, it is evident that the wavefunction cannot be expressed in terms of parabolic functions regular at the origin. Harmin expresses the parabolic wavefunction in terms of a regular function of ξ and a regular and an irregular function of η. Explicitly,[23]

$$\psi \sim \chi_1(\xi)[\chi_2(\eta) \cos \delta_s + \overline{\chi_2}(\eta) \sin \delta_s] \frac{e^{im\theta}}{\sqrt{2\pi}}, \tag{8.22}$$

where $\overline{\chi_2}(\eta)$ is the irregular solution of Eq. (6.22b) for $\chi_2(\eta)$ valid for $r > r_c$. In the classically allowed region $\overline{\chi_2}(\eta)$ is oscillatory, $90°$ out of phase with $\chi_2(\eta)$. Spherical wavefunctions of the form of Eq. (8.21) may be projected onto the parabolic wavefunctions of the form given by Eq. (8.22). There are two points to note regarding Eq. (8.22). First, the phase δ_s is related to the zero field quantum defects by the transformation which transforms the spherical wavefunction of Eq. (8.21) into the parabolic wavefunction of Eq. (8.22). Second, the wavefunctions of Eq. (8.22) are not orthornormal due to the presence of both regular and irregular functions $\chi_2(\eta)$ and $\overline{\chi_2}(\eta)$.. When the wavefunctions analogous to the one of Eq. (8.22) are orthonormalized, and the spherical wavefunctions re-expressed in terms of them, the transformation includes interference terms between the orthonormalized final states. These terms, which vanish if all the quantum defects are zero, lead to the interferences observed in the excitation spectra of nonhydrogenic atoms.

References

1. D. A. Park, *Z. Phys.* **159**, 155 (1960).
2. D. A. Harmin, in *Atomic Excitation and Recombination in External Fields*, eds. M. H. Nayfeh and C. W. Clark (Gordon and Breach, New York, 1985).
3. E. Luc Koenig and A. Bachelier, *J. Phys. B* **13**, 1769 (1980).
4. U. Fano, *Phys. Rev. A* **24**, 619 (1981).
5. D. A. Harmin, *Phys. Rev. A* **24**, 2491 (1981).
6. R. R. Freeman, N. P. Economou, G. C. Bjorklund, and K. T. Lu, *Phys. Rev. Lett.* **41**, 1463 (1978).

7. R. R. Freeman and N. P. Economou, *Phys. Rev. A* **20**, 2356 (1979).
8. D. ter Haar, *Problems in Quantum Mechanics* (Pion, London, 1975).
9. E. Luc-Koenig and A. Bachelier, *Phys. Rev. Lett.* **43**, 921 (1979).
10. A. R. P. Rau, *J. Phys. B* **12**, L193 (1979).
11. W. P. Reinhardt, *J. Phys. B* **16**, 635 (1983).
12. J. Gao, J. B. Delos, and M. C. Baruch *Phys. Rev. A* **46**, 1449 (1992).
13. W. Sandner, K. A. Safinya, and T. F. Gallagher, *Phys. Rev. A* **23**, 2448 (1981).
14. H. A. Bethe and E. A. Salpeter, *Quantum Mechanics of One and Two Electron Atoms* (Academic Press, New York, 1957).
15. H. Rottke and K. H. Welge, *Phys. Rev. A* **33**, 301 (1986).
16. W. L. Glab, K. Ng, D. Yao, and M. H. Nayfeh, *Phys. Rev. A* **31**, 3677 (1985).
17. M. L. Zimmerman, M. G. Littman, M. M. Kash, and D. Kleppner, *Phys. Rev. A* **20**, 2251 (1979).
18. U. Eichmann, K. Richter, D. Wintgen, and W. Sandner, *Phys. Rev. Lett.* **61**, 2438 (1988).
19. M. G. Littman, M. M. Kash, and D. Kleppner, *Phys. Rev. Lett.* **41**, 103 (1978).
20. U. Fano, *Phys. Rev.* **124**, 1866 (1961).
21. S. Feneuille, S. Liberman, J. Pinard, and A. Taleb, *Phys. Rev. Lett.* **42**, 1404 (1979).
22. T. S. Luk, L. DiMauro, T. Bergeman, and H. Metcalf, *Phys. Rev. Lett.* **47**, 83 (1981).
23. D. A. Harmin, *Phys. Rev. A* **26**, 2656 (1982).
24. W. E. Cooke and C. L. Cromer, *Phys. Rev. A* **32**, 2725 (1985).

9

Magnetic fields

The effects of magnetic fields on Rydberg atoms have been studied for more than 50 years. In fact, the experiments of Jenkins and Segre on atomic diamagnetism were among the first in which the large size of the Rydberg atoms was exploited,[1] and when this topic was revisited by Garton and Tomkins 30 years later they observed the unexpected quasi–Landau resonances.[2] Today an atom in a magnetic field is one of the best systems in which to search for the quantum analogue of the onset of classical chaos.

Diamagnetism

We can write the Hamiltonian for a H atom in a magnetic field B in the z direction as

$$H = \frac{p^2}{2} - \frac{1}{r} + A(r)\,\mathbf{L}\cdot\mathbf{S} + \frac{\mathbf{L}\cdot\mathbf{B}}{2} + \mathbf{S}\cdot\mathbf{B} + \frac{1}{8}(x^2 + y^2)B^2 \qquad (9.1)$$

where B is given in units of 2.35×10^5 T, or 2.35×10^9 G, $A(r)$ is a function describing the spin orbit coupling, x, y, and z are the Cartesian coordinates of the electron relative to the proton, and the electron–proton distance $r = \sqrt{x^2 + y^2 + z^2}$. Since $\langle x^2 + y^2 \rangle \propto n^4$, we can see that the ratio of the diamagnetic term to the linear magnetic field term is $\sim n^4 B$, and for a field of 1 T $n^4 B = 1$ at $n = 22$. If $n^4 B \ll 1$ we may safely ignore the quadratic diamagnetic term of Eq. (9.1). In this case the magnetic effects are of the same size as they are in a low lying excited state of the same angular momentum, with one small difference. Since the fine structure interval decreases as $1/n^3$, due to the inverse r^3 dependence of $A(r)$, the Paschen–Back regime in which $\mathbf{L}$ and $\mathbf{S}$ are uncoupled is reached at lower B fields in Rydberg states than in low lying states.

If $n^4 B \geq 1$ we cannot ignore the quadratic term of Eq. (9.1). In this case it is reasonable to assume that $\mathbf{L}$ and $\mathbf{S}$ are completely decoupled, in which case $\mathbf{S}$ can often be completely ignored, as can the spin orbit interaction $A(r)\mathbf{L}\cdot\mathbf{S}$. The resulting Hamiltonian,

$$H = \frac{p^2}{2} - \frac{1}{r} + \frac{\mathbf{L}\cdot\mathbf{B}}{2} + \frac{1}{8}(x^2 + y^2)\,B^2 \qquad (9.2)$$

has rotational symmetry about the z axis and preserves parity. Since m and parity are conserved, there are fewer coupled states than in an electric field, in which

143

parity is not conserved, but the problem is non separable, and, as a result, not amenable to analytic solution.

If we rewrite the diamagnetic term as $H_D = 1/8\, r^2 \sin^2 \theta B^2$, it is apparent that the non zero matrix elements of H_D are those for which $\Delta\ell = 0, \pm 2$, and $\Delta m = 0$.[3] There is, however, no restriction on Δn. Explicit forms of the matrix elements are[4]

$$\langle n\ell m \mid r^2 \sin^2 \theta \mid n'\ell m\rangle = 2\frac{(\ell^2 + \ell - 1 + m^2)}{(2\ell - 1)(2\ell + 3)} \langle n\ell|r^2|n'\ell\rangle \qquad (9.3a)$$

and

$$\langle n\ell m|r^2 \sin^2 \theta|n(\ell + 2)m\rangle = \left[\frac{(\ell + m + 2)(\ell + m + 1)(\ell - m + 2)(\ell - m + 1)}{(2\ell + 5)(2\ell + 3)^2(2\ell + 1)}\right]^{1/2}$$

$$\times \langle n\ell|r^2|n'(\ell + 2)\rangle. \quad (9.3b)$$

The diagonal matrix element of r^2 is the expectation value of $\langle r^2\rangle$, given by[5]

$$\langle n\ell|r^2|n\ell\rangle = \langle r_{n\ell}^2\rangle = \frac{n^2}{2}[5n^2 + 1 - 3\ell(\ell + 1)]. \qquad (9.4)$$

In H all ℓ states of the same n are degenerate, and since the diamagnetic interaction couples all the ℓ states of the same parity by means of the $\Delta\ell = 2$ matrix elements, ℓ is not a good quantum number for any non-zero field. If we consider an atom other than H, in which the low ℓ states are energetically removed from the high ℓ states, the first evidence of a diamagnetic effect comes from the diagonal matrix element of H_D, and according to Eq. (9.2) we can expect a Rydberg level to be displaced from its zero field energy by[3]

$$\Delta W = \frac{mB}{2} + \frac{B^2}{8} \langle n\ell m|r^2 \sin^2 \theta|n\ell m\rangle. \qquad (9.5)$$

If we approximate $\langle r_{n\ell}^2\rangle$ by $5n^4/2$, for the p Rydberg states, which are optically accessible from the ground state, ΔW is given by[3]

$$\Delta W = \frac{mB}{2} + \frac{B^2}{8}(1 + m^2)n^4. \qquad (9.6)$$

The $m = \pm 1$ states are split by the linear shift and have twice the diamagnetic shift of the $m = 0$ state.

In an alkali atom the expressions of Eqs. (9.5) and (9.6) for the shift are not valid for very high fields because the $\Delta\ell = 2$ matrix elements couple states of different ℓ, leading to higher order shifts. From second order perturbation theory it is clear that the shifts scale as $n^{11}B^4$, indicating that the regime in which Eqs. (9.5) and (9.6) are likely to be valid is limited. Historically, the regime in which the diamagnetic effect has produced eigenstates of mixed ℓ is called the ℓ mixing regime. In H it starts at $B=0$ and in an alkali, where it starts depends upon ℓ. When the field is further increased, so that the diamagnetic shifts exceed the n spacing the n mixing regime is reached, in which the eigenstates have contributions from different n as well as different ℓ.[1,3]

The first observations of diamagnetism were made by Jenkins and Segre,[1] who observed the absorption spectra of Na and K using a 30 inch column of vapor located between the poles of a cyclotron magnet. The magnetic field was 27 kG, enough to see clear diamagnetic effects for $n \approx 20$. They only obtained good results with Na. With K they could not observe the absorption of high series members unless the pressure was so high that it broadened the lines. The Cooper minimum of K is at the ionization limit, reducing the absorption of the high np states.[6] In Na they were able to observe the $m = 0$ and ± 1 shifts and the $m = \pm 1$ splittings given by Eq. (9.6) up to $n = 25$, at which point the shifts became larger than those given by Eq. (9.6) due to ℓ mixing. The Na np states lie just above the $n-1$ high ℓ states and are forced up in energy by the $\Delta \ell = \pm 2$ diamagnetic couplings. Although they were not able to resolve the other magnetic states, for the higher n states they observed, $n > 25$, there were clearly visible tails on the long wavelength sides of the lines indicative of other magnetic states lying just below the np states.

In experiments using laser techniques it has been possible to resolve the magnetic levels and display the detailed structure of the levels. An excellent example is the work of Zimmerman *et al.*,[7] who excited Na Rydberg states in the field of a superconducting solenoid. In their experiment a thermal beam of Na moves in the field direction and is crossed by two pulsed laser beams. The first is polarized perpendicular to the field and is tuned to excite the atoms from the ground $3s_{1/2}$ state to the $3p_{3/2}$ $m_j = 3/2$ state, which contains only $m = 1$. The second laser, polarized parallel to the field, drives transitions to even parity $m = 1$ Rydberg levels. The Rydberg atoms are ionized by a 2 kV/cm electric field pulse applied 1 μs after laser excitation, and the liberated electrons are accelerated to 10 keV and detected with a surface barrier detector. Spectra are recorded at fixed magnetic field strengths by scanning the wavelength of the second laser while recording the electron signal. Their spectra, at fields of up to 6 T are shown in Fig. 9.1. Several features are apparent in the data of Fig. 9.1. First, the shifts of the observed transition frequencies are quadratic in the magnetic field. There is no linear shift since both the 3p state and $m = 1$ Rydberg states have the same linear shift. Second the agreement between the observed energy level positions and those calculated by matrix diagonalization is excellent. Finally, the higher energy members of the set of diamagnetic states for a given n have the largest oscillator strength and therefore the largest amount of d character. Kleppner *et al.*[4] have suggested that these two properties are likely to be found together. The low ℓ states have the largest classical outer turning points and thus are likely to be prominent in the states with the largest diamagnetic shifts. Fig. 9.1 also serves to bring out a final point. The observed $m = 1$ even parity levels of different n cross, at least on the scale of Fig. 9.1(b), while the $m = 0$ even parity levels, shown in Fig. 9.1(c), do not. The $m = 0$ levels contain the s state with its large quantum defect. The crossing of the levels in the nearly H like $m = 1$ even parity states suggests that there is a symmetry or conserved quality similar to the H levels in an electric field.

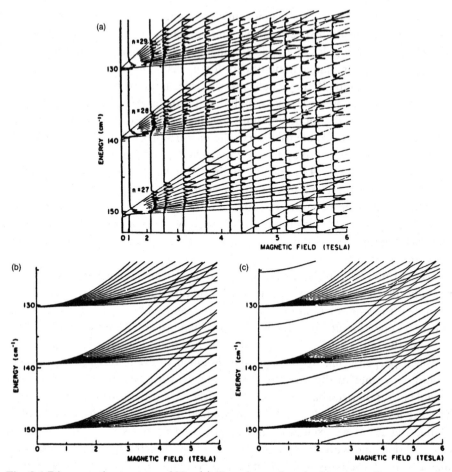

Fig. 9.1 Diamagnetic structure of Na. (a) Experimental excitation curves for even parity levels, $m = 1$, $m_s = \frac{1}{2}$, in the vicinity of $n = 28$. A tunable laser was scanned across the energy range displayed. The zero of energy is the ionization limit. Signals generated by ionizing the excited atoms appear as horizontal peaks. The horizontal scale is quadratic in field. Calculated levels are overlaid in light lines. Some discrepancies are present due to nonlinearity of the lasers. (b) Calculated excitation curves, displayed linearly in field. (c) Same as (b), but for even parity $m = 0$ states. Note the large effect on anticrossings due to the presence of the nondegenerate s states (from ref. 7).

As pointed out by Solov'ev,[8] if the magnetic field is low enough that the coulomb force is dominant, then there exist approximate constants of the motion in addition to L_z and parity. The first, Λ, is given by[8–10]

$$\Lambda = 4A^2 - 5A_z{}^2, \tag{9.7}$$

where $\mathbf{A}$ is the Runge–Lenz vector $\mathbf{A} = \mathbf{p} \times \mathbf{L} - \mathbf{r}/r$, which is directed along the semimajor axis of the classical elliptical orbit. The second, Q, is given by[8–10]

$$Q = \frac{L^2}{1 - A^2}. \tag{9.8}$$

In a pure coulomb potential, i.e. in H or a high m state of an alkali atom, the degeneracy of the orbits is so high that it is relatively easy to form orbits which have approximately constant Λ and Q from the many degenerate coulomb orbits.[9] For $\Lambda > 0$ the motion of **A** is primarily a rotation about the z axis, with **A** lying near the x,y plane. In contrast, for $\Lambda < 0$ the motion of **A** is a librational motion about the z axis with **A** pointing near the z axis. The larger the value of Λ the more the motion is concentrated in the x,y plane and the larger the diamagnetic energy shift. For a specific n state the transition from $\Lambda > 0$ to $\Lambda < 0$ is marked by two observable features. First, the spacing between levels, which is approximately proportional to $|\Lambda|$, is a minimum for $\Lambda \sim 0$. Second, for $\Delta m = 0$ transitions the oscillator strength from lower lying states is lowest to states of $\Lambda \sim 0$ and highest to the states with the highest and lowest values of Λ, corresponding to **A** lying in the x,y plane and along the z axis, respectively. These two points are shown clearly by calculations of Clark and Taylor[11] and Cacciani *et al.*[10] and by the experiment of Cacciani *et al.*[10] They observed single photon transitions from the ground state of Li to high lying Li $m = 0$ odd parity states in a field of 1.94 T using a laser beam polarized parallel to **B**. Their experimental approach was in essence the same as that of Zimmerman *et al.* except that they drove one single photon transition from the ground state rather than two. The largest quantum defect of the Li odd parity $m = 0$ states is that of the np states, which is 0.05, so these states are reasonably hydrogenic. The experimental spectrum is shown in Fig. 9.2 along with the calculated hydrogenic and Li spectra. Conventionally, the label K is attached to the magnetic states, with $K = 1$ being the highest energy state and $K = 15$ being the lowest energy state in this case. For $K = 1$ $\Lambda = 3.5$, for $K = 15$ $\Lambda = -1.0$, and for $K = 12$ $\Lambda = 0$. In Fig. 9.2 it is evident that the lines are weakest and closest together at $K \cong 12$, corresponding to $\Lambda = 0$.

Quasi Landau resonances

Thirty years after the experiment of Jenkins and Segre,[1] Garton and Tomkins[2] observed the absorption spectrum of Ba vapor between the poles of a 25 kG electromagnet. In zero field they were able to resolve np levels as high as 75, and in the presence of the magnetic field they were able to see individual diamagnetic levels as shown in Fig. 9.1. They were also able to verify that for fields below the ℓ mixing regime that the np $m = 0$ and ± 1 shifts were given by Eq. (9.6). The most interesting aspect of their work, however, was that they were able to see quasi-Landau resonances above the ionization limit. An example is shown in Fig. 9.3, in which it is clear that there is a modulation in the σ spectrum, in which the light is polarized perpendicular to the magnetic field, extending well above the zero field limit. The energy spacing between the peaks is

$$\Delta W = \frac{3}{2} \hbar \omega_c \tag{9.9}$$

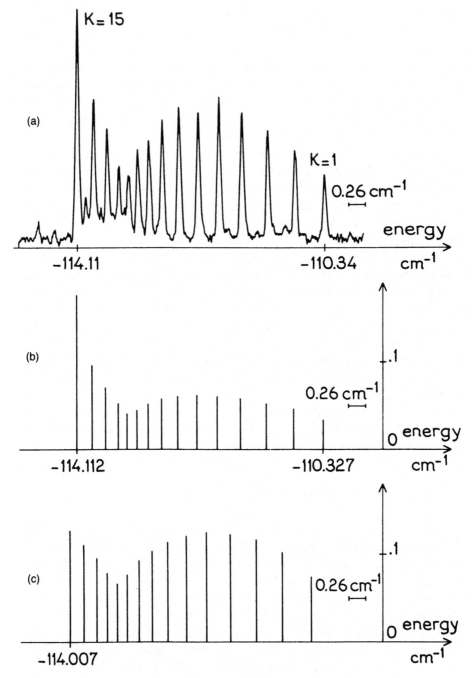

Fig. 9.2 Diamagnetic structure of the $n=31$ multiplet in π excitation ($B=1.94$ T): (a) experimental recording for Li; (b) calculated diamagnetic spectrum for Li; (c) corresponding calculated spectrum for H (from ref. 10).

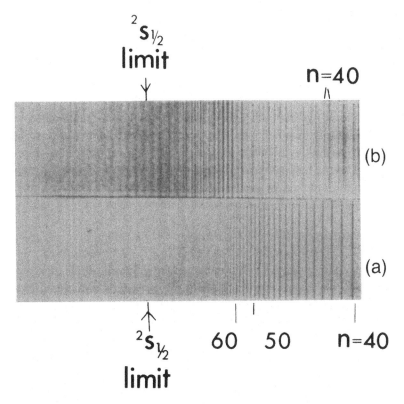

$^2S_{1/2}$
limit

n=40

(b)

(a)

$^2S_{1/2}$
limit

60 50 n=40

Fig. 9.3 Diamagnetic Zeeman effect in Ba: (a) zero-field, (b) σ polarization with resonances extending into the ionization continuum of (a) (from ref. 2).

where ω_c is the cyclotron frequency eB/m. This spacing is 50% higher than the
spacing of the Landau levels of a free electron in a magnetic field and came as a
surprise.

While direct diagonalization of the Hamiltonian matrix works well for situations in which there is a finite number of states, as in Fig. 9.1, it is clearly hopeless
to try it in this case. A useful WKB approach was proposed by Edmonds[12] and
refined by Starace.[13] Using the fact that azimuthal symmetry exists, Starace writes
the wavefunction of the spinless Rydberg electron in cylindrical coordinates as[13]

$$\psi(\rho,\phi,z) = \frac{1}{\sqrt{\rho}} f(\rho,z) e^{im\phi}. \qquad (9.10)$$

Using this wavefunction in the Schroedinger equation leads to an equation for
$f(\rho,z)$,

$$\frac{\partial^2 f(\rho,z)}{\partial \rho^2} + \frac{\partial^2 f(\rho,z)}{\partial z^2} + 2[W' - V(\rho,z)]f(\rho,z) = 0, \qquad (9.11)$$

where

$$W' = W - \frac{mB}{2},$$

and

$$V(\rho,z) = \frac{1}{2}\frac{(m^2 - 1/4)}{\rho^2} - \frac{1}{\sqrt{\rho^2 + z^2}} + \frac{B^2\rho^2}{8}.$$

As pointed out by Edmonds and Starace,[12,13] the atoms are excited near the origin and can only escape in the $\pm z$ directions. The motion in the x,y plane is bound and is most likely to be the source of the quasi Landau resonances. To find the locations of the resonances it is adequate to ignore the z motion entirely and simply compute the energy spectrum of the motion in x,y plane. Applying the Bohr–Sommerfeld quantization condition leads to

$$\int_{\rho_1}^{\rho_2} \left(2W' - \frac{m^2}{\rho^2} + \frac{2}{\rho} - \frac{B^2\rho^2}{4}\right)^{1/2} d\rho = \left(n + \frac{1}{2}\right)\pi, \tag{9.12}$$

where n is an integer and $m^2 - 1/4$ in the centrifugal term of Eq. (9.11) has been replaced by m^2 to account for the fact that the problem is two dimensional.[14] In Eq. (9.12) ρ_1 and ρ_2 are the inner and outer classical turning points. The spacing of the quasi–Landau resonances is obtained by differentiating Eq. (9.12). Explicitly,

$$\frac{dn}{dW} = \frac{1}{\pi} \int_{\rho_1}^{\rho_2} \left(2W' - \frac{m^2}{\rho^2} + \frac{2}{\rho} - \frac{B^2\rho^2}{4}\right)^{-1/2} d\rho. \tag{9.13}$$

Evaluating Eq. (9.13) to obtain the resonance spacing, $\Delta W = dW/dn$, shows that at $W' = 0$

$$\Delta W = \frac{dW}{dn} = \frac{3}{2}B. \tag{9.14}$$

As shown by Fig. 9.4, as the energy is raised above the ionization limit the spacing between the resonances slowly drops from $(3/2)\hbar\omega_c$ to $\hbar\omega_c$

A most interesting question is how to connect the quasi–Landau resonances of Fig. 9.3 to the diamagnetic structure at lower energies, where it is understood in terms of individual magnetic field states, as shown by Figs. 9.1 and 9.2. This question was addressed by both Fonck *et al.*[15,16] and Economou *et al.*[17] using laser excitation of atomic vapor in a magnetic field and detection of the ions produced by collisional ionization of the Rydberg atoms. Using two photon excitation Fonck *et al.*[15,16] observed clear $m = 0$ resonances in Sr similar to the ones shown in Fig. 9.3 and less clear ones in Ba. Economou *et al.*[17] observed the one photon Rb $5s \rightarrow np$ and two photon Rb $5s \rightarrow nd$ transitions with circularly polarized light. In Fig. 9.5 we show the two photon excitation spectrum of the Rb $m = -2$ states, the $5s \rightarrow nd$ series, in a field of 55.5 kG. At $n = 25$ there is one readily identifiable resonance. As n increases to 35 the single resonance splits into several peaks, which still fall into recognizable groups. Finally at $n = 42$ the quasi Landau resonances at the ionization limit are observed. Fig. 9.5 shows clearly that each of the quasi Landau resonances is part of a series of resonances each associated with

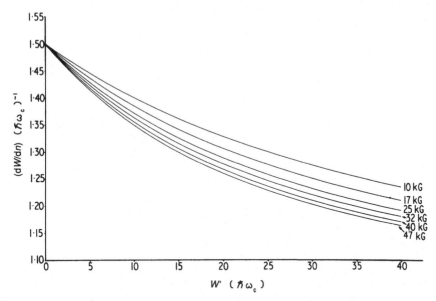

Fig. 9.4 Plot of the energy separation dW/dn against energy W' above the threshold for six values of the magnetic field, $B = 10\,\text{kG}$, $17\,\text{kG}$, $25\,\text{kG}$, $32\,\text{kG}$, $40\,\text{kG}$, $47\,\text{kG}$. Both dW/dn and W' are in units of the cyclotron energy, $\hbar\omega_c = \hbar(eH/mc)$. As $W' \to \infty$, $(dW/dn)/\hbar\omega_c \to 1$ (from ref. 13).

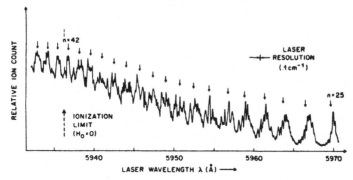

Fig. 9.5 Data showing the $m = -2$, even parity sublevels of Rb at 55.5 kG ($5s \to nd$ series at $B = 0$). Laser wavelengths are plotted on the horizontal axis. The downward arrows indicate the resonance positions predicted by a WKB model (from ref. 17).

a single n state. By modifying the WKB expression of Eq. (9.12) to give the correct coulomb energies, Economou *et al.*[17] were able to reproduce the energies of the observed peaks from $350\,\text{cm}^{-1}$ below (corresponding to $n = 25$) to $20\,\text{cm}^{-1}$ above the zero field ionization limit.

These experiments showed clearly that n states evolve into the quasi–Landau resonances. In higher resolution experiments, Gay *et al.*[18] and Castro *et al.*[19] showed that it was in fact the highest energy diamagnetic states which evolved into

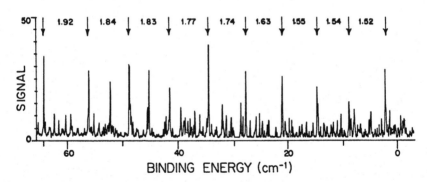

Fig. 9.6 Quasi-Landau spectrum observed by laser excitation and field ionization of even parity, $m = -2$ states of Na in a magnetic field of 4.2 T. The arrows indicate the quasi-Landau levels, the highest energy magnetic states of each principal quantum number. The intermediate peaks are due to other levels. The numbers between the arrows give the level separation in units of $\hbar\omega_c$. A WKB analysis predicts a spacing of 1.5 at $W = 0$, increasing with binding energy in agreement with the data. Relative intensities are not reliable due to laser fluctuations (from ref. 19).

the locations of the quasi–Landau resonances. Gay *et al.*[18] studied the nearly hydrogenic $m=3$ states of Cs using a thermionic diode in the magnetic field of a superconducting solenoid. Castro *et al.*[19] recorded the spectra of even parity Na $m = -2$ states using the same technique used to produce Fig. 9.1, except that both lasers were circularly polarized in the same sense. In Fig. 9.6 their 4.2 T spectrum is shown. The arrows show the highest lying magnetic energy levels of successive n, starting with $n = 35$. These are clearly the dominant features of the spectrum, and the interval between them, given in units of the cyclotron frequency, clearly converges to 3/2, the quasi–Landau resonance interval, at the zero field ionization limit. In contrast to the $m = -2$ spectrum shown in Fig. 9.6, in the analogous $m = -1$ spectra Castro *et al.*[19] observed that many magnetic levels of each n were excited approximately equally. Both of these observations agree with the calculations of Clark and Taylor.[11]

If we now return to the original experiment of Garton and Tomkins, we can ask why did they not see resonances with π polarization. More generally, when are quasi–Landau resonances visible? The requirement is apparently similar to the requirement for seeing strong field mixing resonances in an electric field.[20] The energy ranges covered by the magnetic levels from each n are overlapped at the ionization limit, and only if the oscillator strength is concentrated in a few of the high energy magnetic states are the Landau resonances visible. Exactly how the oscillator strength is distributed varies from atom to atom and depends strongly on the quantum defects of the levels involved. However, for H like systems it is more likely that the requirement will be fulfilled for $|\Delta m| = 1$ than $\Delta m = 0$ single photon transitions.

Quasi classical orbits

While the two dimensional treatment of Edmonds[12] and Starace[13] can be used to explain the development of the quasi–Landau resonances from the bound states it does not give any information beyond the locations of the resonances. For example, it does not tell us how wide the resonances are because motion in the z direction is neglected. Furthermore, using this approach we cannot tell whether or not there exist resonances corresponding to motion outside the x,y plane. With these questions in mind Reinhardt[21] suggested using the wave packet approach to the problem described in Chapter 8. The wave packet approach successfully reproduces the quasi–Landau resonances,[21] but more important, provides a way of looking for resonances which correspond to motions not lying in the x,y plane. These wave packet notions proved to be invaluable in understanding the later observations of Holle *et al.*,[22] who observed the quasi–Landau spectrum of H using a H beam propagating along a magnetic field of up to 6 T. They excited H 1s atoms to well defined 2p $m = 0$ or 1 states with a vacuum ultraviolet laser and then excited final states of $m = 0,1,$ or 2 with a second ultraviolet laser with a linewidth of 0.1–0.3 cm$^{-1\cdot}$ The laser beams crossed the atomic beam at a right angle. The H atoms were excited in an electric field of 1 V/cm parallel to the magnetic field, and the electrons freed by the Rydberg atoms drifted out of the interaction region and were accelerated to 6 keV to register on a surface barrier detector.

When they observed final states of $m = 0$, with both lasers polarized along the field direction, they observed the familiar quasi–Landau resonances spaced at $3\omega_c/2$. In contrast, when they excited the $m = 1$ final state via the intermediate 2p $m = 1$ state these resonances were absent. Only resonances spaced by $0.64\omega_c$ were observed. In the final $m = 0$ states the oscillations at $3\omega_c/2$ are evident, but in the final $m = 1$ states the oscillations spaced by $0.64\omega_c$ are less so, although they are clearly evident if the data are smoothed over 2 cm^{-1}. The periodicity is even more apparent in the Fourier transform of the spectrum, shown in Fig. 9.7. In Fig. 9.7(b) the peak corresponding to the resonance spacing of $0.64\omega_c$ is readily apparent.

The origin of this resonance was identified by extending Reinhardt's[21] wave packet notion. Realizing that the wave packet evolves along the classical trajectories of the electron, Holle *et al.* searched for classical trajectories in which the electron leaving the origin returned in a time of 9.5 ps, $1/2\pi$ $(0.64\omega_c)$. The orbit they discovered which has this return, or recurrence, time does not lie in the x,y plane, and thus cannot be predicted by the Edmonds–Starace approach.[12,13]

It is clear that it is only possible to observe these resonances if their recurrence times are shorter than the coherence time of the exciting laser. With this point in mind, Main *et al.*[23] made a five fold improvement in their spectral resolution and were able to see new resonances in the Fourier transform spectrum with longer recurrence times, as shown in Fig. 9.8. T_c is the recurrence time for a cyclotron orbit. The spectrum vs tuning energy is not shown since it is composed of

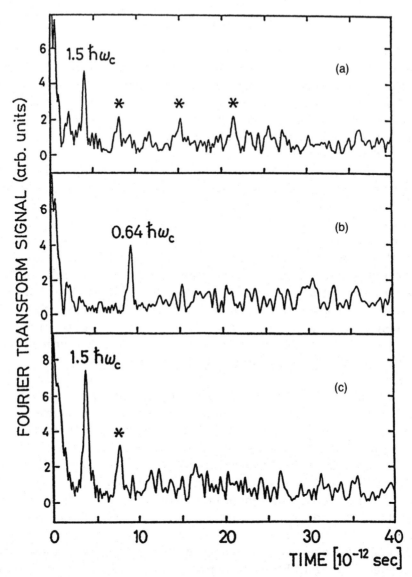

Fig. 9.7 Fourier transforms of two photon, resonant, excitation spectra of the H atom Balmer series around the ionization limit in a magnetic field of strength $B = 6$ T, excited through individually selected magnetic substates $m = 0$ and $m = +1$ of the $n = 2$ state to final m states of even parity: (a) $m = 0$, (b) $m = +1$ (c) $m = +2$, plus some admixture ($\sim 25\%$) of $m = 0$. The resolution is ≈ 0.3 cm^{-1} (from ref. 22).

apparently random lines. Nonetheless, the Fourier transform spectrum of Fig. 9.8 shows clear peaks which may be correlated with classical orbits which return to the origin. In Fig. 9.9 we show the orbits which correspond to the peaks in the Fourier spectrum of Fig. 9.8. As shown by Fig. 9.9 the original quasi–Landau resonances, $v = 1$, are the only ones which have orbits in the x, y plane.

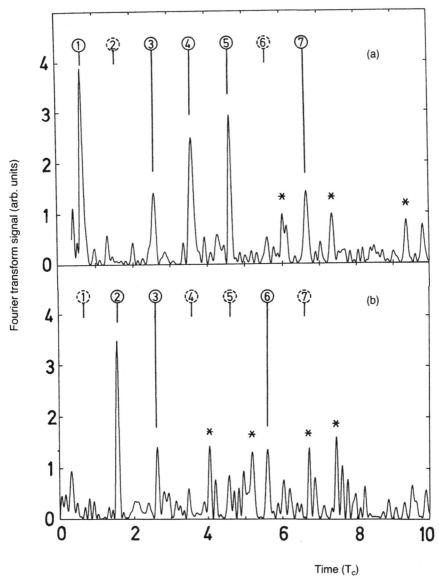

Fig. 9.8 Fourier transforms of H spectra obtained in a magnetic field of 5.96 T with resolution 0.07 cm^{-1}: (a) initial state 2p $m = 0$, final state $m = 0$ even parity states; (b) initial state 2p $m = -1$ final state $m = -1$ even parity states. The squared value of the absolute value is plotted in both cases. The circled numbers correspond to the classical orbits depicted in Fig. 9.9 (from ref. 23).

An interesting hybrid way of calculating the spectrum is one used by Du and Delos.[24] They start with a quantum mechanical wave packet at the origin and let it propagate to $r = 50a_0$, where they use the normals of phase fronts of the wave packets to define classical electron trajectories. These classical trajectories are then followed. Some of the trajectories are reflected back to the origin, and when

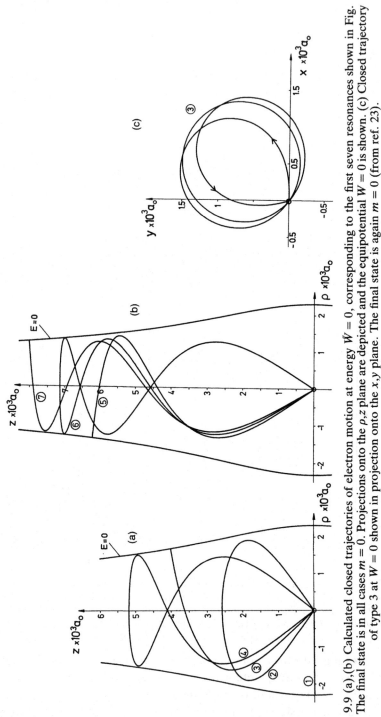

Fig. 9.9 (a),(b) Calculated closed trajectories of electron motion at energy $\bar{W} = 0$, corresponding to the first seven resonances shown in Fig. 9.8. The final state is in all cases $m = 0$. Projections onto the ρ, z plane are depicted and the equipotential $W = 0$ is shown. (c) Closed trajectory of type 3 at $W = 0$ shown in projection onto the x, y plane. The final state is again $m = 0$ (from ref. 23).

they come back to $r = 50a_0$ they are used to construct incoming wave packet phase fronts. The resulting incoming wave packet can interfere constructively or destructively with the outgoing wave packet at the origin, leading to total wavefunctions of varying amplitude, and photoionization cross sections which vary accordingly. Their results are in excellent agreement with the experimental results of Main *et al.*[23]

The fact that the resonances in the Fourier transform spectrum of Fig. 9.8 can be predicted by finding classical orbits which return to the origin suggests a better way of looking for the classical orbits. Wintgen[25] and Wintgen and Friedrich[26] suggested that if the classical Hamltonian is given by

$$H = \frac{p^2}{2m} - \frac{1}{r} + \frac{1}{8}B^2\rho^2,$$
(9.15)

then using the scaled variables

$$\tilde{\mathbf{r}} = \gamma^{2/3}\mathbf{r}$$
(9.16a)

and

$$\tilde{\mathbf{p}} = \gamma^{-1/3}\mathbf{p},$$
(9.16b)

where $\gamma = B/2.35 \times 10^5$ T (γ is B in atomic units),

$$H(\tilde{\mathbf{r}},\tilde{\mathbf{p}},\gamma) = \gamma^{2/3}H(\mathbf{r},\mathbf{p},\gamma = 1).$$
(9.17)

Applying the semiclassical quantization condition to the action around a closed classical orbit yields[25–27]

$$\frac{1}{2\pi}\oint \tilde{p}_\rho \, d\tilde{\rho} + \tilde{p}_z \, d\tilde{z} = n\gamma^{1/3} = C_i,$$
(9.18)

where C_i is the scaled action of the closed orbit. Since the action scales with the scaled energy, as $\tilde{W} = W\gamma^{-2/3}$, if the scaled energy is fixed, the spectrum of possible actions is simply an integer times $\gamma^{1/3}$.

Holle *et al.*[27] observed the scaled energy spectrum by keeping $\tilde{W} = W\gamma^{-2/3}$ fixed while scanning $B^{-1/3}$ linearly, or equivalently $\gamma^{-1/3}$ linearly. Their experimental spectrum, when Fourier transformed, leads to the Fourier transform spectrum of Fig. 9.10, in which clear peaks separated by $\gamma^{-1/3}$ are visible. This spectrum is at a scaled energy $\tilde{W} = -0.45$, corresponding to an energy range $-77.7 \text{ cm}^{-1} \leq W \leq -54.3 \text{ cm}^{-1}$ and $5.19 \text{ T} \geq B \geq 3.03 \text{ T}$. At this scaled energy the classical orbits are regular and the spectrum is well behaved. At higher scaled energies, $\tilde{W} \sim -0.1$, the peaks of Fig. 9.10 break into several peaks and the regularity of Fig. 9.10 disappears, which is often termed the onset of classical chaos.

While the success of scaled energy spectroscopy suggests that the behavior of atoms does become classically chaotic near the ionization limit, higher resolution reveals a surprisingly orderly structure. Iu *et al.*[28] have studied the odd parity Li $m = 0$ states in a beam travelling in the direction of the magnetic field. They

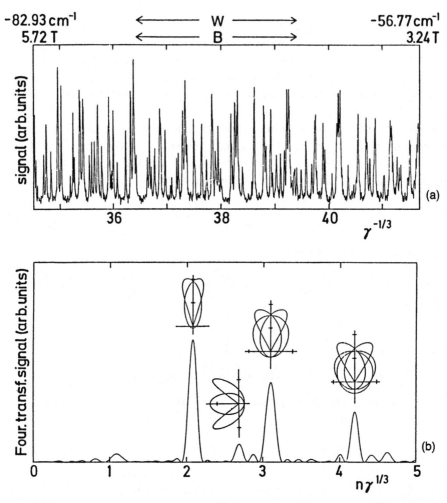

Fig. 9.10 (a) Scaled energy spectrum at $\bar{W} = -0.45$ as a function of $\gamma^{-1/3}$. Range of excitation energy $-77.7\ \mathrm{cm}^{-1} \leq W \leq -54.3\ \mathrm{cm}^{-1}$ and field strength $5.19\ \mathrm{T} \geq B \geq 3.03\ \mathrm{T}$. (b) Fourier transformed action spectrum of (a); closed orbits correlated to respective resonances in ρ,z projection; z coordinate vertical (from ref. 27).

excited the Li atoms from the ground 2s state to the 3s state using two photon excitation, and then drove the single photon 3s → np transitions with 30 MHz resolution. They observed surprisingly regular structure in the vicinity of the ionization limit, a regime which is classically chaotic according to the work of Holle *et al.*[27] In Fig. 9.11, we show their spectra, taken from 1 cm^{-1} below the ionization limit to 3 cm^{-1} above the limit. As shown by Fig. 9.11, the spectra consist of strong lines, which increase rapidly in energy with field, and weaker lines which increase less rapidly with field. The simple fact that the spectrum is so apparently regular is remarkable. To explain the spectrum Iu *et al.* proposed that the adiabatic model of Friedrich[29] could be used. The essential idea is that the

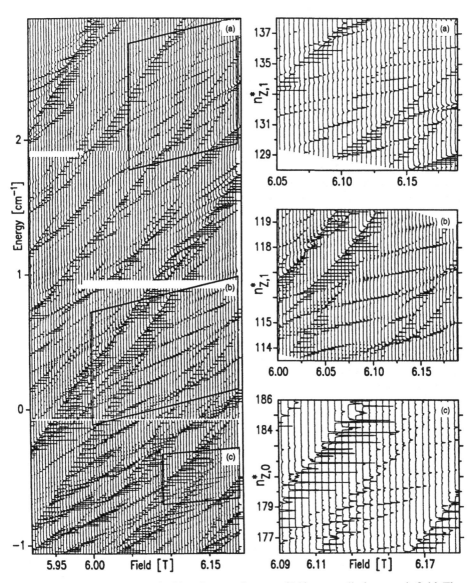

Fig. 9.11 Energy level map of odd parity $m = 0$ states of Li in an applied magnetic field. The maps are created by scanning a laser at successively higher fields: Horizontal peaks are field ionization signals. (The gaps are regions which were skipped during data collection.) The structure is most easily seen by viewing close to the plane from the left. Reduced-term-value plots for the outlined regions are shown at the right. (a),(b) $n_p = 1$; (c) $n_p = 0$. With these coordinates a progression of Rydberg states appears as a series of lines with unit separation. The ionization threshold in the field occurs at approximately $+2.8\,\mathrm{cm}^{-1}$ (from ref. 28).

electron motion is separated into the rapid motion in the x,y plane and a slow motion along the z axis. With this separation the energies are given by[28]

$$W(n_\rho, \nu_z) = (n_\rho + 1/2)B - \frac{1}{2\nu_z^2}, \qquad (9.19)$$

where n_ρ is the quantum number of the Landau level and ν_z is the effective quantum number of the motion in the one dimensional coulomb potential in the z direction.[30] Although Friedrich developed the model for higher magnetic fields, it seems to match the observed spectrum remarkably well in some cases. In the sections on the right side of Fig. 9.11 the energy of the Landau level, corresponding to $n_\rho = 0$ or 1 has been removed, leaving only the energy of the coulomb motion in the z direction. The fact that the levels match the coulomb spacing is apparent, although the quantum defect seems to decrease substantially with magnetic field. While this theoretical description is qualitative, it offers the promise that the spectrum in high magnetic fields can be explained quantum mechanically.

References

1. F. A. Jenkins and E. Segre, *Phys. Rev.* **55** 52 (1939).
2. W. R. S. Garton and F. S. Tomkins, *Astrophys. J.* **158**, 839 (1969).
3. L. I. Schiff and H. Snyder, *Phys. Rev. A* **55**, 59 (1939).
4. D. Kleppner, M. G. Littman, and M. L. Zimmerman, in *Rydberg States of Atoms of Molecules*, eds. R.F. Stebbings and F.B. Dunning (Cambridge Univ. Press, Cambridge, 1983.)
5. H. A. Bethe and E. A. Salpeter, *Quantum Mechanics of One and Two Electron Atoms* (Plenum, New York, 1977.)
6. W. Sandner, T. F. Gallagher, K. A. Safinya, and F. Gounand, *Phys. Rev. A* **23**, 2732 (1981).
7. M. L. Zimmerman, J. C. Castro, and D. Kleppner, *Phys. Rev. Lett.* **40**, 1083 (1978).
8. E. A. Solov'ev, *JETP Lett.* **34**, 265(1981).
9. J. C. Gay and D. Delande, *Comm At. Mol. Phys.* **13**, 275(1983).
10. P. Cacciani, E. Luc-Koenig, J. Pinard, C. Thomas, and S. Liberman, *Phys. Rev. Lett.* **56**, 1124 (1986).
11. C. W. Clark and K. T. Taylor, *J. Phys B* **15**, 1175(1982).
12. A. R. Edmonds, *J. Phys. (Paris)* **31**, C4 (1970).
13. A. F. Starace, *J. Phys. B* **6**, 585 (1973).
14. O. Akimoto and H. Hasegawa, *J. Phys. Soc. Jpn.* **22**, 181 (1967).
15. R. J. Fonck, D. H. Tracy, D. C. Wright, and F. S. Tomkins, *Phys. Rev. Lett.* **40**, 1366 (1978).
16. R. J. Fonck, F. L. Roesler, D. H. Tracy, and F. S. Tomkins, *Phys. Rev. A* **21**, 861 (1980).
17. N. P. Economou, R. R. Freeman, and P. F. Liao, *Phys. Rev. A* **18**, 2506 (1978).
18. J. C. Gay, D. Delande, and F. Biraben, *J. Phys. B* **13**, L729 (1980).
19. J. C. Castro, M. L. Zimmerman, R. G. Hulet, D. Kleppner, and R. R. Freeman, *Phys. Rev. Lett.* **45**, 1780 (1980).
20. E. Luc-Koenig and A. Bachelier, *Phys. Rev. Lett.* **43**, 921 (1979).
21. W. P. Reinhardt, *J. Phys. B* **16**, 635 (1983).
22. A. Holle, G. Wiebusch, J. Main, B. Hager, H. Rottke, and K. H. Welge, *Phys. Rev. Lett.* **56**, 2594 (1986).
23. J. Main, G. Wiebusch, A. Holle and K. H. Welge, *Phys. Rev. Lett.* **57**, 2789 (1986).

24. M. L. Du and J. B. Delos, *Phys. Rev. Lett.* **58**, 1731 (1987).
25. D. Wintgen, *Phys. Rev. Lett.* **58**, 1589 (1987).
26. D. Wintgen and H. Friedrich, *Phys. Rev. A* **36**, 131 (1987).
27. A. Holle, J. Main, G. Wiebusch, H. Rottke, and K. H. Welge, *Phys. Rev. Lett.* **61**, 161 (1988).
28. C. Iu, G. R. Welch, M. M. Kash, L. Hsu, and D. Kleppner, *Phys. Rev. Lett.* **63**, 1133 (1989).
29. H. Friedrich, *Phys. Rev. A* **26**, 1827 (1982).
30. R. Loudon, *Am. J. Phys.* **27**, 649 (1959).

10

Microwave excitation and ionization

In our encounter with ionization by static and quasi–static fields we saw that there are substantial differences between the ionization of H or H like atoms and nonhydrogenic atoms. These differences arise from the fact that in an atom other than H the presence of the ionic core couples states which are not coupled in H. This coupling results in the avoided crossings of bound Stark energy levels and the coupling of nominally stable Stark levels to the continuum of red shifted Stark levels of higher n states. Differences exist in both static and pulsed field ionization, and it is hardly a surprise that microwave ionization of hydrogenic atoms differs from that of nonhydrogenic atoms. For $\omega < 1/n^3$, where ω is the microwave frequency, hydrogenic atoms of $|m| \ll n$ ionize at a field of $1/9n^4$, as the red Stark states do in a static field.[1] Low m states of alkali atoms, however, ionize at the radically different microwave field given by[2,3]

$$E = 1/3n^5. \tag{10.1}$$

We begin by briefly summarizing the two experimental techniques used to study microwave excitation and ionization. The first technique is the use of fast beams of H and He Rydberg atoms. An apparatus used to study the microwave ionization of H and He is shown in Fig. 10.1.[3] A distribution of Rydberg states is produced by charge exchange of a continuous beam of H^+ or He^+ ions, and high lying states, above $n \simeq 10$, are removed by field ionization to produce what is labelled the atom beam in Fig. 10.1. Selected single Rydberg states are then populated by excitation of atoms in typically an $n = 7$ state, first to an $n = 10$ state and then to a higher

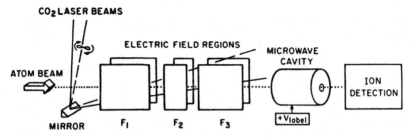

Fig. 10.1 Schematic drawing of a fast beam apparatus. A fast atomic beam enters from the left and is excited sequentially by two different CO_2 lasers in electric field regions F_1 and F_3, respectively. F_2 avoids a zero field region between them. Ions produced by highly excited atoms being ionized in the biased microwave cavity are energy selected and detected by a Johnston particle multiplier (not shown). The output signal is detected in phase with the mechanically chopped F_1 laser beam (from ref. 3).

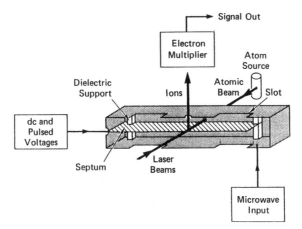

Fig. 10.2 Major components of a thermal atomic beam apparatus for microwave ionization experiments,the atomic source, the microwave cavity, and the electron multiplier. The microwave cavity is shown sliced in half. The Cu septum bisects the height of the cavity. Two holes of diameter 1.3 mm are drilled in the side walls to admit the collinear laser and Na atomic beams, and a 1 mm hole in the top of the cavity allows Na^+ resulting from a field ionization of Na to be extracted. Note the slots for pumping (from ref. 4).

Rydberg state using two CO_2 lasers. The Rydberg atoms pass through a microwave cavity, where they are exposed to a strong microwave field. Due to the high velocities of the energetic, $\sim$10 keV, H and He beams, the atoms pass through the microwave field region in a short time. For example a 10 keV H beam has a velocity of 1.3×10^8 cm/s and passes through a 5 cm long cavity in 40 ns. Downstream from the cavity the beam is analyzed, usually by applying an electric field, to see if ionization or transitions to other states have occurred. In these experiments the excitation, interaction with the microwave field, and final state analysis are separated spatially, as shown in Fig. 10.1.

The second approach is to use thermal beams of alkali atoms as shown in Fig. 10.2.[4] A beam of alkali atoms passes into a microwave cavity where the atoms are excited by pulsed dye lasers to a Rydberg state. A 1 μs pulse of microwave power is then injected into the cavity. After the microwave pulse a high voltage pulse is applied to the septum, or plate, inside the cavity to analyze the final states after interaction with the microwaves. By adjusting the voltage pulse it is possible to detect separately atoms which have and have not been ionized or to analyze by selective field ionization the final states of atoms which have made transitions to other bound states.

As we shall see, microwave ionization can be thought of as a multiphoton absorption or as a process driven by a time varying field. We first discuss microwave ionization of alkali atoms, which can be described by the notions used to describe pulsed field ionization. To show the connection between the time varying field point of view and the photon absorption point of view we then discuss

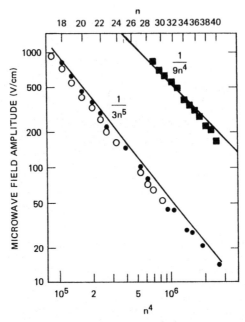

Fig. 10.3 15 GHz microwave ionization fields for the $|m| \leq 1$ components of the Na $(n + 1)$s (○) and nd (●) states. Ionization fields of the $|m| = 2$ components of the nd states (■) (from ref. 4).

microwave multiphoton resonance experiments, which allow us to refine our understanding of microwave ionization. We then discuss ionization of H like atoms. Finally, ionization by circularly polarized fields is described.

Microwave ionization of nonhydrogenic atoms

As noted earlier, nonhydrogenic atoms ionize at a microwave field of $E \sim 1/3n^5$, a field substantially less than the static field required for ionization, $E = 1/16n^4$. This point is made quite graphically in Fig. 10.3, a plot of the observed fields for 50% ionization of the Na nd $|m| = 0$ and 1 states by a 15 GHz microwave field pulse lasting 500 ns.[2] Also shown in Fig. 10.3 are the very different ionization fields of the approximately hydrogenic Na $|m| = 2$ states. Although these fields are often called microwave ionization thresholds they are not nearly as sharp as the pulsed field thresholds shown in Fig. 7.4, a point which is made by Fig. 10.4, a plot of the microwave ionization threshold of the Na 20s state.[4] The complementary signals of Fig. 10.4 are obtained by detecting the atoms which have and have not been ionized by the microwaves. In addition to Na, experiments have been done in other atoms, He, Li, and Ba,[1,5,6] and these experiments also show that ionization generally occurs when $E \sim 1/3n^5$.

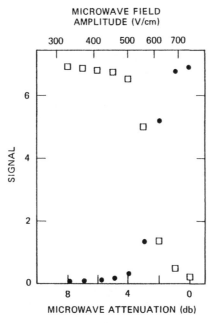

MICROWAVE FIELD
AMPLITUDE (V/cm)

Fig. 10.4 Field ionization signal (□) and 15 GHz microwave ionization signal (●) for the Na 20s state showing the ionization threshold as both the disappearance of the field ionization signal and the appearance of the microwave ionization signal with increasing microwave power (decreasing attenuation). For convenience the microwave-field amplitude is also given (from ref. 4).

The fact that atoms can be ionized by a field so far below the classical field required for ionization, as shown by Fig. 10.3, suggests that the microwave field induces transitions to higher lying states which can be directly field ionized by the microwave field, and that what we observe as the microwave ionization threshold field is in fact the field required to drive the rate limiting step. The fact that the threshold field is so close to $E = 1/3n^5$, the field of the avoided level crossing between the extreme n and $n + 1$ Stark levels, suggests that the rate limiting step is a Landau–Zener transition between these two levels at their avoided crossing.

It is straightforward to generalize the notions previously used to describe pulsed field ionization to understand microwave ionization as a process driven by a temporally varying field. As an example we consider the ionization of an atom initially excited to the Na 20d state with $|m| = 0$. In Fig. 10.5 we show the energy levels of the Rydberg states in positive and negative electric fields.[2] For clarity, we have not shown the s and p states, with quantum defects 1.35 and 0.85, nor have we shown all the Stark energy levels in detail, but only shaded the field energy region in which they lie. Consider applying a pulse of microwaves to atoms initially excited to the Na 20d state. In the presence of the field the good quantum states are the Stark states, and the microwave field rapidly induces transitions

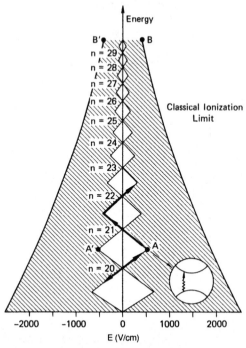

Fig. 10.5 Energy level diagram showing the mechanism by which an $n = 20$ atom is ionized by a microwave field of 700 V/cm amplitude. An atom initially excited to the 20d state is brought to the point at which the $n = 20$ and 21 Stark manifolds intersect. At this point the atom makes a Landau–Zener transition, as shown by the inset, to the $n = 21$ Stark manifold and on subsequent microwave cycles makes further upward transitions as shown by the bold arrow. The process terminates when the atom reaches a sufficiently high n state that the microwave field itself ionizes the atom. This occurs at point B in this example (from ref. 4).

between the n = 20 $|m| = 0$ Stark states at the avoided crossings at zero field. Thus roughly 5% of the atoms are in each Stark state at any given time. If the microwave field reaches $E = 1/3n^5$, atoms in the highest energy $n = 20$ Stark state are brought to the avoided crossing with the lowest lying $n = 21$ Stark state at $E = 1/3n^5$, and these atoms can make a Landau–Zener transition to the lowest $n = 21$ Stark state at the avoided crossing as shown in the inset of Fig. 10.5. We shall shortly return to consider the Landau–Zener transition in more detail, but for a moment we assume that the Landau–Zener transition does occur. It is evident that a field adequate to drive the $n \rightarrow n + 1$ transition is adequate to drive the $n + 1 \rightarrow n + 2$ transition and a sequence of similar transitions to higher lying states, culminating in ionization when it is classically allowed, as shown by Fig. 10.5. On the other hand the field is not strong enough to drive the transition to $n = 19$, so atoms do not make transitions to lower n. In sum the microwave field drives transitions to higher n, resulting ultimately in ionization, and the ionization field is determined by the rate limiting step, the $n \rightarrow n + 1$ transition.

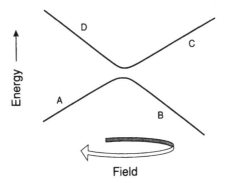

Fig. 10.6 Possible outcomes of a double traversal of an avoided crossing starting from A. Very slow change in field A → B → A. Very fast charge in field A → C → A. Intermediate rate of change A → B and C → D and A.

The Landau–Zener transition can be understood by considering the example shown in Fig. 10.6 in which, at its peak, the microwave field goes slightly past the avoided crossing. If the field traverses the crossing too slowly an atom initially in the lower state A remains at all times on the lower energy level reaching point B at the peak of the field then returning to A when the field returns to its original value. If the field traverses the crossing too rapidly an atom makes a diabatic transition to point C on the upper level on the increase of the field and returns to A when the field returns to its initial value. On the other hand, if the crossing is traversed in a time comparable to the inverse of its magnitude, when the field increases to its peak there is a probability of the atom's being in both the upper and lower states, at points B and C, and after the field returns through the crossing there is a probability of the atom's having made the transition to point D.

To achieve a high, 50%, transition probability the field must reach the avoided crossing, but the transitions still occur, with lower probability, even if the field does not reach the crossing. Since there are many field cycles in the microwave pulse the transition probability need not be high on a single cycle, and the field does not actually need to reach the crossing field, $E = 1/3n^5$. Pillet *et al.*,[4] using the numerical method of Rubbmark *et al.*,[7] have calculated transition probabilities between extreme Stark states due to a single half cycle of the microwave field. At 15 GHz these calculations indicate that the $n \to n + 1$ transition probability is ~1% when $E = 85\%$ of the crossing field, i.e. $E = 1/3.5n^5$, in reasonable agreement with the experimental observations shown in Fig. 10.3. These calculations also show that, if the field amplitude is fixed at $1/3n^5$, the maximum transition probability occurs when the microwave frequency ω is comparable to ω_0, the size of the avoided crossing between the n and $n + 1$ levels. However, the transition probability is negligible if $\omega \gg \omega_0$ or $\omega \ll \omega_0$. The requirement $\omega \sim \omega_0$ for the Landau–Zener transition is consistent with the fact that ionization at $E = 1/3n^5$ does not occur in H like states for which the avoided crossings are vanishingly small, i.e. $\omega \gg \omega_0$.

While the above Landau–Zener description gives an appealing picture of ionization by means of the avoided crossings, it is an oversimplification of the actual process in that it is based on a single microwave cycle and a single pair of levels. Considering only a single cycle ignores the possibility of coherence between field cycles and the contributions of nonextreme Stark states with avoided crossings at $E > 1/3n^5$.

Microwave multiphoton transitions

The field at which ionization occurs in nonhydrogenic systems when $\omega \ll 1/n^3$ is determined by the rate limiting $n \rightarrow n + 1$ transition, which we have described as a Landau–Zener transition. Since this transition is between real states it should, in principle, be possible to observe it as resonant photon absorption. However, in microwave ionization it is difficult to separate the rate limiting step from the ensuing ionization. Instead, it is useful to consider transitions between an isolated pair of levels. An example is provided by the K atom. As shown by Fig. 10.7 for $n = 16$, the K $(n + 2)$s states intersect the n Stark manifolds at the field $E_C \approx 1/10n^5$, far below the field required for microwave ionization.[8,9] Thus it is in principle possible to examine only the transition from the (n + 2)s states to the (n,k) Stark states without having the succession of transitions to higher n as well. For simplicity we shall use the convention of labelling each (n,k) Stark state by the zero field $n\ell$ state to which it is adiabatically connected. Thus the (16,3) state is adiabatically connected to the 16f state and is the lowest member of the $n = 16$ Stark manifold for $E < 1/10n^5$. There are two additional attractive features of these K transitions. First, the locations and widths of the avoided crossings between the $(n + 2)$s and (n,k) states have been measured by anticrossing spectroscopy.[9] Second, the static Stark shifts of both levels are nearly linear, so the avoided crossing of these two levels is a good approximation to the $n \rightarrow n + 1$ avoided level crossing at $E = 1/3n^5$.

The experiment is done using the apparatus shown in Fig. 10.2. Two tunable dye lasers are used to excite K atoms in a beam to the $(n + 2)$s state. A static field can be applied to vary the separation between the 18s and $(16,k)$ states. The atoms are exposed to a microwave pulse, and subsequently to a field ionization pulse which ionizes atoms in the (n,k) Stark state, but not those in the $(n + 2)$s state. The (n,k) field ionization signal is then monitored as either the microwave field amplitude or the static field is swept over many shots of the laser. To the extent that both Stark shifts are linear, the static field alters the energy separation between the two states, but not their wavefunctions.

The similarity of the K $(n + 2)$s $\rightarrow (n,k)$ transitions to the $n \rightarrow n + 1$ transitions of microwave ionization is verified by measuring the number of atoms making the former transition as a function of the microwave field amplitude.[10] As the

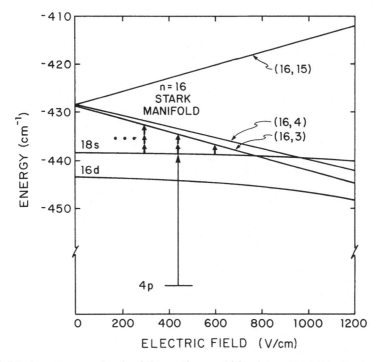

Fig. 10.7 Relevant energy levels of K near the $n = 16$ Stark manifold. The Stark manifold levels are labeled (n,k), where k is the value of ℓ to which the stark state adiabatically connects at zero field. Only the lowest two and highest energy manifold states are shown. The laser excitation to the 18s state is shown by the long vertical arrow. The 18s $\rightarrow$ (16,3) multiphoton rf transitions are represented by the bold arrows. Note that these transitions are evenly spaced in static field, and that transitions requiring more photons occur at progressively lower static fields. For clarity, the rf photon energy shown in the figure is approximately 5 times its actual energy (from ref. 8).

microwave field amplitude is increased, a sharp increase in the number of atoms making the $(n + 2)$s $\rightarrow (n,k)$ transition is observed when the microwave field amplitude reaches a threshold field near the field, E_C, of the avoided crossing between the $(n + 2)$s and lowest (n,k) Stark state. For example, when K 18s atoms are exposed to a 1 μs pulse of 9.3 GHz microwaves of variable amplitude, the threshold is at 775 V/cm, close to the avoided crossing field, $E_C = 753$ V/cm. For $15 \leq n \leq 20$, the observed microwave threshold fields vary from $1.0E_C$ to $1.6E_C$. Although the variation from $1.0E_C$ to $1.6E_C$ is hard to reconcile with the Landau–Zener theory, a threshold field near E_C is expected. There is one exception, the K 19s $\rightarrow (17,k)$ transition, in which a non-monotonic increase in the transition probability is observed with 9.3 GHz field amplitude, beginning at fields well below the avoided crossing field, $E_C = 546$ V/cm, as shown by Fig. 10.8. Similar observations were made in He by van de Water *et al.*[11] This observation is, of course, incompatible with the Landau–Zener picture but can be explained with a multiphoton description.

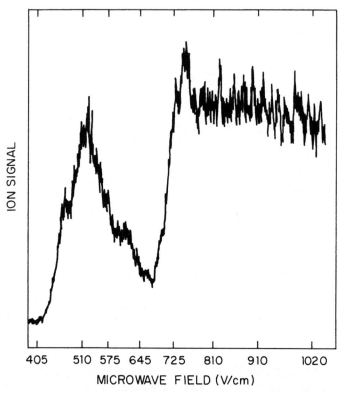

Fig. 10.8 The K 19s → (17,*k*) transition as a function of microwave field. The 19s state is excited by the lasers, a microwave field is applied, and atoms which have made the transition to the (17,*k*) states are detected by field ionization. The static field avoided crossing is at 546 V/cm (from ref. 10).

Sweeping the static field, which alters the energy spacing between the levels, while keeping the microwave field fixed leads to the observation of resonant multiphoton transitions. An example, shown in Fig. 10.9, is the set of the 18s → (16,*k*) transitions observed by sweeping the static field with different strengths of the 10.35 GHz microwave field.[8] The sequence of 18s → (16,3) transitions 25 V/cm apart is quite evident. At the top of Fig. 9 there is a scale in terms of the number of 10.35 GHz photons required to drive the 18s → (16,3) transition.

At low microwave fields transitions involving only a few photons are observed, at high static fields, near the field of the avoided crossing. As the microwave field is increased more transitions are observed, at progressively lower static fields, until at the highest microwave fields the sequence of resonant transitions extends to zero static field. As shown by the scale at the top of Fig. 10.9 the 18s → (16,3) transition nearest zero static field corresponds to the absorption of 28 photons.

Careful inspection of Fig. 10.9 reveals three interesting features. First, most of the 18s → (16,3) resonances corresponding to the absorption of a given number of

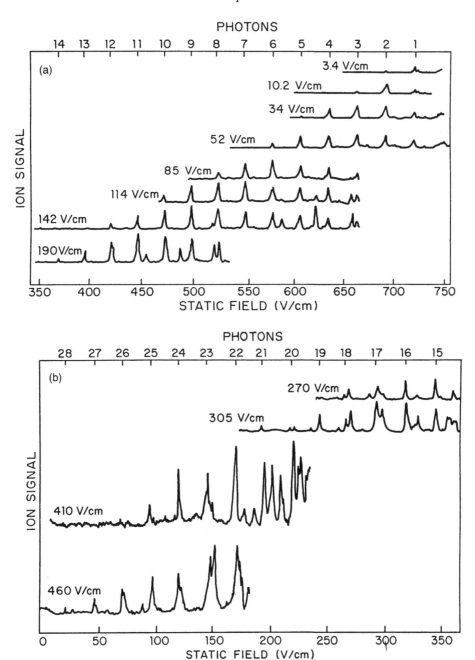

Fig. 10.9 (a) K 18s → (16,3) one to fourteen photon transitions observed as the static field is scanned from 350–750 V/cm for the 10.353 GHz microwave fields indicated above each trace (3.4–190 V/cm). The regularity of the progression is quite apparent. Note the extra resonances in the 142 V/cm microwave field trace. These are due to 18s → (16,4) transitions. (b) 18s → (16,3) 15–28 photon transitions observed as the static field is scanned from 0 to 350 V/cm. Note the congestion of the 410-V/cm trace at static fields above ~200 V/cm, due to many overlapping 18s → (16,k) transitions (from ref. 8).

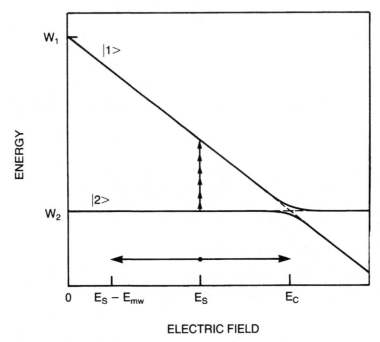

ELECTRIC FIELD

Fig. 10.10 Multiphoton resonant transition at an avoided crossing from the photon and field points of view. The solid curves are the avoiding levels and the dashed lines are the levels which cross when the coupling is ignored. The static field E_S gives rise to a six photon resonant transition, indicated by the stacked arrows. The range of the electric field variation is shown for the case in which the peak field $E_S + E_{mw}$ exactly reaches the crossing is exaggerated for clarity (from ref. 10).

microwave photons are not shifted in static field as the microwave field is increased. Only the resonances at very low static fields, requiring the highest microwave powers, shift as the microwave field is increased, moving to higher static field. Second, in the spectra at microwave fields of 114 V/cm and higher there are resonances which do not match the 25 V/cm interval of the 18s → (16,3) sequence. The "extra" resonances at microwave fields of 114–190 V/cm appear at intervals of 34 V/cm and are the 18s → (16,4) transitions. Especially for high microwave fields, as the static field is increased, resonances to (16,k) states of $k > 3$ are observed, and these overlapping resonances form an effective threshold field for the 18s → (16,k) transitions. Third, beyond the two photon transition, the number of 18s → (16,3) transitions observed increases linearly with the microwave field.

The data of Fig. 10.9 show clearly that the K $(n + 2)$s → (n,k) transitions are multiphoton transitions. On the other hand, most of the data obtained by sweeping the microwave field agree qualitatively with the Landau–Zener description. To reconcile these apparently different descriptions we consider the problem shown in Fig. 10.10, in which there are two states 1 and 2, with a linear Stark shift and no Stark shift respectively.[10] States 1 and 2 are coupled by V, the core interaction, which is time independent. Without V levels 1 and 2 would cross at the field E_C, but due to the coupling V there is an avoided crossing of size ω_0 at the static field E_C.

First, we shall approach the problem as a transition driven by the time variation of a quasi-static field, in other words, in the Landau–Zener terms we used to describe microwave ionization. We can write the Schroedinger equation for this problem as

$$H\psi(\mathbf{r},t) = \frac{\mathrm{i}\, \partial\psi(\mathbf{r},t)}{\partial t}.$$

(10.2)

As we have described the problem above, the full Hamiltonian is given by

$$H = H_H + V + H_{E_s} + H_{E_{mw}}$$

(10.3)

where H_H is the Hamiltonian for the H atom in zero field, V is the core interaction and H_{E_s} and $H_{E_{mw}}$ represent the interactions with the static and microwave fields. If we ignore for a moment the core interaction, V, and imagine that the field is static, i.e. use as the Hamiltonian in Eq. (10.2), $H = H_H + H_{E_s}$, there are two solutions to Eq. (10.2) corresponding to the two eigenstates 1 and 2;

$$\psi_1(\mathbf{r}, t) = \psi_1(\mathbf{r})\mathrm{e}^{-\mathrm{i}(W_1 - kE_s)t}$$

(10.4a)

$$\psi_2(\mathbf{r},t) = \psi_2(\mathbf{r})\mathrm{e}^{-\mathrm{i}W_2 t},$$

(10.4b)

where W_1 and W_2 are the zero field energies of states 1 and 2 and k is the permanent dipole moment of state 1. Following common usage we have used k as both a state label and as a permanent dipole moment. The total wavefunction is given by

$$\psi(\mathbf{r},t) = a\psi_1(\mathbf{r},t) + b\psi_2(\mathbf{r},t)$$

(10.5)

where a and b are time independent and $a^2 + b^2 = 1$.

In this approximation the energy levels cross at $E_C = (W_1 - W_2)/k$. If we add the coupling V to the Hamiltonian, leading to $H = H_H + H_{E_s} + V$, the coupling lifts the degeneracy at E_C, and the two levels are separated in energy by ω_0, defined by

$$\frac{\omega_0}{2} = \langle 2|V|1 \rangle = \int \psi_2^*(\mathbf{r})V\psi_1(\mathbf{r})\, \mathrm{d}\mathbf{r}.$$

(10.6)

Now imagine an experiment with states 1 and 2 analogous to the K experiments. Initially the atoms are in state 2 in a static field, the microwave pulse is applied, and afterwards the atoms can be in state 1 or state 2. During the microwave pulse we must use the full Hamiltonian of Eqs. (10.4). We can no longer represent the wavefunction by Eq. (10.5), but rather by

$$\psi(\mathbf{r},t) = T_1(t)\psi_1(\mathbf{r}) + T_2(t)\,\psi_2(\mathbf{r}).$$

(10.7)

Using the wavefunction of Eq. (10.7) in the Schroedinger equation, Eq. (10.2), leads to two coupled equations for $T_1(t)$ and $T_2(t)$. Explicitly

$$\mathrm{i}\,\dot{T}_1 = [W_1 - kE(t)]T_1 + b\,T_2$$

(10.8a)

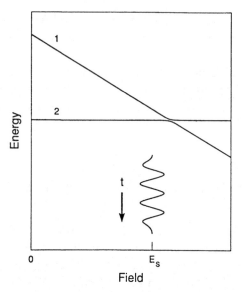

Fig. 10.11 Schematic picture of a multicycle Landau–Zener transition. In combined static and microwave fields the oscillating field brings the atom to the avoided crossing on successive cycles, and the transition amplitudes due to successive cycles add, leading to interference, or resonances (from ref. 12).

$$i\,\dot{T}_2 = bT_1 + W_2T_2. \tag{10.8b}$$

The total electric field is given by

$$E(t) = E_S + E_{mw}\cos\omega t, \tag{10.9}$$

where E_S, and E_{mw} are the static and microwave field amplitudes. To obtain a solution to Eqs. (10.8), they can be integrated numerically, using the method of Rubbmark *et al.*[7] Letting the time evolve from $\omega t = -\pi/2$ to $\pi/2$ yields results analogous to the half cycle results of Pillet *et al.*[4] When the time interval is extended to more than three cycles, resonances appear in the transition probability when the static field has values for which the energy difference between the two states in the static field alone is an integral multiple of the microwave frequency.[10] After many cycles the calculated lineshapes are almost identical to the Rabi lineshapes calculated for two level systems.[10] A pictorial view of the effect of many cycles is given in Fig. 10.11.[12] On successive cycles the atom samples the level crossing at the turning point of the field, and the transition amplitudes of many cycles add coherently, leading to destructive or constructive interference. In other words, there are resonances, and the Landau–Zener description goes over smoothly to the resonant photon absorption picture.

The other way of describing the transitions is based on a Floquet approach.[13–16] To illustrate the essential ideas we use a simple model to calculate the Rabi frequencies for the problem depicted in Fig. 10.10. There are two states: 1, which has a linear Stark shift, and 2, which has no Stark shift. They are coupled by the time independent core coupling V. We treat the problem by first ignoring the core coupling and introducing it later as a perturbation. The unperturbed Hamiltonian, $H = H_H + H_{E_S}$ is used to describe the atom in the static field E_S, and the time dependent wavefunctions of states 1 and 2 are given by Eq. (10.4). Adding a microwave field in the same direction, $E_{mw} \cos \omega t$, corresponding to the Hamiltonian $H = H_H + H_{E_S} + H_{E_{mw}}$, does not couple states 1 and 2. Thus the wavefunction for state 1, for example, may be written as $\psi_1(\mathbf{r},t) = T_1(t)\psi_1(\mathbf{r})$.

It is a solution of the time dependent Schroedinger equation

$$i \dot{T}_1(t)\psi_1(\mathbf{r}) = (W_1 - kE_S - kE_{mw} \cos \omega t)T_1(t)\psi_1(\mathbf{r}) \qquad (10.10)$$

which has as a solution for $T_1(t)$

$$T_1(t) = e^{-i\int_{t_0}^{t} (W_1 - kE_S - kE_{mw} \cos \omega t') \, dt'}. \qquad (10.11)$$

Integrating Eq. (10.11) yields

$$T_1(t) = e^{-i\left(W_1 - kE_S)t - k\frac{E_{mw}}{\omega} \sin \omega t\right)}, \qquad (10.12)$$

where we have dropped the constant phase from the initial time t_0. Finally,

$$\psi_1(\mathbf{r},t) = \psi_1(\mathbf{r})\, e^{-i\left(W_1't - \frac{kE_{mw}}{\omega} \sin \omega t\right)}, \qquad (10.13)$$

where $W_1' = W_1 - kE_S$. When the wavefunction is written in this way it shows that adding the microwave field modulates the energy around the value W_1' at the angular frequency ω. In the same way that frequency modulation of a radio wave leads to sidebands, described by a Bessel function expansion, modulation of the atomic wavefunction also leads to sidebands. Using the fact that[17]

$$e^{ix \sin \omega t} = \sum_{n=-\infty}^{\infty} J_n(x)e^{in\omega t}, \qquad (10.14)$$

we can write the wavefunction $\psi_1(\mathbf{r},t)$ by expressing Eq. (10.12) in terms of the Bessel function expansion of Eq. (10.14). Explicitly,

$$\psi_1(\mathbf{r},t) = T_1(t)\psi_1(\mathbf{r}) = \left\{ e^{-iW_1't} \sum_{-\infty}^{\infty} J_n\left(\frac{kE_{mw}}{\omega}\right) e^{(-in\omega t)} \right\} \psi_1(\mathbf{r}). \qquad (10.15)$$

We can identify W_1' as the carrier energy of state 1, and there are sidebands at energies $W_1' + n\omega$ for all n. There is in principle an infinite number of terms in the expansion. However, for large arguments $J_n(x) \approx 0$ if $|n| > x$,[17] and as a result, the sideband amplitudes are only significant for $kE_{mw} > |n|\omega$. In other words, in a microwave field the sidebands with nontrivial amplitude extend as far in energy, $\pm kE_{mw}$, as the energy modulation produced by the microwave field. In Fig. 10.12 we show the energies of the sidebands of state 1 as a function of the static field.

State 2 has no Stark shift, and thus under the Hamiltonian $H = H_H + H_{E_S} + H_{E_{mw}}$ its wavefunction is given by Eq. (10.4b), and it has no sidebands with nonzero amplitude. With the Hamiltonian $H = H_H + H_{E_S} + H_{E_{mw}}$, i.e. in the

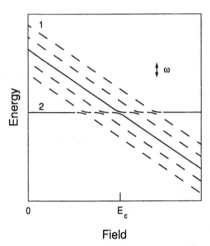

Fig. 10.12 Energy level diagram for two states 1 and 2, showing the sidebands (– – –) of 1 produced by the microwave field of angular frequency ω (from ref. 12).

absence of the coupling V, the Nth sideband of state 1 is degenerate with and crosses state 2 at

$$E_S = E_C + \frac{N\omega}{k}. \tag{10.16}$$

When the core coupling V is introduced as a perturbation, the crossing of state 1 and its sidebands with state 2 become anticrossings, as shown by Fig. 10.12. The magnitude of the avoided crossing, ω_N, between the Nth sideband of state 1 and state 2 at the static field $E_S = E_C + N\omega/k$ is obtained from the time independent part of the coupling matrix element $\langle\psi_2(\mathbf{r},t)|V|\psi_1(\mathbf{r},t)\rangle>$. Using the wavefunctions of Eqs. (10.4b) and (10.15), the time independent part of matrix element is given by

$$\langle\psi_2(\mathbf{r})|V|\psi_1(\mathbf{r})\rangle_N = \langle 2|V|1\rangle J_N(kE_{mw}/\omega). \tag{10.17}$$

Accordingly,

$$\omega_N = 2|\langle 2|V|1\rangle J_N(kE_{mw}/\omega)|$$
$$= \omega_0|J_N(kE_{mw}/\omega)|. \tag{10.18}$$

In essence, the magnitude of the avoided crossing ω_N is the static field avoided crossing ω_0 multiplied by the fractional amplitude of state 1 in the Nth sideband. The significance of the magnitude of the avoided crossing is that it is equal to the frequency at which the atom makes transitions back and forth between 1 and 2, the Rabi frequency. A familiar example showing the equivalence of the avoided crossing size and the Rabi frequency is provided by the analogous anticrossing at E_C between 1 and 2 in a purely static field. At E_C the eigenstates are $(\psi_1 \pm \psi_2)/\sqrt{2}$. If the two eigenstates are excited coherently, in a time $\ll 1/\omega_0$, the population

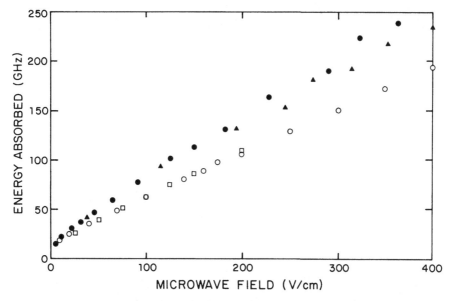

Fig. 10.13 Total energy (number of photons times photon energy) absorbed in the K19s → (17,3) transition vs microwave field: Experimental points; 9.27 GHz, ($\triangle$), and 10.353 GHz ($\bullet$); theoretical points; Landau–Zener calculations ($\square$) and two level Floquet calculations ($\bigcirc$) (from ref. 10).

oscillates between 1 and 2 at frequency $\omega_0 = 2b$, and this oscillation may be detected in a quantum beat experiment.

There are many forms in which the Floquet approach can be more rigorously implemented.[13–16] One which has been used often is the infinite matrix approach of Shirley.[14] It corresponds roughly to the infinite set of sidebands. An alternative more compact approach has been described by Sambe[15] and Christiansen-Dalsgaard.[16]

When Landau–Zener and Floquet calculations are performed to match the K 19s → (17,3) transitions, with parameters $W_1 = 220$ GHz, $W_2 = 0$, $k = 400$ MHz/ (V/cm), and $\omega_0 = 800$ Mhz, the Floquet and multicycle Landau–Zener calculations predict that the same microwave field is required to produce a given Rabi frequency for an N photon transition.[10] To compare calculated transition strengths to the experimental observations requires that we decide what is an observable Rabi frequency. In Fig. 10.9 the observed resonances are ~1 GHz wide while the exposure time of the atoms to the microwaves is 1 μs. Making the questionable assumption of homogeneous broadening leads to a required Rabi frequency of ~30 MHz, and this Rabi frequency was chosen to compare to the experimental results. While the calculations agree with each other, they do not exactly match the experiments, as shown by Fig. 10.13, a plot of the total energy absorbed from the 9 or 10 GHz microwave field as a function of microwave field strength. In Fig. 10.13, absorbing 150 GHz corresponds to fifteen 10 GHz photons or seventeen 9 GHz photons. The calculated fields required to drive the n photon transitions are systematically 25% higher than the experimental results, a difference which is probably due to oversimplifications in the model. For example, the dipole coupling between the levels, is ignored. It is interesting to note that Fig. 10.13 also demonstrates graphically the fact that the number of photons absorbed increases linearly with the microwave field amplitude. Most important, however, it is apparent that the multiphoton resonances can be described equally well in two different ways.

We have thus far treated the K $(n + 2)$s states as having no Stark shifts. While they have no first order Stark shift, they do have a second order shift due to their dipole interaction with the p states, which are removed from the s states by energies large compared to the microwave frequency. The microwave field does not produce appreciable sidebands of the s state since it has no first order Stark shift. However, it does induce a Stark shift to lower energy. Not surprisingly the Stark shift produced by a low frequency microwave field of amplitude E is the same as the second order Stark shift produced by a static field $E_S/\sqrt{2}$, they have the same value of $\langle E^2 \rangle$. Careful inspection of Fig. 10.9 reveals that the resonances observed with high microwave powers shift with power.

In fact, the second order Stark shift of the $(n + 2)$s state due to the microwave field can bring the $(n + 2)$s $\rightarrow (n,k)$ transitions into resonance with no static field. This Stark shift is the explanation of the anomalous K 19s $\rightarrow (17,k)$ threshold shown in Fig. 10.8. The transitions at microwave fields from 500 to 600 V/cm, i.e. fields below the avoided crossing field, $E_C = 546$ V/cm, are due to the fact that the 19s state is shifted into 27 photon 19s–$(17,k)$ resonances in this field range, and the Rabi frequencies of the transitions to a few, three or four, of the $(17,k)$ states, $k \sim 3$, are adequate to observe the transitions in this field range. At higher fields, ~700 V/cm, the transition becomes nonresonant, and is not observed again until E ~ 750 V/cm, corresponding to the 28 photon resonances. At this field, the Rabi frequencies of many more 19s $\rightarrow (17,k)$ transitions are adequate to allow observation of the resonances, and what is observed as a threshold is really many overlapping resonant transitions. In retrospect, it is now clear why the threshold fields for the $(n + 2)$s $\rightarrow (n,k)$ transitions observed with microwaves alone are scattered from $1.0E_C$ to $1.6E_C$, contrary to the prediction of a single cycle Landau–Zener model. Both the resonance condition and Rabi frequency condition must be fulfilled, and the field at which the resonance condition is fulfilled is rather random.

Having considered the connection between the multiphoton resonances and the microwave threshold field for the K $(n + 2)$s $\rightarrow (n,k)$ transitions, it is now interesting to return to the analogous $n \rightarrow n + 1$ transitions which are responsible for microwave ionization and consider them from this point of view. We start with a two level description based on the extreme n and $n + 1$ $|m| = 0$ Stark states, a description which is the multiphoton resonance counterpart to the single cycle Landau–Zener model presented earlier. The problem is identical to the problem depicted in Fig. 10.10 except that both states have Stark shifts and sidebands. For simplicity we label these two extreme Stark states as n and $n + 1$. The Hamiltonian for this problem is identical to that of Eq. (10.2) except that there is no static field, i.e. $H = H_H + H_{E_{mw}} + V$. Using $H = H_H + H_{E_{mw}}$ we find the solutions to Eq. (10.2) analogous to the solution given in Eq. (10.15), i.e. [16,17]

$$\psi_n(\mathbf{r},t) = \psi_n(\mathbf{r})\, e^{\frac{it}{2n^2}} \sum_k J_k\left(\frac{k_n E_{mw}}{\omega}\right) e^{-ik\omega t} \qquad (10.19a)$$

and

$$\psi_{n+1}(\mathbf{r},t) = \psi_{n+1}(\mathbf{r})\, e^{\frac{it}{2(n+1)^2}} \sum_{k'} J_{k'}\left(\frac{k_{n+1} E_{mw}}{\omega}\right) e^{-ik'\omega t}. \qquad (10.19b)$$

We wish to calculate the Rabi frequency for the N photon transition at resonance, i.e. $N\omega = 1/2n^2 - 1/2(n + 1)^2 \approx 1/n^3$. The matrix element of V coupling the two states of Eqs. (10.19) is given by

$$\langle \psi_{n+1}(\mathbf{r},t)|V|\psi_n(\mathbf{r},t)\rangle = \langle n + 1|V|n\rangle\, e^{\frac{it}{n^3}} \sum_{k'k} J_{k'}\left(\frac{k_{n+1} E_{mw}}{\omega}\right) J_k\left(\frac{k_n E_{mw}}{\omega}\right) \times e^{i(k'-k)\omega t}. \qquad (10.20)$$

At the N photon resonance, when $N\omega = 1/n^3$, the n and $n + 1$ states are degenerate and are coupled by the time independent part of the coupling matrix element of Eq. (10.20), which is

$$\langle\psi_{n+1}|V|\psi_n\rangle_N = \langle n + 1|V|n\rangle \sum_k J_{k-N}\left(\frac{k_{n+1}E_{mw}}{\omega}\right) J_k\left(\frac{k_n E_{mw}}{\omega}\right). \qquad (10.21)$$

Using the relation[17]

$$J_N(x - y) = \sum_k J_{k-N}(x)J_k(y) \qquad (10.22)$$

Eq. (10.21) may be written as

$$\langle\psi_{n+1}|V|\psi_n\rangle_N = \langle n + 1|V|n\rangle J_N\left(\frac{(k_{n+1} - k_n)E_{mw}}{\omega}\right). \qquad (10.23)$$

Eq. (10.23) can be applied to any pair of n and $n + 1$ Stark states. If we consider the extreme $m = 0$ states, for which $k_n = 3n(n - 1)/2$ and $k_{n+1} = -3n(n + 2)/2$, Eq. (10.23) reduces to

$$\langle\psi_{n+1}|V|\psi_n\rangle_N = \langle n|V|n + 1\rangle J_N\left(\frac{3n^2 E_{mw}}{\omega}\right). \qquad (10.24)$$

Eq. (10.23) implies that it is only the difference in the Stark shifts which is important.[10,17] Again the fact that for large N, $J_N(x) \to 0$ for $N > x$ leads to the conclusion that $E_{mw} \geq 1/3n^5$ to have a nonnegligible coupling of the extreme Stark

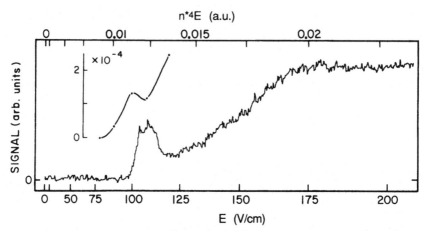

Fig. 10.14 Ionization signal of He 28s ^{3}S atoms as a function of the peak electric field inside the cavity. The size of the vertical scale corresponds to approximately 1/3 of the saturated signal. Inset: the calculated transition probability from the 28s ^{3}S to the 29s ^{3}S state after one field cycle (from ref. 3).

states. Slightly higher fields are required to have significant coupling between states other than the extreme ones.

Of course the resonance condition must also be met. For ionization by a fixed frequency microwave field tuning must come from the AC stark shifts, which are sufficient. If only the extreme Stark states participated in the transitions, sharp resonances might be observed. However, in the $n \to n + 1$ transition there are $\sim n$ possible initial states in the n Stark manifold as well as $n + 1$ final states in the $n + 1$ Stark manifold, leading to $n(n + 1)$ transitions in a tuning range of ω. For a frequency of 10 GHz at $n = 20$ there is a resonance every 25 MHz on the average. When the microwave field reaches $1/3n^5$ some of these resonances begin to have observable Rabi frequencies, and the onset of these closely spaced, usually overlapping, resonances produce what appears to be a threshold field for the transition. Since the resonance criterion is not an issue, the primary criterion is that the Rabi frequency be high enough, leading to the observed $E \sim 1/3n^5$ dependence.

Due to the $n(n + 1)$ possible $n \to n + 1$ transitions it is in general difficult to observe resonance effects in microwave ionization as obvious as those shown in Fig. 10.9. Nonetheless several experiments show clearly the importance of multiphoton resonance in microwave ionization. In Ba and in He the observed microwave ionization thresholds are structured by resonances[3,6]. An excellent example is the microwave ionization probability of the He 28 ^{3}S state shown in Fig. 10.14. In He the ^{3}S states intersect the Stark manifold at fields approaching $1/3n^5$, and as a result making transitions from the energetically isolated ^{3}S state requires a field comparable to the field required to drive $n \to n + 1$ transition. The structure in Fig. 10.14 is quite similar to the structure in Fig. 10.8, which is not

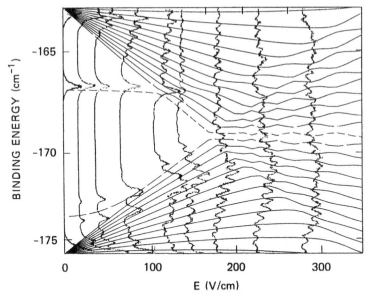

Fig. 10.15 Observed Na 3p → n = 25,26 spectra in varying strengths of microwave field overlaid on an energy level diagram of the Na |m| = 0 states. The baseline of each spectrum is located at the amplitude of the microwave field. Note that only odd sidebands of the p states occur and that the predominant effect of higher microwave fields is to add more sidebands (from ref. 18).

surprising since most of tuning comes from the second order Stark shifts of the s states in both cases.

In alkali atom experiments no explicit resonances have been observed in microwave ionization. However, there are indirect confirmations of the multiphoton resonance picture. First, according to the multiphoton picture the sidebands of the extreme n and $n + 1$ Stark levels should overlap if $E = 1/3n^5$. In the laser excitation spectrum of Na Rydberg states from the $3p_{3/2}$ state in the presence of a 15 GHz microwave field van Linden van den Heuvell *et al.* observed sidebands spaced by 15.4 GHz, as shown in Fig. 10.15.[18] The extent of the sidebands increases linearly with the microwave field, as shown in Fig. 10.15, and the $n = 25$ and $n = 26$ sidebands overlap at microwave fields of 150 V/cm or higher, matching the observation that the 25d state has an ionization threshold of 150 V/cm in a 15 GHz field.

Pillet *et al.* observed that adding small static fields dramatically reduces the microwave fields required for the ionization of Li.[19] For example the application of a static field of 1 V/cm lowers the 15 GHz ionization threshold of the Li 42d state from ~200 V/cm, to a broad threshold centered at 20 V/cm, a field only slightly in excess of $E = 1/3n^5 = 13$ V/cm. The threshold field 200 V/cm corresponds to the hydrogenic threshold field of $1/9 n^4$, which will be described shortly. A small field has virtually no effect in a single cycle Landau–Zener model, but its dramatic effect is readily explained in a multilevel resonance picture. The small static field lifts the degeneracy of sidebands from different Stark states, transforming them into a quasi-continuum of states, ensuring that the resonance condition is met.

Finally, Mahon *et al.*[5] have observed that even at frequencies as low as 670 MHz that ionization of Na occurs at fields near $1/3n^5$. At such low frequencies it is impossible to explain the ionization on the basis of incoherent single cycle Landau–Zener transitions. Rather the coherent effect of many cycles of the field is required.

Hydrogen

Microwave ionization of H and H like atoms differs radically from the ionization of nonhydrogenic atoms. For our present purposes, an atomic system is H like if it has small quantum defects. For example the Na $|m| = 2$ states are composed of $|m| \geq 2$ states all of which have quantum defects $<1.5 \times 10^{-2}$. With small quantum defects the avoided crossings between the Stark levels of adjacent n are negligibly small, and there is negligible coupling between n levels due to the core interaction. As this coupling is responsible for the ionization of nonhydrogenic atoms at $E = 1/3n^5$, it is not surprising that H like atoms do not exhibit ionization at $E = 1/3n^5$. It is worth noting that atoms may appear to be hydrogenic in one case but not in another. For example, the Na $|m| = 2$ states are ionized by 15 GHz fields of $1/9n^4$, as shown by Fig. 10.3, but can be ionized at lower fields, $\sim 1/3n^5$, when a static field is applied to produce a quasi-continuum of levels.[20] Ionization at $E = 1/3n^5$ does not occur in H under the same conditions. Thus, even quantum defects of 10^{-2} are adequate to produce nonhydrogenic behavior in some circumstances.

The first measurements of microwave ionization in any atom were carried out with a fast beam of H by Bayfield and Koch[1], who investigated the ionization of a band of approximately five n states centered at $n = 65$. Using microwave and rf fields with frequencies of 9.9 GHz, 1.5 GHz, and 30 MHz, to ionize the atoms they found that the same field was required at 30 MHz and 1.5 GHz to ionize the atoms, but that a smaller field was required at 9.9 GHz. The measurements showed that at $n = 65$ frequencies up to 1.5 GHz are identical to a static field. Later, more systematic measurements have confirmed the initial measurements and have allowed significant refinements of our understanding. In Fig. 10.16 we show the ionization threshold fields (in this case the field at which there is 10% ionization) of H in a 9.9 GHz field.[21] The ionization fields are plotted as n^4E vs $n^3\omega$, and they bring out two factors. First, at low frequencies the field required is $\sim 1/9n^4$, the static field required to ionize the red n Stark state of $|m| \ll n$. Second, as shown by the scaling of the horizontal axis, the required field drops below $1/9n^4$ as the microwave frequency approaches the interval between adjacent n states, $1/n^3$.

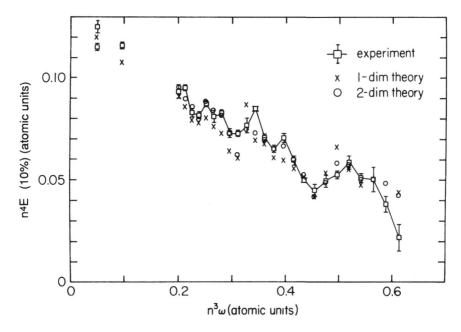

Fig. 10.16 Scaled microwave ionization field, n^4E, for H plotted against the scaled microwave frequency $n^3\omega$: experimental ($\square$); one dimensional theory (x); two dimensional theory ($\bigcirc$). $n^3\omega = 0.05$ corresponds to $n = 32$ and $n^3\omega = 0.6$ corresponds to $n = 73$. Note the decline from $n^4E = 1/9$ at $n = 30$ to progressively lower values as n approaches 60 (from ref. 21).

Qualitatively similar observations have been made for $\omega < 1/3n^3$ for the H like Na and Li $|m| = 2$ states with 15 GHz fields.[22]

When $\omega \ll 1/n^3$, the field required for ionization is $E = 1/9n^4$, and as ω approaches $1/n^3$ it falls to $E \cong 0.04n^{-4}$. These observations can be explained qualitatively in the following way. At low n, so that $\omega \ll 1/n^3$, the microwave field induces transitions between the Stark states of the same n and m by means of the second order Stark effect. With only a first order Stark shift a state always has the same dipole moment and wavefunction, as indicated by the constant slope dW/dE of the energy level curve. Thus when the field reverses, $E \rightarrow -E$, the Rydberg electron's orbit does not change. With a second order Stark shift as well, the slope dW/dE is not the same at E and $-E$, and as a result the dipole moment and wavefunction are not the same. If the field is reversed suddenly a single Stark state in the field E is projected onto several Stark states of the same n and m when $E \rightarrow -E$. Since all the Stark states of the same n make transitions among themselves they ionize once the field is adequate to ionize one of them, the red one, at $E = 1/9n^4$ for $|m| \ll n$.

As $\omega \rightarrow 1/n^3$, or more commonly, as n is increased with fixed ω, the required field falls below $1/9n^4$ as transitions to higher n occur, allowing ionization at lower fields. Since many energy levels are coupled this development has been explained as the onset of classical chaos.[23] It can also be explained in terms of a rate limiting step. We here present a development along the latter line, due to Christiansen-Dalsgaard,[24] to calculate the hydrogenic microwave ionization fields for ω approaching but not exceeding $1/n^3$. We assume that ionization is always possible if $E > 1/9n^4$, as outlined above. On the other hand as $\omega \rightarrow 1/n^3$ the field required to drive the rate limiting $n \rightarrow n + 1$ transition falls below $1/9n^4$, and ionization occurs by transitions to higher n states followed by ionization from the higher n states.

The central problem is to calculate the field required to drive the $n \rightarrow n + 1$ transition via an electric dipole transition. In the presence of an electric field, static or microwave, the natural states to use are the parabolic Stark states. While there is no selection rule as strict as the $\Delta \ell = \pm 1$ selection rule for angular momentum eigenstates, it is in general true that each n Stark state has strong dipole matrix elements to only the one or two $n + 1$ Stark states which have approximately the same first order Stark shifts. Red states are coupled to red states, and blue to blue. Explicit expressions for these matrix elements between parabolic states have been worked out,[25] and, as pointed out by Bardsley *et al.*[26], the largest matrix elements are those between the extreme red or blue Stark states. These matrix elements are given by $\langle n |z|n + 1 \rangle = n^2/3$.[26]

Since the extreme n and $n + 1$ Stark states have the largest coupling matrix elements, it seems reasonable to assume that the $n \rightarrow n + 1$ transitions occur through these states and calculate the field required to drive the transition between this pair of levels. While this is an approximation, it is useful, just as considering the extreme Stark states gives a reasonable description of the ionization of Na.

Following the approach of Christiansen-Dalsgaard,[24] we can calculate the $n \rightarrow n + 1$ Rabi frequency. We write the Hamiltonian as

$$H = H_{\mathrm{H}} + H_{E_{\mathrm{mw}}} + H_{\Delta n E_{\mathrm{mw}}}. \tag{10.25}$$

H_{H} has the same definition as before, $H_{E_{\mathrm{mw}}}$ includes only the diagonal matrix elements, and $H_{\Delta n E_{\mathrm{mw}}}$ includes the off diagonal elements. $H = H_{\mathrm{H}} + H_{E_{\mathrm{mw}}}$ has as its eigenfunctions the linear hydrogenic Stark states. The dominant effect of $H_{\Delta n E_{\mathrm{mw}}}$ is the dipole coupling between n states. If we use $H = H_{\mathrm{H}} + H_{E_{\mathrm{mw}}}$ the wavefunctions for the n and $n + 1$ blue Stark states are given by Eq. (10.19) with $k_n = 3n(n - 1)/2$ and $k_{n+1} = 3n(n + 1)/2$.

At the n photon resonance $N\omega = 1/2n^2 - 1/2(n + 1)^2 \cong 1/n^3$, and we can calculate the Rabi frequency starting from the dipole coupling matrix element,[24,27]

$$\langle \psi_{n+1}(\mathbf{r},t)|H_{\Delta n E_{\mathrm{mw}}}|\psi_n(\mathbf{r}, t) \rangle = \langle \psi_{n+1}(\mathbf{r},t)|zE \cos \omega t|\psi_n(\mathbf{r},t) \rangle. \tag{10.26}$$

Using the Bessel function forms of the wavefunctions given by Eqs. (10.19), writing $\cos \omega t$ in exponential form and using Eqs. (10.19), we identify the time independent part of Eq. (10.26) as

$$\langle\psi_{n+1}|H_{\Delta nE_{mw}}|\psi_n\rangle_N = \frac{\langle n+1|z|n\rangle}{2}\left\{\sum_k J_{k+1-N}\left(\frac{k_{n+1}E_{mw}}{\omega}\right)J_k\left(k_n\frac{E_{mw}}{\omega}\right)\right.$$

$$\left. + J_{k-1-N}\left(k_{n+1}\frac{E_{mw}}{\omega}\right)J_k\left(k_n\frac{E_{mw}}{\omega}\right)\right\}. \quad (10.27)$$

Using Eq. (10.22), Eq. (10.27) becomes

$$\langle\psi_{n+1}|H_{\Delta nE_{mw}}|\psi_n\rangle_N = \frac{\langle n+1|z|n\rangle}{2}\left\{J_{N+1}\left(\frac{(k_{n+1}-k_n)E_{mw}}{\omega}\right)\right.$$

$$\left. + J_{N-1}\left(\frac{(k_{n+1}-k_n)E_{mw}}{\omega}\right)\right\}. \quad (10.28)$$

Using[17]

$$J_{v-1}(x) + J_{v+1}(x) = \frac{2v}{x}J_v(x) \quad (10.29)$$

allows us to write Eq. (10.28) as

$$\langle\psi_{n+1}|H_{\Delta nE_{mw}}|\psi_n\rangle_N = \frac{\langle n+1|z|n\rangle N\omega}{k_{n+1}-k_n}J_N\left(\frac{(k_{n+1}-k_n)E_{mw}}{\omega}\right). \quad (10.30)$$

For the extreme Stark states $\langle n+1|z|n\rangle = n^2/3$, and $k_{n+1} - k_n = 3n$. At the N photon resonance $N\omega \cong 1/n^3$, and the N photon matrix element is given by

$$\langle\psi_{n+1}|H_{\Delta nE_{mw}}|\psi_n\rangle_N = \frac{1}{9n^2}J_N\left(\frac{3nE}{\omega}\right), \quad (10.31)$$

and the Rabi frequency is twice as large. In this form it is clear that it is predominantly the variation of the Bessel function which determines the field required to achieve a useful Rabi frequency.

If we assume that many photons are required for the $n \to n+1$ transition, N is large and $J_N(3nE/\omega) \to 0$ unless $3nE/\omega > N$. Since $N\omega \cong 1/n^3$, this requirement can be restated as

$$E \geq \frac{1}{3n^4}, \quad (10.32)$$

a field above the static ionization field of $E = 1/9n^4$. Evidently, for $\omega \ll 1/n^3$ ionization does not occur by transitions to higher states but by direct field ionization as described earlier. On the other hand when N is a small number, <5, the above reasoning for large arguments of the Bessel function does not apply. Recall that in the K $(n+2)s \to (n,k)$ transitions the first few transitions require very small fields. Similarly, when $n < 5$, the Rabi frequency can be appreciable even for $E \ll 1/3n^4$. Explicit evaluation of Eq. (10.31) yields immediately the Rabi frequency for the transitions.

An upper limit to the required Rabi frequency is found by requiring that it be equal to the detuning. Using this requirement with Eq. (10.31) Christiansen-

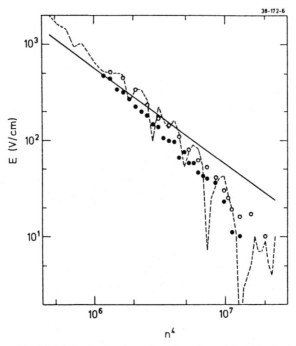

Fig. 10.17 The critical field, for ionization, E, as function of the principal quantum number n of the lower state for dipole transitions in H (–––). Experimental points for Na (●) and Li (○) are shown as well as the critical field for tunneling (——) (from ref 24).

Dalsgaard has computed the hydrogenic fields for ionization by a 15 GHz field.[24] The results of this two state model are compared to the experimental Na and Li $|m| = 2$ threshold fields in Fig. 10.17. The computed threshold fields are in general agreement with the observations, although they exhibit more structure than do the experimental points. For example the two photon resonance at $n = 60$ is apparent in the theoretical curve but not the experimental points. The discrepancies are due to the simplicity of the model. In the theoretical model the second order Stark shifts and the existence of other states were ignored. Only the extreme n and $n + 1$ Stark states were considered, but just as in the Na or K cases the other Stark states also make $n \rightarrow n + 1$ transitions although, usually at slightly higher fields due to their smaller $n \rightarrow n + 1$ matrix elements. However, the fact that the carrier energies of different Stark states of the same n are different, due to their different second order Stark shifts, means that it is not obvious which transitions between n and $n + 1$ Stark states are more likely to be resonant. When the many n and $n + 1$ Stark states and possible variations of the microwave field with time are taken into account, it is apparent that sharp resonances are less likely to be as apparent as they are in the two state theoretical curve of Fig. 10.17.

If the microwave ionization can be described by a resonance multiphoton picture, it should be possible to observe other manifestations of resonance

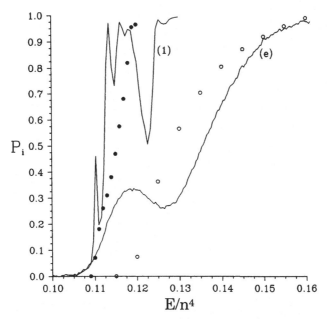

Fig. 10.18 Ionization probability P_i of H n =36 atoms by 9.92 GHz field experimental (curve (e)), one dimensional adiabatic quantum calculation (curve (1)), one dimensional classical calculation (●), three dimensional classical calculation (○) (from ref. 28).

phenomena, and such is the case. Ionization thresholds in H show, in some cases, marked structure. An example of this phenomenon is shown in Fig. 10.18, the structured ionization curves of H n = 36 atoms in a 9.92 GHz field.[28] The origin of the pedestal at $E \sim 0.12 \, n^{-4}$ is due, in the terms used above, to a resonance. It is similar to the K and He resonances shown in Figs. 10.8 and 10.14.

The ionization curve of Fig. 10.18 is obtained in the same way as the data shown in Fig. 10.14, by exciting atoms in zero field and then exposing them to a strong microwave field. When atoms are excited in the presence of a static field, to a single Stark state, and held in single Stark state by the continued application of the field, resonances became more apparent when a microwave field in the same direction is applied. Bayfield and Pinnaduwage have observed transitions from the extreme red H n = 60, m = 0 Stark state to other nearby extreme Stark states in static fields of 5–10 V/cm.[29] As shown by Fig. 10.19 resonances corresponding to the four photon transition to the extreme red n = 61 Stark state and four and five photon transitions to the extreme red n = 59 Stark state are visible. These experiments are similar to the K and He multiphoton resonance experiments described earlier, but are inherently simpler because the extreme red n = 60 Stark state is only coupled to the extreme n = 59 Stark state. In contrast, the K $(n + 2)$s state is coupled to all the (n,k) Stark states.

As shown by Fig. 10.16, as the microwave frequency approaches $1/n^3$ the scaled field $n^4 E$ required for ionization drops steadily, raising the question of what

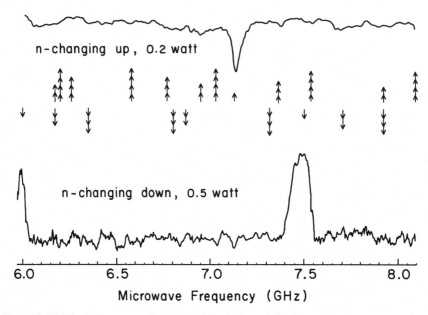

Fig. 10.19 The microwave frequency dependence of the n changing signals at low microwave power, where n changes up or down only by 1. Resonant multiphoton transitions are observed near the expected static field Stark shifted frequencies indicated. These resonances involve the absorption of four or five microwave photons. The down n changing atom production curve was obtained with the state analyzer field E_A set at 50.0 V/cm, while up n changing was studied as $n = 60$ atom loss with $E_A = 45.5$ V/cm. The locations of resonances for larger direct (not stepwise) changes in n are indicated along with their order k (from ref. 29).

happens when ω exceeds $1/n^3$. Reaching the regime $\omega > 1/n^3$ requires higher n, higher frequency, or both. For example, at 10 GHz $\omega = 1/n^3$ at $n = 87$. Galvez *et al.*[30] have observed the ionization of H $n = 40$ to $n = 80$ states by a 36 GHz microwave field. They took data in two modes; the quench mode, observing the atoms remaining in the initial state, and the ionization mode, observing those which have been ionized. Static fields downstream from the microwave field region were used to analyze the final states, and in the ionization mode the microwave ionization signal also included bound states of $n > n_c$, where $160 < n_c < 190$. Their results, obtained in the ionization mode are shown in Fig. 10.20.

As shown by Fig. 10.20, when ω exceeds $1/n^3$ the scaled field, n^4E, required to ionize H no longer drops sharply but in fact increases very slightly with n. At 36 GHz $n^3\omega = 0.4$ and 2.8 correspond to $n = 42$ and 80. The measured ionization fields are not inconsistent with a constant value of n^4E for $n^3\omega > 1$. Jensen *et al.*[31] have developed a simple quantum mechanical model for the regime, $1/n^3 \ll \omega \ll 1/n^2$ which predicts an n independent ionization field. In their model ionization occurs by a sequence of single photon transitions through higher lying

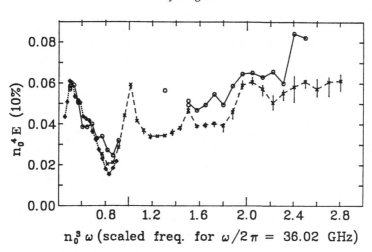

Fig. 10.20 H 36 GHz microwave ionization fields (10% ionization): experimental ionization mode results (o); three dimensional classical calculations ($\times$, $\Diamond$) (from ref. 30).

states. The sequence of transitions, and ionization, only occurs if the Rabi frequencies of the transitions equal or exceed the detunings. If the energy difference between two states of principal quantum numbers n and n' is given by $\Omega = 1/2n'^2 - 1/2n^2$, the electric dipole matrix element connecting these two states is given by[31]

$$\langle n|z|n'\rangle = 0.4(nn')^{-3/2}\Omega^{-5/3}$$
$$= 0.4n^{-3}\Omega^{-5/3} \tag{10.33}$$

provided that $n,n' \gg 1$ and $|n - n'| \ll n$. If the two requirements for Eq. (10.33) are fulfilled, the dipole matrix element for the $n \to n'$ transition scales as n^{-3}, just as separation between n states and the detuning from resonance. As a result, requiring that the Rabi frequency equal the maximum detuning, $\sim 1/2n^3$ leads to the requirement

$$E = 2.5\Omega^{5/3}. \tag{10.34}$$

Since resonance $\Omega \cong \omega$, the microwave frequency, the predicted ionizing field is n independent. For a microwave frequency of 36 GHz the field predicted by Eq. (10.34) is 10 V/cm, in reasonable agreement with Fig. 10.17.

As ω increases further above $1/n^3$ a single photon drives the initially populated state closer and closer to the ionization limit, and ionization occurs with the absorption of fewer photons. Few photon processes are well described by lowest order perturbation theory, which shows that the rates are proportional to E^{2N}, where N is the number of photons absorbed. For small N such processes are not well described by a threshold field, and it is not meaningful to discuss ionization threshold fields in this case.

The discussion of microwave multiphoton processes given here is largely a Floquet description, i.e. a steady state description. We have implicitly assumed, for example, that the microwave fields are turned on and off rapidly compared to the Rabi frequencies and that the time during which the microwaves are turned on or off is so short that it contributes nothing to the transition. It is clear that these assumptions are not always met, and interesting dynamic effects due to tuning through resonances by second order Stark shifts have been predicted and observed.[32-34] Similarly, we have not considered the effects of noise as anything other than a source of line broadening. In fact noise can have a profound effect, especially in a broadband microwave system where broadband noise almost ensures coincidences with relevant atomic transitions.[35]

Ionization by circularly polarized fields

While ionization by linearly polarized fields has been well studied, there is only one report of ionization by a circularly polarized field, the ionization of Na by an 8.5 GHz field.[36] In the experiment Na atoms in an atomic beam pass through a Fabry–Perot microwave cavity, where they are excited to a Rydberg state using two pulsed tunable dye lasers tuned to the 3s → 3p and 3p → Rydberg transitions at 5890 Å and ~4140 Å respectively. The atoms are excited to the Rydberg states in the presence of the circularly polarized microwave field which is turned off 1 μs after the laser pulses. Immediately afterwards a pulsed field is applied to the atoms to drive any ions produced by microwave ionization to a microchannel plate detector. To measure the ionization threshold field the ion current is measured as the microwave power is varied.

The experimental approach to producing the circularly polarized field is to use a Fabry–Perot cavity which supports two orthogonally linearly polarized 8.5 GHz fields 90° out of phase. The major experimental problem is to minimize the interaction between these two nominally degenerate cavity modes, as any inter-action lifts the degeneracy. To minimize the interaction Fu *et al.* fed the two modes from orthogonally polarized waveguides through irises in the mirrors.[36] The two modes were offset by 2 MHz, and they had Qs of 2000 and 2100, so the linewidths of the modes were 4 MHz. Due to the fact that the Qs were not identical circular polarization could exist in steady state, but not as the cavities modes were filled with the microwave field. To minimize the exposure of the atoms to elliptically polarized microwave fields Fu *et al.* excited the atoms to a Rydberg state in the microwave field.

A much higher circularly polarized field is required to ionize the atoms than a linearly polarized field, as shown by Fig. 10.21, a plot of the threshold fields, where 50% ionization occurs, for linearly and circularly polarized 8.5 GHz fields. As shown by Fig. 10.21, the circularly polarized microwave ionization threshold field is very nearly $E = 1/16n^4$, the same as the the static field required to ionize a Rydberg Na atom and much higher than the field required for ionization by

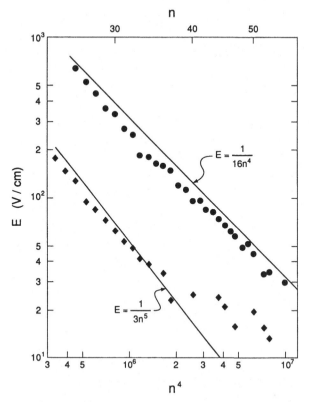

Fig. 10.21 Ionization threshold fields for linear ($\blacklozenge$) and circular ($\bullet$) polarization as a function of n when Na atoms are excited in the 8.5 GHz microwave field (from ref. 36).

linearly polarized microwaves, $E = 1/3n^5$. Furthermore there is a sharp dependence of the observed ionization signal on the relative phase between the two polarizations in the cavity. When E is 20% below $1/16n^4$, so that ionization by a circularly polarized field does not occur, a 10° phase deviation from the 90° phase shift required to produce circular polarization leads to nearly complete ionization. At the same microwave power, a small ellipticity in the polarization produces ionization, while pure circular polarization does not.

If we transform the problem to a frame rotating with the microwave field, it is static and cannot induce transitions. The transformation to the rotating frame, often used to describe two level magnetic resonance experiments, is discussed by Salwen[37] and Rabi *et al.*[38].

If we have a microwave field given by

$$\mathbf{E} = \hat{\mathbf{x}}E \cos \omega t + \hat{\mathbf{y}}E \sin \omega t, \tag{10.35}$$

and we transform the problem to a frame rotating about the z axis at angular frequency ω, the field becomes a static field in the x direction in the rotating frame. If we use as eigenstates in the laboratory frame the usual spherical $n\ell m$ states, a rotation through angle ϕ about the z axis transforms the wave function $\psi_{n\ell m}$ to $e^{im\phi}\psi_{n\ell m}$. For uniform rotation $\phi = \omega t$ and $\psi_{n\ell m} \rightarrow e^{im\omega t}\psi_{n\ell m}$, i.e., the energy is shifted by $-m\omega$. Transforming the Na $n\ell m$ wave functions to the rotating frame only adds $-m\omega$ to their energies. In principle, it is then a simple matter to diagonalize the new Hamiltonian matrix containing a field in the x direction to find the energy levels in the rotating frame. In practice, the fact that there are $n^2/2$ coupled levels for each principal quantum number complicates the problem. To develop an understanding Fu *et al.* diagonalized the Na Hamiltonian matrix, for $n = 5, 6$, and 7; in a frame rotating at 1500 GHz, 2% of the $n = 4$–5 interval, to correspond to the fact that 8.5 GHz is 2% of the $n = 25$ to 26 interval. When the Stark manifolds of adjacent n overlap there are avoided crossings, just as in a static field. In the limit of very low frequency the Stark manifolds overlap at $E = 1/3n^5$, but the overlap occurs at lower fields as the frequency is raised.

In analogous calculations for H the levels cross, as in a static field. Thus the avoided crossings in Na are due to the core coupling not the transformation to the rotating frame. In static fields $1/3n^5 < E < 1/16n^4$ the core coupling is manifested in avoided level crossings, and in fields $E > 1/16n^4$ the same coupling between discrete states and the underlying Stark continuum of ionized red Stark states leads to ionization.[39] In other words field ionization of Na at $E > 1/16n^4$ is really autoionization, and in the rotating frame the situation is presumably the same. There are avoided crossings for $E < 1/16n^4$, presumably due to the $m = 0$ and ± 1 parts of the wavefunction, and ionization when $E > 1/16n^4$. In other words, ionization by a circularly polarized microwave field is field ionization in the rotating frame.

The pronounced effect of ellipticity in the polarization can also be understood with the rotating frame description. When the polarization is elliptical, in the rotating frame there is a field oscillating at 2ω superimposed on the static field. If the static field in the rotating frame exceeds $1/3n^5$, the atom is in the field regime in which there are many level crossings, and even a very small additional oscillating field can drive transitions to higher lying states via these level crossings, leading ultimately to ionization.

Nauenberg[40] has approached the problem of ionization by a circularly polarized field classically, by transforming the problem to a rotating frame. In the rotating frame the potential is depressed by the addition of a term $-\omega^2(x^2 + y^2)/2$. At low n the ionization field is given by $E = 1/16n^4$, but at high n the required ionization field falls below $1/16n^4$. As an example, at 8.5 GHz for $n = 30$, $E = 1/16n^4$, but for $n = 50$ the classically calculated field is about a factor of five below $E = 1/16n^4$. As shown by Fig. 10.21, the experimental results exhibit a $1/16n^4$ dependence up to $n = 60$. While ionization of $n = 50$ atoms is energetically allowed at a much lower field, it does not happen. A possible impediment is an angular momentum constraint.[41] At the threshold for ionization the electron barely escapes over the saddle point in the rotating frame. In the laboratory frame the same electron has a large angular momentum. For example for $n = 50$ it has an angular momentum of

$60\hbar$. In quantum mechanical terms acquiring this much angular momentum requires ten $\Delta n = 1$ transitions, which is unlikely.

References

1. J. E. Bayfield and P. M. Koch, *Phys. Rev. Lett.* **33**, 258, (1974).
2. P. Pillet, W. W. Smith, R. Kachru, N. H. Tran, and T. F. Gallagher, *Phys. Rev. Lett.* **50**, 1042 (1988).
3. D. R. Mariani, W. van de Water, P. M. Koch, and T. Bergeman, *Phys. Rev. Lett.* **50**, 1261 (1983).
4. P. Pillet, H. B. van Linden van den Heuvell, W. W. Smith, R. Kachru, N. H. Tran, and T. F. Gallagher, *Phys. Rev. A* **30**, 280 (1984).
5. C. R. Mahon, J. L. Dexter, P. Pillet, and T. F. Gallagher, *Phys. Rev. A* **44**, 1859 (1991).
6. U. Eichmann, J. L. Dexter, E. Y. Xu, and T. F. Gallagher, *Z. Phys. D* **11**, 187 (1989).
7. J. Rubbmark, M. M. Kash, M. G. Littman, and D. Kleppner, *Phys. Rev.* **23**, 3107 (1981).
8. L. A. Bloomfield, R. C. Stoneman, and T. F. Gallagher, *Phys. Rev. Lett* **57**, 2512 (1986).
9. R. C. Stoneman, G. R. Janik, and T. F. Gallagher, *Phys. Rev. A* **34**, 2952 (1986).
10. R. C. Stoneman, D. S. Thomson, and T. F. Gallagher, *Phys. Rev. A* **37**, 1527 (1988).
11. W. van de Water, S. Yoakum, T. van Leeuwen, B. E. Sauer, L. Moorman, E. J. Galvez, D. R. Mariani and P. M. Koch, *Phys. Rev. A* **42**, 872 (1990).
12. T. F. Gallagher, in *Atoms in Intense Laser Fields*, ed. M. Gavrila (Academic Press, Cambridge, 1992).
13. S. H. Autler, and C. H. Townes, *Phys. Rev.* **100**, 703 (1955).
14. J. Shirley, *Phys. Rev.* **138**, B979 (9165).
15. H. Sambe, *Phys. Rev. A* **7**, 2203 (1973).
16. B. Christiansen-Dalsgaard, unpublished, (1990).
17. M. Abramowitz and I. A. Stegun, *Handbook of Mathematical Functions*, (U.S. GPO, Washington, DC., (1964).
18. H. B. van Linden van den Heuvell, R. Kachru, N. H. Tran, and T. F. Gallagher, *Phys. Rev. Lett.* **53**, 1901 (1984).
19. P. Pillet, C. R. Mahon, and T. F. Gallagher, *Phys. Rev. Lett.* **60**, 21 (1988).
20. G. A. Ruff and K. M. Dietrick (unpublished).
21. K. A. H. van Leeuwen, G. V. Oppen, S. Renwick, J. B. Bowlin, P. M. Koch, R. V. Jensen, O. Rath, D. Richards, and J. G. Leopold, *Phys. Rev. Lett.* **55**, 2231 (1985).
22. T. F. Gallagher, C. R. Mahon, P. Pillet, P. Fu and J. B. Newman, *Phys. Rev. A* **39**, 4545 (1989).
23. R. V. Jensen, S. M. Susskind, and M. M. Sanders, *Phys. Rept.* **201**, 1 (1991).
24. B. Christiansen-Dalsgaard (unpublished).
25. H. A. Bethe and E. A. Salpeter, *Quantum Mechanics of One and Two Electron Atoms* (Academic Press, New York, 1957).
26. J. N. Bardsley, B. Sundaram, L. A. Pinnaduwage, and J. E. Bayfield, *Phys. Rev. Lett.* **56**, 1007 (1986).
27. M. A. Kmetic and W. J. Meath, *Phys. Lett.* **108A**, 340 (1985).
28. D. Richards, J. G. Leopold, P. M. Koch, E. J. Galvez, K. A. H. van Leeuwen, L. Moorman, B. E. Sauer and R. V. Jensen, *J. Phys. B.* **22**, 1307 (1989).
29. J. E. Bayfield and L. A. Pinnaduwage, *Phys. Rev. Lett.* **54**, 313 (1985).
30. E.J. Galvez, B. E. Sauer, L. Moorman, P. M. Koch, and D. Richards, *Phys. Rev. Lett.* **61**, 2011 (1988).
31. R. V. Jensen, S. M. Susskind, and M. M. Sanders, *Phys. Rev. Lett.* **62**, 1476 (1989).
32. H. P. Breuer, K. Dietz, and M. Holthaus, *Z. Phys D* **10**, 13 (1988).
33. M. C. Baruch and T. F. Gallagher, *Phys. Rev. Lett.* **68**, 3515 (1992).
34. S. Yoakum, L. Sirko, and P. M. Koch, *Bull. Am. Phys. Soc.* **37**, 1105 (1992).

35. R. Blumel, R. Graham, L. Sirko, U. Smilansky, H. Walther, and K. Yamada, *Phys. Rev. Lett.* **62**, 341 (1987).
36. P. Fu, T. J. Scholz, J. M. Hettema, and T. F. Gallagher, *Phys. Rev. Lett.* **64**, 511 (1990).
37. H. Salwen, *Phys. Rev.* **99**, 1274 (1955).
38. I. I. Rabi, N. F. Ramsey, and J. Schwinger, *Rev. Mod. Phys.* **26**, 167 (1955).
39. M. G. Littman, M. M. Kash, and D. Kleppner, *Phys. Rev. Lett.* **41**, 103 (1989).
40. M. Nauenberg, *Phys. Rev. Lett.* **64**, 2731 (1990).
41. T. F. Gallagher, *Mod. Phys. Lett. B* **5**, 239 (1991).

11

Collisions with neutral atoms and molecules

The large size and low binding energies, scaling as n^4 and n^{-2}, of Rydberg atoms make them nearly irresistible subjects for collision experiments. While one might expect collision cross sections to be enormous, by and large they are not. In fact, Rydberg atoms are quite transparent to most collision partners.

Collisions involving Rydberg atoms can be broken into two general categories, collisions in which the collision partner, or perturber, interacts with the Rydberg atom as a whole, and those in which the perturber interacts separately with the ionic core and the Rydberg electron. The difference between these two categories is in essence a question of the range of the interaction between the perturber and the Rydberg atom relative to the size of the Rydberg atom. A few examples serve to clarify this point. A Rydberg atom interacting with a charged particle is a charge–dipole interaction with a $1/R^2$ interaction potential, and the resonant dipole–dipole interaction between two Rydberg atoms has a $1/R^3$ interaction potential. Here R is the internuclear separation of the Rydberg atom and the perturber. In both of these interactions the perturber interacts with the Rydberg atom as a whole. On the other hand when a Rydberg atom interacts with a N_2 molecule the longest range atom–molecule interaction is a dipole–induced dipole interaction with a potential varying as $1/R^6$. This interaction is vanishingly small and negligible for R greater than the orbital radius of the Rydberg atom. As a consequence, only when the N_2 molecule actually penetrates the Rydberg electron's orbit is there any appreciable interaction. Once the N_2 molecule is inside the Rydberg electron's orbit it can interact separately with the charged ionic core and the Rydberg electron. In this chapter the focus is primarily on collisions of this type. The discussion of collisions due to long range interactions is deferred to later chapters.

This chapter begins with a brief summary of the basic physical notions behind short range Rydberg atom–neutral scattering, followed by an outline of the theoretical connection between Rydberg atom scattering and free electron scattering.[1-5] Commonly used experimental techniques are described, and a summary of the experimental results is presented. The closely related line broadening and shift measurements are described in the chapter immediately following this one.

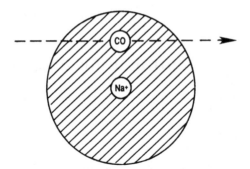

Fig. 11.1 A CO molecule colliding with a Na Rydberg atom which is composed of the Na^+ core and the diffuse electron cloud indicated by the shaded area (from ref. 6).

A physical picture

An illustrative example is the collision of a Na Rydberg atom with CO, and in Fig. 11.1 we show a picture of the CO passing through the Na electron cloud.[6] There are three interactions:

$$e^- - Na^+ \qquad e^- - CO \qquad Na^+ - CO \qquad (11.1)$$

The $e^- - Na^+$ interaction leads to the energy levels of the Na atom. As long as the characteristic $e^- - CO$ and $Na^+ - CO$ interaction lengths are small compared to the size of the Rydberg atom, we can describe collision processes as being due to the sum of the $e^- - CO$ interaction and the $Na^+ - CO$ interaction, ignoring any correlation between the two. This approximation is excellent for high n states, but is unlikely to be good for $n = 5$ states.

How does the electron, for example, interact with the CO? It is useful to introduce the distance ρ between the Rydberg electron and the CO. First, there is the short range interaction characterized by the elastic scattering length a and a $\delta(\rho)$ dependence.[1] Second, the electron interacts with the electric dipole and quadrupole moments of the polar CO molecule, yielding ρ^{-2} and ρ^{-3} dependences for the interactions. Finally, the electron interacts with induced multipole moments of the CO. The largest of these is the electron–induced dipole interaction, which scales as ρ^{-4}. The interactions of the Na^+ with the CO are analogous except for the short range scattering length interaction. All the above mentioned interactions are of short range, compared to the size of a $n = 20$ Rydberg atom, so for them to play an important role, the CO must come near the electron or the Na^+. Since the Na^+ is localized at a point, while the electron can be found anywhere in the electron cloud of Fig. 11.1, the $e^- - CO$ interaction is usually, but not always, the dominant interaction.

Above we have considered collisions with CO. If we now consider collisions with N_2, for example, the electron–dipole interaction is absent, and the longest range e^-—N_2 interaction is the electron–quadrupole interaction. Since this interaction is of shorter range than the electron–dipole interaction in CO, it is not unreasonable to expect to see a difference between Rydberg atom collisions with CO and N_2. If we substitute a rare gas atom, or probably any atom, for the CO, both the electron–dipole and electron–quadrupole interactions are absent, leaving only the polarization interaction and the delta function interaction.

If we assume that the electron–perturber interaction is dominant, what can we expect to observe in collisions between Rydberg atoms and different collision partners? A ground state rare gas atom has no energetically accessible states, and as a result collisions between the electron and the rare gas atom are elastic collisions, involving only the exchange of translational energy. Since an electron scattering from an atom is roughly comparable to a ping pong ball scattering from a bowling ball, very little energy is exchanged. Specifically, the typical energy exchanged, ΔW, is given by

$$\Delta W = 4kTv/V \tag{11.2}$$

where k is Boltzman's constant, T is the temperature, v is the velocity of the Rydberg electron, and V is the relative velocity of the rare gas atom and the Rydberg atom. Using $v = 1/n$, for collisions between $n = 20$ Rydberg atoms and He at 300 K the expression of Eq. (11.2) gives $\Delta W = 3$ cm^{-1}, which is small compared to $kT = 200$ cm^{-1}. We may thus expect that collision processes in which the Rydberg atom changes energy by more than $\Delta W = kT/100$ to be unlikely. While this notion is a good general rule, it is important to bear in mind that in obtaining this value of ΔW we have used a typical value of $v \sim 1/n$. However, the Rydberg electron's velocity is larger near the ionic core, so collision processes requiring a larger change in the energy of the Rydberg state are possible, albeit with smaller cross sections.

In collisions with molecules the fact that there are energetically accessible vibrational and rotational states alters the picture significantly. An electron scattering from a molecule can induce rotational transitions in the molecule by means of dipole or quadrupole transitions in the molecule. The dipole transitions are of course stronger, but only occur in polar molecules, such as CO or NH_3. In such electron induced transitions the energy of the molecule must change by the amount separating the initial and final rotational states, and as a result the Rydberg electron's energy must change by the same amount. In other words this process is necessarily resonant, although the resonant behavior may not be readily apparent if there are many rotational states populated. Vibrational excitation is also possible, and molecular vibrational transitions, which are easily induced by electron impact, play a role in the depopulation of low lying Rydberg states. In sum, the presence of the rotational and vibrational states of molecules allows inelastic collisions of the Rydberg electrons with the molecule, and this new channel vastly increases the chance of substantially changing the energy of a Rydberg state in a collision.

Theory

As we have already noted, many Rydberg atom collision processes are due to the interaction of primarily the Rydberg electron with the perturber, and much theoretical work, beginning with that of Fermi,[1] has been devoted to connecting electron scattering to Rydberg atom scattering. For example, detailed reviews have been given by Omont,[2] Matsuzawa,[3] Hickman *et al.*[4] and Flannery.[5] While it is not possible to give an exhaustive treatment of the theory here, it is useful to present a brief outline which shows the connection between electron scattering and Rydberg atom scattering.

We are interested in collisions for which the range of the interaction between the Rydberg electron and the perturber is small compared to $n^2 a_0$, the size of the Rydberg atom, for we are assuming that the Rydberg electron and ion core scatter independently from the perturber. Furthermore we shall in many cases ignore the interaction of the Rydberg ion core with the perturber. For low energy electron scattering from rare gas atoms the scattering length gives a reasonable estimate of the range of the electron–rare gas interaction. Since typical scattering lengths are $\sim 3a_0$,[7] it is evident that for any Rydberg state of $n > 2$ this criterion is met. On the other hand, electron–polar molecule scattering is characterized by cross sections of 10^3 Å^2,[8] so interaction lengths of $\sim 30a_0$ are not uncommon. Thus for polar molecules only for $n > 5$ can we ignore the interaction of the perturber with the Rydberg ion core.

Consider very fast collisions, in which the collision velocity V between the Rydberg atom and the perturber is much larger than the Rydberg atom's orbital velocity, v, explicitly, $V \gg v$. For collisions with 1 keV H atoms this criterion is met for $n \gg 10$. In such a collision the orbital velocity of the Rydberg electron is negligible compared to V. In effect, scattering from a Rydberg atom with velocity V is identical to the scattering from a nearly static electron cloud plus the scattering from the Rydberg ion core. Explicitly, for fast collisions,[3,9,10]

$$\sigma_{\text{Ryd–perturber}} = \sigma_{e^- \text{-perturber}} + \sigma_{\text{ion–perturber}}. \tag{11.3}$$

If we focus on the other extreme, collisions at thermal energies, the relative velocity of the colliding atoms is $\sim 10^{-3}$, and the typical velocity of the Rydberg electron is $v = 1/n$. Therefore for $n \ll 1000$ the Rydberg electron's velocity substantially exceeds the velocity of the perturber, and a useful way of approximating the collisions is as electron scattering from static perturbers. This approach is valid when both the requirements of the range being smaller than the orbital radius and $v \gg V$ are met. Even for the relatively restrictive case of polar molecules they are met for $10 < n < 1000$.

Consider a perturber which collides with a Rydberg atom as shown in Fig. 11.2(a). Before the collision the perturber is in state β and has momentum **K**. After the collision the perturber is in state β' and has momentum **K**'. The

(a)

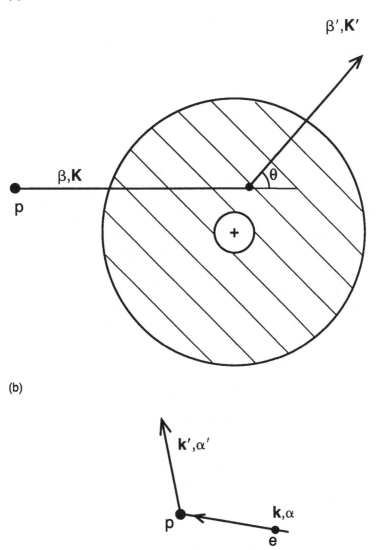

(b)

Fig. 11.2 (a) Schematic diagram of a perturber p initially in state β and having momentum **K** scattering from a Rydberg atom to produce final state β' and momentum **K'**. (b) Magnified view of the Rydberg electron colliding with the perturber. The Rydberg electron is initially in state α and has momentum **k**. After the collision these are α' and **k'**.

perturber moves slowly through the Rydberg atom until it is hit by the more rapidly moving Rydberg electron, which causes an abrupt change in the momentum and state of the perturber to **K'** and β' respectively.

In Fig. 11.2(b) we show the trajectory of the Rydberg electron just before and after hitting the perturber. Prior to the collision the Rydberg atom is in state α, and the electron has momentum **k**; after the collision these are α' and **k'**, respectively. In Fig. 11.2(a) we have drawn the scattering angle θ as a fairly large angle, for clarity, but, since the Rydberg electron cannot exchange much momentum with the perturber, θ is in reality quite small. Accordingly, we begin by treating the problem using the Born approximation. If we define the momentum transfer $\mathbf{Q} = \mathbf{K} - \mathbf{K}'$, the Born scattering amplitude for the collision of Fig. 11.2 is given by[3,11]

$$f(\alpha, \beta, \mathbf{K} \to \alpha', \beta', \mathbf{K}') = \frac{-\mu}{2\pi} \int e^{i\mathbf{Q} \cdot \mathbf{R}} \, \psi^*_{n'\ell'm'}(\mathbf{r}) U(\beta,\beta',\mathbf{r},\mathbf{R}) \psi_{n\ell m}(\mathbf{r}) \, d^3r \, d^3R,$$

(11.4)

where U is the interaction potential for initial and final perturber states β and β', **r** and **R** are the positions of the Rydberg electron and the perturber relative to the Rydberg ion core, and μ is the reduced mass of the Rydberg atom and the perturber. In writing Eq. (11.4) we have chosen the usual spherical $n \, \ell \, m$ states for the α Rydberg states. Throughout this section we shall assume that the α states are spherical states, but this is not the only possible choice. For example, for hydrogen in a field the parabolic states would be a better choice.[12] Since we are considering the electron–perturber interaction, in general $U(\beta, \beta', \mathbf{r}, \mathbf{R})$ must be a function of $\mathbf{r} - \mathbf{R}$, so we introduce $\boldsymbol{\rho} = \mathbf{r} - \mathbf{R}$ and rewrite Eq. (11.4) substituting $\boldsymbol{\rho} + \mathbf{r}$ for **R**. This procedure yields

$$f(\alpha,\beta,\mathbf{K} \to \alpha,\beta,\mathbf{K}') = \frac{-\mu}{2\pi} \int e^{i\mathbf{Q} \cdot \mathbf{r}} \psi^*_{n'\ell'm'}(\mathbf{r}) U(\beta,\beta',\boldsymbol{\rho}) \, e^{-i\mathbf{Q} \cdot \boldsymbol{\rho}} \psi_{n\ell m}(\mathbf{r}) \, d^3r \, d^3\rho.$$

(11.5)

Examining Fig. 11.2(b), we can see that by conservation of momentum

$$\mathbf{K} + \mathbf{k} = \mathbf{K}' + \mathbf{k}'.$$

(11.6)

Thus $\mathbf{Q} = \mathbf{K} - \mathbf{K}'$ implies that $\mathbf{Q} = \mathbf{k}' - \mathbf{k}$. In other words, **Q** is the momentum transfer to the electron. If we assume that there is no particular orientation of the perturber during the collision we can replace $U(\beta, \beta', \boldsymbol{\rho})$ by the isotropic potential $U(\beta, \beta', \rho)$. With this approximation in Eq. (11.5) we recognize the Born approximation to the electron scattering amplitude

$$\frac{2\pi}{m} f_{eB}(\beta, \beta', \mathbf{Q}) = -\int e^{-i\mathbf{Q} \cdot \boldsymbol{\rho}} U(\beta, \beta', \rho) \, d^3\rho,$$

(11.7)

and the form factor

$$F(\alpha, \alpha', \mathbf{Q}) = \langle \psi_{n'\ell'm} | e^{i\mathbf{Q} \cdot \mathbf{r}} | \psi_{n\ell m} \rangle = \int \psi^*_{n'\ell'm'}(\mathbf{r}) \, e^{i\mathbf{Q} \cdot \mathbf{r}} \psi_{n\ell m}(\mathbf{r}) \, d^3r \quad (11.8)$$

In Eq. (11.7) m is the mass of the electron, i. e. 1 in atomic units. Using Eqs. (11.7) and (11.8) we may rewrite Eq. (11.5) as

$$f(\alpha, \beta, \mathbf{K} \to \alpha', \beta', \mathbf{K}') = \frac{\mu}{m} F(\alpha, \alpha', \mathbf{Q}) f_{eB}(\beta, \beta', \mathbf{Q}).$$

(11.9)

For polar molecule perturbers the Born electron scattering amplitude is quite accurate and Eq. (11.9) is immediately useful. As an example, the squared Born scattering amplitude for $J \to J - 1$ rotational deexcitation of a polar diatomic molecule is given by[3]

$$|f_{eB}(\beta, \beta', \mathbf{Q})|^2 = \frac{4}{3}\left(\frac{D}{ea_0}\right)^2 \frac{J}{(2J+1)Q^2},$$ (11.10)

where $\beta = J, \beta' = J-1$, D is the permanent dipole moment of the molecule, and e is the electron's charge. The Born amplitudes for other polar molecules have similar forms and are given by Matsuzawa.[3] The presence of $1/Q^2$ in Eq. (11.10) reflects the long range of the electron–dipole interaction.

In contrast, the Born amplitude for electron scattering is not always accurate, especially for low energy scattering from rare gases, in which the scattering angles are large. A natural approach is to allow the replacement of Born amplitude of Eq. (11.7) by a more accurate electron scattering amplitude $f_e(\beta\beta'\mathbf{Q})$.[13] In other words, we express the scattering amplitude of Eq. (11.5) as

$$f(\alpha, \beta, \mathbf{K} \to \alpha', \beta', \mathbf{K}') = \frac{\mu}{m} F(\alpha, \alpha', \mathbf{Q}) f_e(\beta, \beta, '\mathbf{Q}).$$ (11.11)

Including the electron mass explicitly in Eq. (11.7) yields the ratio μ/m in Eq. (11.11).

The electron–rare gas scattering amplitude is given by[6]

$$f_e(\beta, \beta, \mathbf{Q}) = \frac{e^{i\eta(k)} \sin \eta(k)}{k},$$ (11.12)

where η is the s wave phase shift, which is related to the scattering length a by[6]

$$\tan \eta(k) = -ak.$$ (11.13)

If we assume that η is small, $f_e(\beta, \beta, \mathbf{Q}) \cong -a$, and the total electron–perturber elastic scattering cross section is given by

$$\sigma_e(Q) = 2\pi \int_0^\pi |f_e(\beta, \beta, \mathbf{Q})|^2 \sin \theta \, d\theta$$ (11.14)

$$= 4\pi a^2.$$

For a rare gas atom it is useful to write out the coupling between states α and α' in a fashion analogous to the scattering amplitude of Eq. (11.4). Explicitly,

$$\langle \alpha' | U(\beta, \beta, \mathbf{r}, \mathbf{R}) | \alpha \rangle = \int \psi^*_{n'\ell'm'}(\mathbf{r}) U(\beta, \beta, \mathbf{r}, \mathbf{R}) \psi_{n\ell m}(\mathbf{r}) \, d^3 r.$$ (11.15)

If we substitute the Fourier transforms of the spatial wavefunctions and use the fact that $U(\beta, \beta, \mathbf{r}, \mathbf{R}) = U(\beta, \beta, \rho)$, we can rewrite Eq. (11.15) as

$$\langle \alpha' | U | \alpha \rangle = \int e^{-i\mathbf{k}' \cdot \mathbf{r}} G^*_{\alpha'}(\mathbf{k}') U(\beta, \beta, \rho) G_\alpha(\mathbf{k}) e^{i\mathbf{k} \cdot \mathbf{r}} \, d^3 r \, d^3 k \, d^3 k'$$ (11.16)

substituting $\mathbf{R} + \rho$ for $\mathbf{r}$ yields

$$\langle \alpha' | U | \alpha \rangle = \int e^{-i(\mathbf{k}' - \mathbf{k}) \cdot \mathbf{R}} G^*_{\alpha'}(\mathbf{k}') G_\alpha(\mathbf{k}) e^{i(\mathbf{k}' - \mathbf{k}) \cdot \rho} U(\beta, \beta, \rho) \, d^3\rho \, d^3 k \, d^3 k'.$$ (11.17)

The integral over ρ is the Born amplitude for electron scattering multiplied by -2π. Using $f_e(\beta, \beta, Q) \cong -a$, replacing the integral over ρ of Eq. (11.17) by $2\pi a$, and integrating over $\mathbf{k}$ and $\mathbf{k}'$ yields

$$\langle a'|U|a\rangle = \langle \psi_{n'\ell'm'}|U|\psi_{n\ell m}\rangle = 2\pi a\, \psi^*_{n'\ell'm'}(\mathbf{R})\psi_{n\ell m}(\mathbf{R}) \qquad (11.18)$$

Inspecting Eq. (11.18), we can see that we can equivalently express $U(\beta,\beta, \mathbf{r}, \mathbf{R})$ as

$$U(\beta,\beta,\mathbf{r}, \mathbf{R}) = 2\pi a\delta(\mathbf{r} - \mathbf{R}). \qquad (11.19)$$

The form factor of Eq. (11.8) is an integral over three factors, two wavefunctions and an oscillatory plane wave term, $e^{i\mathbf{Q}\cdot\mathbf{r}}$. Due to the plane wave term the magnitude of the form factor decreases with increasing Q, and it may be evaluated using either the methods of Gounand and Petitjean[14] or Cheng and van Regemorter.[15] In an alternative approach[16] used by Hickman,[17] the form factor is replaced by a Bessel function expansion.

To compute the total cross section we integrate the squared scattering amplitude over the scattering angle θ. Explicitly,

$$\sigma(a, \beta \rightarrow a', \beta') = 2\pi \int |f(a, \beta, \mathbf{K} \rightarrow a', \beta', \mathbf{K}')|^2 \sin\theta\, d\theta. \qquad (11.20)$$

Using the fact that

$$Q^2 = K^2 + K'^2 - 2KK' \cos\theta, \qquad (11.21)$$

we can replace the integral over the scattering angle θ by an integral over Q, i.e.[3,11]

$$\sigma(a, \beta \rightarrow a', \beta') = \frac{2\pi}{KK'} \int_{Q_{min}}^{Q_{max}} |F(a, a',Q)|^2 f_e(\beta\beta'Q)|^2 Q\, dQ, \qquad (11.22)$$

where $Q_{min} = |\mathbf{K} - \mathbf{K}'|$ and $Q_{max} = |\mathbf{K} + \mathbf{K}'|$. Since almost all the scattering is near the forward direction Q_{max} can be replaced by ∞ with minimal error. Energy conservation determines Q_{min}. Explicitly,[3,11]

$$\frac{K^2}{2\mu} + W_a + W_\beta = \frac{K'^2}{2\mu} + W'_a + W'_\beta, \qquad (11.23)$$

where W_a, $W_{a'}$, W_β, and $W_{\beta'}$ are the internal energies of the Rydberg atom and the perturber before and after the collision. Solving Eq. (11.23) for $|\mathbf{K} - \mathbf{K}'|$ in the approximation $K \approx K'$ yields

$$Q_{min} = |\mathbf{K} - \mathbf{K}'| = \frac{\mu}{K}|(W_a - W_{a'}) - (W_\beta - W_{\beta'})|. \qquad (11.24)$$

There are two important points to note regarding Q_{min}, and Eq. (11.24). First, the right hand side contains the difference between the internal energy lost by the Rydberg atom and that gained by the perturber. If this difference is zero, so that the collision is resonant and no energy goes into translation, $Q_{min} = 0$. Second, for an appreciable cross section Q_{min} should be small, since the form factor decreases with Q, and in the case of polar molecule perturbers the scattering amplitude, f_{eB}, does as well.

Consider two examples. For a rare gas perturber there can be no change in the state of the perturber, so $\beta = \beta'$ and Q_{min} is determined entirely by the energy difference, $W_\alpha - W_{\alpha'}$, between the initial and final Rydberg states. On the other hand, if the perturber is a diatomic molecule which is rotationally deexcited in the collision so that $W_{\beta'} < W_\beta$ and the Rydberg atom's energy gain equals the molecule's energy loss, i.e. $W_{\alpha'} - W_\alpha = W_\beta - W_{\beta'}$, $Q_{min} = 0$ in spite of the fact that $W_\alpha \neq W_{\alpha'}$.

The cross section of Eq. (11.22) is for a single initial and final state. Assuming that we have no control over the direction of the collision velocities, to compute a cross section we must, at a minimum, average over m of the initial state and sum over m' of the final state. This procedure leads to

$$\sigma(n, \ell, \beta \to n', \ell', \beta')$$

$$= \frac{2\pi}{KK'} \int_{Q_{min}}^{\infty} \frac{1}{(2\ell + 1)} \sum_{mm'} |F(n\ell m, n'\ell'm', \mathbf{Q})|^2 |f_e(\beta, \beta', \mathbf{Q})|^2 Q \, dQ \tag{11.25}$$

Eq. (11.25) is in some ways the minimally useful cross section, from one $n\ell$ state to another $n'\ell'$ state. If we are interested in the total depopulation of the $n\ell$ state we must sum over the possible $n'\ell'$ final states. The summation over $n'\ell'$ includes, implicitly, an integration over the continuum, although including the continuum is usually unnecessary. In the continuum $|F(n\ell m, n'\ell'm', \mathbf{Q})|^2 n'^{-3}$ is replaced by $d|F(n\ell m, W'\ell'm', \mathbf{Q})|^2/dW'$, where

$$\frac{d|F(n\ell m, W'\ell'm', \mathbf{Q})|^2}{dW'} = \lim_{n' \to \infty} \frac{|F(n\ell m, n'\ell'm', \mathbf{Q})|^2}{n'^3}. \tag{11.26}$$

The squared form factor is similar to the oscillator strength in that it passes smoothly across the ionization limit. If we are interested in ionization we simply sum over ℓ' and integrate over the continuum using Eq. (11.26).

The approach outlined above has been used extensively by Matsuzawa[3] to calculate cross sections for excitation and ionization of Rydberg atoms by polar molecules and by Hickman[4] to calculate cross sections for state changing by rare gas atoms colliding with Rydberg atoms.

Another approach to the problem of rare gas scattering is to replace the spatial wavefunctions of Eq. (11.4) with their Fourier transforms, the momentum space wavefunctions. These wavefunctions represent the velocity distributions of the electron in the Rydberg states. Proceeding along these lines, we rewrite Eq. (11.4) as[11]

$$f(\alpha, \beta, \mathbf{K} \to \alpha', \beta', \mathbf{K}') = \frac{-\mu}{2\pi} \int e^{i\mathbf{Q} \cdot \mathbf{R}} G_{\alpha'}^*(\mathbf{k}')$$

$$\times e^{i(\mathbf{k} - \mathbf{k}') \cdot \mathbf{r}} U(\beta, \beta', \mathbf{r}, \mathbf{R}) G_\alpha(\mathbf{k}) \, d^3k \, d^3k' \, d^3r \, d^3R \tag{11.27}$$

Replacing **r** by $\boldsymbol{\rho}$ + **R**, assuming $U(\beta, \beta', \mathbf{r}, \mathbf{R}) = U(\beta, \beta', \boldsymbol{\rho})$, and integrating over **R** yields

$$f(\alpha, \beta, \mathbf{K} \rightarrow \alpha', \beta', \mathbf{K}') = \frac{-\mu}{2\pi} \int \delta(\mathbf{k}' - \mathbf{k}) G_{\alpha'}^*(\mathbf{k}') G_\alpha(\mathbf{k})$$

$$\times e^{-i\mathbf{k}' \cdot \boldsymbol{\rho}} U(\beta, \beta', \boldsymbol{\rho}) e^{i\mathbf{k} \cdot \boldsymbol{\rho}} d^3\rho \, d^3k \, d^3k' \quad (11.28)$$

In addition to showing the fact that the momentum transferred to the collision partner comes from the electron, this expression also contains, in the integral over ρ, the Born scattering amplitude, given by Eq. (11.7). If we again replace it by the correct amplitude $f_e(\beta, \beta', \mathbf{k}' - \mathbf{k})^{12}$ and integrate over $\mathbf{k}'$ we find

$$f(\alpha, \beta, \mathbf{K} \rightarrow \alpha', \beta', \mathbf{K}') = \frac{\mu}{m} \int G_{\alpha'}^*(\mathbf{k} + \mathbf{Q}) G_\alpha(\mathbf{k}) f_e(\beta, \beta', \mathbf{Q}) \, d^3k. \tag{11.29}$$

This expression for the scattering amplitude, given in terms of the momentum, or velocity, distribution of the Rydberg electrons, is usually termed the impulse approximation. Examination of Eq. (11.29) shows the similarity of the integral over the momentum to the form factor of Eq. (11.8).

It is useful to consider rare gas scattering, for which $\alpha = \alpha'$ and $\beta = \beta'$. Using the optical theorem we can relate the imaginary part of the forward scattering amplitude to the total cross section.[18] Explicitly, we can apply the optical theorem to the scattering amplitudes on both the left hand and right hand sides of Eq. (11.29). This procedure yields

$$K \frac{\sigma_T}{4\pi}(\alpha, \beta, K) = \frac{\mu}{m} \int |G_\alpha(\mathbf{k})|^2 \frac{k}{4\pi} \sigma_{eT}(\beta, k) \, d^3k. \tag{11.30}$$

Using $K/\mu = V$ and $k/m = v$, Eq. (11.30) reduces to

$$V \sigma_T(\beta, K) = \int |G_\alpha(\mathbf{k})|^2 v \sigma_{eT}(\beta, k) \, d^3k. \tag{11.31}$$

The physical significance of this expression is that the rate constant for elastic perturber–Rydberg atom collisions is the same as the rate constant for elastic perturber–electron collisions averaged over the Rydberg electron's velocity distribution, in spite of the fact that $V \ll$ v.

Eq. (11.30) is strictly applicable only to elastic collisions, in which $\alpha' = \alpha$, and is thus of limited utility. However, it is physically appealing to assume that the cross section $\sigma(\alpha, \alpha', \beta, \beta')$ for an inelastic process $\alpha' \neq \alpha$ and $\beta' \neq \beta$ can be written as the integral of the electron scattering cross section $\sigma_e(\beta, \beta', q)$ over the velocity distribution of the Rydberg electron in the initial state α. Making this notion explicit, we write[19]

$$V \sigma_T(\alpha, \alpha', \beta, \beta',) = \int_{k_{min}} |G_\alpha(\mathbf{k})|^2 v \sigma(\beta, \beta', k) \, d^3k, \tag{11.32}$$

where k_{min} accounts for the fact that not all momenta present in the initial state wavefunction contribute to the scattering. Explicitly,

$$k_{\min} = \begin{cases} \dfrac{2m}{\hbar^2}(W_\beta{}' - W_\beta) & \text{if } W_\beta{}' > W_\beta \\ 0 & \text{if } W_\beta{}' < W_\beta \end{cases} \tag{11.33}$$

As is the case for Eq. (11.31), Eq. (11.32) states explicitly that the rate constant for Rydberg atom scattering is the same as the rate constant for free electron scattering averaged over the Rydberg electron velocity distribution.

It is instructive to consider applying Eq. (11.32) to processes in which the electron–perturber scattering is short range and long range. When the interaction is short range, as in elastic electron–rare gas scattering, the electron–rare gas scattering cross section σ_{eT}, has no velocity dependence and is simply given by $4\pi a^2$. In this case the cross section for Rydberg atom scattering given by Eq. (11.32) can be written as

$$\sigma_T(\alpha, \alpha', \beta, \beta') \cong 4\pi a^2 \frac{\langle v \rangle}{V}, \tag{11.34}$$

where the expectation value of the electron velocity, $\langle v \rangle$ is given by

$$\langle v \rangle = \int |G_\alpha(\mathbf{k})|^2 v \, d^3k. \tag{11.35}$$

The cross section of Eq. (11.34) scales as $1/n$. In contrast, for a long range interaction, such as electron-polar molecule scattering, the cross section σ_{eT} has a $1/v$ dependence and the electron scattering rate constant $v\sigma_{eT}$ is independent of v, leading to an n independent cross section $\sigma_T(\beta, K)$ and rate constant $V\sigma_T(\beta, K)$. The notion that the Rydberg atom scattering rate constant is equal to the electron scattering rate constant averaged over the Rydberg state velocity distribution has proven to be very useful in, for example, treating collisions with attaching targets.[20,21] It also forms the basis for the method of de Prunele and Pascale for treating ℓ mixing collisions.[22]

Experimental methods

There are three methods which have been used to study collisions of Rydberg atoms with neutrals. They are direct measurement of collisionally induced population changes, line shift and broadening measurements, and photon echo measurements.[23] In this chapter we describe the first of these. The last two are described in the chapter immediately following.

The most commonly used method is the direct measurement of a decay rate by pulsed excitation and time resolved detection. The most straightforward example of this technique is laser induced fluorescence applied to alkali Rydberg atoms. Alkali atoms are typically contained in a glass cell, which also holds a known pressure of perturber gas. The alkali atoms are excited to the Rydberg state at time $t = 0$ and the time resolved fluorescence from the Rydberg atoms is detected at $t > 0$. Assuming that the alkali ground state atoms are at a low enough pressure that they do not depopulate the Rydberg state, the intensity of the fluorescence with no perturbing gas is given by

$$I = I_0 \, e^{-\gamma_0 t}, \tag{11.36}$$

where γ_0 is the inverse of the radiative lifetime at the cell temperature, which usually includes a contribution due to black body radiation, as discussed in Chapter 5. When a perturbing gas is added the fluorescence intensity is given by

$$I = I_0 \, e^{-(\gamma_0 + \gamma_c)t}, \tag{11.37}$$

where γ_c is the rate at which collisions remove atoms from the state excited by the laser. The collision rate depends on the density, or pressure, of the perturbing gas. Explicitly,

$$\gamma_c = nk = n\sigma V, \tag{11.38}$$

where n is the number density of the perturbing gas, σ and k are the cross section and rate constant for depopulation, and V is the average relative velocity of the Rydberg atom and the perturber. In a cell of temperature T, $V = \sqrt{8kT/\pi\mu}$, k is Boltzman's constant and μ is the reduced mass of the collision partners. Since the number density is proportional to the pressure, the collisional contribution to the decay rate increases linearly with pressure, and measuring the decay rate as a function of pressure, as shown in Fig. 11.3, yields the cross section.[24] Such a measurement really yields the rate constant k, but since it is more physically appealing to compare a cross section to the size of an atom we shall most often use cross sections derived using Eq. (11.38). However it should be borne in mind that the cross section so obtained is an average over thermal velocities.

The approach described above is the simplest variant of time resolved fluorescence detection for the measurement of collision processes. If we choose to excite state A and detect the fluorescence from a second state B, the fluorescence intensity has the form

$$I = I_0(e^{-\gamma_s t} - e^{-\gamma_f t}), \tag{11.39}$$

where γ_s and γ_f are the slower and faster of the decay rates of the two states A and B.

Eq. (11.37) is written assuming that after atoms are removed from the state excited by the laser, state A, they never return. In contrast, if the atoms initially excited to state A are collisionally transferred to a longer lived state R and do return to A, the fluorescence from state A exhibits a two exponential decay which is of the form[25]

$$I = I_0(\alpha \, e^{-\gamma_A t} + \beta \, e^{-\gamma_R t}), \tag{11.40}$$

where α and β depend on the decay rates γ_A and γ_R. Whenever γ_A and γ_R are at all close to each other it is difficult to extract good values from even slightly noisy data.

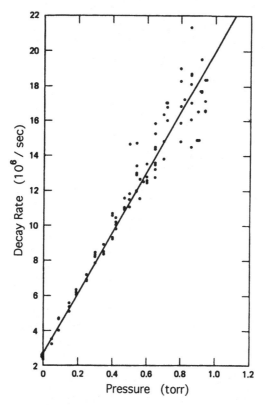

Fig. 11.3 Decay rate of the Na 8s state vs N₂ pressure, obtained by observing the time resolved 8s–3p fluorescence subsequent to pulsed laser excitation (from ref. 24).

Fluorescence detection is simple, and if the detected fluorescence is well resolved in wavelength there is no doubt about which state is being observed. On the other hand fluorescence detection does have its limitations. As n increases, the number of Rydberg atoms excited decreases as n^{-3} as does the spacing of the levels, so that increasingly higher spectral resolution is required. Although the time integrated intensity only declines as n^{-3}, the intensity of the fluorescence signals decreases as n^{-6}, and background noise becomes more of a problem with increasing n. In addition, it is rarely possible to collect even 10% of the fluorescence emitted at any given wavelength. Finally, fluorescence radiated on unobserved transitions goes undetected, so the branching ratio for the transition monitored must be favorable. In light of the inherent difficulties of fluorescence detection at high n, it is not surprising that fluorescence detection has only been used for $n \leq 22$.[26]

The other commonly used technique is selective field ionization. The atoms can be in a beam or in a cell of the type shown in Fig. 11.4, in which the detector is in a separately pumped region connected to the interaction region, which may have a high pressure, $\sim 10^{-3}$ torr, of added gas.[27] The basic principle is to use the known

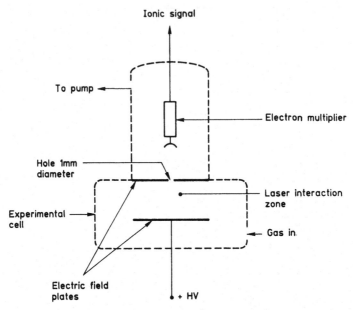

Fig. 11.4 Two chambered cell allowing the use of field ionization to study collision processes. The lower chamber contains a relatively high pressure and the upper chamber, containing the electron multiplier, is at a lower pressure (from ref. 27).

values of the fields at which states ionize to connect, using the principles outlined in Chapter 7, a signal at a given ionizing field to a specific state. If a single state A is excited, and the field ionization ramp is applied at a variable time t later, the Rydberg states present at time t may be determined by simply noting the times, and thus the fields, at which signals are observed. Repeating this procedure on many shots of the laser for different values of t allows one to build up a record of the time resolved populations of the Rydberg states. The attractions of field ionization are two. First, it is efficient. All the Rydberg atoms are ionized, and all the resulting ions or electrons can be detected with high efficiency. Second, all the atoms present are ionized at once, so the signals have an n^{-3} scaling, and background noise is usually not a problem. On the other hand, the successful use of field ionization is always dependent upon correctly connecting the ionization field to a state, as described in Chapter 7.

Collisional angular momentum mixing

One of the more well studied collision processes involving Rydberg atoms is collisional angular momentum mixing, or ℓ mixing, the collisional transfer of population among the nearly degenerate ℓ states of the same n.[28] The process has

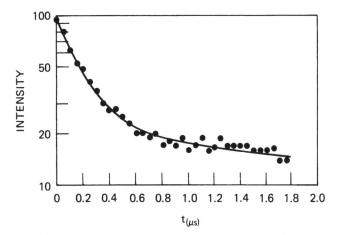

Fig. 11.5 Semilogarithmic plot of the initial portion of observed decay of the Na 10d state in the presence of 0.027 Torr Ne (●). This plot shows the entire fast decay but only the beginning of the slow decay. The solid line is the computer fit of two exponentials to the data. The decay times of the fast and slow component from the computer fit are 0.19 and 3.9 μs, respectively (from ref. 25).

been studied in alkali atoms and Xe, and it is limited to ℓ states with quantum defects near zero. For example, rapid ℓ mixing has been observed from the Li nd,[29] Na nd,[25,30,31] Rb nf,[32] and Xe nf[33,34] states to the higher ℓ states of the same n.

Most of these experiments have been done by laser induced fluorescence techniques. Typical of the observations is the decay of the Na 10d–3p fluorescence in the presence of Ne, shown in Fig 11.5.[25] The decay clearly contains more than one exponential, and, in fact, is fit well by a two exponential decay. The fast initial component is the decay of the initially populated 10d state, and the slow component is the apparently pressure independent decay of the mixture of $n = 10$, $\ell \geq 2$ states. The identification of this process as ℓ mixing was based on two observations. First, the same rare gas pressure had no observable effect on the Na ns states which have a quantum defect of 1.35. Second, measuring the pressure independent slow decay rate, such as the one shown in Fig. 11.5 gives a lifetime $\tau_n \propto n^{4.5(6)}$. This n dependence is in agreement with the expected $n^{4.5}$ scaling of the average lifetime of all ℓ,m states of the same n.[35,36] When the missing ns and np states as well as the difference between the radiative lifetimes of the Na and H nd states are taken into account the observed values of τ_n are in excellent agreement with the calculated values.[25] In retrospect, the agreement is probably somewhat fortuitous. Thermal radiation shortens the lifetime of the mixture of $\ell \geq 2$ states populated by collisions of rare gas atoms with the initially excited nd states. However, the most likely transitions from an $n\ell$ state are to $n\pm1$, $\ell \pm 1$ states. Since the Na nd–3p fluorescence was observed through a filter which would

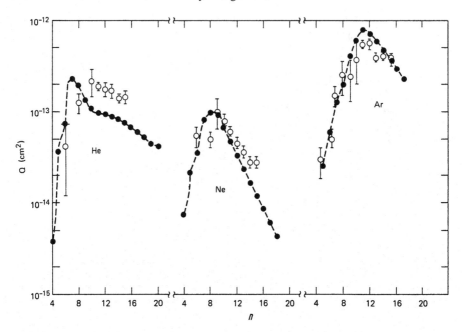

Fig. 11.6 Observed (○) (ref. 25) and calculated (●) (ref. 37) cross sections for the collisional ℓ mixing of the Na nd states by He, Ne, and Ar. The calculated cross sections are based on the two state model of Olson (from ref. 37).

transmit the $(n\pm1)$d–3p fluorescence as well, the effect of black body transitions between ℓ states of similar n was not noticeable.

The ℓ mixing cross sections are obtained by measuring the decay rate of the initial fast decay as a function of perturbing gas pressure. In Fig. 11.6 we show the experimental Na nd ℓ mixing cross sections for the rare gases He, Ne, and Ar,[25] and the cross sections calculated by Olson,[37] using a two state model in which the coupling arises from the scattering length and the polarizability of the rare gas. As shown, the agreement between the calculated and observed cross sections is excellent. The calculated cross sections of Fig. 11.6 are representative of calculations of ℓ mixing carried out by de Prunele and Pascale,[22] Gersten,[38] Derouard and Lombardi,[39] and Hickman[40,41] as well. As shown by Fig. 11.6 the cross sections rise approximately as the geometric cross section, n^4, at low n, reach a peak, and then decline at high n. It is apparent that the location of the peak and how high it becomes depends on the rare gas, and that the peak ℓ mixing cross section increases from Ne to Ar. Not surprisingly, the cross sections for ℓ mixing of the Na nd states by Xe measured by Kachru *et al.*[30] rise to an even higher peak value of 4×10^{-12} cm^2 at $n = 18$.

In classical terms ℓ mixing can be thought of in a simple way. When an Ar atom, for example, enters the Rydberg electron's orbit the electron can scatter elastically from the Ar. The reorientation of the orbit corresponds to a new ℓ. If we pursue this notion a little further we can understand qualitatively the dependence on n and the identity of the rare gas. The coupling between the ℓ states which leads to ℓ mixing is provided predominantly by the short range interaction between the electron and the rare gas atom. Thus, to a good approximation, the coupling is given by Eq. (11.18) i.e.

$$\langle \alpha' | U | \alpha \rangle = 2\pi a \int \psi_{\alpha'}^*(\mathbf{r}) \, \delta(\mathbf{r} - \mathbf{R}) \psi_\alpha(\mathbf{r}) \, \mathrm{d}^3 r, \qquad (11.41)$$

where α and α' are the initial and final states, $\mathbf{r}$ and $\mathbf{R}$ are the locations of the electron and the Ar relative to the Na^+, and a is the Ar scattering length. In other words, the strength of the coupling depends on the rare gas through its scattering length. If the Ar atom passes through any part of a low n Rydberg atom the strength of the coupling is enough that ℓ mixing occurs, and the ℓ mixing cross section equals the geometric cross section of the Rydberg atom. On the other hand, as n increases the coupling decreases because the wave function becomes more dilute, and the mere presence of the Ar atom inside the Rydberg electron's orbit does not ensure an ℓ changing collision. In this regime the ℓ mixing cross section falls farther and farther below the geometric cross section and begins to decline with increasing n. The n at which the maximum cross section occurs increases with the absolute value of the scattering length of the rare gas. For He, Ne, Ar, and Xe the scattering lengths are $1.19a_0$, $0.24a_0$, $-1.70a_0$, and $-6.5a_0$.[6] As expected, the maximum cross section for Ne occurs at the lowest n and the maximum cross section for Xe occurs at the highest n. At high n the ℓ mixing cross section is proportional to the product of the probability of finding the electron at any point in its orbit, $\sim n^{-6}$, and the length of the path of the Ar atom through the Rydberg atom, $\sim n^2$, yielding a cross section which decreases as n^{-4}.

In addition to the measurements of total ℓ mixing cross sections, it has been possible to analyze the final states subsequent to ℓ mixing. As suggested by the simple classical picture and the coupling matrix element of Eq. (11.41), there is no reason to expect any ℓ selection rule for ℓ mixing. Using field ionization Gallagher et al.[42] observed that the ℓ mixing of the Na 15d state by Ar results in an apparently even mixture of $n = 15$ $\ell > 2$ states of $m \leq 2$. The distribution of $|m| > 2$ states was not observed. Since the Na $nd - nf$ interval is roughly five times greater than the $nf - ng$ interval the possibility of a slow, rate limiting $15d \rightarrow 15f$ transfer followed by a rapid $15f \rightarrow 15\ell > 3$ transfer cannot be excluded on the basis of the above experiment alone. However, using resonant microwave fields to form 50–50 mixtures of Na nd and nf states Gallagher et al. observed that the ℓ mixing rate of the $nd - nf$ mixture was nearly unchanged from that of the nd state, indicating that the ℓ mixing cross section of the Na nf states is very nearly the same as that of the Na nd states, in spite of the small $nf-ng$ energy separation.[42] Later measurements by Kachru et al.[30] and Slusher et al.[43] have confirmed the lack of a $\Delta\ell$ selection rule.

The experimental evidence also seems to indicate that there is no strong Δm selection rule, although there may be a slight propensity for Δm to be small. First, the fact that the observed pressure independent lifetime of the mixture of ℓ states scales as $n^{4.5}$ implies that all $|m|$ values are eventually populated. However, several collisions could be required to reach an even distribution over $|m|$. More direct evidence that there is no Δm selection rule is provided by the work of Slusher *et al.*[44] First, they observed that the diabatic field ionization signal subsequent to ℓ mixing of the Xe nf states matched the distribution expected for all $|m|$ states, not simply low $|m|$ states. Second, in another experiment they allowed the Xe 31f state to be ℓ mixed by CO_2 molecules, producing distinct adiabatic $|m| \leq 3$ and diabatic $|m| > 3$ field ionization signals. They then selectively field ionized the low $|m|$ atoms responsible for the adiabatic field ionization signal and observed the rate at which the low $|m|$ states were repopulated from the high $|m|$ states, i.e. the rate of Δm collisions. They found the rate constant $k_{\Delta m} = 2 \times 10^{-7}$ cm^3/s, virtually the same rate as $k_{\Delta \ell} = 1.5 \times 10^{-7}$ cm^3/s.

In the ℓ mixing of the Na $18d_{3/2}$ state by Xe Kachru *et al.*[30] found that when half the atoms were removed from the initially populated 18d state roughly 30% of the atoms removed were in states which ionized at fields lower than the ionization field of the $18d_{3/2}$ state. These are states of $|m| \leq 2$. The remaining 70% of the atoms ionized at fields in excess of the ionization field of the $18d_{3/2}$ state. These atoms must be in states of $|m| \geq 3$. Counting the $|m|$ states of $n = 18$ and $\ell \geq 2$ reveals that 23% of the $\ell \geq 2$ states have $|m| \leq 2$, and 77% have $|m| > 2$, so the observed distribution, with 30% of the atoms having $|m| \leq 2$, shows at most a slight propensity for small Δm but no evidence for any strong Δm selection rule.

The effects of electric fields on ℓ mixing

The ℓ mixing of both the Na nd and Xe nf states has been studied in electric fields. Both these states are the lowest states of nearly degenerate high ℓ states, and both are thus adiabatically connected to the lowest member of the Stark manifold. The experiments are done by exciting the Na nd or Xe nf states in zero field then applying a field which rises slowly to a chosen value and then remains constant while collisions are allowed to occur. In this way only the lowest member of the Stark manifold is excited. After the 1–10μs period during which collisions occur, the time resolved field ionization signal from this initially populated level is observed as a function of perturbing gas pressure to determine the total cross section. As the field is raised from zero to approximately $1/3n^5$ the cross section drops steadily. Kachru *et al.*[30] show a factor of 2 decline for a field of $1/30n^5$, and Slusher *et al.*[43] and Chapelet *et al.*[31] show that the cross section continues to decrease with increasing field, the decrease reaching a factor of 4 for a field of $1/3n^5$.

Equally as interesting as the size of the total cross section is the distribution of the final states subsequent to ℓ mixing. Examining the adiabatic field ionization signals of Kachru *et al.*,[30] it appears that only the lowest Stark states nearest to the one initially populated are populated by ℓ mixing. In the data of Slusher *et al.*[43] this point is very clear. As shown by Fig. 11.7, when the Xe 31f state is allowed to remain in zero field the diabatic field ionization signal of the products shown in Fig. 11.7(a) exhibits the form characteristic of having contributions of all ℓ and *m* states. Recall from Fig. 6.10 that the lowest field portion of the signal comes from low $|m|$ redmost Stark states. As shown by Fig. 11.7 raising the field removes the higher field portion of the field ionization signal, indicating that the presence of the field restricts the final states to the low lying Stark states of low *m* adjacent to the one initially populated.

There are two effects which come into play in an electric field. First, the Stark states are polarized along the field axis, the red and blue Stark states having their wavefunctions primarily in the downfield or upfield directions. The consequence of this spatial character is that the matrix elements of the short range e^-–Xe interaction given by Eq. (11.41) are only appreciable between relatively similar Stark states. This effect alone would certainly restrict the number of final states with appreciable populations, as observed. Hickman's quantitative theoretical description of collisions in electric fields, using parabolic Stark states, also indicates that only relatively similar Stark states are populated.[12] The final state restriction might or might not alter the total cross section depending upon whether or not the spatial overlap with nearby Stark states was improved enough to offset the smaller number of final states. However, this effect cannot explain the steady decrease in cross section with increasing field, for the Stark wavefunctions are field independent.

This realization brings us to the second important effect of the field, the limitation of kinetic energy transfer. According to our previous argument regarding kinetic energy transfer between an electron and a He atom, we expect typical energy transfers of ≤ 1 cm^{-1}. Since the energy between adjacent Stark states is 3nE, this constraint reduces the number of final states as the field is increased, and it is for this reason that the cross section decreases monotonically with field, at least for $E \leq 1/3n^5$.[43]

ℓ mixing by molecules

As discussed above, molecular collision partners have energetically accessible degrees of freedom and longer range interactions. What effect do these have on ℓ mixing? While not as much effort has gone into the measurement of ℓ mixing by molecules, a reasonable range of measurements has been carried out. For nonpolar molecules and only slightly polar molecules such as CO, the behavior is qualitatively the same as a rare gas atom's. The cross sections for ℓ mixing of the Na *n*d states by CO and N_2 have been measured and are nearly identical to the cross sections for Ar.[45] The vibrational and rotational degrees of freedom play no role, because the Na Rydberg ℓ states being mixed are degenerate on the scale of

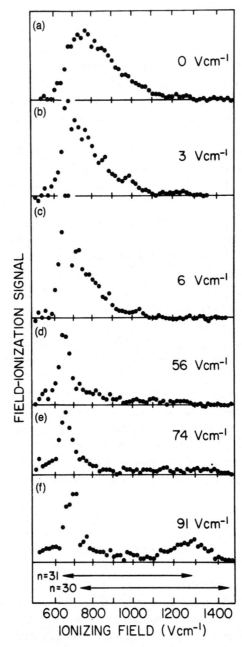

Fig. 11.7 Collisionally induced diabatic ionization features observed following collisions of laser-excited Xe 31f atoms with Xe target gas at a pressure of 10^{-5} Torr for 8 μs in the presence of several applied static fields. The arrows indicate the ranges of ionizing field strengths over which $n = 30$ and 31 states are expected to ionize diabatically (from ref. 43).

molecular rotational and vibrational frequencies. Since no rotational or vibrational transitions of the molecule are induced, the long range electron–dipole and electron–quadrupole interactions are unimportant.

Highly polar molecules, such as HCl[46] and NH_3[34] have very large ℓ mixing cross sections, as shown by Stebbings *et al.* It is not clear whether the polar molecules interact with only the electron of the Rydberg atom or with the atom as a whole.

Fine structure changing collisions

In the heavier alkali atoms, Rb and Cs, the fine structure intervals of the nd states can be resolved well enough that population transfers between these states can be observed. Typically these measurements have been done by time and wavelength resolved laser induced fluorescence, for example populating the $nd_{3/2}$ state and observing the fluorescence from either the $nd_{3/2}$ or $nd_{5/2}$ state. Since the nd states of Rb and Cs are energetically well removed from the high ℓ states the depopulation of the initially populated nd_j state is usually due to fine structure changing collisions, which have large cross sections. For example, as shown by Deech *et al.*[47] and Pendrill[48], the fine structure changing cross section for the Cs nd states of $n = 9$–15 by ground state Cs atoms increases as n^{*4}, the geometric cross section. This is the same dependence as the ℓ mixing cross sections at low n. The n^{*4} dependence of the Cs nd fine structure changing collisions was confirmed by Tam *et al*[49] using a rather different technique. They used continuous wave dye laser excitation to excite Cs $6p_{1/2}$ atoms to the $nd_{3/2}$ level and measured the ratio of the fluorescence radiated on the $nd_{3/2} \rightarrow 6p_{3/2}$ and $nd_{5/2} \rightarrow 6p_{3/2}$ transitions. Aside from the 6d state, which has an anomalously large cross section, the cross sections for the process $nd_{3/2} \rightarrow nd_{5/2}$ follows an n^{*4} dependence, rising from $3.7(7) \times 10^{-14}$ cm^2 at 7d to $29.8(60) \times 10^{-14}$ cm^2 at 10d.[49]

A set of measurements similar to those of Deech *et al.*[47] was carried out by Hugon *et al.*,[50] who measured the Rb $nd_{3/2} \rightarrow nd_{5/2}$ cross section for $n = 9,10$ and 11, in collisions with He. They found that the cross section decreased from $5.1(10) \times 10^{-14}$ cm^2 at $n = 9$ to $2.2(6) \times 10^{-14}$ cm^2 at $n = 11$, behavior similar to the He–Rb nf ℓ mixing cross section in the same n range. Furthermore the cross section is smaller than the Rb nf ℓ mixing cross sections by a factor of 2–3.[32] In contrast to the readily observed fine structure transitions in the Rb and Cs nd states, when Gallagher and Cooke attempted to observe them in the Na nd states using rare gas collision partners they were unable to do so. Specifically, they attempted to observe the Na $nd_{5/2} \rightarrow nf_{7/2}$ microwave transition when the Na $nd_{3/2}$ state was populated and rare gas added. However they were never able to observe the transition due to its being obscured by ℓ mixing collisions.[51]

All of the above observations are consistent with interpreting fine structure changing collisions in Rydberg atoms as elastic e$^-$–perturber scattering leading to

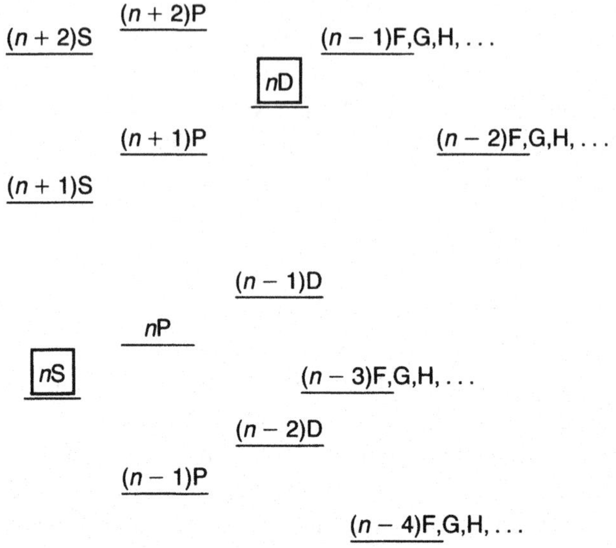

$(n+2)$S $\underline{\quad(n+2)\text{P}\quad}$ $(n-1)$F,G,H, . . .

$\boxed{n\text{D}}$

$\underline{\quad(n+1)\text{P}\quad}$ $(n-2)$F,G,H, . . .

$(n+1)$S

$\underline{\quad(n-1)\text{D}\quad}$

$\underline{\quad n\text{P}\quad}$

$\boxed{n\text{S}}$ $(n-3)$F,G,H, . . .

$\underline{\quad(n-2)\text{D}\quad}$

$\underline{\quad(n-1)\text{P}\quad}$

$(n-4)$F,G,H, . . .

Fig. 11.8 Rb energy level diagram showing the proximity of the $(n+3)$s state to the $n\ell, \ell \geq 3$ states and the relative isolation of the np and nd states (from ref. 50).

m but not ℓ changing, which has been shown to have basically the same cross section as ℓ mixing.[42] Similar n dependences of the cross sections are observed in both cases. The only real difference is that in fine structure changing collisions the number of possible final states is reduced. However, the overlap of the radial wave functions is 100%, or, equivalently, the form factor is 1.

n changing collisions by rare gases

We consider first the simplest case, n changing collisions with rare gas atoms, where by n changing collisions we mean collisions in which the initial Rydberg state is not approximately degenerate with other nearby states as is the case in ℓ mixing collisions. Since there are no energetically accessible internal states of an atomic collision partner, it is reasonable to expect the cross sections to be small since energy must go into, or come from, translational kinetic energy.

Many n changing cross sections have been measured. For example, the depopulation of the Na ns states by several rare gases has been studied over a wide range of n values.[52–54] To convey the essential ideas, though, it is useful to consider a single set of measurements, the depopulation cross sections of Rb Rydberg states with He. In Fig. 11.8 we show the energy levels of Rb.[50] The s, p, and d states all have substantial quantum defects, and the f states have a quantum

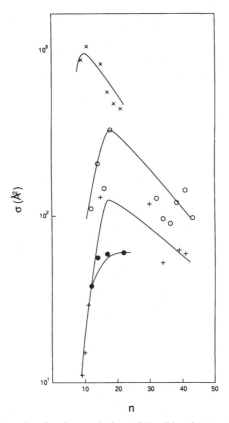

Fig. 11.9 Cross sections for the depopulation of the Rb $n\ell$ states by He vs n; ns(○) (ref. 50,55), np(●) (ref. 26), nd(+) (refs. 50,55), and nf(×) (ref. 32).

defect of 0.05. We begin with the Rb nf states, which undergo primarily ℓ mixing collisions populating the degenerate $n\ell > 3$ states. The He ℓ mixing cross sections, shown in Fig. 11.9, rise to a maximum of 1050 Å² at $n = 11$.[32]

The cross sections for depopulation of the mixed pair of nd fine structure levels are much smaller, as shown by Fig. 11.9. The low n, <15, cross sections, measured by fluorescence detection, rise to their maximum of 130 Å² at $n = 15$,[50] and the higher n cross sections, measured using field ionization, decrease from 118 Å² at $n = 30$ to 59 Å² at $n = 41$.[55] Although there are no measurements for $15 < n < 30$, from the data in Fig 11.9 it is hard to imagine that the cross section ever exceeds 200 Å².

The np state depopulation cross sections were all measured using fluorescence detection, and as a result the highest n observed is $n = 22$.[26] As shown by Fig 11.9, the cross sections rise to a plateau of 60 Å², a value distinctly smaller than the cross sections for any other $n\ell$ states.

Finally the Rb ns cross sections for $12 \leq n \leq 18$ were measured by laser induced fluorescence and show a sharp increase with n, to 320 Å at $n = 18$.[49] For $n = 32$ to n

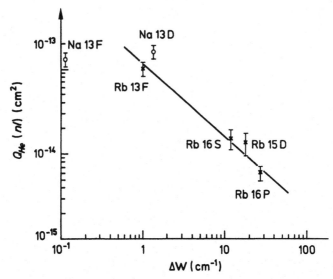

Fig. 11.10 Log–log plot of the cross sections for quenching of Rb ($\times$) and Na ($\bigcirc$) levels having approximately $n^* \sim 13$ by He against the energy defect ΔW between the level of interest and its closest state (from ref. 32).

$= 45$ the cross sections were measured using time resolved selective field ionization,[55] and in this n range the cross section decreases from 125 Å^2 to 82 Å^2. Inspection of Fig. 11.9 suggests that the 320 Å cross section of $n = 18$ may well represent the peak cross section.

It is useful to make some general observations about all the cross sections of Fig. 11.9. For each $n\ell$ series the cross section increases sharply with n at low n, reaches a peak at $n \sim 10$–20, and then declines. Furthermore, it seems possible that all the $n\ell$ states might have the same cross section at high n, although this point is not clear from the data of Fig 11.9. Finally, in the region of the maxima in the cross sections there are pronounced differences in the cross sections of different ℓ states. The difference in the cross sections is directly related to how energetically far removed an $n\ell$ state is from possible final states. This point is made explicitly by Fig. 11.10, which is a plot of the depopulation cross sections of Rb $n\ell$ states of effective quantum number $n^* = 13$ by He.[32] For comparison the Na 13d and 13f ℓ mixing cross sections are also shown. The cross sections are plotted as a function of the energy to the nearest state, ΔW, and, as shown, the cross sections are inversely proportional to ΔW for large values of ΔW, but independent of ΔW for small values of ΔW. The Na nd, Na nf, and Rb nf states all have similar cross sections followed in size by the Rb ns states which are energetically close to the $(n-3)$ $\ell \geq 3$ states. McIntire *et al.*[53] present a graph similar to Fig. 11.10 making the same point.

In Fig. 11.10 the energy separation used for the d state, the d–s interval, suggests that the d state is depopulated by collisional transitions to primarily the s

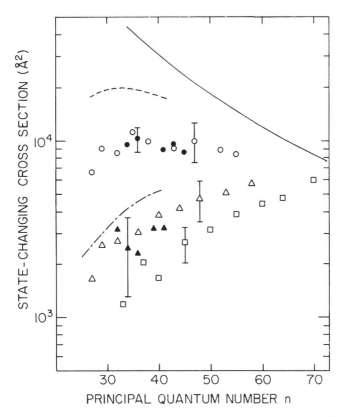

Fig. 11.11 Cross sections for state changing in Rb $n\ell$ − Xe collisions. Rb ns(○) Rb np(□), Rb nd(△) (ref. 56); Rb ns(●), Rbnd(▲) (ref. 55) Born approximation calculations for the Rb ns (- - -) and nd(· · · ·) states (ref. 55), degenerate initial and final state calculation (—) (ref. 12) (from ref. 56).

state. In this case one might expect the cross section to be much smaller than the depopulation cross section of the s state which can go to any high ℓ final state. However, it does not seem to be. This observation and the fact that fine structure changing cross sections are comparable to ℓ mixing cross sections suggest that the relatively good spatial overlap of states of similar or identical ℓ more than offsets the availability of a greater number of states of quite different ℓ. Stated another way these observations imply a propensity for small $\Delta\ell$, in contrast to the observation that ℓ mixing seems to have no $\Delta\ell$ selection rule.

In Fig. 11.9 there is the suggestion that at high n all the Rb $n\ell$ depopulation cross sections coalesce. Goeller *et al.*[56] have demonstrated this point explicitly using Xe to depopulate Rb ns, np and nd states of $27 < n < 70$. Their cross sections, shown in Fig. 11.11 show clearly that at high n these three $n\ell$ series have depopulation cross sections which converge to the same value.

In summary, n changing collisions are typically collisions which transfer an initially excited Rydberg atom to the nearest manifold of degenerate high ℓ states, perhaps via an intermediate state. These collisions can be understood as arising predominantly from the short range electron–perturber interaction leading to elastic e⁻—He scattering. The Rb—He n changing cross sections have the same qualitative dependence as the ℓ mixing cross sections. They increase sharply at low n to a plateau and then fall as n is further increased. There is only a quantitative difference from ℓ mixing collisions, more energy must be transferred to or from the Rydberg electron, requiring that the perturber be closer to the ionic core where the Rydberg electron's velocity is higher. Stated another way these collisions rely on the high energy tail of the Rydberg electron's velocity distribution. At very high n, higher than investigated in this type of experiment, the e⁻–perturber interaction should fall to the point where the ion–perturber scattering is dominant.

n changing by alkali atoms

While the cross sections for depopulation of excited alkali Rydberg atoms by rare gas atoms depend heavily on the quantum defects of the Rydberg states, the same is not true of the depopulation by ground state alkali atoms. In Figure 11.12 we show the cross sections for the depopulation of the Rb ns, np, nd, and nf states in collisions with ground state Rb,[26,32,57] as well as the $nd_{3/2}$ fine structure mixing cross sections.[57] As can be seen in Fig. 11.12, the cross sections for all of these processes are near the geometric cross section $5\pi n^{*4}/2$. The similarity of the cross sections is surprising in that the states under consideration range from the nf states, which are essentially degenerate with the higher ℓ states, to the nd and np states which are well isolated, as shown by Fig. 11.8. In the same n range the depopulation cross sections by He are very different for these $n\ell$ series, as shown by Fig. 11.9. A point about Fig. 11.12 which is worth noting is that the cross sections for the $nd_{3/2}$ states represent primarily fine structure changing collisions, the lone nd depopulation point (13d) represents depopulation of the mixture of 13d fine structure states to other levels. Note that this cross section is a factor of 5 smaller than the fine structure changing collision. In a more extensive set of measurements, Tam et al.[49] measured the ratio of the fine structure changing to depopulation cross sections for the Cs nd states. Their ratio is 2.4(4) independent of n, for $6 \leq n \leq 10$, with the exception of $n = 7$, which may be more rapidly depopulated due to an accidental resonance.

Precisely why the alkali atoms exhibit self quenching behavior so different from the rare gases is not fully understood, although it is no doubt due to the longer range interaction of the alkali atom with the Rydberg electron, and perhaps the Rydberg ion core as well.

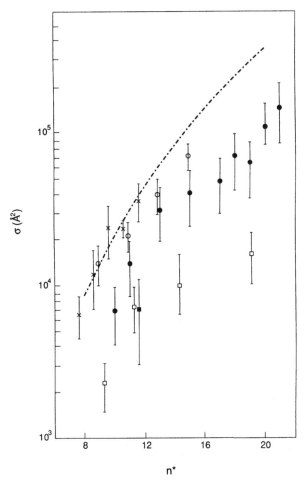

Fig. 11.12 Plot of the depopulation cross sections of the Rb *nℓ* states by ground state Rb: *n*f(●) (ref. 32); *n*s (○), *n*d$_{3/2}$(×), *n*d (■) (ref. 57); *n*p (□) (ref. 26). The cross sections are plotted vs effective quantum number *n**. For reference the geometric cross section $(5/2)\pi n^4 a_0^2$ is also shown (– · –).

n changing by molecules

When Rydberg atoms collide with molecules, the vibrational and rotational degrees of freedom of the molecule allow the molecule to absorb or release internal energy in the collision. In addition, the fact that the rotational and vibrational states are energetically accessible introduces interactions of longer range than the polarization and short range interaction of the Rydberg electron with an atomic perturber. Specifically, the electron can interact with the multipole moments of the molecule. For a homonuclear diatomic molecule, such as N_2, the e$^-$–quadrupole interaction is the longest range interaction, while for a polar molecule such as CO, HF, or NH_3 the dominant interaction is the e^-–dipole interaction. Typically it is the rotational dipole moment which is important. However, it is possible to observe the interaction between the Rydberg electron and the dipole moment of infrared active vibrational transitions.

The only nonpolar molecule which has been studied extensively is N_2. As we have already noted, ℓ mixing by N_2 is not different from ℓ mixing by Ar or CO. However, n changing is quite different. The cross sections for depopulation of the Na ns, $n<10$, states by N_2 have been measured both by time and wavelength resolved laser induced fluorescence and by sensitized fluorescence in Hg–Na–N_2mixtures.[24,58,59] The depopulation cross section of the Na ns states section is roughly constant at ~90 $Å^2$ up to the 8s state at which point it begins to increase slightly.[24,58] This cross section is very different from the rare gas depopulation cross sections for the same low n Na ns states. For example the cross sections for the depopulation of the 6s and 7s states by Ar are <0.12 $Å^2$ and 0.48 $Å^2$ respectively.[52] This difference is directly attributable to the fact that in collisions with rare gas atoms energy lost by the Rydberg electron must go into translation, while in collisions with N_2 it can go into vibration and rotation.

The N_2 depopulation cross sections of Na ns states of $n \leq 10$ can be described with a variant of the molecular curve crossing model used by Bauer *et al.*[60] to describe the depopulation of the Na 3p state by thermal collisions with N_2. At higher n we would expect that a better description of Rydberg atom–N_2 scattering would be given by picture based on e^-–N_2 scattering. While such experiments have not been done with Na, they have with Rb,[61] and these experiments provide an example of a case in which the free electron picture does seem to work well. Specifically, collisions of Rb ns and nd states with N_2 have been studied using time resolved field ionization for detection. To obtain the cross section the decay of the adiabatic feature corresponding to the initially populated Rb ns or nd state was observed as a function of time after laser excitation for different gas pressures. The results obtained for the Rb ns states are shown in Fig. 11.13. As shown by Fig 11.13 the cross section is largest, 305 $Å^2$, at $n = 30$ and decreases to 170 $Å^2$ at $n = 46$. These measurements agree with cross sections based on the elastic scattering of the free electron from the N_2. More than 90% of the computed cross section is due to collisional transfers from the ns states to the nearby $(n-3)\ell,\ell \geq 3$ states. These collisions occur because of the short range e^-–N_2 interaction. Less than 10% is due to n changing collisions due to the long range interaction of the electron with the N_2 quadrupole moment to produce $\Delta J = \pm2$ rotational transitions. In other words, N_2 behaves almost like a rare gas atom.

In collisions with Rb nd states the fact that N_2 is a molecule becomes more apparent. The Rb nd state cross sections, measured in the same way, are smaller, with cross sections of ~150 $Å^2$, independent of n for 24 < n < 46. These cross sections can also be compared to those calculated using the free electron model. Consider the 35d state as a typical example. The computed cross sections for transfer to the nearby 35$\ell \geq 3$ and 34$\ell \geq 3$ states by the short range e^-–N_2

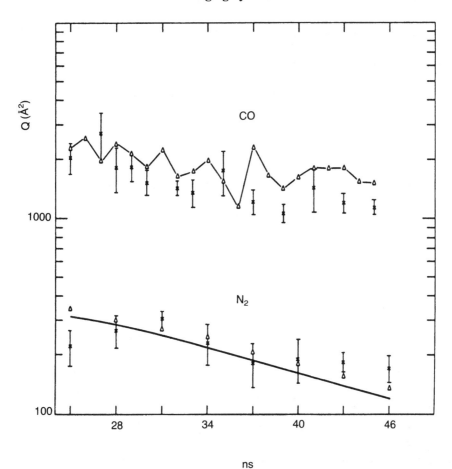

Fig. 11.13 Quenching cross sections for the Rb *ns* levels by N_2 (ref. 61) and CO (ref. 62) vs. the principal quantum number *n*. Experimental values ($\times$). The smooth curve represents the calculated ℓ mixing cross sections by N_2. The ℓ mixing cross section by CO, which is not shown in the figure, is a factor of 2 larger. The triangles ($\triangle$) represent the total calculated cross sections including ℓ mixing, *n* changing and ionization. For CO the line joining the triangles ($\triangle$) is drawn for clarity (from ref. 61).

interaction are 62 and 18 Å^2 respectively. These cross sections are far smaller than the measured cross section of 130(25) Å^2. The long range e^-–N_2 interactions leading to the rotational excitation and deexcitation of N_2 lead to cross sections of 12, 5, and 4 Å^2 for excitation, deexcitation and ionization of the initially populated 35d Rydberg state. The total computed cross section of 99 Å^2 is still substantially below the measured value, but it is clear that the short range interactions are by no means solely responsible for the observed cross sections.

It is also instructive to compare the *n* dependences of the depopulation cross sections of the Rb *ns* and *nd* states. The *nd* cross sections are smaller and *n* independent while the ns cross sections are larger and decrease with *n*. From the discussion of the theory we expect that cross sections due to the short range e^-–N_2

interactions should decrease with n whereas those due to longer range interactions should not. This expectation also suggests that the ns, but not the nd, depopulation is due to the short range interaction.

If we consider a polar molecule, such as CO, we introduce the possibility of the longer range electron–dipole coupling, leading to collisions in which the molecule undergoes $\Delta J = \pm 1$ transitions. The similarity of the Na nd ℓ mixing cross sections for N_2 and CO shows the similarity of the short range e^-–N_2 and e^-–CO interactions.[45] Unfortunately, no measurements of n changing by CO have been made at low n. However, measurements of the Rb ns and nd depopulation cross sections by CO have been carried out by Petitjean *et al.*, using the same field ionization techniques described above for Rb–N_2 collisions.[62] The total depopulation was measured by observing the decay with time of the adiabatic field ionization feature of the initially excited state.

The cross sections for the depopulation of the Rb ns states are much larger than the corresponding cross sections for N_2. As shown by Fig. 11.13, the peak cross section is 2000 Å^2 at $n = 25$, and the cross section decreases to $\sim$1000 Å^2 at $n = 45$. An experimental indication of the reason for the huge difference from N_2 is obtained by a qualitative inspection of the field ionization signals. With rare gases, and presumably with N_2, there is negligible signal observed at field ionization voltages lower than the onset of the adiabatic signal from the state populated by the laser. In contrast, with CO added there is a very obvious, approximately continuous, signal from zero field up to the threshold ionization field of the initially populated state. This signal has been interpreted as the diabatic field ionization signal from higher n states, but whether it is diabatic or adiabaic ionization, it is clear that a significant amount of population is being transferred to high n states.

Using the free electron model the cross sections are readily calculated. The short range e^-–CO interaction leads to the $ns \rightarrow (n-3)\ell$, $\ell \geq 3$ transition, which has a calculated cross section a factor of 2–5 smaller than the observed cross section. The long range e^-–CO interaction, principally the electron–dipole interaction, which leads to $J \rightarrow J \pm 1$ transitions in CO, provides most of the cross section, as shown by Fig. 11.13. For substantial cross sections these transitions must be resonant energy transfers between the CO and the Rydberg electron, i.e. no energy goes into translation. Rotational deexcitation of the molecule is more favorable for two reasons. First, the densely packed higher n states make the resonance more likely. Second, rotational excitation of the CO requires that energy be removed from the Rydberg electron, and this process is less likely to occur near its outer turning point than at smaller orbital radii.

Although the intrinsic width of the resonances is computed to be $\sim$1 cm^{-1}, the manifestations of the resonant character are not particularly striking in the cross section for two reasons. First, at room temperature there are about 30 rotational states of CO populated, all of which have different resonance frequencies. Second, there is a substantial $ns \rightarrow (n-3)\ell$, $\ell \geq 3$ component to the cross section, which makes observing small variations due to resonances more difficult. A most interesting aspect of Fig. 11.13 is that the total cross section declines as n increases from 25 to 40, reflecting the decrease in cross section for the transitions from the ns to $(n-3)\ell$, $\ell \geq 3$ states. These transitions are due to the short range e^-–CO interaction.

The depopulation cross sections of the Rb nd states of $25 < n < 40$ are ~ 1000 Å^2, which is the same as the cross section of the Rb ns state if the $ns \rightarrow (n-3)\ell$, ℓ ≥ 3 contribution is subtracted. For the Rb nd states the calculated contribution of the scattering of the nd state to $n\ell \geq 3$ and $(n-1)\ell \geq 3$ states with no change in the rotational state of the CO is $<100 \text{ Å}^2$, so 90% of the cross section is due to the inelastic transitions leading to rotational excitation. Presumably it is because the resonant transfer accounts for 90% of the observed cross section that the structure in the cross section is more visible in the nd cross sections than in the ns cross sections. For both the ns and nd states minimal collisional ionization is observed and calculated in this n range, principally because there are too few CO molecules with energetic enough $\Delta J = -1$ rotational transitions. For example, only CO $J > 18$ states can ionize an $n = 42$ Rydberg state by a $\Delta J = -1$ transition, and only 3% of the rotational population distribution is composed of $J > 18$ states.

When molecules such as HBr, or NH_3 are used instead of CO there are two important changes. First, due to the larger dipole moments of these molecules the electron–dipole interaction has a longer range, and the collisions are of longer duration. Second, due to the larger rotational constant of these H bearing molecules, the rotational transition frequencies are higher and more widely spaced. These changes have two effects. First they make it possible to observe explicitly resonances in the transfer of energy from molecular rotation to electronic energy of the Rydberg atom. Second, they make collisional ionization possible. Both these effects have been observed in collisions between Xe Rydberg atoms and NH_3.[63,64] The experiments were done using selective field ionization, and there are typically three features in the field ionization spectrum subsequent to collisions; an adiabatic peak corresponding to the initially populated Xe nf state, a diabatic peak corresponding to the Xe $\ell \geq 3$ states of the same n, and resolved diabatic peaks corresponding to specific higher n states.

The most striking feature of these experiments is the clearly visible rotational to electronic energy transfer first observed by Smith *et al.*[63] Their field ionization data showing the final Rydberg states of Xe 27f–NH_3 collisions, shown in Fig. 11.14, exhibit clear peaks due to diabatic field ionization of higher n states lying above the initial $n=27$ level by the energies of the $J \rightarrow J-1$ rotational transitions, for $J \leq 6$. Note that the lowest J transition, $J=2 \rightarrow 1$, leads to a diabatic ionization field higher than the adiabatic ionization field for the initially populated Xe 27f state.

Kellert *et al.*[64] both measured the total depopulation rates of the initially populated Xe nf states and analyzed the final bound states resulting from these collisions. The total depopulation rates of the Xe nf states were measured by

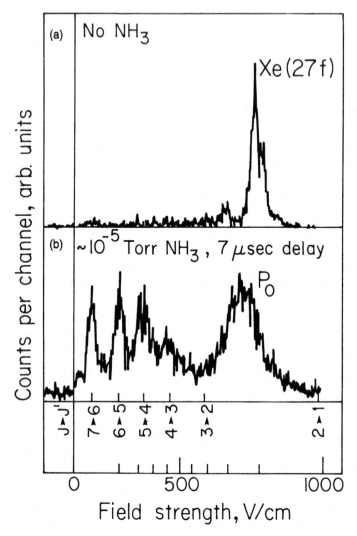

Fig. 11.14 Field ionization signal when the Xe 27f state is populated as a function of ionizing field strength (a) without NH_3, (b) with 10^{-5} Torr NH_3. The resonant rotational NH_3 transitions are indicated beneath each field ionization feature (from ref. 63).

monitoring the dependence of the adiabatic field ionization signal from the nf states as a function of time after the laser excitation for different NH_3 pressures. For $n = 25$–40 the rate constants are $\sim 2 \times 10^{-6}$ cm^3/s. Analysis of the final states under single collision conditions, shortly after the laser excitation shows that, not surprisingly, the depopulation is dominated by ℓ mixing collisions, but that the resonant n changing collisions to higher n states comprise $\sim 1/3$ of the total depopulation rate. At long times after the laser excitation it is clear that the atoms transferred to the high ℓ states will contribute to the resonant transfer signals shown in Fig. 11.14. Since the high ℓ states are, for all practical purposes, degenerate with the nf states their participation does not affect the locations of the resonances of Fig. 11.14.

The Xe nf states are nearly degenerate with the $n\ell > 3$ states. The Rb ns states, however, are more isolated, having a quantum defect of 3.13. Kalamarides et al.[65] studied collisions of Rb ns atoms, $40 \leq n \leq 48$, with HF and observed that the $ns \rightarrow (n-3)\ell$, $\ell \geq 3$ rate constants in this case only comprise 30% of the total depopulation rate constants, which are $10(2) \times 10^{-7}$ cm^3/s for $40 \leq n \leq 48$. Roughly 60% is due to the n changing $(\Delta n > 0)$ transitions resonant with $J = 1 \rightarrow 0$ rotational transition in HF, and 10% is due to n changing, $\Delta n < 0$, transitions resonant with the HF $J = 0 \rightarrow 1$ transition.

Although we have discussed n changing collisions first, ionization of He, Ne, and Ar Rydberg atoms by molecules was observed before n changing collisions by Hotop and Niehaus,[66] who, not having selective methods of preparing or detecting the Rydberg states, could not detect changes in n, but only ionization. They observed ionization cross sections of $\sim 10^4$ Å^2 for ionization by H_2O, NH_3, SO_2, C_2H_5OH, and SF_6, noting that in the case of SF_6 the production of a positive ion led also to the production of SF_6^-. They were not able to detect ionization in collisions with the molecules H_2, O_2, N_2, NO, or CH_4. Of these only NO is polar and has dipole allowed rotational transitions, but the dipole moment is small. More important, the NO rotational constant is small, and only the highest rotational states can ionize a Rydberg atom by a $\Delta J = -1$ rotational transition. The situation is the same as for the collisions with CO described earlier. The fact that only rapidly rotating polar molecules caused collisional ionization was also demonstrated by Kocher and Shepard,[67,68] who showed that this same class of molecules led to collisions with large changes in n as well.

A good example of a rapidly rotating molecule is NH_3, and, as shown in Fig. 11.14, NH_3 molecules in states of $J > 7$ can ionize Xe 27f atoms by a $\Delta J = -1$ transition. If we go to higher n states lower rotational states of NH_3 can ionize the initially populated state, and the ionization rate should increase accordingly. Kellert et al.[64] measured the ionization rates in two ways. One of these was to use a low field pulse to collect the Xe$^+$ ions which had been formed in short enough times after laser excitation, $\sim 5\,\mu s$, that single collision conditions prevailed. Comparing the number of ions produced to the number of atoms remaining in the initially populated nf state gives the cross section, or rate constant. This method is, in essence, the same as the one used to determine how much of the total depopulation cross section is due to ℓ mixing as opposed to n changing collisions. The measured rate constants of Kellert et al.[64] for ionization and total depopulation are shown in Fig. 11.15 along with the calculated ionization rate constants[11,69,70] of the nf state. While measured and calculated ionization rate constants have qualitatively similar behavior, the measured rate constants are roughly a factor of 4 larger than the calculated ones. An interesting aspect of Fig. 11.15 is the step structure in the calculated ionization rate constant, reflecting the fact that as n increases lower rotational states are added to the number capable of ionization.

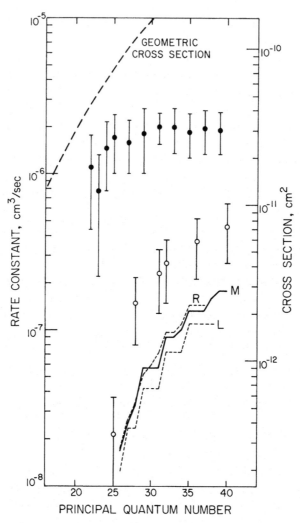

Fig. 11.15 Principal quantum number dependence of the experimental rate constants for the total depopulation of the Xe nf states by NH_3 (●) and for collisional ionization (○). Also shown is the calculated ionization rate constants of Rundel (R) (ref. 69), Latimer (L) (ref. 10), and Matsuzawa (M) (ref. 70) (from ref. 64).

The data of Fig. 11.15 are too sparse to reveal such structure, but it has been observed in collisions between Kr Rydberg atoms and HF, which has a larger rotational constant than NH_3.[71] As we shall see shortly SF_6 readily removes the electron from a Rydberg atom and is hence an excellent detector of Rydberg atoms. Using first a mixture of SF_6 and Kr, Matsuzawa and Chupka scanned the wavelength of a vacuum ultraviolet source across the Kr Rydberg series and observed the SF_6^- ions produced from the Rydberg atoms, which remain in the observation volume for microseconds.[71] Below the ionization limit they observed the Kr Rydberg series, and above the limit they observed no signal since the

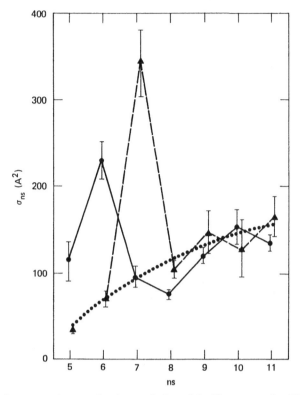

Fig. 11.16 Total cross sections σ_{ns} for depopulation of the Na ns states by CH_4 (●,——) and CD_4 (▲,– – –). The smooth dotted curve (. . . .) shows the expected depopulation cross section in the absence of resonant e^-–v transfer (from ref. 72).

photoelectrons rapidly leave the observation region. When HF was added the SF_6^- signal from the high n states was substantially depressed. As the wave length is increased from the ionization limit the SF_6^- signal can be seen to increase in steps coincident with the energies of the HF $J \to J - 1$ transitions.

The clear evidence for resonant transfer of energy from molecular rotation to Rydberg atoms prompts us to ask if an analogous process might occur between molecular vibration and Rydberg atoms. In general, vibrational frequencies are higher than rotational frequencies, so we would expect to see evidence for resonant transfer at relatively low n, <10, and, in the vicinity of room temperature, we would expect to see energy transferred from the atom to the molecule. In the depopulation of the Na ns states by N_2 there is no sign of any resonant transfer, the depopulation cross section increases monotonically with n, consistent with a curve crossing model.[24] On the other hand, laser induced fluorescence measurements of the depopulation cross sections of the Na ns states by methane, CH_4, and deuterated methane, CD_4, show evidence of resonant behavior similar to that seen for Xe nf states with NH_3.[72] In Fig. 11.16 we show the depopulation cross sections of the Na ns states by CH_4 and CD_4. In Fig. 11.16 the dotted line shows

Table 11.1. *Observed resonant transfers from Na to methane and deuterated methane.*[a]

Na transition	Na frequency (cm^{-1})	Molecule	Vibrational mode, rotational branch	Branch center frequency (cm^{-1})	Branch width (cm^{-1})
$5s \to 4p$	2930	CH_4	v_3, P	2940	60
$6s \to 5p$	1331	CD_4	v_4, R	1340	60
$7s \to 5d$	975	CD_4	v_4, R	965	50

[a](from ref. 71)

the "resonance free" behavior, as seen with N_2. It is evident that the 5s and 6s CH_4 cross sections and the 7s CD_4 cross section lie above the smooth dotted curve.

The increased cross sections for these three states are attributed to resonant electronic to vibrational energy transfer. Table 11.1 identifies the three atomic transitions and the resonant molecular transitions in CH_4 and CD_4. For example the rapid depopulation of the Na 7s state by CD_4 is attributed to the Na $7s \to 5d$ transition. To verify this assignment the cross section for the $7s \to 5d$ transfer was measured for both CH_4 and CD_4 by observing the 5d–3p fluorescence as well as the 7s–3p fluorescence. The $7s \to 5d$ cross sections are 215 $Å^2$ for CD_4 and 15 $Å^2$ for CH_4. As shown by Fig. 11.16, the 7s CD_4 cross sections is ~240 $Å^2$ above the smooth dotted curve in good agreement with the $7s \to 5d$ cross section. Similar confirmations were carried out for the other two resonant collisional transfers.

The observed resonant energy transfer cross sections can be described in terms of a free electron scattering from the CH_4 or CD_4. First, both the v_3 and v_4 modes are infrared active, and there is a long range electron–dipole interaction.[73–75] Second, electron scattering measurements have shown that the cross section for electron impact excitation of the v_3 and v_4 modes is high at threshold.[76] As a result of the second point the Rydberg electron is able to excite the methane at the largest orbital radius at which it is energetically possible, i.e. just within the outer turning point of the final Na states.

Electron attachment

Collisions of some halogen bearing molecules, such as SF_6, with Rydberg atoms result in attachment of the Rydberg electron to the molecule to form a negative molecular ion. Simple attachment, dissociative attachment, and attachment followed by autodetachment have all been observed. Collisions with attaching molecules are an excellent example of a process dominated by the electron molecule interaction, for at high n the rate constants for attachment of the electrons in a Rydberg atoms are identical to those for attachment of free electrons, as predicted by Eq. (11.32).[77,78] However, at low n the rate constants fall below the free electron value.

Measurements of the cross sections for attachment are most often done using time resolved selective field ionization. In collisions of Rydberg states of $n > 20$ with attaching targets no n changing collisions have been observed, but ℓ mixing collisions have.[79] Often they are negligible, but even the largest reported ℓ mixing rates, for the Xe nf states with molecules, are ~30% of the collisional attachment rates. The most straightforward way of measuring the attachment rate is to observe the decay of the sum of the populations in the initially populated state and those populated by ℓ mixing for short times after laser excitation. By short times we mean those during which there is not much depopulation of the initially populated state. At short times this decay rate, γ, is approximately[79]

$$\gamma = \gamma_{0i} + \gamma_{att}, \tag{11.42}$$

but at long times it becomes

$$\gamma = \gamma_{0\ell} + \gamma_{att}, \tag{11.43}$$

where γ_{0i} and $\gamma_{0\ell}$ are the radiative decay rates of the initially populated state and the mixture of ℓ states respectively. Measuring the decay rate as a function of the molecular gas pressure gives the attachment rate constant since ℓ mixing and ionization are the only two significant processes. In Fig. 11.17 are shown the free electron attachment cross sections for a series of halogen bearing molecules derived from Rydberg atom measurements[78] and swarm measurements.[80] The cross sections of Fig. 11.17 are obtained from the Xe nf Rydberg states of $25 < n < 40$ using a variant of Eq (11.32),

$$\sigma_{att} = \frac{k}{v_{rms}} \tag{11.44}$$

where k is the measured attachment rate constant and v_{rms} is the rms value of the electron's velocity, which scales as $1/n$. We use the fact that the rate constants for attachment of a Rydberg electron and a free electron of the same average rms velocity are the same, as shown in Eq. (11.32). In Fig. 11.17 the most striking feature is the continuity of the cross sections from the swarm measurements at higher energies[80] to the Rydberg atom measurements at low energies. This continuity, in essence, verifies the applicability of Eq. (11.44).

As noted above, attachment can occur in several ways, and analyzing the negatively charged products allows us to differentiate between them. Specifically, negatively charged products can be identified by their flight times to a detector.[79] Representative examples of attachment processes are shown in Fig. 11.18, the time resolved ion signals subsequent to the collisions of Xe 26f atoms with attaching targets in the presence of a 2 V/cm field in the interaction region. For all three molecules shown in Fig. 11.18 the Xe$^+$ signal exhibits the expected

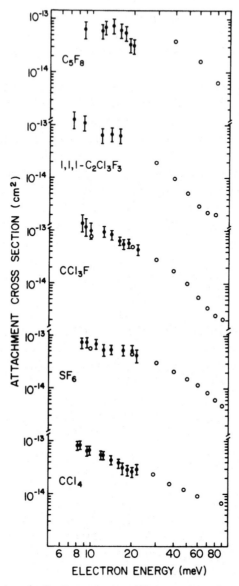

Fig. 11.17 Cross sections for Rydberg electron (●) (ref. 78) and free electron (○) (ref. 80) attachment to C_5F_8, 1,1,1-$C_2Cl_3F_3$, CCl_3F, SF_6, and CCl_4. The Rydberg electron attachment cross sections are calculated using Eq. (11.42) (from ref. 78).

exponential decay. The temporal displacement of the beginning of the ion signal from the laser pulse is due to the Xe^+ flight time to the detector. When Xe collides with CH_3I, to produce I^-, the flight time is consistent with that expected for I^-, shown by the arrow. It is not inconsistent with the flight time of CH_3I^-, so the most stringent statement that can be made on the basis of this measurement alone is that the negatively charged particle contains I^-. The assignment as I^- is due to the

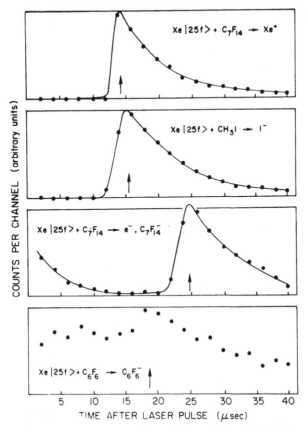

Fig. 11.18 Arrival time spectra of the products of the collisional ionization of Xe 26f high Rydberg atoms by C_7F_{14}, CH_3I, and C_6F_6. As shown, collisions with CH_3 lead only to I^-. Collisions with C_7F_{14} lead to both $C_7F_{14}^-$ and e^-, as shown by the large signal at early times due to electrons. C_6F_6 leads to the production of a long lived autodetaching state of $C_6F_6^-$ which produces a nearly continuous electron signal at early times (from ref. 79).

higher resolution mass spectrometric work of Stockdale *et al.*[81] who only observed I^- when free electrons were attached to CH_3I. In contrast to CH_3I, C_7F_{14} produces both electrons and $C_7F_{14}^-$. The electrons have essentially zero transit time on the scale of Fig. 11.18, and the $C_1F_{14}^-$ has a transit time of 25 μs. Experimentally it is not clear if the electron signal comes from simple collisional ionization or collisional attachment followed by very rapid autodetachment. In Fig. 11.18 the electron and $C_7F_{14}^-$ signals are comparable in size. Although the relative sensitivity of the detector to electrons and $C_7F_{14}^-$ is not known, if we assume the sensitivities are equal, implying that the two processes have comparable rates, we can reconcile the fact that the rate constant for Xe^+ production from Rydberg states exceeds the attachment rate constant obtained from the swarm experiments of Davis *et al.*[82] by a factor of 2–3. Finally, in Fig. 11.18 Xe 26f atoms colliding with C_6F_6 leads to a continuous signal followed by a more or less

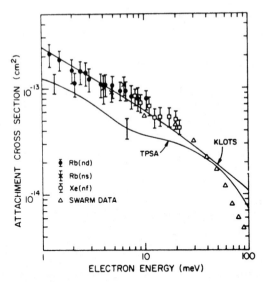

Fig. 11.19 Cross sections for Rydberg electron, Rb nd (●), Rb ns (×) (ref. 84), Xe nf (□) (ref. 85), and free-electron (△) (ref. 88) attachment to SF_6. The solid lines show the theoretical result of Klots (ref. 86) and the results of threshold photoelectron spectroscopy by electron attachment(TPSA) studies (ref. 87) (from ref. 84).

normal exponential decay starting 18 μs after the laser pulse. The normal decay starting at 18 μs is due to the detection of $C_6F_6^-$. The earlier signal is attributed to autodetachment electrons from long lived $C_6F_6^-$ states. This assignment is consistent with previous observations of Naff *et al.*[83] that $C_6F_6^-$ produced by attachment of low energy electrons has a 12 μs autodetachment lifetime.

One of the reasons for using Rydberg atoms is that it is in principle possible to observe electron scattering processes at very low electron energies. An elegant example of this notion is the extension of the measurements of attachment by SF_6 to very high n, by Zollars *et al.*[84] They used the inherently high resolution of a single mode cw dye laser to resolve the high lying, $n > 100$, Rydberg states of Rb by two photon excitation. Their measured rate constants are approximately the same, 4×10^{-7} cm^3/s, as the rate constants for lower n states.[85] These measurements agree with calculated[86] and measured[87,88] free electron attachment rate constants. The n independence of the rate constant indicates that the free electron attachment cross section scales as the inverse of the electron velocity, as expected. This point is shown explicitly by Fig. 11.19, a plot of the cross section as a function of the electron energy. As shown by Fig. 11.19, the measurements go down to electron energies of 1 meV, an energy unattainable by conventional means.

At low n the rate constants for free electron attachment and attachment of Rydberg electrons are no longer the same. In some cases a dependence of the attachment rate constant on the ℓ of the Rydberg state has been observed,[89,90] and at low n the rate constant falls below the free electron value.[90–92] For example, the rate constant for attachment of the electron in Rydberg K nd states by SF_6[91] falls precipitously from its asymptotic value of 4×10^{-7} cm^3/s below $n = 20$. There are two reasons for the discrepancy between the free electron attachment rate constant and the Rydberg atom rate constant at low n. The first is that the attachment rate constant is n independent only in the limit that the attachment cross section is small compared to the geometric cross section of the Rydberg atom. In other words on any given passage of the attaching molecule through the Rydberg atom the probability of attachment is assumed to be much less than 1. When n decreases to the point that the geometric cross section approaches the high n attachment cross section, the attachment cross section and rate constant fall.

A second effect which should further suppress attachment at low n is that the negative molecular ion may not have enough energy to escape from the positive ion of the Rydberg atom.[91] If the molecule captures the electron at a distance R from the Rydberg ion core, it must then overcome the attractive coulomb potential $-1/R$ to escape. As a result, free positive and negative ions are only possible if the relative velocity of the Rydberg ion and the molecule is enough that the relative translational energy exceeds $1/R$. If this condition is not met, attachment can occur, but the result is an orbiting pair of ions which most likely recombine to produce deactivation of the initially excited Rydberg state or a chemical reaction with neutral products. This point is made by Fig. 11.20.[93] For high n the K nd states have total depopulation and attachment rate constants which are very nearly the same, but low n K nd states have depopulation rate constants substantially larger than the negative ion formation rate constants,[91,93] suggesting that temporary attachment may occur, but coulomb trapping prohibits the formation of a free negative ion.[93] The calculated rate constant of Fig. 11.20 is obtained by requiring that the relative velocity of the colliding atom and molecule exceed $1/R$. For a thermal velocity distribution this requirement leads to a rate constant for negative ion formation scaling as $e^{-W/kT}$ where W is the binding energy of the Rydberg state.[91,93]

Beterov *et al.*[92] have measured the electron attachment rate constants for Na np–SF_6 collisions, extending the measurements to slightly lower values of n than those of Fig. 11.20. By comparing SF_6^- production to black body photoionization Harth *et al.*[90] have measured the rate constants for Ne ns, nd–SF_6 electron attachment collisions down to effective quantum number $n^* = 5$. The measured Rydberg atom–SF_6 rate constants are shown in Fig. 11.21.[90] At high n^* the results are in substantial agreement with each other and the free electron attachment rate constant. At low n^* the rate constant does not exhibit an $e^{-W/kT}$ scaling. Such a scaling would lead to unobservably small rate constants for $n < 10$.

Two suggestions have been advanced as to why the rate constant does not vanish at low n. First, Beterov *et al.*[92] have suggested that before the electron is captured by the SF_6 n increasing collisions with SF_6 lower the binding energy to the point that coulomb trapping does not eliminate SF_6^- production. In their

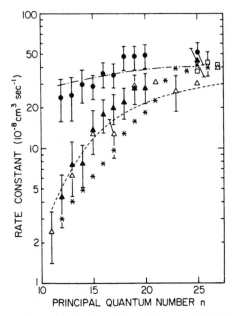

Fig. 11.20 The n dependence of the rate constants for collisional depopulation of K nd atoms by SF_6 (●) (ref. 93) and for formation of free SF_6^- ions in collisions of SF_6 with K nd (▲) (ref. 93), K nd (△) (ref. 91), Na np(∗) (ref. 92), Xe nf (□) (ref. 85). Also shown are rate constants calculated with (- - -) and without (– · – ·) taking into account postattachment electrostatic interactions (from ref. 93).

picture the SF_6 is treated as being structureless. Using this model Harth *et al.*[90] have been able to reproduce the rate constants of Fig. 11.21. Second, some of the internal energy of the SF_6^- can be converted into translational energy after the electron is captured. If a constant fraction of the available internal energy is assumed to be converted to translational energy, it is not possible to reproduce the n^* dependence of the rate constant shown in Fig. 11.21. However, assuming a $1/n^{*4}$ scaling of the fraction of internal SF_6^- energy transferred Harth *et al.*[90] have been able to match their experimental rate constants over the entire range of Fig. 11.21.

The $1/n^{*4}$ scaling of the energy transfer fraction implies that the transfer occurs more readily for low than high Rydberg states. This notion is supported by the work of Harth *et al.* who observed CS_2^- resulting from Ne ns, nd collisions with CS_2.[90] At high n the rate constant for the production of Ne^+ is a constant, indicating that the Rydberg electron is easily removed from the Ne. In contrast, the rate constant for CS_2^- production peaks at $n^* = 18$, not at high n^*, as shown by Fig. 11.22. This unusual n dependence of the rate constant suggests that only when the CS_2^- is formed near enough to the residual Ne^+ can it dispose of enough internal energy to live long enough, 35 μs, to be detected.[90]

A graphic illustration of the effects of the interaction of the two ions is obtained by observing the spatial location of the parent Xe^+ ion after attachment by SF_6.[84]

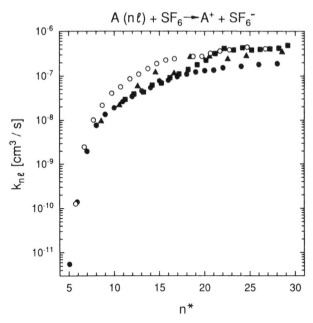

$$A (n\ell) + SF_6 \rightarrow A^+ + SF_6^-$$

Fig. 11.21 Rate constants for SF_6^- formation in collisions of Ne, K, and Na Rydberg atoms with SF_6 vs effective quantum number n^*, Ne ns (○), Ne nd (●) (ref.90), K nd (▲) (ref.91), and Na np (■) (ref. 92) (from ref. 90).

A beam of metastable Xe atoms is excited to a Rydberg state or photoionized and the product Xe^+ ions are expelled by a field pulse from 2–5 μs later and impinge upon a position sensitive detector. When the metastable Xe is photoionized the pattern reflects the geometry of the metastable Xe beam, and the same is true when Xe^+ is produced by attachment of the electron from a Xe 60f state to SF_6. Apparently, there is no deflection of the Xe^+ by the SF_6^-. However, at low n, $n <$ 40, the pattern of Xe^+ detected is twice as broad as the metastable beam, indicating deflection of the Xe^+ ion by the SF_6^-. As expected, the deflections of the Xe^+ become more pronounced at lower n.

If we consider dissociative attachment, the dissociation of the molecular ion provides translational energy to the fragments, and this energy may overcome the coulomb trapping. Walter *et al.*[94] studied the attachment of the electron of a high n, $n \approx 55$, Rydberg atom to CH_3I to form CH_3I^-, which dissociates rapidly, yielding I^-. They found that most of the available energy, ~1/2 eV, from forming the negative ion went into translational energy of the molecular fragments CH_3 and I^-. By observing the spatial distribution of I^- from dissociative attachment of electrons in K nd Rydberg states by CH_3I at low n Kalamarides *et al.*[95] observed a very clear illustration of coulomb trapping. The K atoms in a beam are excited by a laser as shown by Fig. 11.23, and the I^- ions are detected by a position sensitive detector. When the K 26d state is excited the I^- distribution is apparently isotropic, as shown by Fig. 11.23. As n is lowered to 15 and 9 the I^- distribution

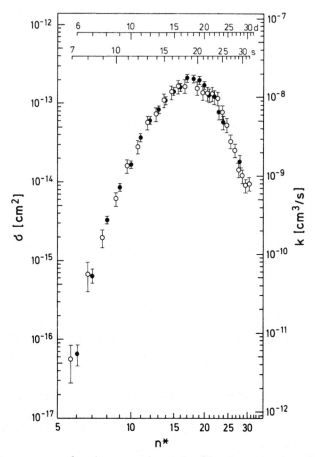

Fig. 11.22 Rate constants k and cross sections σ for CS_2^- formation in collisions of Ne ns (O) and nd (●) Rydberg atoms with CS_2 molecules vs n^*. The decrease of the rate constant and cross section for $n^* > 20$ is due to the reduced probability of stabilizing Ne^+—CS_2^- encounters (from ref. 90).

becomes more sharply peaked in the direction opposite the K beam's. This angular distribution is a reflection of the fact that at $n = 9$ only if the dissociation of CH_3^- produces I^- moving in the direction opposite to the K^+ motion is there enough relative energy for the K^+ and I^- to overcome the coulomb barrier and separate.

The use of the weakly bound electron in a Rydberg atom to measure low energy electron attachment rate constants has proven to be one of the more useful applications of Rydberg atoms. Measurements have been refined to the point of measuring the lifetimes of negative ions formed by attachment,[96] and it is likely that further developments will follow.

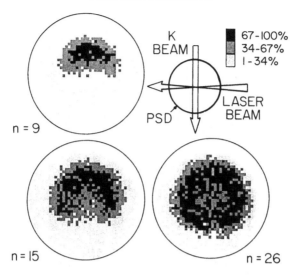

Fig. 11.23 Spatial distribution of I^- ions detected following K nd–CF_3I collisions at intermediate n. The inset indicates the directions of the laser and potassium atom beams (from ref. 95).

Associative ionization

Associative ionization is the process in which an excited atom collides with either a ground state atom or molecule to form a molecular ion and a free electron. For example, the associative ionization of a K Rydberg atom by collision partner M is represented by

$$Kn\ell + M \rightarrow KM^+ + e^-. \tag{11.45}$$

Ignoring the translational kinetic energy of the collision partners, associative ionization is in fact the only way in which a collision with a ground state atom can ionize a Rydberg atom.

If we consider collisions between alkali Rydberg atoms and ground state atoms, associative ionization is possible as long as the energy of the Rydberg state exceeds the minimum energy of the dimer ion, as measured from the energy of the ground states of two separated atoms. This notion is shown graphically in Fig. 11.24, and Lee and Mahan[97] used this requirement to measure the well depth of the alkali dimer ions. Using a lamp and a monochromator they excited the np levels of alkali atoms in a cell and measured the monomer and dimer ions produced. They were able to discriminate between, for example, Cs^+ and Cs_2^+ by their mobilities in Cs vapor (Cs^+ has a lower mobility due to resonant charge transfer). They observed that when np levels of $n < 12$ were excited, Cs_2^+ was the dominant ion, whereas above $n = 12$, Cs^+ was dominant, indicating that below $n = 12$ associative ionization is by far the dominant ionization mechanism.

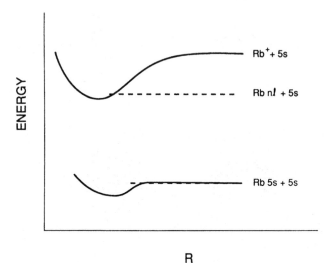

R

Fig. 11.24 Adiabatic potential curves for the Rb–Rb system. The initially populated Rb $n\ell$
Rydberg state can lead to associative ionization if its energy at $R = \infty$ exceeds the minimum
of the Rb$^+$ + Rb 5s potential well, as shown.

Associative ionization was also observed between atoms in Sr $5sn\ell$ states and
the ground Sr $5s^2$ state in a beam of Sr by Worden *et al.*[98] They used multi step
resonant laser excitation of the Sr $5sn\ell$ states, and at Sr number densities of 10^{13}
cm^{-3}, they observed Sr$_2^+$. They were able to determine the well depth of Sr$_2^+$ to be
0.77 eV.

Systematic measurements of associative ionization have been done in Na,[99,100]
Rb,[101,102] and Ne,[103] by several different methods. Measurements of the associative ionization rate constants of high lying Na $n\ell$ states were made using a beam of
Na, which was excited by two pulsed dye lasers from the ground state to the 3p
state and then to an $n\ell$ state.[99] Stark switching allowed the population of the np
states and a mixture of $\ell \geq 2$ states. The excited atoms were allowed to collide for 5
μs after which the Na$^+$ and Na$_2^+$ products were accelerated out of the interaction
region toward the detector by a pulsed extraction field. Na$^+$ and Na$_2^+$ were
distinguished by their flight times to the detector. To determine the associative
ionization rate constant, the densities of the ground and $n\ell$ Rydberg states must
be known, as well as the Na$_2^+$ production. The number of Rydberg atoms was
determined by field ionizing the Rydberg atoms on alternate shots of the laser.
The number of Rydberg atoms, together with the laser beam geometry, gives their
density. The density of the 3s atoms was determined by measuring the lengthening
of the 3p lifetime due to radiation trapping. Specifically, the variation in the
population of the Rydberg state was measured as a function of the time delay
between the two lasers, a method which works well for Na densities in excess of
10^{10} cm^{-3}. The results of the experimental measurements of associative ionization
of Na Rydberg state atoms with ground state Na atoms are shown in Fig. 11.25. As

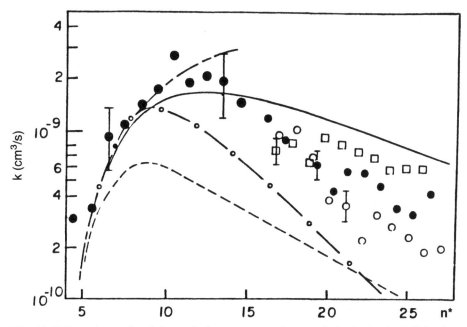

Fig. 11.25 Experimental and theoretical rate constant for associative ionization of Na $n\ell$ + Na 3s atoms at $T = 1000$ K plotted against effective principal quantum number. Results for $17 \leq n^* \leq 27$ (ref. 99) are scaled to absolute results obtained by Boulmer *et al.* for $5 \leq n \leq 15$ (ref. 100). Experimental data; *ns* levels ($\square$); *np* levels ($\bullet$); $n\ell \geq 2$ levels ($\bigcirc$). Theoretical curves: associative ionization of *ns* levels (– – –), *np* levels (——), $n\ell \geq 2$ levels (— o —), associative plus Penning ionization (– – – –) (from ref. 99).

shown, the rate constant, or the cross section, peaks at $n = 11$ and decreases at higher n.

Extensive experiments based on rather different methods were used to study the Rb $n\ell$ + Rb 5s associative ionization. Klucharev *et al.*[101] used a lamp and a monochromator, an approach not unlike that of Lee and Mahan,[97] and Cheret *et al.*[102] used cw dye lasers. The results are very similar to those shown in Fig. 11.25, and are in good agreement with the rate constants calculated by Mihajlov and Janev.[104] By comparing black body photoionization and associative ionization signals Harth *et al.*[103] measured the associative ionization of Ne ns and nd Rydberg atoms with both Ne and He ground state atoms. Using Ne they found results almost identical to those shown in Fig. 11.25, but using He the cross sections were observed to be almost two orders of magnitude smaller.

The alkali associative ionization cross sections can be understood using a model proposed by Janev and Mihajlov.[105] In Fig. 11.26, we show the potential curves of the Rb–Rb system. R is the internuclear separation. At small R the potentials connecting to the same $R = \infty$ states are split by the exchange interaction. The initial state of associative ionization is Rb $n\ell$ + Rb 5s, which is shown by a broken line. If the two Rb atoms approach to within small enough R that the exchange splitting is significant, the molecular Rb $n\ell$–Rb 5s state converging to the nearby

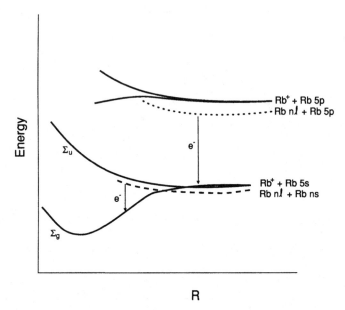

Fig. 11.26 Rb–Rb potential curves showing the origin of the differing rates for Penning and associative ionization. In associative ionization the initial state Rb $n\ell$ + Rb 5s only is above the lower Σ_g ionic state at small R where the $\Sigma_g - \Sigma_u$ exchange splitting is large. Only at small R does autoionization to the ionic molecular state occur. In contrast, in Penning ionization the initial state Rb $n\ell$ + Rb 5p always lies above the ionic final state, and autoionization can occur at any R.

Σ_u molecular ion state can autoionize into the degenerate continuum of the lower Σ_g molecular ion state. This picture suggests that the peak cross section should come from the value of R at which the autoionization rate becomes significant, $\sim 10\,\text{Å}^2$. In other words the peak cross section should be $\sim 100\,\text{Å}^2$, which for Rb at 450 K corresponds to a rate constant of $0.5 \times 10^{-9}\,\text{cm}^3/\text{s}$, approximately the highest rate constant of Fig. 11.25. The model also implies that the cross section should decrease roughly as n^{-3}, a factor reflecting the scaling of autoionization rates due to the normalization of Rydberg electron's wave function at the ion core. As shown by Fig. 11.25, the rate constants fall sharply at high n. In Fig. 11.26 the associative ionization of Na $n\ell$–Na collisions was explained in terms of autoionization from the repulsive upper Σ_u molecular state to the attractive lower Σ_g state. In the rare gas systems the lower state is a Σ state but the upper state is a Π state, precluding autoionization by the ejection of an s wave electron.[103] For this reason the Ne $n\ell$–He cross section is quite small, and one might expect the Ne $n\ell$–Ne cross section to be also. However, for the Ne $n\ell$–Ne case the presence of resonant charge transfer channels enables the autoionization.[103]

Associative ionization measurements have been done with molecules as well. Specifically, associative ionization of K nd states with a variety of molecules has been studied.[106,107] K atoms in a thermal beam were excited with pulses formed from a cw laser beam at a high repetition rate, 10 kHz. The K nd atoms were allowed to collide with a mixture of SF_6 and, for example, H_2S for a period of 1–10 μs, after which the positive and negative ions formed were collected. Using the known rate constants for electron attachment by SF_6, the observed KH_2S^+ and SF_6^- signals immediately imply the ratio of the H_2S associative ionization cross section to the SF_6 electron attachment cross section. To offset the much larger size of the SF_6 attachment cross section a higher density of H_2S was used than SF_6, with typical densities 10^{11}–10^{12} cm^{-3} and 3 x 10^{10} cm^{-3}, respectively.

The results of these measurements, carried out for H_2S, H_2O, CH_3OCH_3, and other similar polyatomic molecules, are uniformly small rate constants, ~10^{-11} cm^3/s, for the $n = 9$ to $n = 15$ K nd states. These rate constants are two orders of magnitude smaller than the rate constants for associative ionization of Rydberg atoms with alkali ground state atoms. Why are the cross sections so small? A possibility suggested by Kalamarides *et al.*[107] is that the dipole moments of these polar molecules may produce more ℓ mixing of the states of the Rydberg electron than do alkali atoms. On average, ℓ mixing puts the Rydberg electron into a higher ℓ state in which it is further from the core and thus less likely to autoionize, in the terms of the model of Janev and Mihajlov.[105]

Penning ionization

Penning ionization occurs when the colliding pair of atoms has enough electronic energy to create a free atomic ion, i.e. the sum of the electronic energies must exceed the ionization potential of one of the atoms. The one case which has been studied is Rb $n\ell$ + Rb 5p $\rightarrow$ Rb$^+$ + Rb + e$^-$. This system has been studied by Barbier and Cheret using two step cw laser excitation of Rb in a cell coupled with mass spectroscopic detection of the product ions.[108] They obtained the Penning ionization rate constants shown in Fig. 11.27. Comparing Figs. 11.27 and 11.25, it is apparent that the Penning ionization rate constants exceed those for associative ionization by two orders of magnitude. The reason for this, pointed out by Barbier *et al.*,[109] is shown schematically in Fig. 11.26. The entrance channel for Penning ionization is Rb $n\ell$ + Rb 5p, which is shown as a dotted line just below the excited state molecular ion potential curve. At virtually any value of R, not only R small enough for a significant exchange splitting, it can undergo the resonant dipole–dipole energy exchange Rb $n\ell$ + Rb 5p$\rightarrow$Rb $\varepsilon\ell\pm1$+Rb 5s.

Ion–perturber collisions

While most collisions involving Rydberg atoms are those in which the perturber interacts primarily with the outer electron, in some cases the dominant interaction

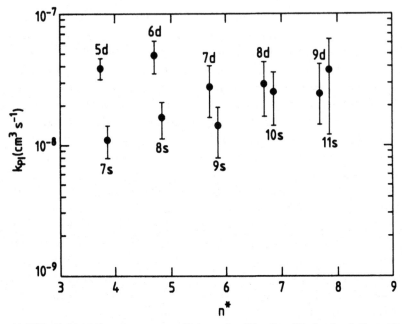

Fig. 11.27 Individual Penning rate coefficients for Rb $n\ell$ + Rb 5p, $k_{PI}(n\ell)$, against the effective quantum number n^* (from ref. 108).

is with the Rydberg ion core. Clear evidence for Rydberg ion–perturber inter-action is the deflection of a beam of Li Rydberg atoms by collisions with gases observed by Kocher and Smith,[110] since only interaction of the perturber with the Rydberg ion core can lead to observable deflection of the Li atomic beam.

They excited a thermal beam of Li atoms to Rydberg states using a pulsed electron beam which delivered $10\,\mu$s bursts at an 850 Hz repetition rate. The beam of Rydberg atoms was collimated and passed through 35 cm of rare gas at a variable pressure. They verified that the rare gas deflected the Rydberg atoms out of the collimated beam, and by measuring the attenuation of the on axis beam they determined the rate constants given in Table 11.2. Since the Rydberg atom beam was pulsed they could measure the rate constant as a function of the atomic beam velocity and found it to be independent of it, implying a cross section with a $1/v$ dependence. By changing the ionizing field in the detector they could change the lower bound of n detected from 35 to 65, and no change in the rate constant was detected.

Both the n independence and the $1/v$ dependence of the cross section agree with a description based on the perturber's scattering from the Li$^+$ core. Since the electron is not involved, the cross section is n independent, and the charge-induced dipole interaction leads to a $1/v$ dependence of the cross section. Furthermore, the measured rate constants are in good agreement with those computed for the scattering of Li$^+$ from the perturbers, as shown in Table 11.2.

Table 11.2. *Experimental and calculated rate constants for deflection of Li Rydberg atoms by five target gases.*[a]

Gas	Rate constant	
	experimental (10^{-9} cm^3/s)	theoretical (10^{-9} cm^3/s)
He	1.97 (40)	1.75
Ne	1.88 (45)	2.42
Ar	2.37 (65)	4.94
H$_2$	4.00 (65)	3.47
N$_2$	2.12 (60)	5.12

[a](from ref. 110)

For typical thermal velocities, the rate constants of Table 11.2 correspond to cross sections of ~100 Å^2.

Another clear illustration of scattering from the Rydberg ionic core is the isotopic exchange observed by Boulmer *et al.*[111] In a He afterglow composed of 50% ^{3}He and 50% ^{4}He they populated, using laser excitation from the 2s state, either ^{3}He or ^{4}He np states and observed the fluorescence from both ^{3}He and ^{4}He Rydberg states. They observed that the 300 K rate constant for the process.

$$^3\text{He } n = 9 + {}^4\text{He 1s} \rightarrow {}^3\text{He 1s} + {}^4\text{He } n = 9 \tag{11.46}$$

is $5.7(10) \times 10^{-10}$ cm^3/s. The results are unchanged if the roles of ^{3}He and ^{4}He are interchanged. This process is interpreted as charge exchange between the Rydberg ion core and the perturbing ground state He atom with the Rydberg electron remaining a spectator. This interpretation is supported by several observations. First, the total exchange rate constant, including processes in which there is a change in n of the Rydberg electron, is only slightly larger, $6.8(20) \times 10^{-1}$ cm^3/s. Second, the theoretical 300 K ion–atom charge exchange rate constant is 5.3×10^{-10} cm^3/s.[112,113] Finally, the extrapolation of the experimental[114] ion-atom rate constants leads to good agreement with these observations.

Fast collisions

All the collision processes we have discussed in any detail are thermal collisions. We would now like to return to a point made early in the discussion of the theory. If the collision velocity is high compared to the Rydberg electron's velocity, the Rydberg atom–perturber cross section should be equal to the sum of electron–perturber and the Rydberg ion–perturber cross section at the same velocity. A

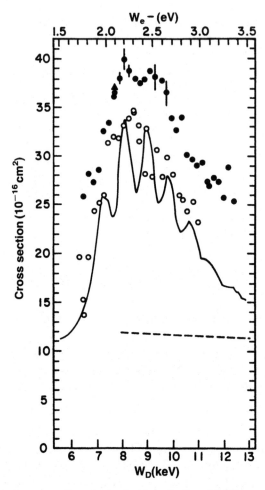

Fig. 11.28 Measured cross section for ionization of $n^* = 46$ ($35 \leq n \leq 50$) D atoms in collisions with N_2 vs the kinetic energy W_D of the deuterium atom ($\bigcirc$), measured cross section for destruction of $n^* = 46$ D atoms ($\bullet$) and $n^* = 71$ D atoms ($\triangle$) in collisions with N_2 (ref. 116). For comparison, Kennerly's measured total cross section for free electron-N_2 scattering vs W_e (——) (ref. 118) is also shown. The W_D and W_e scales have been arranged to correspond to the same collision velocity. The measured cross section σ for electron transfer in H^+–N_2 collisions is plotted vs $W_D = 2W_H$ (– – –) (ref. 117). The destruction cross section equals the sum of the electron transfer and ionization cross sections (from ref. 116).

convincing demonstration of this notion was made by Koch,[115,116] who examined the deexcitation, to $n \leq 28$, and ionization or excitation to $n \geq 61$, of fast D $n = 46$ atoms colliding with N_2. The 6–13 keV D_2 beam energies gave the range of velocity ratio $6 < V/v < 13$. The sum of the excitation, ionization, and deexcitation cross sections is the destruction cross section, which corresponds to the total e^-–N_2 scattering cross section. As shown in Fig. 11.28, if we add the electron transfer cross section for H^+–N_2 collisions,[117] to account for the Rydberg ion core, to the $e^- - N_2$ scattering cross section,[118] the sum is in excellent agreement with the Rydberg atom destruction cross section. In particular, the N_2^- resonance in the $e^- - N_2$ scattering is reproduced well in the Rydberg atom data.

References

1. E. Fermi, *Nuovo Cimento* **11**, 157 (1934).
2. A. Omont, *J. Phys.* (Paris) **38**, 1343 (1977).
3. M. Matsuzawa, in *Rydberg States of Atoms and Molecules*, eds. R. F. Stebbings and F. B. Dunning (Cambridge University Press, New York, 1983).
4. A. P. Hickman, R. E. Olson, and J. Pascale, in *Rydberg States of Atoms and Molecules*, eds. R.F. Stebbings and F. B. Dunning (Cambridge University Press, New York, 1983).
5. M. R. Flannery, in *Rydberg states of Atoms and Molecules*, eds. R. F. Stebbings and F. B. Dunning (Cambridge University Press, New York, 1983).
6. T. F. Gallagher, *Rep. Prog. Phys.* **51**, 143 (1988).
7. T. F. O'Malley, *Phys. Rev.* **130**, 1020 (1963).
8. N. F. Lane, *Rev. Mod. Phys.* **52**, 29 (1980).
9. M. Matsuzawa, *J. Phys. B* **13**, 3201 (1980).
10. M. R. Flannery, *Phys. Rev. A* **22**, 2408 (1980).
11. M. Matsuzawa, *J. Chem. Phys.* **35**, 2685 (1971).
12. A. P. Hickman, *Phys. Rev. A* **28**, 111 (1983).
13. A. P. Hickman, *Phys. Rev. A* **19**, 994 (1979).
14. F. Gounand, and L. Petitjean, *Phys. Rev. A* **30**, 61 (1984).
15. L. Y. Cheng and H. van Regemorter, *J. Phys. B* **14**, 4025 (1981).
16. A. M. Arthurs and A. Dalgarno, *Proc. R. Soc. London* **256**, 540 (1960).
17. A. P. Hickman, *Phys. Rev. A* **18**, 1339 (1978).
18. D. A. Park, *Introduction to the Quantum Theory* (McGraw-Hill, New York, 1964).
19. M. Matsuzawa, *J. Phys. B* **12**, 3743 (1979).
20. F. B. Dunning and R. F. Stebbings in *Rydberg States of Atoms and Molecules*, eds. R. F. Stebbings and F. B. Dunning (Cambridge University Press, New York, 1983).
21. C. J. Latimer, *J. Phys. B* **10**, 1889 (1977).
22. E. de Prunele and J. Pascale, *J. Phys. B* **12**, 2511 (1979).
23. F. Gounand and J. Berlande, in *Rydberg States of Atoms and Molecules*, eds. R. F. Stebbings and F. B. Dunning (Cambridge University Press, New York, 1983).
24. L. M. Humphrey, T. F. Gallagher, W. E. Cooke, and S. A. Edelstein, *Phys. Rev. A* **18**, 1383 (1978).
25. T. F. Gallagher, S. A. Edelstein, and R. M. Hill, *Phys. Rev. A* **15**, 1945 (1977).
26. F. Gounand, P. R. Fournier, and J. Berlande, *Phys. Rev. A* **15**, 2212 (1977).
27. M. Hugon, P. R. Fournier, and F. Gounand, *J. Phys. B* **12**, 1207 (1979).
28. T. F. Gallagher, S. A. Edelstein, and R. M. Hill, *Phys. Rev. Lett.* **35**, 644 (1975).
29. M. Harnafi and B. Dubreuil, *Phys. Rev. A* **31**, 1375 (1985).
30. R. Kachru, T. F. Gallagher, F. Gounand, K. A. Safinya, and W. Sandner, *Phys. Rev. A* **27**, 795 (1983).
31. M. Chapelet, J. Boulmer, J. C. Gauthier, and J. F. Delpech, *J. Phys. B* **15**, 3455 (1982).
32. M. Hugon, F. Gounand, P. R. Fournier and J. Berlande, *J. Phys. B* **12**, 2707 (1979).
33. F. G. Kellert, K. A. Smith, R. D. Rundel, F. B. Dunning and R. F. Stebbings, *J. Chem. Phys.* **72**, 6312 (1980).
34. C. Higgs, K. A. Smith, F. B. Dunning, and R. F. Stebbings, *J. Chem. Phys.* **75**, 745 (1981).
35. H. A. Bethe and E. A. Salpeter, *Quantum Mechanics of One and Two Electron Atoms* (Academic Press, New York, 1957).
36. E. S. Chang, *Phys. Rev. A* **31**, 495 (1985).

37. R. E. Olson, *Phys. Rev. A* **15**, 631 (1977).
38. J. L. Gersten, *Phys. Rev. A* **14**, 1354 (1976).
39. J. Derouard and M. Lombardi, *J. Phys. B* **11**, 3875 (1978).
40. A. P. Hickman, *Phys. Rev. A* **18**, 1339 (1978).
41. A. P. Hickman, *Phys. Rev. A* **23**, 87 (1981).
42. T. F. Gallagher, W. E. Cooke, and S. A. Edelstein, *Phys. Rev. A* **17**, 904 (1978).
43. M. P. Slusher, C. Higgs, K. A. Smith, F. B. Dunning, and R. F. Stebbings, *Phys. Rev. A* **26**, 1350 (1982).
44. M. P. Slusher, C. Higgs, K. A. Smith, F. B. Dunning, and R. F. Stebbings, *J. Chem. Phys.* **76**, 5303 (1982).
45. T. F. Gallagher, R. E. Olson, W. E. Cooke, S. A. Edelstein, and R. M. Hill, *Phys. Rev. A* **16**, 441 (1977).
46. R. F. Stebbings, F. B. Dunning, and C. Higgs, *J. Elec. Spectr. and Rad. Phen.* **23**, 333 (1981).
47. J. S. Deech, R. Luypaert, L. R. Pendrill, and G. W. Series, *J. Phys. B* **10**, L137 (1977).
48. L. R. Pendrill, *J. Phys. B* **10**, L469 (1977).
49. A. C. Tam, T. Yabuzaki, S. M. Curry, M. Hou, and W. Happer, *Phys. Rev. A* **17**, 1862 (1978).
50. M. Hugon, F. Gounand, P. R. Fournier, and J. Berlande, *J. Phys. B* **13**, 1585 (1980).
51. T. F. Gallagher and W. E. Cooke, *Phys. Rev. A* **19**, 820 (1979).
52. T. F. Gallagher and W. E. Cooke, *Phys. Rev. A* **19**, 2161 (1979).
53. J. P. McIntire, G. B. McMillian, K. A. Smith, F. B. Dunning and R. F. Stebbings, *Phys. Rev. A* **29**, 381 (1984).
54. J. Boulmer, J. F. Delpech, J. C. Gauthier, and K. Safinya, *J. Phys. B* **14**, 4577 (1981).
55. M. Hugon, B. Sayer, P. R. Fournier, and F. Gounand, *J. Phys. B* **15**, 2391 (1982).
56. L. N. Goeller, G. B. McMillian, K. A. Smith, and F. B. Dunning, *Phys. Rev. A* **30**, 2756 (1984).
57. M. Hugon, F. Gounand, and P. R. Fournier, *J. Phys. B* **13**, L109 (1980).
58. T. F. Gallagher, W. E. Cooke, and S. A. Edelstein, *Phys. Rev. A* **17**, 125 (1977).
59. M. Czajkowski, L. Krause, and G. M. Skardis, *Can. J. Phys.* **51**, 1582 (1973).
60. E. Bauer, E. R. Fisher, and F. R. Gilmore, *J. Chem. Phys.* **51**, 4173 (1969).
61. L. Petitjean, F. Gounand, and P. R. Fournier, *Phys. Rev. A* **30**, 736 (1984).
62. L. Petitjean, F. Gounand, and P. R. Fournier, *Phys. Rev. A* **30**, 71 (1984).
63. K. A. Smith, F. G. Kellert, R. D. Rundel, F. B. Dunning and R. F. Stebbings, *Phys. Rev. Lett.* **40**, 1362 (1978).
64. F. G. Kellert, K. A. Smith, R. D. Rundel, F. B. Dunning, and R. F. Stebbings, *J. Chem. Phys.* **72**, 3179 (1980).
65. A. Kalamarides, L. N. Goeller, K. A. Smith, F. B. Dunning, M. Kimura, and N. F. Lane, *Phys. Rev. A* **36**, 3108 (1987).
66. H. Hotop and A. Niehaus, *J. Chem. Phys.* **47**, 2506 (1967).
67. C. A. Kocher and C. L. Shepard, *J. Chem. Phys.* **74**, 379 (1981).
68. C. L. Shepard and C. A. Kocher, *J. Chem. Phys.* **78**, 6620 (1983).
69. R.D. Rundel (unpublished).
70. M. Matsuzawa, *J. Phys. B* **55**, 2685 (1971).
71. M. Matsuzawa and W. A. Chupka, *Chem. Phys. Lett.* **50**, 373 (1977).
72. T. F. Gallagher, G. A. Ruff, and K. A. Safinya, *Phys. Rev. A* **22**, 843 (1980).
73. G. Herzberg, *Spectra of Diatomic Molecules* (Van Nostrand, New York, 1950).
74. A. H. Nielsen and H. H. Nielsen, *Phys. Rev.* **48**, 864 (1934).
75. A. H. Nielsen and H. H. Nielsen, *Phys. Rev.* **14**, 118 (1938).
76. K. Rohr, *J. Phys. B* **13**, 4897 (1980).
77. W. P. West, G. W. Foltz, F. B. Dunning, C. J. Latimer, and R. F. Stebbings, *Phys. Rev. Lett.* **36**, 854 (1976).
78. B. G. Zollars, K. A. Smith, and F. B. Dunning, *J. Chem. Phys.* **81**, 3158 (1984).
79. G. F. Hildebrandt, F. G. Kellert, F. B. Dunning, K. A. Smith, and R. F. Stebbings, *J. Chem. Phys.* **68**, 1349 (1978).
80. L. Christophorou, D. L. McCorkle, and J. G. Carter, *J. Chem. Phys.* **54**, 253 (1971).

81. J. A. Stockdale, F. J. Davis, R. N. Compton, and C. E. Klots, *J. Chem. Phys.* **60**, 4279 (1974).

82. E. J. Davis, R. N. Compton, and D. B. Nelson, *J. Chem. Phys.* **59**, 2324 (1973).

83. W. T. Naff, C. D. Cooper, and R. N. Compton, *J. Chem. Phys.* **49**, 2784 (1968).

84. B. G. Zollars, C. Higgs, F. Lu, C. W. Walter, L. G. Gray, K. A. Smith, F. B. Dunning, and R. F. Stebbings, *Phys. Rev. A* **32**, 3330 (1985).

85. G.W. Foltz, C.J. Latimer, G.F. Hildebrandt, F.G. Kellert, K.A. Smith, W.P. West, F.B. Dunning, and R.F. Stebbings, *J. Chem. Phys.* **67** 1352 (1977).

86. C.E. Klots, *Chem. Phys. Lett.* **38**, 61 (1976).

87. A. Chutjian and S. H. Alajajian, *Phys. Rev. A* **31**, 2885 (1985).

88. R.Y. Pai, L.G. Christophorou, and A.A. Christadoulides, *J. Chem. Phys.* **70**, 1169 (1979).

89. H. S. Carman, Jr., C. E. Klots, and R. N. Compton, *J. Chem. Phys.* **90**, 2580 (1989).

90. K. Harth, M.-W. Ruf, and H. Hotop, *Z. Phys. D* **14**, 149 (1989).

91. B. G. Zollars, C. W. Walter, F. Lu, C. B. Johnson, K. A. Smith, and F. B. Dunning, *J. Chem. Phys.* **84**, 5589 (1986).

92. I.M. Beterov, F.L. Vosilenko, I.I. Riabstev, B.M. Smirnov, and N.V. Fateyev, *Z. Phys. D* **7**, 55 (1987).

93. Z. Zheng, K. A. Smith, and F. B. Dunning, *J. Chem. Phys.* **89**, 6295 (1988).

94. C. W. Walter, K. A. Smith, and F. B. Dunning, *J. Chem. Phys.* **90**, 1652 (1989).

95. A. Kalamarides, C. W. Walter, B. G. Lindsay, K. A. Smith, and F. B. Dunning, *J. Chem. Phys.* **91**, 4411 (1989).

96. A. Kalamarides, R. W. Marawar, M. A. Durham, B. G. Lindsay, K. A. Smith, and F. B. Dunning, *J. Chem. Phys.* **93**, 4043 (1990).

97. Y. T. Lee and B. H. Mahan, *J. Chem. Phys.* **42**, 2893 (1965).

98. E. F. Worden, J. A. Paisner, and J. G. Conway, *Opt. Lett.* **3**, 156 (1978).

99. J. Weiner and J. Boulmer, *J. Phys. B* **19**, 599 (1986).

100. J. Boulmer, R. Bonanno, and J. Weiner, *J. Phys. B* **16**, 3015 (1983).

101. A. N. Klucharev, A. V. Lazavenko and V. Vujnovic, *J. Phys. B* **31** 143 (1980).

102. M. Cheret, L. Barbier, W. Lindinger, and R. Deloche, *J. Phys. B* **15**, 3463 (1982).

103. K. Harth, H. Hotop, and M.-W. Ruf, in *International Seminar on Highly Excited States of Atoms and Molecules, Invited Papers*, eds. S. S. Kano and M. Matsuzawa (Chofu, Tokyo, 1986).

104. A. Mihajlov and R. Janev, *J. Phys B* **14**, 1639 (1981).

105. R. K. Janev and A. A. Mihajlov, *Phys. Rev. A* **21**, 819 (1980).

106. B. G. Zollars, C. W. Walter, C. B. Johnson, K. A. Smith, and F. B. Dunning, *J. Chem. Phys.* **85**, 3132 (1986).

107. A. Kalamarides, C. W. Walter, B. G. Zollars, K. A. Smith, and F. B. Dunning, *J. Chem. Phys.* **87**, 4238 (1987).

108. L. Barbier and M. Cheret, *J. Phys. B* **20**, 1229 (1987).

109. L. Barbier, A. Pesnelle, and M. Cheret, *J. Phys. B* **20**, 1249 (1987).

110. C. A. Kocher and A. J. Smith, *Phys. Rev. Lett.* **39**, 1516 (1977).

111. J. Boulmer, G. Baran, F. Devos, and J. F. Delpech, *Phys. Rev. Lett.* **44**, 1122 (1980).

112. D. Rapp and W. E. Francis, *J. Chem. Phys.* **37**, 2631 (1962).

113. D. P. Hodgkinson and J. S. Briggs, *J. Phys. B* **9**, 255 (1976).

114. R. D. Rundel, D. E. Nitz, K. A. Smith, M. W. Geis, and R. F. Stebbings, *Phys. Rev. A* **19**, 33 (1979).

115. P. M. Koch, *Phys. Rev. Lett.* **41**, 99 (1978).

116. P. M. Koch, in *Rydberg States of Atoms and Molecules* eds. R. F. Stebbings and F. B. Dunning (Cambridge University Press, New York, 1983).

117. H. Tawara and A. Russek, *Rev. Mod. Phys.* **45**, 178 (1973).

118. R. E. Kennerly, *Phys. Rev. A* **21**, 1876 (1980).

12

Spectral line shifts and broadenings

The first measurements of collision properties of Rydberg atoms were the pressure shift measurements of Amaldi and Segre,[1] which prompted the description of the shifts in terms of free electron scattering by Fermi.[2]

Lineshift and broadening measurements provide information complementary to that obtained from the conventional collision measurements described in the previous chapter. Just as the tunable laser has made possible many of the collision experiments described in the previous chapter, it has allowed much more sensitive line broadening measurements. In this chapter we connect the normal description of lineshapes to the Rydberg atom collision processes, briefly describe the two modern experimental techniques, and describe the results of the experiments.

Theory

If we consider the intensity of a transition from the ground state to a Rydberg state, it has the form[3]

$$I(\omega) = \frac{\gamma/2\pi}{(\omega - \omega_0 + \Delta)^2 + (\gamma/2)^2},$$ (12.1)

irrespective of whether it is a single photon transition or a two photon transition. Our primary interest here is the impact regime in which the Rydberg atom is colliding with one perturber at a time. In this case the shift and broadening rates Δ and γ are related to the shift and broadening cross sections by[3]

$$\gamma = 2N\langle V\sigma^b\rangle$$ (12.2a)

and

$$\Delta = N\langle V\sigma^s\rangle,$$ (12.2b)

where σ^b and σ^s are the broadening and shift cross sections, V is the relative collision velocity, and N is the perturber number density.

We are particularly interested in transitions from the ground state to the Rydberg states, in which case the effects of collisions on the ground state are negligible, and only the effects of collisions on the Rydberg states need be considered. First we consider the broadening cross section σ^b. It arises from all collisions which disrupt the atomic phase during the emission or absorption of radiation. The more often the atomic phase is changed, the less well defined is the atomic frequency and the broader the emission or absorption line. Equivalently, the coherence between the upper and lower levels is destroyed.[4] Its destruction is observed not only in line broadening experiments, but in photon echo experiments as well.

For transitions between the ground state and a Rydberg state of $n > 10$ there are three contributions to σ^b. The first is inelastic collisions of the Rydberg electron with the perturber, i.e. state changing collisions, which remove population from the Rydberg state under study and halt the coherent absorption or emission of radiation. Second, elastic collisions of the Rydberg electron with the perturber contribute to the broadening if the collisionally induced phase shift is large enough, $\sim \pi$. Third, collisions of the Rydberg ion with the perturber contribute if the induced phase shift is large enough. Elastic electron–perturber collisions obviously make no contributions to state changing collisions, nor do the Rydberg ion–perturber collisions make a large contribution, thus these cannot be observed in conventional depopulation experiments, but only in line broadening or echo measurements. We can write the broadening cross section explicitly as[5]

$$\sigma^b = \frac{1}{2}\left(\sigma_{el}^b + \sigma_{inel} + \sigma_{ion}^b\right),\qquad(12.3)$$

where σ_{el}^b is the Rydberg electron–perturber elastic scattering broadening cross section, σ_{inel} is the depopulation cross section due to the inelastic Rydberg electron–perturber scattering, and σ_{ion}^b is the broadening cross section for Rydberg ion–perturber scattering. We ignore the ground state contribution to the broadening, and the factor of 1/2 is due to the fact that only the Rydberg state contributes to the cross section.

The elastic scattering cross section for a perturber in state β incident on a Rydberg atom in state α with momentum $\mathbf{K}$ is obtained from the optical theorem. Explicitly[3]

$$\sigma_{el}^b = \frac{4\pi}{K}\,\text{Im}\,[f(\alpha,\beta,\mathbf{K} \to \alpha,\beta,\mathbf{K})].\qquad(12.4)$$

We have already computed this cross section, multiplied by the relative collision velocity V, in Eq. (11.31). Explicitly

$$V\sigma_{el}^b = \int_0^\infty |G_\alpha(\mathbf{k})|^2\sigma(k)vd^3k,\qquad(12.5)$$

where $\mathbf{k}$ is the momentum of the Rydberg electron. Using Eq. (11.14), $\sigma(k) = 4\pi a^2$, where a is the scattering length, and comes out of the integral of Eq. (12.5), yielding

$$\sigma_{el}^b = \frac{4\pi a^2}{V}\int |G_\alpha(\mathbf{k})|^2vd^3k = 4\pi a^2\frac{\langle v\rangle}{V}.\qquad(12.6)$$

The inelastic cross section is simply the total depopulation cross section to all other states, i.e.

$$\sigma_{inel} = \sigma_{depop}. \tag{12.7}$$

Finally, the cross section for the Rydberg ion–perturber scattering is obtained by computing the phase shift η due to the polarization interaction during the collision. Explicitly

$$\eta(b) = \frac{1}{\hbar} \int_{-\infty}^{\infty} \left(\frac{-\alpha_p}{2R^4} \right) dt, \tag{12.8}$$

where b is the impact parameter of the collision between the Rydberg ion and the perturber, α_p is the polarizability of the perturber, and R is the separation between the ionic core of the Rydberg atom and the perturber. Assuming a straight line trajectory, the phase shift may be computed to be

$$\eta(b) = \frac{\pi \alpha_p}{4\hbar V b^3}, \tag{12.9}$$

and the cross sections for both the polarization broadening and shift can be calculated using the method proposed by Anderson.[6] Specifically,[4,6,7]

$$\sigma_{ion}^b = 2\pi \int_0^{\infty} [1 - \cos \eta(b)]b\, db, \tag{12.10}$$

$$\sigma_{ion}^s = 2\pi \int_0^{\infty} \sin \eta(b) b\, db \tag{12.11}$$

Using Eqs. (12.6), (12.7), and (12.10) we can assemble the total broadening cross section given by Eq. (12.3). The Anderson approximation bypasses the ambiguous question of precisely how big a phase shift constitutes a line broadening collision, or equivalently, precisely how the Weiskopf radius[4] is defined, removing an inherent ambiguity of the theory.

The pressure shift cross section has two terms, one from the elastic Rydberg electron–perturber interaction and one from the ion–perturber interaction. Explicitly,

$$\sigma^s = \sigma_{el}^s + \sigma_{ion}^s. \tag{12.12}$$

In some cases it is also convenient to break the shift itself into two pieces,

$$\Delta = \Delta_{el} + \Delta_{ion}$$
$$= N\langle V\sigma_{el}^s \rangle + N\langle V\sigma_{ion}^s \rangle. \tag{12.12a}$$

We have already computed σ_{ion}^s in Eq. (12.11). In the line broadening literature the electron contribution is usually described as[3]

$$\sigma_{el}^s = \frac{2\pi}{K} \text{Re} \left[f(\alpha,\beta,\mathbf{K} \to \alpha,\beta,\mathbf{K}) \right]. \tag{12.13}$$

We can compute this cross section following the same approach used to compute σ_{el}^b. Explicitly,

$$\sigma_{\text{el}}^{\text{s}} = \frac{2\pi}{K} \frac{(-\mu)}{2\pi\hbar^2} \int |G_\alpha(\mathbf{k})|^2 \left(\frac{-2\pi}{m}\right) \text{Re} \left[f_e(\alpha,\beta,\mathbf{k} \to \alpha,\beta,\mathbf{k})\right] \text{d}^3k, \quad (12.14)$$

where μ is the reduced mass of the two colliding atoms and m is the mass of the electron. Using the fact that $\text{Re}[f_e(\alpha,\beta,\mathbf{k})] = -a$ from Eq. (11.12), we find that

$$\sigma_{\text{el}}^{\text{s}} = \frac{2\pi\mu(-a)}{Km} = \frac{-2\pi a}{mV} \quad (12.15)$$

or

$$V\sigma_{\text{el}}^{\text{s}} = \frac{-2\pi a}{m}.$$

This result was first derived in an elegant manner by Fermi.[2] It can also be derived in a pedestrian way using the form of the interaction given by Eq. (11.17). The derivation is implicitly statistical, in that the motion of the perturber is ignored, which at first seems to imply that it should not match the results of an impact theory. However, we are only ignoring the thermal motion of the perturber relative to the electron motion. It is still true that the Rydberg electron interacts with one perturber at a time, thus the requirement of the impact regime is met.

The energy shift due to a rare gas atom at $\mathbf{R}$ due to the electron–rare gas interaction is given by

$$\Delta_{\text{el}} = \int \psi_{n\ell m}^*(\mathbf{r})U(\beta, \beta,\mathbf{r},\mathbf{R})\psi_{n\ell m}(\mathbf{r}) \, \text{d}^3r. \quad (12.16)$$

We can connect this energy shift per perturbing atom to an absolute average shift by multiplying by the rare gas number density N and integrating over the volume. If we also replace $U(\beta,\beta,\mathbf{r},\mathbf{R})$ using Eq. (11.19), Eq. (12.16) becomes

$$\Delta_{\text{el}} = N \int \psi_{n\ell m}^*(\mathbf{r})2\pi a\delta(\mathbf{r} - \mathbf{R}) \, \psi_{n\ell m}(\mathbf{r}) \, \text{d}^3r\text{d}^3R$$

$$= 2\pi aN. \quad (12.17)$$

In other words the shift is n independent and depends only on the rare gas scattering length a and density N. It is often termed the Fermi shift.

For completeness, we also present the method used by Fermi[2] to calculate the shift due to the Rydberg ion polarization interaction in the statistical regime. If there are rare gas atoms located at points $\mathbf{R}_i$, the energy shift of a single Rydberg ion is given by[8]

$$\Delta_{\text{ion}} = -\sum_i^{\infty} \frac{\alpha_{\text{p}}}{2R_i^4}. \quad (12.18)$$

If there are N perturber atoms per unit volume, and they are distributed uniformly, there will be one perturber per cube of volume $1/N$. If we assume that the perturbers are arranged in spherical shells starting with radius R_1 such that

$$\frac{4\pi R_1{}^3}{3} = \frac{1}{N},\tag{12.19}$$

we can convert the sum of Eq. (12.18) to an integral. Explicitly,

$$\begin{aligned}\Delta_{\text{ion}} &= -N\int_{R_1}^{\infty}\frac{\alpha_{\text{p}}}{2R_i^4}2\pi R_i^2\mathrm{d}R_i\\ &= -N\alpha\pi\left(\frac{4\pi N}{3}\right)^{1/3}.\end{aligned}\tag{12.20}$$

This result, first obtained by Fermi,[2] has an $N^{4/3}$ dependence, markedly different from the linear density dependence characteristically found in the impact regime.

Experimental methods

Historically, the shift and broadening of spectral lines was the first way in which collisions of Rydberg atoms were studied, by Amaldi and Segre.[1] Much of the classical line broadening work has been summarized by Allard and Kielkopf,[9] and we shall not describe it here. Using classical spectroscopic techniques it was necessary to introduce enough perturbing gas that the shifts and broadenings were visible above the Doppler broadening, ~1 GHz. The development of Doppler free two photon spectroscopy makes it a straightforward matter to observe lines which are less than 10 MHz wide, reducing the required pressure by two orders of magnitude.[10] The reduction in required pressure ensures that measurements can be done in the impact regime, not the quasi–static regime. In the latter case there are so many perturbing atoms that simply considering their presence gives an adequate representation of the observations. In the former case the motion of the perturbing atoms is important.

Conventional lineshift and broadening measurements are similar to absorption spectroscopy, in which the atoms never need to be detected at all. All the information is in the absorption of the light. If the atoms are detected, as is usually the case in two photon spectroscopy, all that is necessary is discrimination between the ground state and a Rydberg state, a trivial proposition. In many of the two photon, Doppler free measurements the atoms are detected either by fluorescence detection[11] or, more often, using a thermionic diode[12] in which collisions convert the Rydberg atoms to ions. As an example, Fig. 12.1 depicts the thermionic diode and single mode cw dye laser used by Weber and Niemax to study self broadening in Cs.[10] Results typical of such line broadening experiments are shown in Fig. 12.2, in which the broadening and shift of the two photon Rb $5s \rightarrow 24s$ transition with Ar pressure are quite evident.[11] It is also interesting to note that the shifts observed in this work would have been marginally detectable, and the broadening unobservable, using conventional absorption spectroscopy.

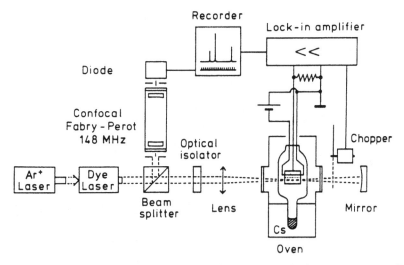

Fig. 12.1 Thermionic diode and cw dye laser used by Weber and Niemax to measure Cs self broadening (from ref. 10).

The final technique which has been used extensively to measure line broadening is the tri-level echo technique.[12] The principle is shown in Fig. 12.3 as applied to the Na $ns_{1/2}$ or $nd_{3/2}$ states. At time t_1 a forward travelling laser beam of frequency ω_1 excites the atoms from the $3s_{1/2}$ state to the $3p_{1/2}$ state, and a beam propagating in the reverse direction, with frequency ω_2, excites these atoms to an $ns_{1/2}$ or $nd_{3/2}$ state at time t_2. This process forms a coherent superposition of the ground state and the Rydberg state, and the decay of this coherence is dominated by processes affecting the Rydberg state. At a later time t_3 the 3s–Rydberg coherence is transferred to a 3s–3p coherence by a laser pulse of frequency ω_2 propagating in the reverse direction, and at time $t_4 = (t_3 - t_2)\omega_2/\omega_1 + t_1$ an intense echo pulse at ω_1 is emitted in the forward direction. The decay of the echo as the time interval t_3–t_2 is increased reflects the decay of the 3s–Rydberg coherence. Echo measurements are inherently Doppler free, and all the information is contained in the well directed optical echo beam. The atoms never need to be detected at all. On the other hand the echo beam emerges from the sample cell collinear with the ω_1 beam. For a sensitive detector to see the echo pulse without being at least temporarily saturated by the ω_1 laser pulse requires an electro optic shutter with an extinction of 10^7 when closed.[13]

Measurements of shifts and broadening

Extensive measurements of shifts and broadening of transitions to alkali Rydberg states by rare gases have been made. A representative recent example is the use of

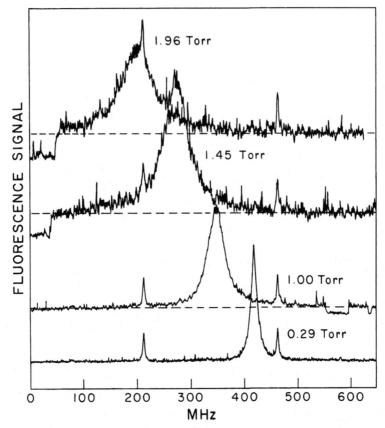

Fig. 12.2 Recorder trace of Rb $5^2s_{1/2} \rightarrow 24^2s_{1/2}$ signal at pressures of 0.29, 1.00, 1.45, and 1.96 Torr of argon, with different detector sensitivities, as a function of laser detuning ν. The sharp spikes are superimposed signals from a 250 MHz reference cavity (from ref. 11).

two photon, Doppler free spectroscopy to measure both the shift and broadening of Rb 5s → ns and 5s → nd transitions. In Fig. 12.4 we show the shift and broadening of Rb 5s → ns and 5s → nd transitions by He observed by Weber and Niemax.[14] In Figs. 12.4 and 12.5 the shift coefficients are given in wave numbers per number of atoms per cubic centimeter, or $cm^{-1}/cm^{-3} = cm^2$. As shown by Fig. 12.4, the shifts of the transitions to both the ns and nd states are virtually identical, and the shifts rise from zero at low n to a plateau at high n. Aside from the fact that the shift at high n is independent of n and ℓ, the fact that it is positive is of historical interest. As we have already pointed out, the shift has two components, the polarization shift and the Fermi shift. The polarization shift is always negative, but the Fermi shift can be either positive or negative depending on the sign of the electron scattering length of the rare gas atom. When positive pressure shifts were observed by Amaldi and Segre[1] it was rather surprising as only the negative polarization shifts had been anticipated. As shown by Fig. 12.4, the

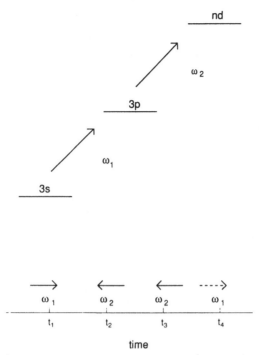

Fig. 12.3 Tri-level echo energy level and timing diagram. The 3s atoms are excited to the nd Rydberg state by two laser pulses of frequencies ω_1 and ω_2 at times t_1 and t_2. The ω_1 beam propagates in the forward direction (to the right) and the ω_2 beam propagates in the reverse direction (to the left). At time t_3 a second beam at ω_2 passes through the sample in the reverse direction, and an echo beam at ω_1 emerges in the forward direction at
$$t_4 = (t_3 - t_2)\omega_2/\omega_1 + t_1.$$

Rb–He pressure shifts agree fairly well with the pressure shifts predicted by Alekseev and Sobelman.[3] Roughly 90% of the shift is due to the Fermi shift and 10% to the polarization shift. The fractional values of these shifts are typical for all the rare gases.

The shifts of the Rb transitions by He are in agreement with expectations. For Xe though, shifts are somewhat different. As shown by Fig. 12.5, the 5s $\rightarrow$ ns and 5s $\rightarrow$ nd shifts both rise from zero and reach similar plateaus at $n > 40$. At high n, >40, the shifts match those calculated on the basis of elastic scattering due to the polarization and Fermi interactions. However, the 5s $\rightarrow$ ns shifts for $n \approx 15$ exceed the high n 5s $\rightarrow$ ns and 5s $\rightarrow$ nd shifts and the 5s $\rightarrow$ nd shifts for $n \approx 15$. Why are the $n \approx 15$ 5s $\rightarrow$ ns shifts larger than the 5s $\rightarrow$ nd shifts for $n \approx 15$, and why do they exceed the high n shifts, which are usually the largest? A possible answer to both these questions is that the Rb ns states of $n = 20$ are responding to the Xe not as separated, noninteracting Rb$^+$ and a Rydberg electron, but as a whole, easily perturbed atom. The Rb ns states have quantum defect 3.13 and lie just below the degenerate high ℓ states. They therefore are much more susceptible

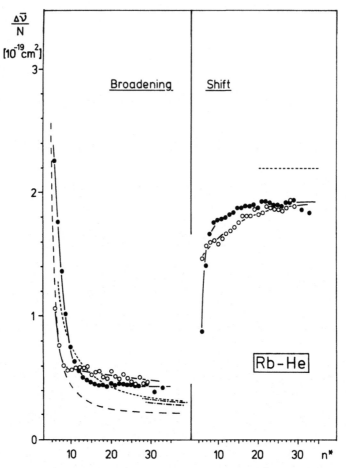

Fig. 12.4 The measured broadening and shift rates of the Rb $n^2s_{1/2}$ (○) and the n^2d_j levels (●) by He as functions of the effective principal quantum number n^*. Broadening: (– – –) theory by Omont (ref. 16), (——) Omont plus the theoretical inelastic contribution by Alekseev and Sobelman (ref. 3) and (– · – · – · –) Omont plus inelastic contributions measured by Hugon *et al.* (ref. 15) (lower curve:nd, upper curve:ns). Shift: (——) theory by Alekseev and Sobelman (ref. 3) (from ref. 14).

to external perturbations, such as the presence of a Xe atom, than are the nd states which have quantum defect 1.35.

In Figs. 12.4 and 12.5 the broadenings of the Rb 5s → ns and 5s → nd transitions are also shown. If we examine first the broadening by He, we can see that the broadening decreases from the lowest states studied to a plateau at high n. The drop in the broadening from $n = 10$ to $n = 20$ is due to the decline in elastic broadening cross section, which is nearly the same for both ns and nd states with He. As shown by Hugon *et al.*,[15] the depopulation, or inelastic, cross sections for Rb ns and nd collisions with He are negligibly small on the scale of Figs. 12.4 and 12.5. The high n plateau is due to the Rydberg ion–He scattering, which is n

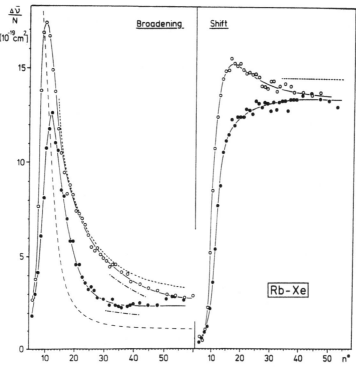

Fig. 12.5 The measured broadening and shift rates of the Rb $n^2s_{1/2}$ (o) and the n^2d_j levels (●) by Xe as functions of n^*. Broadening: (—) Omont's theory (ref. 16), (—) Omont plus inelastic contributions by Alekseev and Sobelman (ref. 3) and (– · – · – · –) Omont plus inelastic contributions measured by Hugon *et al.* (ref. 15) (lower curve: *n*d, upper curve: *n*s). Shift:(—) theory by Alekseev and Sobelman (ref. 3) (from ref. 14).

independent. The broadening is in reasonable agreement with the theory of Omont,[16] and the agreement is improved by adding theoretical[3] or experimental[15] corrections for inelastic collisions.

The broadening of the Rb 5s → *n*s, *n*d transition by Xe is again different. For both *n*s and *n*d states the broadening increases at low *n*, reflecting the geometric cross section. The broadening peaks at $n \sim 20$ for both *n*s and *n*d states and then falls to the same plateau at high *n*, which is due to Rydberg ion–Xe scattering. The major difference from He is that for $n \sim 20$ the broadening is greater for the *n*s than the *n*d states. This difference is a reflection of the larger inelastic Xe scattering cross sections of the Rb *n*s states. In fact, the difference in the broadening cross sections of the *n*s and *n*d states from $n = 32$ to $n = 41$ is the same as the difference in the depopulation cross sections measured by Hugon *et al.*[15] As shown by Fig. 12.5, the agreement with the theoretical predictions is reasonable but not spectacular.

An interesting illustration of the importance of inelastic collisions and ion–rare gas scattering is the broadening of the Na *n*s and *n*d states observed by the tri-level echo technique. In Fig. 12.6 we show the rare gas broadening of the Na *n*s and *n*d

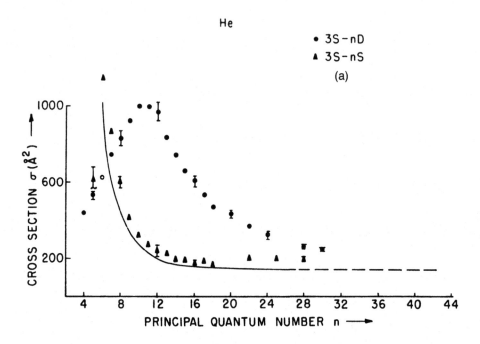

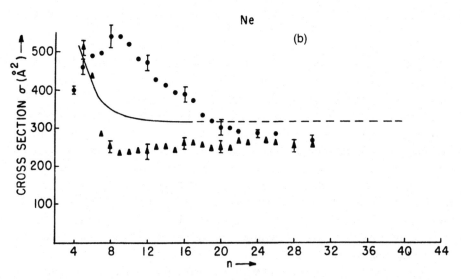

Fig. 12.6 The $3s \rightarrow ns$ and $3s \rightarrow nd_{3/2}$ collisional broadening cross sections derived from tri-level echo decay data are shown plotted vs the principal quantum number of the upper state of the transition. The curved line corresponds to the calculation of Omont (ref. 16). The flat part of the line is also calculated by Alekseev and Sobelman (ref. 3). The errors shown are typical and represent the statistical error in the data (from ref. 7).

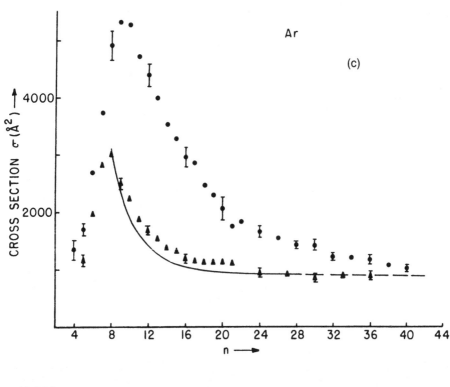

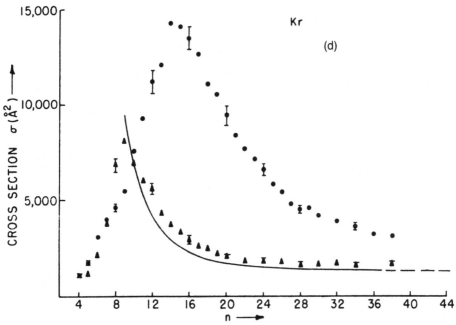

Fig. 12.6 *Continued*.

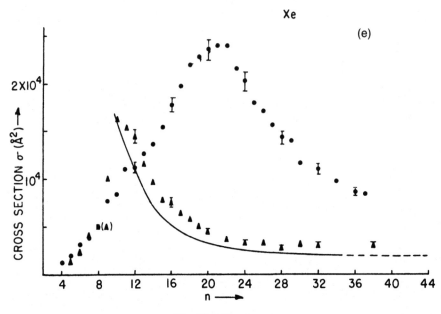

Fig. 12.6 *Continued.*

states observed by Kachru *et al.*[7] The broadening cross sections for both *n*s and *n*d states rise at low *n*, reflecting the geometric cross section of the Rydberg atom. After peaking, the *n*s and *n*d cross sections fall to the same plateau at high *n*, reflecting Na$^+$–rare gas scattering. The evident difference between the *n*s and *n*d states is due to the fact that the *n*d states have large inelastic ℓ mixing cross sections,[17–19] while the depopulation cross sections of the *n*s states are nearly invisible on the scale of Fig. 12.6.[20] If the ℓ mixing cross sections of the Na *n*d states are added to the ns broadening cross sections the resulting cross sections are in reasonable agreement with the *n*d broadening cross sections. An interesting experimental point is that the ℓ mixing cross sections go to zero at high *n* while the broadening cross sections reach a plateau due to the Na$^+$–rare gas scattering. A graphic illustration of the relative importance of the three components of the broadening cross sections of the Na *n*d states by He and Xe is found in the work of Gounand *et al.*[5] and shown in Fig. 12.7. In Fig. 12.7 are shown the calculated elastic e$^-$–rare gas, Na$^+$–rare gas, and inelastic collision cross sections from Hickman's scaling formulae.[21] At low *n* the elastic cross sections can not exceed the geometric cross section $\sim n^4$Å^2. From Fig. 12.7 is apparent that at high *n* it is the Na$^+$–rare gas interaction which dominates and at low *n* the e$^-$–rare gas interaction which dominates. If the same graphs were to be drawn for the *n*s states the inelastic scattering cross sections would by several orders of magnitude smaller since there are nearly no ℓ mixing collisions of the Na *n*s states.

A good illustration of the contribution of inelastic collisions to the broadening cross sections is provided by the alkaline earth atoms, which have perturbations in

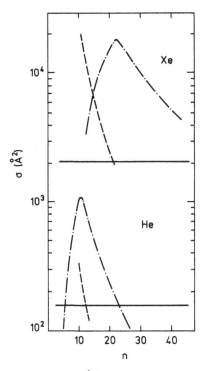

Fig. 12.7 Relative contributions to the σ^b values of the three terms of Eq. (12.3) in the case of Na nd states perturbed by He and Xe. Full line (——) indicates the n independent polarization term, dashed line (– – –) the elastic part due to the (e$^-$–perturber) interaction, and dot-dashed line (– · – · – · –) shows the inelastic contribution calculated according to the scaling formula of Hickman (ref. 21) (from ref. 5).

their quantum defects.[22,23] In particular, in Sr the 5snd states increase their quantum defects by 1 between $n = 10$ and $n = 20$.[22] As a consequence, their energy spacings from other 5s$n\ell$ Rydberg states changes radically over this range. Using two photon absorption spectroscopy Weber and Niemax[22] have measured the broadening of the Sr $5s^2\,^1S_0 \rightarrow 5sns\,^1S_0$, $5s^2\,^1S_0 \rightarrow 5snd\,^1D_2$, and $5s^2\,^1S_0 \rightarrow 5snd$ $^3D_2$, transitions by He and Xe. Where the 5snd quantum defects match those of other Rydberg series increased broadening is observed, due to the larger inelastic cross sections. A theoretical treatment, based on quantum defect theory, of the variation in the broadening with n has been presented by Sun *et al.*[24]

While pressure shift and broadening by rare gases can be characterized as well understood, the same is not true of self shift and broadening of the alkalis. The first two photon measurements of self broadening of Rydberg states were made by Weber and Niemax,[10] who measured the self broadening of the Cs $6s_{1/2} \rightarrow nd$, $11 \leq n \leq 42$ transitions using a thermionic diode. Measuring typically every second or third n state they observed an apparently smooth increase in the broadening rate up to a peak of 13.8(33) MHz cm^3 at $n = 25$, then a slight decrease as n increases to 40. While the self broadening rate is an order of magnitude larger

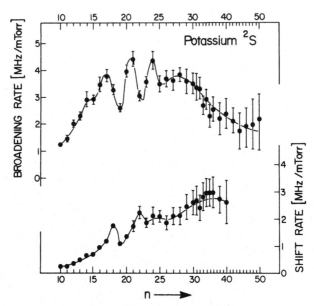

Fig. 12.8 Observed broadening and shift rates for the K 4s → ns transitions by ground state K atoms (from ref. 27).

than for Xe, the most surprising aspect of the self shift and broadening is the oscillatory structure in the alkali self shift and broadening rates at $n \approx 20$, first observed clearly in Rb by Stoicheff and Weinberger,[25] who measured the rates for every n. In retrospect, evidence of the oscillations can be seen in the earlier measurements of Weber and Niemax,[26] but as they did not measure every n, the oscillations are not apparent.

The self shift and broadening of the K ns and nd states provide a good example of all the features mentioned above. Not surprisingly, the broadening and shift rates are large, as shown by Figs. 12.8 and 12.9, plots of the self shift and broadening of the K ns and nd states.[27] At first glance the shifts appear normal. They rise from approximately zero at low n to n independent plateaus at high n. However, the high n value is different for the ns and nd states, an observation unlike any of the shifts observed with rare gases and in stark disagreement with the usual theoretical description. Furthermore, there is not a monotonic increase of the shift with n, rather there is apparent oscillatory structure, as mentioned above. If we examine the broadening of the levels, the gross characteristics of the ns and nd states are the same and similar to rare gas broadening. The broadening of both ns and nd states rises from zero at low n to a maximum at $n \sim 20$ and declines to the same value at high n. These characteristics are not different from those observed in the Rb–Xe experiments shown in Fig. 12.5.

The most striking feature of the broadening is the pronounced oscillatory structure at $n \sim 20$. While the structure is also present in the shift data, it is there not so pronounced. Oscillations in the self shift and broadening similar to those

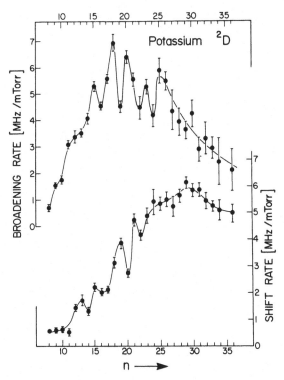

Fig. 12.9 Observed broadening and shift rates for the K 4s → nd transitions by ground state K atoms (ref. 27).

shown in Figs. 12.8 and 12.9 have been observed in K, Rb, and Cs.[25–28] Previously Mazing and Serapinas[29] had observed an oscillation in the self broadening of the Cs 6s–np transitions between $n = 20$ and $n = 30$. However, as shown by Heinke *et al.*,[28] the origin of the oscillation is the variation in proximity of a peak in the quasi-static line wing. The oscillations shown in Figs. 12.8 and 12.9, obtained in the impact regime, are not of the same origin. Matsuzawa[30,31] has suggested that the oscillatory structure may be due to narrow resonances in the e⁻–K scattering. As n is increased the lobes of the momentum distribution of the Rydberg electron pass through the e⁻–K scattering resonances. In addition to describing the locations of the oscillations, this explanation has the attraction of implying oscillations in both the broadening and shift, with the oscillations in the shift being about half as large. Kaulakys[32] compared the oscillations in the shift and broadening to those calculated on the basis of theoretically predicted[32,33] resonances in e⁻–K scattering. Kaulakys[32] and Thompson *et al.*,[27] using the formulation of Matsuzawa, calculated oscillations which matched those observed in the K self shift and broadening experiments. The same structure is observed when Rb is used to produce the broadening of the K Rydberg states.[27,28] suggesting that low energy electron scattering resonances in K⁻ and Rb⁻ have the same energies and widths.

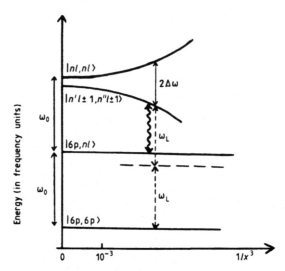

Fig. 12.10 Schematic diagram showing the variation of the energy levels of a system made of two Cs atoms at distance xa_0n^2 from each other vs $1/x^3$ (from ref. 34).

Herman[34] has suggested a more refined approach to this problem in which the correlation between the Rydberg electron's position and momentum is included. Since the negative ion resonance is a p wave resonance, he requires that there be a non-zero derivative of the Rydberg atom spatial wavefunction where the resonant electron momentum occurs. As n is increased from 15 to 25 the orbital radius at which the resonant momentum occurs moves inward from outside the outermost lobe of the radial wavefunction, crossing several lobes of the wavefunction and leading to the observed oscillations in the shift and broadening.

We have, until this point, discussed the broadening of the transitions to Rydberg states by ground state rare gas and alkali atoms. Line broadening due to other Rydberg atoms has also been detected by Raimond *et al.*[35] and Allegrini *et al.*[36] In a Cs beam of ground state atom density 10^{12} cm^{-3}, and a Rydberg atom density at line center of $\sim 10^{10}$ cm^{-3} Raimond *et al.*[35] have observed 30–40 GHz broadening of the transitions from the Cs 6p state to the Cs ns and nd Rydberg states of $n \sim 40$. Although the 6p $\rightarrow$ ns and 6p $\rightarrow$ nd transitions are single photon transitions, the origin of the broadening is due to two photon molecular transitions, an idea which is easily appreciated by examining the molecular energy level diagram of Fig. 12.10. In Fig. 12.10, x is the scaled interatomic distance, the ratio of the interatomic spacing of two nearest neighbor atoms, R, to the size of the Rydberg atom a_0n^2. The molecular states are composed of two Cs atomic states. The low lying state 6p6p is unaffected by variations in x on the scale shown, as is the molecular state 6p$n\ell$. However, the molecular state $n\ell n\ell$ and the nearby $n'\ell \pm 1$, $n''\ell \pm 1$ state are significantly affected. These doubly excited molecular states interact by means of the dipole–dipole interaction,

$$U \cong \frac{\mu'\mu''}{R^3}, \tag{12.21}$$

where μ' and μ'' are the electric dipole matrix elements connecting the $n\ell$ state to the $n'\ell \pm 1$ and $n''\ell \pm 1$ states, and R is the internuclear separation. The two states are shifted in energy by $\pm\Delta\omega$. When the strength of the dipole–dipole interaction of Eq. (12.21) becomes larger than the $R = \infty$ energy spacing between the $|n\ell,n\ell\rangle$ and $|n'\ell \pm 1,n''\ell \pm 1\rangle$ states the energy splitting between the two levels, $2\Delta\omega$, becomes twice the dipole interaction of Eq. (12.21). If we make the approximation, $\mu' = \mu'' = n^2$, and replace R by xn^2 the splitting is given by

$$2\Delta\omega \cong \frac{2}{x^3 n^2}. \tag{12.22}$$

If there were only these two states, their energy levels would split with increasing $1/x^3$ as shown in Fig. 12.10. There are of course not only two neighboring Rydberg states, but many, so that in reality the entire region spanned by the range $2\Delta\omega$, contains the doubly excited $n\ell,n'\ell'$ states. Thus $2\Delta\omega$ should be a good estimate of the expected broadening. For a Cs atom density of 10^{12} cm^{-3} and $n = 40$, $x \cong 7$, which using Eq. (12.22) yields $2\Delta\omega = 3.6 \times 10^{-6} = 24$ GHz, in reasonable agreement with the observed broadening. As shown by Fig. 12.10 the doubly excited $n\ell,n\ell$ states are, in general, excited from the 6p6p state by a two photon molecular transition through a virtual intermediate state near the $6pn\ell$ state. However, near the center of the 6p $\rightarrow$ $n\ell$ transition they may also be excited via the real $6pn\ell$ state.

It is interesting to note that the pressure broadening described here is in essence the non–resonant manifestation of the same interaction responsible for resonant Rydberg atom–Rydberg atom collisional energy transfer.[37]

References

1. E. Amaldi and E. Segre, *Nuovo Cimento* **11**, 145 (1934).
2. E. Fermi, *Nuovo Cimento* **11**, 157 (1934).
3. V. A. Alekseev and I. I. Sobelman, *Zh. Eksp. Teor. Fiz.* **49**, 1274 (1965) [Sov. Phys. – JETP **22**, 882 (1966)].
4. I. I. Sobelman, *Introduction to the Theory of Atomic Spectra* (Pergamon, New York, 1972).
5. F. Gounand, J. Szudy, M. Hugon, B. Sayer, and P. R. Fournier, *Phys. Rev. A* **26**, 831 (1982).
6. P. W. Anderson, *Phys. Rev.* **76**, 647 (1949).
7. R. Kachru, T. W. Mossberg, and S. R. Hartmann, *Phys. Rev. A* **21**, 1124 (1980).
8. H. Margenau and W. W. Watson, *Rev. Mod. Phys.* **8**, 22 (1936).
9. N. Allard and J. Kielkopf, *Rev. Mod. Phys.* **54**, 1103 (1982).
10. K. H. Weber and K. Niemax, *Opt. Comm.* **28**, 317 (1979).
11. W. L. Brillet and A. Gallagher, *Phys. Rev. A* **22**, 1012 (1980).
12. T. Mossberg, A. Flusberg, R. Kachru, and S. R. Hartmann, *Phys. Rev. Lett.* **39**, 1523 (1977).
13. E. Y. Xu, F. Moshary, and S. R. Hartmann, *J. Opt. Soc. Am. B* **3**, 497 (1986).
14. K. H. Weber and K. Niemax, *Z. Phys. A* **307**, 13 (1982).
15. H. Hugon, B. Sayer, P. R. Fournier, and F. Gounand, *J. Phys. B* **15**, 2391 (1982).

16. A. Omont, *J. Phys.* (Paris) **38**, 1343 (1977).
17. T. F. Gallagher, S. A. Edelstein, and R. M. Hill, *Phys. Rev. A* **15**, 1945 (1977).
18. R. Kachru, T. F. Gallagher, F. Gounand, K. A. Safinya, and W. Sandner, *Phys. Rev. A* **27**, 795 (1983).
19. M. Chapelet, J. Boulmer, J. C. Gauthier, and J. F. Delpech, *J. Phys. B* **15**, 3455 (1982).
20. J. Boulmer, J.-F. Delpech, J.-C. Gauthier, and K. Safinya, *J. Phys. Rev. B* **14**, 4577 (1981).
21. A. P. Hickman, *Phys. Rev. A* **23**, 87 (1981).
22. K. H. Weber and K. Niemax, *Z. Phys. A* **309**, 19 (1982).
23. K. S. Bhatia, D. M. Bruce, and W. W. Duley, *Opt. Comm.* **53**, 302 (1985).
24. J.- Q. Sun, E. Matthias, K.- D. Heber, P. J. West, and J. Gidde, *Phys. Rev. A* **43**, 5956 (1991).
25. B. P. Stoicheff, and E. Weinberger, *Phys. Rev. Lett.* **44**, 733 (1980).
26. K. H. Weber and K. Niemax, *Opt. Comm.* **31**, 52 (1979).
27. D. C. Thompson, E. Weinberger, G.-X. Xu, and B. P. Stoichreff, *Phys. Rev. A* **35**, 690 (1987).
28. H. Heinke, J. Lawrenz, K. Niemax, and K.-H. Weber, *Z. Phys. A* **312**, 329 (1983).
29. M. Mazing and P. D. Serapinas, *Sov. Phys. JETP* **33**, 294 (1971).
30. M. Matsuzawa, *J. Phys. B* **10**, 1543 (1977).
31. M. Matsuzawa, *J. Phys. B* **17**, 795 (1984).
32. B. Kaulakys, *J. Phys. B* **15**, L719 (1982).
33. A. L. Sinfailam and R. K. Nesbet, *Phys. Rev. A* **7**, 1987 (1973).
34. R. M. Herman, unpublished (1993).
35. J. M. Raimond, G. Vitrant, and S. Haroche, *J. Phys. B* **14**, L655, (1981).
36. M. Allegrini, E. Arimondo, E. Menchi, C. E. Burkhardt, M. Ciocca, W. P. Garver, S. Gozzini, and J. J. Leventhal, *Phys. Rev. A* **38**, 3271 (1988).
37. K. A. Safinya, J.-F. Delpech, F. Gounand, W. Sandner, and T. F. Gallagher, *Phys. Rev. Lett.* **47**, 405 (1981).

13

Charged particle collisions

Due to the long range interaction between a Rydberg atom and a charged particle, these collisions have by far the largest cross sections of all nonresonant Rydberg atom collision processes. While there have been a few experiments with electrons, most of the the experimental work has been done with ions, because it is possible to make ion beams in which the ion is moving at a speed comparable to the Rydberg electron's. Specifically $\Delta\ell$ and Δn state changing collisions, ionization, and charge exchange have been carefully studied.

State changing collisions with ions

State changing collisions between singly charged positive ions and Rydberg atoms of Na have been studied extensively using crossed beams of atoms and ions, the approach shown in Fig. 13.1.[1] The Na atoms in a thermal beam are excited to a

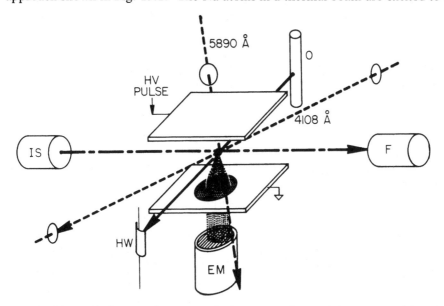

Fig. 13.1 Schematic diagram of apparatus: IS, ion source; O, atomic beam oven; F, Faraday cup; EM, electron multiplier; HW, hot-wire detector. Long and short dashed line, ion beam; solid line, Na beam; dashed lines, laser beams (from ref. 1).

269

Rydberg state by two pulsed dye lasers tuned to the Na 3s → 3p and 3p → nd transitions. The excited Na atoms are bombarded by an ion beam, which has an energy of 40–2000 eV, a current of 10–600 nA, and a diameter of 5 mm. The Rydberg atoms are exposed to the ion beam for times of 1–5 μs, after which the final states of the Na atoms are analyzed by selective field ionization. Since only low–ℓ s, p, and d states can be optically excited, the initial states are of low ℓ and m and ionize adiabatically when exposed to a field ionization pulse with a 1 μs risetime. The final states are most often the higher angular momentum states, which are composed mostly of $|m| > 2$ states, which ionize diabatically with the same field pulse. In all experiments the ion beam axis has been perpendicular to the direction in which the ionizing field was applied, so that the tendency of the collisions to preserve m relative to the ion beam's direction still results in high m states relative to the field ionization axis. In most experiments it has not been possible to observe the difference in the adiabatic field ionization signals of different ℓ states of low m, and collisions are typically detected by the presence of a diabatic field ionization signal.

Na $nd \rightarrow n\ell$ transitions

The most easily observed process with ionic, as with neutral, collisions partners is ℓ mixing. When the Na nd states are exposed to an ion beam, the field ionization signal changes from one which is predominantly adiabatic to one which is predominantly diabatic. By measuring the fraction R of signal transferred from the adiabatic to the diabatic peak of the field ionization signal MacAdam *et al.* measured the depopulation cross section of the Na nd states by He$^+$ ions.[1] In the limit of small values of R the depopulation cross section is given by[1]

$$\sigma = \frac{Re}{JT} \tag{13.1}$$

where J is the ion current density, T is the exposure time and e is the charge of the electron. In Fig. 13.2 we show the values of R, obtained with constant J and T, as a function of n. Equivalently, Fig. 13.2 is a plot of the cross section, in arbitrary units, vs n. Since Fig. 13.2 is plotted on logarithmic scales, the cross sections evidently exhibit a $\sigma \propto n^\beta$ dependence, and $\beta \approx 5$. From the magnitudes of J and T, the absolute value of the cross section for 450 eV He$^+$ is determined to be 2.6×10^8 Å^2 for $n = 28$. The fact that the higher n cross sections at the ion energy of 450 eV fall below the n^5 dependence was later found to be an artifact due to insufficient resolution of the diabatic and adiabatic field ionization signals.[2] In later experiments with other ions the n^5 dependence shown in Fig. 13.2 was also observed.[2,3] The later measurements also verified that the cross section was independent of ion species as long as the ions had the same velocity. Using ions of

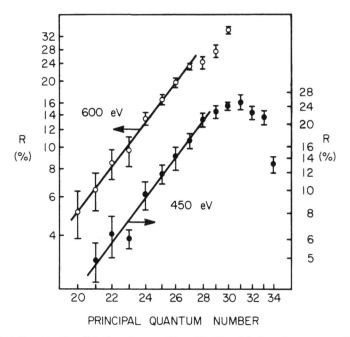

Fig. 13.2 R, defined by Eq. (13.1) and proportional to Na nd ℓ-changing cross sections, vs n for 450 and 600 eV He$^+$ impact. Note the vertical offset of left and right logarithmic scales, introduced for clarity. Different ion-beam currents were used at the two energies. Least-squares fits to the data are shown by the solid lines (from ref. 1).

different mass the velocity dependence of the cross section can be observed over a wide range of velocities, as shown in Fig. 13.3.[2] Fig. 13.3 is a plot of the cross section for depopulation of the Na 28d state as a function of the ratio of the ion velocity v to the Rydberg electron velocity, $v_e = 1/n$, in atomic units. Equivalently, $v_e = ca/n$, and $v_e = 0.78 \times 10^7$ cm/s for $n = 28$. As shown by Fig. 13.3, the cross section decreases as v/v_e increases. The observations are in good functional agreement with the theoretical results of Percival and Richards,[4] Herrick,[5] and Shevelko *et al.*[6] for $v/v_e > 1$. Later calculations of Beigman and Syrkin[7] agree with the experimental cross sections to lower values of v/v_e than do the theoretical results shown in Fig. 13.3.

The final state distributions, which yield additional insights into the $\Delta\ell$ selection rules for the process, are equally as interesting as the cross sections. When ion velocities of $v \approx v_e$ are used, for example 1000 eV Ar$^+$ colliding with Na 28d atoms, there is a clear evolution of the diabatic field ionization signal with increasing exposure of the atoms to the ion beam, as shown in Fig. 13.4.[8] At low current the diabatic field ionization signal has a sharp peak, indicating that only a few states are populated, while at high current the diabatic field ionization signal is quite broad, as would be expected for all the $n = 28, \ell > 2, |m| > 2$ states. The fact that the field ionization signal evolves with ion current indicates that the high current final state distribution is due to multiple collisions. An analysis of the field

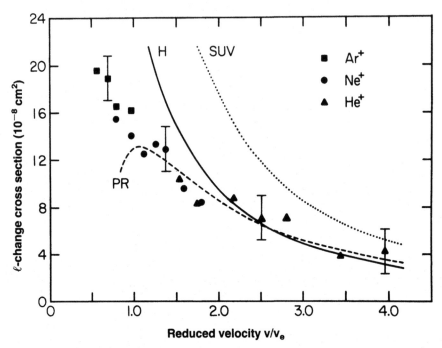

Fig. 13.3 Velocity dependence of the ℓ-changing cross section. Error bars are the same for all points. Uniform 30 nA beams of He$^+$, Ne$^+$, and Ar$^+$ at 400–2000 eV were incident on Na 28d atoms. The data have been normalized to Percival and Richard's (PR) result (ref. 4) at $v/v_e = 1.5$. The results of Herrick (H) (ref. 5) and Shevelko *et al.* (SUV) (ref. 6) are also shown (from ref. 2).

ionization curves shows that the predominant initial transfer is into states of $3 \leq \ell \leq 5$, leading to the very narrow peak in the field ionization signals of Fig. 13.4 for low ion currents.[8]

Measurements with Na$^+$ ions of energies in the 29–590 eV range, corresponding to v/v_e from 0.2 to 0.9, were compared to the diabatic SFI spectra of the Na 28f, 28g, and 28h states observed individually by driving resonant microwave transitions from the 28d state.[9] These detailed comparisons show clearly that for high velocity, $v/v_e \approx 0.9$, 59% of the $28d \to 28\ell$ cross section is to the 28f state, the dipole allowed transition. However, at lower values of v/v_e, nondipole processes play a more important role. For example, at $v/v_e = 0.2$, only 37% of the cross section is due to the $28d \to 28f$ transition.[9] At high velocities the process is predominantly a dipole $\Delta \ell = 1$ process, but at low velocities the dipole selection rule breaks down.

Depopulation of the Na *n*s and *n*p states

The Na *n*s and *n*p states, having quantum defects of 1.35 and 0.85 respectively, are well removed from the hydrogenic $\ell \geq 2$ states, and as a result the cross sections

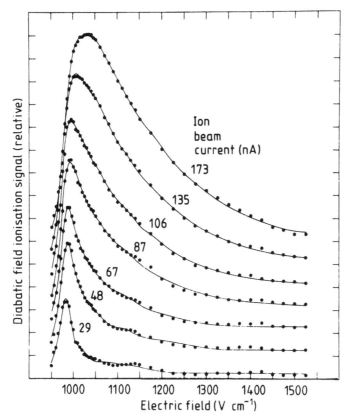

Fig. 13.4 Diabatic field ionization signals for ℓ changing under increasing incident beam intensities. For clarity the successive curves are displaced upward by one scale unit. Data points are taken from the transient digitizer records, and a small sloping background has been subtracted. Full curves are fits to the model of MacAdam *et al.* (from ref. 8).

for collisionally induced transitions from these states are smaller, by a factor of 10–30. The collisions of ns states from 32s to 41s have been studied with both Ar^+ and Na^+.[10] At scaled velocities $v/v_e \sim 0.9$ the cross sections exhibit an n^4 scaling, but at the lowest values of v/v_e studied the cross sections exhibit an n^5 scaling, as do the Na nd states. The np states also exhibit cross sections scaling as n^4 which are approximately one order of magnitude smaller than the nd cross sections. Even so, these cross sections are about two orders of magnitude larger than the geometric cross sections.

One of the most interesting aspects of the study of the Na ns and np states is the distribution of final states. In Fig. 13.5 we show the field ionization signals obtained when the 39p, 40s, 39d, and 40p states are exposed to 43 eV Na^+ ions. [10] There is an initial adiabatic peak and a later broader diabatic feature. The Na^+ current is more than adequate to depopulate the 39d state, and the 39d signal presumably reflects substantial population of the higher ℓ, m states of $n = 39$, due to both non-dipole low velocity collisions and multiple collisions. As shown by

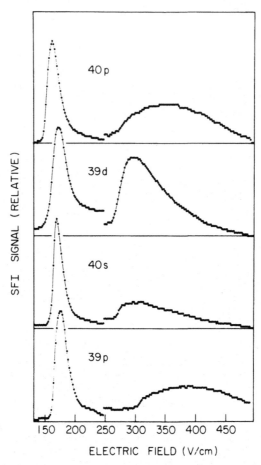

Fig. 13.5 Adiabatic and diabatic selective field ionization (SFI) for ℓ-changed ensembles produced from Na 39p, 40s, 39d, and 40p states by 43 eV Na$^+$ impact. The adiabatic peaks occur at 170–180 V/cm, and the diabatic features occur above 250 V/cm (note change of vertical scale). The diabatic SFI from ℓ-changed 50s targets most closely resembles that from 39d. In contrast, 40p and 39p targets yields SFI that indicates a different distribution of Stark sublevels lying high in the $n = 39$ and 38 manifolds, respectively (from ref. 10).

Fig. 13.5, the 40s diabatic signal is approximately identical in form to the 39d signal, indicating that nearly all of the $n = 39$ high ℓ states are populated from the s state. The extent to which multiple collisions play a role is not so clear, but the final result is clearly similar to that of the 39d state. In contrast, the np signals are very different. Much more of the diabatic signal comes at higher fields. The fact that the s state and p state signals are so different shows that the s state is not depopulated by dipole transitions to the p states. Rather, collisional transfers are directly to high ℓ states. The fact that the p state adiabatic signal differs so markedly from the d and s signals also shows that the states populated from the p states are not readily depopulated by subsequent collisions. This observation suggests that these states might be the highest ℓ states, which have small dipole matrix elements connecting them to other states.

Theoretical descriptions

At high velocities, $v/v_e \geq 1$, these ion–atom collisions can be described by the Born approximation, and this description leads to a reasonable description of the observed Na $nd \rightarrow n\ell$ results.[4] For example, the results of Percival and Richards,[4] shown in Fig. 13.3, are obtained by calculating the Born cross section for the $nd \rightarrow nf$ transition and using a sudden approximation to estimate the redistribution of population among the nearly degenerate higher angular momentum states.

At low velocities, $v/v_e < 1$, the Born approximation breaks down. Beigman and Syrkin[7] have employed a strong coupling model in which the ℓ and $\ell \pm 1$ states are coupled by the field from the passing ion. In its simplest form their model consists of three levels, the initial level ℓ_0, corresponding to the Na nd state, and two degenerate higher ℓ levels $\ell_0 + 1$ and $\ell_0 + 2$. These three states are coupled by the $\ell - (\ell + 1)$ electric dipole matrix elements and the field of the passing ion. Instead of using the zero field ℓ states as a basis they use the three states[7]

$$|\ell_0\rangle$$

$$|+\rangle = \frac{1}{\sqrt{2}}(|\ell_0 + 1\rangle + |\ell_0 + 2\rangle) \tag{13.2}$$

$$|-\rangle = \frac{1}{\sqrt{2}}(-|\ell_0 + 1\rangle + |\ell_0 + 2\rangle).$$

Using this approach the $|+\rangle$ and $|-\rangle$ states are not coupled by the field of the ion, but are only split in energy. At high collision velocities the initial state $|\ell_0\rangle$ is simply projected onto the $|\ell_0 + 1\rangle$ state, a coherent superposition of $|+\rangle$ and $|-\rangle$ states, by the dipole matrix element. However, at lower velocities the change in energy of the $|+\rangle$ and $|-\rangle$ states during the collision allows the $|+\rangle$ and $|-\rangle$ states themselves to be populated rather than only a coherent superposition. The latter feature allows nondipole transitions at lower collision velocities, as observed experimentally.

The $|+\rangle$ and $|-\rangle$ states of Beigman and Syrkin are Stark states, and their description of slow collisions is related to one given originally by Smith *et al.*[3] They proposed that the collisional transfers occur if the field of the passing ion brings the initial state to an avoided crossing with a Stark state of the adjacent Stark manifold. This requirement is easily stated in a quantitative form,

$$\frac{\delta_i}{n^3} = \frac{3}{2}n^2 E_i, \tag{13.3}$$

where δ_i is the smallest absolute value of the quantum defect of the initial state modulo 1; i.e. the energy separation of the initial i state from the nearest high ℓ states is δ_i/n^3. Since E_i is given by $1/R^2$ where R is the distance between the ion and the Rydberg atom, a reasonable estimate of the cross section is obtained by using for R the impact parameter b. Explicitly,

$$\sigma = \pi b^2 = \pi \frac{3n^5}{2\delta_i}. \tag{13.4}$$

Although this simple expression does not allow for any velocity dependence of the cross section, it does agree with the observed zero field nd cross sections and implies that the Na nd cross sections are a factor of 10 larger than the np and ns cross sections, as observed experimentally. Furthermore, this picture suggests that during the collision the Rydberg high ℓ states are better described as Stark states than ℓ states, and that we might expect the $(n + 1)p$ and $(n + 1)s$ states to populate the highest and lowest members of the n Stark manifold preferentially. In the case of the s states there is no evidence to support this notion, but ion collisions with the Na np states produce atoms which ionize with high ionization fields, presumably due to diabatic ionization of either high ℓ states or higher lying Stark states. It is easy to believe that collisions transfer the Na np state atoms to high Stark levels when $R \sim b$ and that as the ion departs the Stark states relax to ℓ states. In sum, this simple picture allows us to describe many, but not all, of the features of state changing ion–atom collisions.

Electron loss

The logical extension of the state changing $\Delta\ell$ and Δn collisions is collisional ionization, for Na Rydberg atoms and Ar^+ the process

$$Na\,n\ell + Ar^+ \rightarrow Na^+ + e^- + Ar^+, \tag{13.5}$$

which is the dominant mechanism by which high velocity, $v > v_e$, ions remove the electron from a Rydberg atom. On the other hand, when $v/v_e \leq 1$ the incident ion passes by the Rydberg atom at a speed comparable to the Rydberg electron's, and the electron is much more likely to attach itself to the incident ion than to escape from both the Na^+ and Ar^+ ions.[11] In other words, charge exchange, the process

$$Na\,n\ell + Ar^+ \rightarrow Na^+ + Ar\,n'\ell' \tag{13.6}$$

is more likely to occur than collisional ionization in slow collisions. The sum of the cross sections for ionization, σ_I, and charge exchange, σ_{CX}, is usually termed the electron loss cross section, σ_{EL}, i.e.

$$\sigma_{EL} = \sigma_I + \sigma_{CX}. \tag{13.7}$$

It corresponds to the cross section for converting the Rydberg atom to an ion with no regard for the fate of the electron.

At high velocities, $v/v_e \gg 1$, the ionization cross sections can be described using the Born approximation, but the cross sections are small. In the region $v \sim v_e$, where the cross sections are large, the Born approximation is not valid, but there are so many open quantum mechanical channels that rigorous coupled channel calculations become impractical. Two theoretical approaches have proven to be useful. One is to apply classical scaling arguments to existing calculations of H^+–H 1s collisions. The essential notion is the following. If the ratio of the velocity of the ion to the Rydberg electron's velocity, v/v_e, is the same as the velocity of the proton relative to the velocity of the H ground state electron, then the ion–Rydberg atom cross section is n^4 times as large as the H^+–H 1s cross section, i.e. it is increased by the ratio of the geometric cross sections.

The second approach is to use classical trajectory Monte Carlo techniques to calculate directly ion–Rydberg atom cross sections.[11,12] In these calculations a set of initial conditions for the Rydberg electron is chosen to correspond to the initial n and ℓ of the Rydberg electron, and the classical trajectory of the electron is computed as the ion passes by. At the end of the collision the electron is either still bound in the initial atom, free, or bound to the incident ion, corresponding to no electron loss, ionization, and charge exchange, respectively. Olson[11] compared his Monte Carlo Results for n changing collisions to the analytic results of Lodge *et al.*[13] and found excellent, 25%, agreement. In Fig. 13.6 we show the calculated charge exchange and ionization cross sections for a Rydberg atom in states of $n = 15$ and $\ell = 2$ and 14 as functions of v/v_e. Although the cross sections differ for $\ell = 2$ and 14, it is apparent that for $v/v_e < 1$ charge exchange dominates, while for $v/v_e > 1$ ionization dominates.[12] As shown by the scale of Fig. 13.6, the peak cross sections for both processes are approximately equal to $\pi n^4 a_0^2$, the geometric cross section of the atom.

The first electron loss measurements were made by Koch and Bayfield[14] using a fast beam technique. They passed a H^+ beam through a charge exchange cell of Xe to produce collinear beams of H^+ and neutral H atoms, in all states. By modulating an axial field between 105 and 171 V/cm they were able to isolate, by field ionization, the signal from a band of about five n states centered at $n = 47$. Prior to entering a high vacuum collision chamber, the H^+ beam was decelerated to approximately the velocity of the H beam by a second axial field. Accelerating plates after the collision chamber accelerated the H^+ ions. The H^+ ions formed from H Rydberg atoms by electron loss collisions had slightly different energies from those of the primary H^+ beam, which allowed them to be detected separately from the ions in the far more intense primary H^+ beam. Using this technique they measured the electron loss cross section for center of mass collision energies from 0.2 to 60 eV, corresponding to v/v_e from 0.3 to 3.3. The cross section decreases from 8×10^{-9} cm^2 to 1×10^{-9} cm^2 over the velocity range examined. For reference, the geometric cross section, $\pi n^4 a_0^2$, is 4.3×10^{-10} cm^2. At high

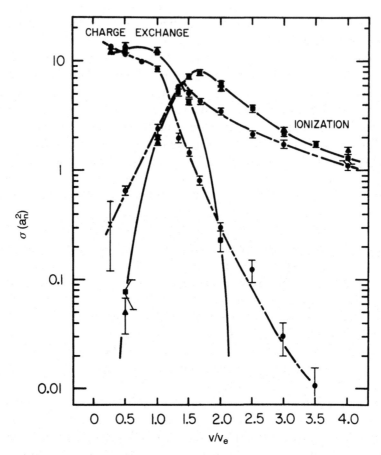

Fig. 13.6 Classical trajectory Monte Carlo (CTMC) ionization and charge transfer cross sections, with statistical standard deviations, for specific initial ℓ plotted against reduced impact speed v/v_e. The cross sections are given in units of a_n^2, where $a_n = n^2 a_0$. Circles are for $\ell = 2$, squares for $\ell = 14$. Also included for comparison are points from approximate CTMC calculations with the Na target core held fixed for $\ell = 14$ (triangles) and $\ell = 2$ (crosses) (from ref. 12).

velocities the observed cross sections have the same velocity dependence as calculated cross sections[15–17] but are a factor of 3.5 higher.

Measurements of electron loss from Na 40d and 30d states were done by MacAdam *et al.*[18] using crossed Na and ion beams, an arrangement similar to the one shown in Fig. 13.1. The resulting electron loss cross sections are shown in Fig. 13.7, along with the H^+–H $n = 47$ results of Bayfield and Koch scaled by $(40/47)^4$. The H^+–H results exhibit the same v/v_e dependence but are a constant factor of about 3.5 larger than the other results. As shown, the cross section for v/v_e is twice the geometric cross section. Several theoretical cross sections for electron loss and ionization are also shown in Fig. 13.7.[15,19–22] The theoretical ionization curves are the two which have cross sections which increase with velocity at low

Equivalent Energy for Stationary H(1s) Target (keV/amu)

Fig. 13.7 $n = 40$ electron loss cross sections of MacAdam *et al.*: solid circles (Xe$^+$), triangles (Ar$^+$), squares (Ne$^+$). Experimental results of ref. 13: open circles. Theoretical curves: (a) Janev (ref. 19); (b) Percival and Richards (ref. 20); (c) CTMC (ref. 21); (d) Born approximation (ionization only) (ref. 14); (e) 35-state close coupling (ionization only) (ref. 22). Geometric cross section is σ_g (from ref. 18).

velocities.[15,22] The experimental and theoretical cross sections are in excellent agreement for $v/v_e \geq 1$. Only the theoretical cross section of Janev[19] goes to low velocities. It is calculated by applying classical scaling laws to classical ionization and charge exchange cross sections. As shown by Fig. 13.7, the theoretical cross section is somewhat larger at low velocities than the experimental cross section. We have only discussed electron loss produced by collisions of singly charged ions with Rydberg atoms. Electron loss in collisions with multiply charged ions has also been observed.[23,24] but has not been studied extensively.

Charge exchange

As we have already mentioned in the discussion of electron loss collisions and shown in Fig. 13.6, ionizing collisions of Rydberg atoms with low velocity ions are

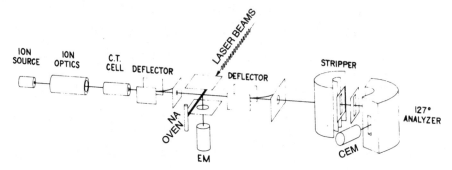

Fig. 13.8 Overall perspective view of the charge exchange apparatus. A thermal Na beam is laser excited in the interaction region to produce a target for Rydberg-to-Rydberg charge transfer experiments. The number of target Rydberg atoms is determined by pulsed field ionization between the parallel plates and by collection of the resultant ions in the electron multiplier EM (from ref. 27).

most likely to result in charge exchange, and this process has been measured directly by detecting the neutral product.[25,26]

The experimental arrangement used to measure the final n state distribution subsequent to charge exchange is shown in Fig. 13.8.[27] A neutral beam of Na atoms is excited to an ns or nd Rydberg state by two pulsed dye lasers. An Ar^+ ion beam, for example, with an energy from 10 to 2000 eV crosses the Rydberg atom target. Some of the ions undergo charge exchange, resulting in the production of fast neutral Ar $n\ell$ Rydberg atoms which travel colinearly with, and at the same speed as, the ion beam. The parent Ar^+ beam is deflected from the fast neutral beam with electrostatic deflector plates, and the fast neutral Ar Rydberg atom products of charge exchange are detected by field ionization.

In measurements of total charge exchange cross sections the stripper and 127° analyzer of Fig. 13.8 are replaced by a field ionizer with a 14 kV/cm field, one adequate to ionize states as low as $n = 15$, ensuring ionization of virtually all the charge exchange products from initial Rydberg states of $n = 25$.[26] The ions produced by field ionization are accelerated to an energy of ~14 keV, pass through a thin C foil, and are detected by a particle multiplier. The C foil blocks the passage of slower (<1 keV) non–Rydberg neutral atoms resulting from charge exchange with the background gas. The number of Rydberg atoms initially produced by laser excitation is determined by field ionization of the target Rydberg atoms, and the ion beam current is monitored with a Faraday cup. Using these three signals, the geometry, and the timing of the experiment, it is straightforward to extract the total charge exchange cross section.

In Fig. 13.9 we show the relative charge exchange cross sections measured for Na^+ ions impinging upon Na atoms initially in the 29s and 28d states, which are nearly degenerate in energy.[28] For $v/v_e > 0.8$ the two cross sections coincide, but for smaller v the 28d cross section falls below the 29s cross section, except at the lowest velocity. Also shown are the classically scaled cross sections for H^+–H 1s

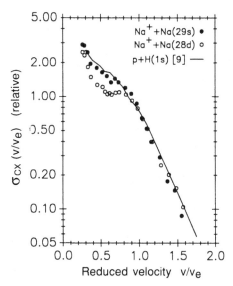

Fig. 13.9 Velocity dependence of charge transfer into all states. Na (28d) and Na (29s) targets. The line represents experimental data of McClure (ref. 29) for $H^+ + H(1s)$ electron capture scaled to match at $v/v_e = 1.0$ (from ref. 28).

charge exchange,[29] which are normalized to the experimental cross sections at $v/v_e = 1$. As can be seen in Fig. 13.9, the 29s cross section agrees almost perfectly with the scaled H ground state cross section, while the 28d cross section obviously does not; it is smaller. Precisely why the 29s cross section agrees with the scaled H^+–H 1s cross section and the 28d cross section does not is unclear, but there are several possibilities. First, the 29s and 28d states do not experience the same amount of collisional ℓ mixing. Presumably the initially populated 28d state has been spread over all the $n = 28$, $\ell \geq 2$ states, while the 29s state is not depopulated to the same extent. The 29s state has zero angular momentum, as does the H 1s state. In contrast, the $\ell \geq 2$ states have angular momentum and thus their orbits have classical outer turning points at smaller radii than the 29s orbit, which would tend to make the charge exchange cross section lower than the 29s cross section.

Equally as interesting as the total cross section is the distribution of the final Rydberg state products of charge exchange. Since the cross section for $v/v_e \sim 1$ is approximately the geometric cross section of the target Rydberg atom, it is not surprising that the final n state distribution is peaked near the n of the target Rydberg atom.[25]

The experiments to measure final state distributions are done using the apparatus of Fig. 13.8. The stripper and 127° analyzer allow the analysis of the final Rydberg states. When an Ar^+ ion beam is used the parent Ar^+ beam is deflected from the neutral Ar $n\ell$ charge exchange products by a pair of deflecting plates, and only the neutrals enter the concentric cylindrical stripper shown in Fig. 13.8. The outer cylinder is grounded, and the inner concentric cylinder is held at the voltage V_S, the stripper voltage. The field increases as $1/r$ between the two cylinders where r is measured from the mutual axis of the two cylinders. A neutral Ar $n\ell$ atom entering through a hole in the outer cylinder ionizes at the field corresponding to its $n\ell$ state and is then accelerated through a small hole in the inner cylinder. The field at which an atom ionizes determines the energy it acquires in passing through the stripper. The ions then pass through the 127° cylindrical energy analyzer set to pass ions with kinetic energies 200 eV less than eV_S. Scanning the stripper voltage over many laser shots yields the distribution of fields at which the final states ionize.

Converting the field ionization signals to an n distribution is a slightly ambiguous process. In principle, charge exchange should result in no change in m along the mutual axis of the ion beam and the field of the stripper, in which case the field ionization signals would be purely adiabatic if the initial state was $\ell \leq 2$. With purely adiabatic ionization it would be possible to make a unique conversion of a field ionization spectrum to an n distribution using $E_n \approx 1/16n^4$. However, the final state distribution is not in only a few low $|m|$ states. This point was verified by adding a 2.5 G magnetic field perpendicular to the ion beam to cause the magnetic moments of the atoms to precess about the field and alter the m values along the ion beam axis. The presence of the magnetic field did not alter the field ionization spectra. Therefore we are forced to conclude that the final state distribution is one consisting of a wide distribution of m states. This observation can be reconciled with an expectation of a $\Delta m = 0$ selection rule for charge exchange by recalling that the deflecting field separating the parent ions from the product Rydberg atoms is transverse, allowing Δm transitions.

To convert the observed field ionization spectra to n distributions two procedures were employed.[30] One, termed the "SFI centroid" approach, was simply to calculate the average field at which each n state is ionized, including both adiabatic, $|m| \leq 2$, and diabatic, $|m| > 2$, contributions. This procedure yields[30]

$$E_n = 3.3601 \times 10^8 n^{-3.8096}. \tag{13.8}$$

The field decreases less rapidly than n^{-4} because of the increasing number of high m states with increasing n. This method allows a unique transformation from the field ionization spectrum to an n distribution.

A second method is to assume that the final state distribution is given by[30,31]

$$P(n) = Cf(n)e^{-\mu(n^*/n-1)^2} \tag{13.9}$$

where the adjustable parameters n^* and μ specify the peak and width of the distribution. C is an arbitrary constant and $f(n)$ is a weighting factor. Choosing $f(n) = n$ corresponds to allowing all ℓ, but only low m, states to be populated, and choosing $f(n) = n^2$ corresponds to allowing all ℓm states to be populated. The maximum value of $P(n)$ occurs at $n = n_{\max}$. If $f(n) = 1$, $n_{\max} = n^*$. However, for the more physically likely cases $f(n) = n$ and $f(n) = n^2$, $n_{\max} > n^*$. For example, for $f(n) = n$ and $n \approx 25$ $n_{\max} \approx n^* + 1$. A synthetic field ionization signal is then computed using[30]

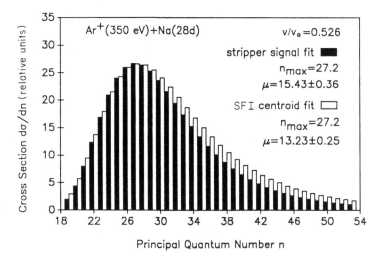

Fig. 13.10 Charge transfer final state n distributions following collisions of 350 eV Ar^+ with Na 28d atoms obtained by the SFI centroid fit and the probability distribution of Eq. (13.9) (from ref. 30).

$$S(E) = \Sigma P(n)\mathcal{E}(n,E), \qquad (13.10)$$

where $\mathcal{E}(n,E)$ is the probability that the state n ionizes at the field E, assuming that the $m \leq 2$ and $m > 2$ states ionize adiabatically and diabatically, respectively. The computed values of $S(E)$ are compared to field ionization spectra to determine $P(n)$. The difficulty is, of course, that the fields at which different n states ionize overlap.

Fortunately, using the two methods of analysis described above it is possible to obtain similar n distributions, as shown in Fig. 13.10. A very interesting aspect of the fit using Eqs (13.9) and (13.10) is that $f(n) = n$ produces a better fit than does $f(n) = n^2$. This choice is consistent with the electron in an initial low ℓ Rydberg state's being captured into a state of any ℓ but low m.

The distribution of the final states depends on the velocity of the ion, as shown by Fig. 13.11. As v/v_e is raised, the peak of the final state distribution first rises then falls. In general, n^* is very near the n of the initial Rydberg state. Also, n^* is systematically higher for the $(n + 1)s$ state than the nd state. Finally, the maximum value of n^* occurs for $v/v_e = 0.8$. The widths of the final state distributions also depend on v/v_e. For example, of the three distributions shown in Fig. 13.11, it is apparent that the one for $v/v_e = 0.751$ is the narrowest. Finally, it is difficult to discern a significant difference between the widths of the final state distributions for the $(n + 1)s$ and nd states.

The observed final state distributions are in reasonable agreement with CTMC calculations of Becker and MacKellar,[12] who calculated the distributions for different initial values of ℓ. In Fig. 13.12 we show the theoretical final n distributions for target Rydberg atoms of $n = 28$ and different values of ℓ for

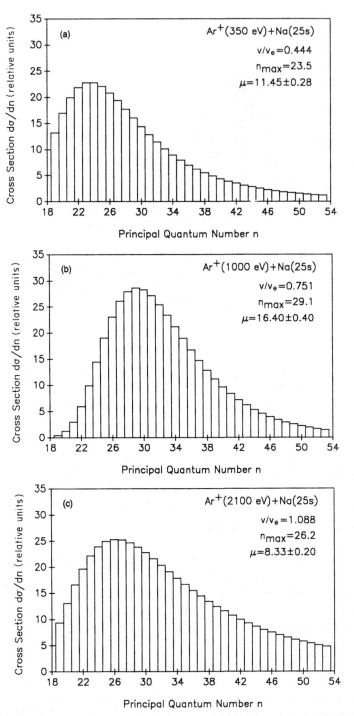

Fig. 13.11 Final state histograms for the process, Ar$^+$ + Na 25s, demonstrating low, intermediate, and high energy behavior. The distributions are obtained by using Eq. (13.9) to fit the data: (a) 350 eV; (b) 1000 eV; (c) 2100 eV (from ref. 30).

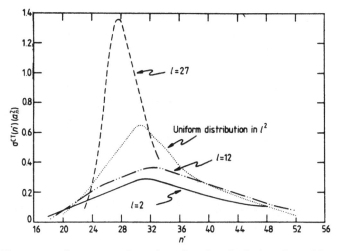

Fig. 13.12 Charge transfer cross sections plotted against final n' at $v/v_e = 1$ for the reaction $\mathrm{Na}^+ + \mathrm{Na}\,(n = 28,\ell) \rightarrow \mathrm{Na}\,(n', \text{all } \ell') + \mathrm{Na}^+$. The cross sections are expressed in terms of $a_n{}^2$, where $a_n = n^2 a_0$. The distributions are shown for $\ell = 2, 12, 27$ and for an initial distribution uniform in ℓ^2 (from ref. 12).

$v/v_e = 1$. As shown by Fig. 13.12, for $\ell = 27$ the theoretical final n state distribution is quite narrow, while for $\ell = 2$ it is very broad. The distribution for all initial ℓ states, however, provides the best match to the experimental data, such as those of Fig. 13.11. Since ℓ mixing by the ions has a much larger cross section than charge exchange, most Rydberg atoms undergoing charge exchange are equally likely to be in all ℓ and m states.

Although extensive charge exchange measurements have only been done with singly charged ions, charge exchange with multiply charged ions is now being explored.[32]

Electron collisions

Electron collisions with Rydberg atoms are similar to ion collisions in that they are of very long range. They differ, however, in several important respects. First, charge exchange is clearly not possible, leaving only state changing and ionizing collisions. As shown by Fig. 13.6, the ionization cross sections by ions peak when $v/v_e \approx 1$. For electron collisions the same criterion applies, and for Rydberg atoms thermal, 0.01–0.1 eV, electrons are the ones which are most effective in inducing collisions. Several indirect indications exist of the effects of electron collisions on Rydberg atoms. Schiavone *et al.*[33,34] produced Rydberg atoms by electron bombardment of ground state atoms and observed atoms with high threshold fields for ionization and such long radiative lifetimes that they must have been high m states of relatively low n. In addition, at very low electron currents they observed a nonlinear dependence of the Rydberg atom production on the electron current, suggesting that the long lived atoms were being produced in two steps. First, low ℓ Rydberg states were produced by electron collisions with ground state atoms, then in subsequent collisions with electrons these atoms were collisionally transferred to high ℓ states. Since the cross sections for transitions between Rydberg states are large compared to the cross sections for exciting the Rydberg states, only at very low electron bombardment currents is the nonlinearity in the electron current visible. They estimated the size of the cross section for ℓ changing of $n > 20$ atoms to be 10^{-10}–10^{-9} cm^2.[33,34] Similar observations of long lived, presumably high ℓ, states were reported by Kocher and Smith,[35] who excited Li atoms to Rydberg states by electron impact. However, they were unable to determine if the high ℓ atoms were the result of single or multiple electron collisions.

Foltz *et al.*[36] have made systematic measurements of electron collisions with laser excited Na nd Ryderg atoms using an apparatus similar to the one shown in Fig. 13.1 but differing in two respects. First, an electron beam is used instead of an ion beam, and second, a magnetic shield encloses the electron beam and interaction region.

Rydberg Na nd atoms are produced by two step pulsed laser excitation, via the 3p state. The atoms are exposed to a 25 eV electron beam of current 1 μA and diameter 1 cm for a time of 6 μs, after which the Na atoms are selectively field ionized. When the atoms have been exposed to the electron beam, two new features appear in the field ionization signal, as shown by Fig. 13.13. The adiabatic feature at an ionization field below the field of the initial nd state indicates the presence of higher ℓ, $m < 2$ states, and the diabatic feature indicates the presence of higher ℓ, $m \geq 2$ states. Comparing these signals to the total Rydberg atom signal allows the cross sections to be determined. The results are given in Table 13.1 and compared to the theoretical results of Percival and Richards[4] and Herrick,[5] two of the theories shown in Fig. 13.3. As can be seen from Table 13.1 and Fig. 13.3 both theories agree fairly well with both sets of experimental data even though there is a difference of several orders of magnitude in the cross sections. The difference in the size of the cross sections is due to the relatively high velocity of the 25 eV electrons. A 25 eV electron has a velocity of 3×10^8 cm/s, more than an order of magnitude higher than the typical ion velocities represented in Fig. 13.3. Since the cross sections at high velocities decrease approximately as $1/v^2$ it is not surprising that the observed electron cross sections are so much smaller than the ion cross sections.

While electrons in conventional beams have velocities too high to have large cross sections, thermal electrons have large cross sections for state changing collisions with Rydberg atoms, and these collisions have been studied in a systematic fashion. Specifically, metastable He atoms in a stationary afterglow have been excited to specific Rydberg states with a laser.[37,38] The populations of

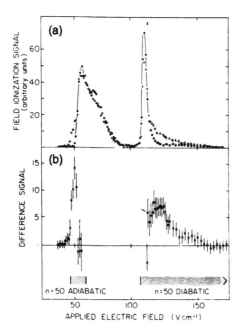

Fig. 13.13 Typical selective field ionization data for laser excited Na 50d atoms: (a) data with electron beam gated off (●), data following collisions with 25 eV electrons (+) corrected for electron-induced background signals; (b) net signal due to electron impact. The horizontal bars indicate the range of field strengths over which $n = 50$ atoms are expected to ionize adiabatically and diabatically (from ref. 36).

Table 13.1. *Measured and calculated cross sections for Na nd ℓ changing collisions with 25 eV electrons.*

		Theoretical $nd \rightarrow nf$	
n	Experimental $nd \rightarrow n\ell^a$ (cm^2)	Percival & Richards[b] (cm^2)	Herrick[c] (cm^2)
35		2.1×10^{-9}	2.0×10^{-9}
36	1.6×10^{-10}		
40	1.5×10^{-9}	3.6×10^{-9}	3.5×10^{-9}
45	3.0×10^{-9}	5.8×10^{-9}	5.7×10^{-9}
50	3.4×10^{-9}	9.0×10^{-9}	8.9×10^{-9}

[a] (from ref. 36)
[b] (from ref. 4)
[c] (from ref. 5)

all ℓ states of the same n were monitored by time-resolved $np \rightarrow 2s$ fluorescence to determine the total population and depopulation rates by thermal electrons. From the results several conclusions may be drawn. First, the collisional mixing of the degenerate ℓm levels of the same n is immeasurably rapid and is therefore at least two orders of magnitude faster than Δn changing rates. From the buildup times of the fluorescence from n levels other than the one populated by the laser the cross sections to those levels can be determined, and they are $\sim 10^{-11}$ cm^2 and decrease rapidly with increasing Δn. This dependence is in agreement with the Monte Carlo calculations of Mansback and Keck[39] and shows clear disagreements with the more analytical approach of Johnson and Hinnov,[40] which is based on more restrictive assumptions, such as dipole transitions, and which predicts a strong $\Delta n = \pm 1$ selection rule.

References

1. K. B. MacAdam, D. A. Crosby, and R. Rolfes, *Phys. Rev. Lett.* **44**, 980 (1980).
2. K. B. MacAdam, R. Rolfes, and D. A. Crosby, *Phys. Rev. A* **24**, 1286 (1981).
3. W. W. Smith, P. Pillet, R. Kachru, N.H. Tran, and T.F. Gallagher, *Abstracts*, ICPEAC 13, eds. J. Eichler, W. Fritsch, I. V. Hertel, N. Stotlerfoht, and U. Wille (North Holland, Amsterdam, 1983).
4. I. C. Percival and D. R. Richards, *J. Phys. B* **10**, 1497 (1977).
5. D. R. Herrick, *Mol. Phys.* **35**, 1211 (1976).
6. V. P. Shelvelko, A. M. Urnov, and A. V. Vinograd, *J. Phys. B* **9**, 2859 (1976).
7. I. L. Beigman and M. I. Syrkin, *Sov. Phys. JETP* **62**, 226 (1986) [Zh.Eksp. Teor. Fiz 89, 400 (1985)].
8. K. B. MacAdam, D. B. Smith, and R. G. Rolfes, *J. Phys. B.***18**, 441 (1985).
9. K. B. MacAdam, R. G. Rolfes, X. Sun, J.Singh, W. L. Fuqua III, and D. B. Smith, *Phys. Rev. A* **36**, 4254 (1987).
10. R. G. Rolfes, D. B. Smith, and K. B. MacAdam, *Phys. Rev. A* **37**, 2378 (1988).
11. R. E. Olson, *J. Phys. B* **13**, 483 (1980).
12. R. L. Becker and A. D. MacKellar, *J. Phys. B* **17**, 3923 (1984).
13. J. G. Lodge, I. C. Percival, and D. Richards, *J. Phys. B* **9**, 239 (1976).
14. P. M. Koch and J. A. Bayfield, *Phys. Rev. Lett.* **34**, 448 (1975).
15. D. R. Bates and G. Griffing, *Proc. Phys. Soc. London* **66**, 961 (1953).
16. R. Abrines and I. C. Percival, *Proc. Phys. Soc. London* **88**, 873 (1966).
17. D. Banks, PhD. thesis, University of Stirling (1972).
18. K. B. MacAdam, N. L. S. Martin, D. B. Smith, R. G. Rolfes, and D. Richards, *Phys. Rev. A* **34**, 4661 (1986).
19. R. K. Janev, *Phys. Rev. A* **28**, 1810 (1983).
20. I. C. Percival and D. Richards, *Adv. Atomic and Molecular Physics* **11**, 1 (1975).
21. R. E. Olson, K. H. Berkner, W. G. Graham, R. V. Pyle, A. E. Schlacter, and J. W. Stearns, *Phys. Rev. Lett.* **41**, 163 (1976).
22. R. Shakeshaft, *Phys. Rev. A* **18**, 1930 (1978).
23. H. J. Kim and F. W. Meyer, *Phys. Rev. Lett.* **44**, 1047 (1980).
24. R. E. Olson, *Phys. Rev. A* **23**, 3338 (1981).
25. K. B. MacAdam and R. G. Rolfes, *J. Phys. B* **15**, L243 (1982).
26. S. B. Hansen, L. G. Gray, E. Hordsal-Pederson, and K. B. MacAdam, *J. Phys. B* **24**, L315 (1991).
27. K. B. MacAdam and R. G. Rolfes, *Rev. Sci. Instr.* **53**, 592 (1982).

28. K. B. MacAdam, *Nucl. Instr. and Methods B* **56/57**, 253 (1991).
29. G. W. McClure, *Phys. Rev.* **148**, 47 (1966).
30. K. B. MacAdam, L. G. Gray, and R. G. Rolfes, *Phys. Rev. A* **42**, 5269 (1990).
31. T. Aberg, A. Blomberg, and K. B. MacAdam, *J. Phys. B* **20**, 4795 (1987).
32. B. D. De Paola, J. J. Axmann, R. Parameswaran, D. H. Lee, T. J. M. Zouros, and P. Richard, *Nucl. Instr. and Methods B* **40/41** 187 (1989).
33. J. A. Schiavone, D. E. Donohue, D. R. Herrick, and R. S. Freund, *Phys. Rev. A* **16**, 48 (1977).
34. J. A. Schiavone, S. M. Tarr, and R. S. Freund, *Phys. Rev. A* **20**, 71 (1979).
35. C. A. Kocher and A. J. Smith, *Phys. Lett.* **61A**, 305 (1977).
36. G. W. Foltz, E. J. Beiting, T. H. Jeys, K. A. Smith, F. B. Dunning, and R. F. Stebbings, *Phys. Rev. A* **25**, 187 (1982).
37. J. F. Delpech, J Boulmer, and F. Devos, *Phys. Rev. Lett.* **39**, 1400 (1977).
38. F. Devos, J. Boulmer, and J. F. Delpech, *J. Phys.* (Paris) **40**, 215 (1979).
39. P. Mansbach and J. C. Keck, *Phys. Rev.* **181**, 275 (1969).
40. L. C. Johnson and E. Hinnov, *Phys. Rev.* **181**, 143 (1969).

14

Resonant Rydberg–Rydberg collisions

Resonant energy transfer collisions, those in which one atom or molecule transfers only internal energy, as oppposed to translational energy, to its collision partner require a precise match of the energy intervals in the two collision partners. Because of this energy specificity, resonant collisional energy transfer plays an important role in many laser applications, the He–Ne and CO_2 lasers being perhaps the best known examples.[1–4] It is interesting to imagine an experiment in which we can tune the energy of the excited state of atom B through the energy of the excited state of atom A, as shown in Fig. 14.1.[5] At resonance we would expect the cross section for collisionally transferring the energy from an excited A atom to a ground state B atom to increase sharply as shown in Fig. 14.1. In general, atomic and molecular energy levels are fixed, and the situation of Fig. 14.1 is impossible to realize. Nonetheless systematic studies of resonant energy transfer have been carried out by altering the collision partner, showing the importance of resonance in collisional energy transfer.[6–8]

The use of atomic Rydberg states, which have series of closely spaced levels, presents a natural opportunity for the study of resonant collisional energy transfer. One of the earliest experiments was the observation of resonant rotational to electronic energy transfer from NH_3 to Xe Rydberg atoms by Smith et al.[9] Resonant electronic to vibrational energy transfer has also been observed, from Rydberg states of Na to CH_4 and CD_4.[10] Both of these experiments, which are described in Chapter 11, correspond to tuning in steps equal to the discrete spacing of the Rydberg levels. In an ideal study of resonant collisions it would be

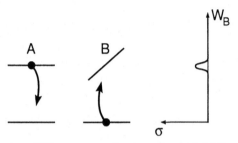

Fig. 14.1 Two atoms, A and B have energy levels as shown. Initially atom A is in its excited state, and atom B is in its ground state. If the energy of the excited state of atom B could be tuned by some means, we would expect the cross section for resonant energy transfer from atom A to atom B to increase at resonance as shown on the right (from ref. 5).

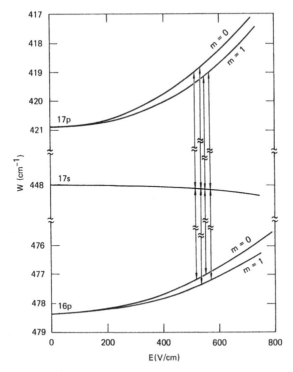

Fig. 14.2 Energy level diagram for the Na 16p, 17s, and 17p states in a static electric field. The vertical lines are drawn at the four fields where the s state is midway between the two p states and the resonant collisional transfer occurs (from ref. 12).

possible to tune continuously, as shown in Fig. 14.1, to observe the collisional resonance explicitly. In fact, due to the large Stark shifts of Rydberg atoms it is possible to tune them through a collisional resonance. A good example is the resonant energy transfer in the collision of two Na ns atoms in an electric field.[11] In Fig. 14.2 we show the energy levels of the Na ns, $(n-1)$p, and np states as a function of electric field. In zero field the ns state lies slightly above midway between the two p states. However, as the field is increased the p states are shifted to higher energies by their Stark shifts. As shown by Fig. 14.2, the field also lifts the degeneracy of the $|m| = 0$ and 1 levels of the p states. Here m is the azimuthal orbital angular momentum quantum number. Due to the lifting of the m degeneracy there are four fields at which the ns state lies halfway between the two p states. At these fields two ns atoms can collide to produce an $(n-1)$p and an np atom by the process

$$\text{Na } ns + \text{Na } ns \rightarrow \text{Na}(n-1)\text{p} + \text{Na } n\text{p}. \qquad (14.1)$$

As shown by Fig. 14.2, there are four collisional resonances for each ns state. We label them by (m_ℓ, m_u), where m_ℓ and m_u are the $|m|$ values of the lower and upper final p states. As shown by Fig. 14.2 in order of increasing field the

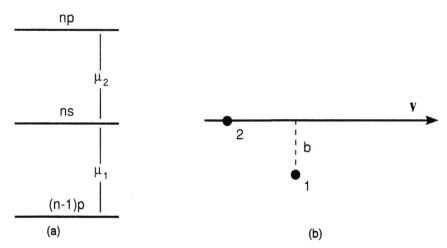

Fig. 14.3 (a) Energy levels for the Na ns, np and $(n-1)$p states showing the dipole matrix elements coupling them. (b) Diagram of the geometry of the collision in which atom 2 passes by atom 1 with velocity **v** and impact parameter b (from ref. 5).

resonances are $(0,0)$, $(1,0)$, $(0,1)$, and $(1,1)$. The Na collision process described in Eq. (14.1) and depicted in Fig. 14.2 is a resonant dipole–dipole process, and a simple picture enables us to determine both the magnitude of the peak cross section and the width of the collisional resonance. We consider the collision of two Na atoms initially in the ns state. During the collision they undergo transitions to the np and $(n-1)$p states, by means of dipole transitions with the dipole matrix elements μ_1 and μ_2, as shown by Fig. 14.3. We assume that the atoms follow undeflected straight line trajectories described by the impact parameter b and collision velocity **v**, as shown in Fig. 14.3. For simplicity, we first consider the special case of the static field's being tuned to the collisional resonance, i.e. to the field at which the energy of the ns state is midway between those of the np and $(n-1)$p states. We wish to calculate the probability of the atoms' making the transition to the np and $(n-1)$p states during the collision. One approach is to treat one atom, atom 2, as providing an oscillating field which drives the transition in the other atom, atom 1.[12] In other words, we treat atom 2 as a classical dipole. Classically, a dipole transition matrix element becomes a dipole of strength equal to the transition matrix element oscillating at the transition frequency. Thus atom 2 can be thought of as a collection of dipoles corresponding to the $ns \rightarrow n'$p dipole transitions allowed from the ns state. Most of these transitions are off resonant when compared to the $ns \rightarrow (n-1)$p transition, and we ignore them. The $ns \rightarrow (n-1)$p transition in atom 2, while degenerate can be ignored on the basis of energy conservation (both atoms cannot make transitions to lower energy states). Thus the only important transition of atom 2 is the $ns \rightarrow np$ transition, in which case the field produced by atom 2 at atom 1 is given by $E_2 \cong \mu_2/r^3$, μ_2 being the ns–np dipole matrix element, **r** the distance vector from atom 1 to atom 2, and *r* the internuclear separation between the two atoms. This field is resonant with the $ns \rightarrow (n-1)$p transition of atom 1 and drives the transition if the interaction is strong enough. Since the dipole field falls quickly with increasing *r*, it is a reasonable approximation to set

$$E_2 = \mu_2/b^3 \qquad r \leqslant \frac{\sqrt{5}}{2} b \qquad (14.2a)$$

$$E_2 = 0 \qquad r > \frac{\sqrt{5}}{2} b. \qquad (14.2b)$$

This choice of cutoff implies that the field from atom 2 is present for a time $\tau = b/v$. In general atom 1 undergoes the $ns \rightarrow (n-1)$p transition if

$$E_2 \cdot \mu_1 \cdot \tau \approx 1, \qquad (14.3)$$

or if

$$\frac{\mu_2 \mu_1}{b^3} \cdot \frac{b}{v} = 1. \qquad (14.4)$$

In Eqs. (14.3) and (14.4) μ_1 is the ns–$(n-1)$p dipole matrix element. Eq. (14.3) makes it apparent that the requirement for driving the transition is that the time integrated interaction be one. Solving Eq. (14.4) for b^2, ignoring factors of π, gives the cross section σ_R for the resonant collisional energy transfer. Explicitly,

$$\sigma_R \approx b^2 = \frac{\mu_1 \mu_2}{v}. \qquad (14.5)$$

Using the impact parameter *b* and the velocity *v* we can also calculate the duration, or time, of the collision,

$$\tau = \frac{b}{v} = \frac{\sqrt{\mu_1 \mu_2}}{v^{3/2}}. \qquad (14.6)$$

For the $ns \rightarrow n$p and $ns \rightarrow (n-1)$p transitions the dipole matrix elements are $\sim n^2$. Accordingly, we insert n^2 for μ_1 and μ_2 in Eqs. (14.5) and (14.6), finding

$$\sigma_R = n^4/v \qquad (14.7)$$

and

$$\tau = n^2/v^{3/2}. \qquad (14.8)$$

As shown by Eq. (14.7), the cross section is given by the geometric cross section divided by the collision velocity. For thermal collisions $v \sim 10^{-4}$, and as a result, $\sigma_R \sim 10^4 n^4$. For $n=20$ this expression yields a cross section of $\sim 10^9 a_0^2$, or $\sim 2 \times 10^8$ Å^2, which is large by atomic standards. Perhaps more interesting than the magnitudes of the cross sections are the collision times. Evaluating Eq. (14.8) for $n = 20$ and $v = 10^{-4}$ gives $\tau = 4 \times 10^8$ or $\sim 10^{-9}$ s, a time far longer than the typical atomic collision times of 10^{-12} s.

The discussion above is focused on dipole–dipole collision processes, but it may easily be extended to higher multipole processes. If atoms 1 and 2 have $2k$ pole moments of n^{2k} and $n^{2k'}$ respectively, when the atoms are separated by a distance *r* the interaction *V* due to these multipoles is

$$V = \frac{n^{2k}n^{2k'}}{r^{1+k+k'}}. \tag{14.9}$$

If we again assume that the atoms collide with an impact parameter b and velocity v, following straight line trajectories , we can again assume that V is only non zero for $r \approx b$ and a time interval $\tau = b/v$. Requiring that $V\tau = 1$ for a significant transition probability leads to

$$\sigma \cong b^2 = \frac{n^4}{v^{2/(k+k')}}. \tag{14.10}$$

For a dipole–dipole collision, $k = k' = 1$, and this result reduces to Eq. (14.5). On the other hand, as k and k' increase, the factor $v^{2/(k+k')}$ approaches 1 from below, and the cross section decreases to the geometric size of the atom. It is worth noting that the assumption of a straight line trajectory is not strictly valid in mixed multipole collisions, or in any collision in which the internal angular momentum of the atoms is not conserved. However, the amount of angular momentum in the translational motion is usually sufficiently high that small changes in it do not significantly alter the trajectories of the colliding atoms.

Two state theory

The expression of Eq. (14.7) for the cross section is adequate for a calculation of the magnitude of the cross section, but it does not take into account effects due to the orientation of the collision velocity **v** relative to the tuning field **E**, nor does it allow the calculation of the lineshape of the collisional resonance. With these thoughts in mind, we present a more refined treatment of the problem which enables us to describe some of the details of the process. We shall consider the resonant collisions of two Na ns atoms to yield an np and an $(n-1)p$ atom, as indicated by Eq. (14.1). The description is, in principle, similar to the treatment of resonant rotational energy transfer between polar molecules given by Anderson.[8] It differs, however, in that the presence of the electric field, which lowers the symmetry, must be taken into account. Treatments specific to the problem of dipole–dipole collisions in a field, such as the process depicted by Fig. 14.2, have been given by Gallagher *et al.*,[12] Fiordilino *et al.*,[13] and Thomson *et al.*[14]

We assume again that the atoms follow straight line trajectories, and we calculate the transition probability, $P(b)$, from the initial to the final state in a collision with a given impact parameter, b. We then compute the cross section by integrating over impact parameter, and, if necessary, angle of **v** relative to **E** to obtain the cross section. The central problem is the calculation of the transition probability $P(b)$. The Schroedinger equation for this problem has the Hamiltonian

$$H = H_0 + V, \tag{14.11}$$

where $H_0 = H_1 + H_2$ is the Hamiltonian for two non–interacting atoms in the static field, and V is the dipole–dipole interaction

$$V = \frac{\boldsymbol{\mu}_1 \cdot \boldsymbol{\mu}_2}{r^3} - \frac{3(\boldsymbol{\mu}_1 \cdot \mathbf{r})(\boldsymbol{\mu}_2 \cdot \mathbf{r})}{r^5}. \tag{14.12}$$

We construct the molecular product states which have the spatial wavefunctions

$$\psi_A = \psi_{1ns} \otimes \psi_{2ns}$$
$$\psi_B = \psi_{1np} \otimes \psi_{2(n-1)p}, \tag{14.13}$$

where $\psi_{1n\ell}$ and $\psi_{2n\ell}$ are the atomic wavefunctions of atoms 1 and 2 in the $n\ell$ state. These product states are the time independent solutions to H_0, having energies $W_A = W_{ns} + W_{ns}$ and $W_B = W_{np} + W_{(n-1)p}$.

The dipole–dipole interaction V has a diagonal term, which slightly shifts the energies, and an off diagonal term, which induces the collisional transitions. Since V depends on r, V is time dependent, and the wavefunction is given by the solution to the time dependent Schroedinger equation

$$H\psi = i\partial\psi/\partial t. \tag{14.14}$$

When the two possible states are those given by Eq. (14.13) the general form of the solution to Eq. (14.14) is

$$\psi = C_A(t)\psi_A + C_B(t)\psi_B, \tag{14.15}$$

in which all of the time dependence resides in the coefficients $C_A(t)$ and $C_B(t)$. Using the wavefunction of Eq. (14.15) in the time dependent Schroedinger equation, Eq. (14.14), we find two coupled equations for $C_A(t)$ and $C_B(t)$,

$$W_A C_A(t) + V_{AB}C_B(t) + V_{AA}C_A(t) = i\dot{C}_A(t), \tag{14.16a}$$
$$W_B C_B(t) + V_{BA}C_A(t) + V_{BB}C_B(t) = i\dot{C}_B(t), \tag{14.16b}$$

where $V_{BA} = \langle \psi_B | V | \psi_A \rangle = \int \psi_B^* V \psi_A \, d\tau_1 \, d\tau_2$ is the matrix element of V between the two spatial wavefunctions.

In Eqs. (14.16) the matrix elements of V are time dependent, since the internuclear distance is a function of time, as are the coefficients $C_A(t)$ and $C_B(t)$. In general, Eqs. (14.16) cannot be solved analytically, but in several special cases analytic solutions can be obtained.

Box interaction strength approximation

If we set $V_{AA} = V_{BB} = 0$ and use Eq. (14.2), i.e. assume that V_{AB} is a constant, $= \mu_1\mu_2/b^3$, for $r < \sqrt{5}b/2$ and zero for $r > \sqrt{5}b/2$, as given in Eq (14.2), the resulting approximation is equivalent to the usual molecular beam magnetic resonance treatment.[15] If we assume that $r = \sqrt{b^2 + v^2 t^2}$ and that initially both atoms are in the ns state so that $\psi(t_0) = \psi_A$, $C_A(t_0) = 1$ and $C_B(t_0) = 0$ for $t_0 < b/2v$. After the collision, $t > b/2v$, the probability of finding the atoms in the p states, $\psi = \psi_B$, is given by

$$P(b) = |C_B(t)|^2 = \frac{|V_{BA}|^2}{\Omega^2} \sin^2\left(\frac{\Omega b}{v}\right), \tag{14.17}$$

where $\Omega = \sqrt{(W_A - W_B)^2 + 4|V_{BA}|^2}/2$. Eq. (14.17) yields Lorentzian resonances with linewidths $\sim(\sqrt{v^3/\mu_1\mu_2})$, in agreement with Eq. (14.6). While the widths are meaningful, the lineshape, which depends on the form of the interaction during the collision, is only approximate.

At resonance $W_A = W_B$, and Eq. (14.17) reduces to

$$P(b) = \sin^2\left(\frac{\mu_1\mu_2}{b^2 v}\right), \tag{14.18}$$

in which we have used the explicit form $V_{AB} = \mu_1\mu_2/b^3$.

Exact resonance approximation

Another case which may be treated analytically is the case of exact resonance, $W_A = W_B$, if in addition, $V_{AA} = V_{BB} = 0$ and $V_{AB} = V_{BA}$, i.e. the coupling matrix element is real. In this case Eqs. (14.16) are readily decoupled, leading to two identical uncoupled equations. The equation for $C_A(t)$ is

$$\ddot{C}_A(t) = \frac{\dot{V}_{AB}}{V_{AB}} \dot{C}_A(t) - V_{AB}^2 C_A(t). \tag{14.19}$$

The equation for $C_B(t)$ has identical coefficients. If we again assume that initially, at $t = -\infty$, $C_A(-\infty) = 1$ and $C_B(-\infty) = 0$ the solutions to Eq. (14.19) which satisfy these boundary conditions are[12,13]

$$C_A(t) = \cos\left(\int_{-\infty}^{t} V_{AB}(t')dt'\right) \tag{14.20a}$$

and

$$C_B(t) = \sin\left(\int_{-\infty}^{t} V_{AB}(t')dt'\right). \tag{14.20b}$$

The probability of making a transition is therefore given by

$$P(\infty) = C_B^2(\infty) = \sin^2\left(\int_{-\infty}^{\infty} V_{AB}(t')dt'\right). \tag{14.21}$$

To obtain an explicit form for the matrix element V_{AB} given in Eq. (14.12), requires that we define the geometry of the collision. While the collision velocity is normally the logical choice of quantization axis for field free collisions, the

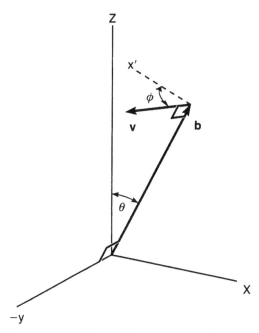

Fig. 14.4 Diagram of the geometry of the collision. The static field is in the z direction. The impact parameter vector **b** lies in the x,z plane, at an angle θ from the z axis. The velocity vector **v** lies in the plane perpendicular to **b** and makes an angle ϕ with the x' axis, which is perpendicular to **b** and in the x,z plane (from ref. 5).

presence of the tuning electric field makes the **E** field direction the logical choice, since the eigenstates of H_0 are easily described with this choice of quantization axis. If we define the field direction as the z axis, so that $\mathbf{E}\|\hat{z}$, then the impact parameter vector **b** makes an angle θ with the z axis as shown in Fig. 14.4. We may assume, with no loss of generality, that **b** lies in the x,z plane. The collision velocity **v**, which is a constant vector due to the assumption of straight line trajectories, is normal to **b** and makes an angle ϕ with the x' axis, which is in the x,z plane and perpendicular to **b**, as shown in Fig. 14.4. If the closest approach of the two atoms occurs at $t = 0$, the coordinates of atom 2 relative to atom 1 are given by

$$y = -vt \sin \phi$$
$$x = b \sin \theta - vt \cos \phi \cos \theta \qquad (14.22)$$
$$z = b \cos \theta + vt \cos \phi \sin\theta.$$

Using these coordinate values we may now evaluate the matrix elements of Eq. (14.12) by substituting for the dipole moments the dipole matrix elements between the initial and final states. This procedure yields explicitly time dependent matrix elements $V_{AB}(t)$. It is particularly interesting to consider the (0,0) resonances, for two reasons. First, the (0,0) resonances have no further splitting due to the spin orbit interaction and are therefore good candidates for detailed experimental study. Second, since these resonances only involve the matrix elements of μ_z, which are real, it is possible to evaluate the transition probability at resonance analytically using Eq. (14.21). The matrix element $V_{AB}(t)$ is given by

$$V_{AB}(t) = \mu_{z_1}\mu_{z_2}\left(\frac{1}{r^3} - \frac{3b^2\cos^2\theta}{r^5} - \frac{6bvt\cos\theta\cos\phi}{r^5} - \frac{3v^2t^2\cos^2\phi\sin^2\theta}{r^5}\right), \quad (14.23)$$

where $r = \sqrt{b^2 + v^2t^2}$.

In Fig. 14.5 we show graphs of $V_{AB}(t)$ for two cases; $\mathbf{v}\|\mathbf{E}$, and one example of $\mathbf{v}\perp\mathbf{E}$, for which $\theta = 0$ and $\phi = \pi/2$. Note that the specification $\mathbf{v}\|\mathbf{E}$, specifies both θ and ϕ. However $\mathbf{v}\perp\mathbf{E}$, only specifies $\phi = \pi/2$; θ can take any value.

The form of $V_{AB}(t)$ given in Eq. (14.23) can be integrated analytically to find the transition probability at resonance. Carrying out the integration yields

$$\int_{-\infty}^{\infty} V_{AB}(t')dt' = \frac{2\mu_{z_1}\mu_{z_2}}{vb^2}(1 - 2\cos^2\theta - \cos^2\phi\sin^2\theta). \quad (14.24)$$

Evaluating Eq. (14.24) for $\mathbf{v}\|\mathbf{E}$, $\theta = \pi/2$ and $\phi = 0$, reveals that the integral vanishes, as might have been anticipated from a close examination of Fig. 14.5(a). The average value of V_{AB} in Fig. 14.5(a) is clearly near zero. On the other hand, for $\mathbf{v}\perp\mathbf{E}$, $\theta = 0$ and $\phi = \pi/2$, the integral of Eq. (14.24) does not vanish, since the average value of $V_{AB}(t)$ does not vanish in this case, as can be seen in Fig. 14.5(b). Correspondingly, at resonance, the cross section must vanish for $\mathbf{v}\|\mathbf{E}$ but not for $\mathbf{v}\perp\mathbf{E}$.

Numerical calculations

Normally one might expect that if the transition probability vanishes on resonance it also vanishes off resonance. However, such is not the case. When the transition probability is calculated off resonance, by numerically solving Eqs. (14.16) using a Taylor expansion method, it is nonzero for both $\mathbf{v}\|\mathbf{E}$ and $\mathbf{v}\perp\mathbf{E}$.[14,16] In Fig. 14.6 we show the transition probabilities obtained using two different approximations for $\mathbf{v}\|\mathbf{E}$, and $\mathbf{v}\perp\mathbf{E}$ for the 17s (0,0) collisional resonance.[16] To allow direct comparison to the analytic form of Eq. (14.21) we show the transition probabilities calculated with $V_{AA} = V_{BB} = 0$. For these calculations the parameters $\mu_{z_1} = \mu_{z_2} = 156.4\ ea_0$, $b = 10^4a_0$, and $v = 1.6 \times 10^{-4}$ au have been used. The resulting transition probability curves are shown by the broken lines of Fig. 14.6. As shown by Fig. 14.6 these curves are symmetric about the resonance position. The $\mathbf{v}\perp\mathbf{E}$ curve of Fig. 14.6(b) has an approximately Lorentzian form, but the $\mathbf{v}\|\mathbf{E}$ curve of Fig. 14.6(a), while it vanishes on resonance as predicted by Eq. (14.24), has an unusual double peaked structure.

Since the transition probabilities of Fig. 14.6 are obtained numerically, it is not appreciably more difficult to include the diagonal matrix elements V_{AA} and V_{BB} arising from the permanent dipole moments of the atomic states in the field. The

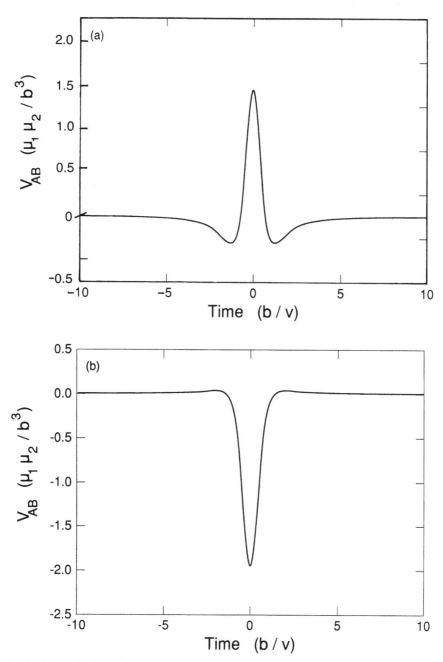

Fig. 14.5 Dipole–dipole interaction potential for (a) the case in which $\mathbf{v}\,\|\,\mathbf{E}$, (b) the case in which $\mathbf{v}\perp\mathbf{E}$ with $\theta = 0$ and $\phi = \pi/2$ (from ref. 14).

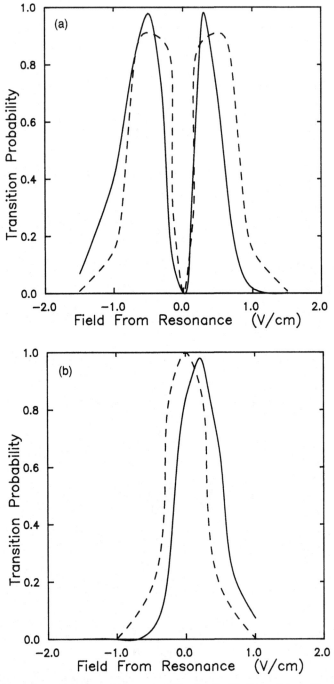

Fig. 14.6 Calculations of the transition probability for the Na 17s + 17s → 17p + 16p, (0,0) transition for (a) **v** ∥ **E**, and (b) **v** ⊥ **E**. Solid lines are for calculations which include the effects of the permanent dipole moments; dashed lines indicate that the permanent dipole moments have been neglected (from ref. 16).

solid lines of Fig. 14.6 are the transition probabilities calculated with the permanent dipole moments at the field of the collisional resonance taken into account. The permanent dipole moments of the 16p, 17p, and 17s states are $\mu_{16p} = 113.3ea_0$, $\mu_{17p} = 182.3\ ea_0$, and $\mu_{17s} = -16.5\ ea_0$, respectively. As shown by Fig. 14.6, when the permanent moments are taken into account the location of the resonance shifts slightly, by approximately a resonance linewidth, as expected from the fact that the permanent moments are nearly as large as the transition moments. In addition, the lineshapes become slightly asymmetric, due to the fact that the energy spacing of the atomic states now changes with internuclear separation.

Intracollisional interference

The unusual lineshape of Fig. 14.6(a) for $\mathbf{v} \parallel \mathbf{E}$ has approximately the same form and origin as the Ramsey separated oscillatory field pattern when there is a 180° phase shift between the two oscillatory fields.[14] To see the similarity we first examine $V_{AB}(t)$, shown in Fig 14.5(a). As shown by Fig. 14.5(a), for $\mathbf{v} \parallel \mathbf{E}$ V_{AB} changes sign twice during the collision, so that $\int_{-\infty}^{\infty} V_{AB}(t')dt' = 0$. A sign change of $V_{AB}(t')$ is equivalent to a 180° phase shift between the oscillatory fields in a separated oscillatory field experiment, and the $\mathbf{v} \parallel \mathbf{E}$ collisions are thus approximately equivalent to doing an rf resonance experiment in which there are three sequential rf fields with phase changes of 180° between successive fields. At resonance, the contributions to the transition amplitude from $V_{AB}(t') < 0$ cancel those from $V_{AB}(t') > 0$ so that, integrated over the whole collision, the transition probability vanishes. Off resonance, however, the cancellation is not complete.

Calculation of cross sections

To convert the transition probabilities to cross sections we must integrate over impact parameter using

$$\sigma = \int_0^\infty P(b)2\pi b db. \tag{14.25}$$

It is useful to consider the case of exact resonance, which may be worked out analytically. At resonance, using either Eq. (14.18) or Eq. (14.21) the transition probability is given by

$$P(b) = \sin^2\left(\frac{D}{b^2 v}\right), \tag{14.26}$$

where $D = \mu_1\mu_2$ if we use Eq. (14.18) and $D = 2\mu_{z_1}\mu_{z_2}$ $(1 - 2\cos^2\theta - \cos^2\phi\sin^2\theta)$ if we use Eq. (14.21). In either case D is proportional to the product of the two dipole moments. Examining Eq. (14.26) we see that if we define b_0 such that $D/b_0^2v = \pi/2$, as b increases from 0 to b_0, $P(b)$ oscillates between 0 and 1, and as b increases from b_0 to infinity $P(b)$ decreases smoothly to zero. If we insert the transition probability of Eq. (14.26) into Eq. (14.25), the integral can be done analytically, yielding

$$\sigma = \frac{\pi^3}{4}b_0^2 = \frac{\pi^2}{2}\frac{D}{v}. \tag{14.27}$$

Recalling that $D \sim \mu_1\mu_2$, we see that Eq. (14.27) resembles Eq. (14.5) which was derived in a very simple way.

If we are interested in the cross section for $v \parallel E$, the angles θ and ϕ are uniquely defined, $\theta = \pi/2$ and $\phi = 0$. On the other hand, to calculate the observed $v \perp E$ cross section Eq. (14.27) must be averaged over θ for $\phi = \pi/2$. In either case the calculated cross sections must be averaged over the appropriate distribution of collision velocities.

In an analogous fashion the numerically obtained transition probabilities shown in Fig. 14.6 can be converted to cross sections by integrating over impact parameter and averaging over the possible collision velocities. Collisions with $v \perp E$ do not have a single allowed value of θ. However, since the lineshape of Fig. 14.6(b) is simple, some averaging over the possible values of θ has no appreciable effect. For $v \parallel E$, θ and ϕ are fixed, at 90° and 0°, so the integration over impact parameter is only over b, and the calculated $v \parallel E$ cross section is very similar to Fig. 14.6.

The lineshapes of Figs. 14.6(a) and 14.6(b) are very different. However, this difference is only apparent in an experiment with high resolution. At low resolution the $v \parallel E$ and $v \perp E$ collisional resonances can be expected to exhibit the same instrumentally determined shape. Furthermore, both cross sections are approximately the same size. The calculated peak cross sections for the Na $17s + 17s \rightarrow 16p + 17p$ collisional resonances are $6.0 \times 10^8 a_0^2$ and $1.0 \times 10^9 a_0^2$ for $v \parallel E$ and $v \perp E$ respectively.[14] When integrated over the tuning field, to allow comparison to low resolution experiments, the two integrated cross sections are $8.3 \times 10^8 a_0^2$ V/cm and $1.2 \times 10^9 a_0^2$ V/cm.[14]

Experimental approach

The basic principle of the experimental approach is easily understood by considering the Na collisions of Eq. (14.1) and Fig. 14.2 as a concrete example. As shown by Fig. 14.7, atoms effuse from a heated oven in which the Na vapor pressure is $\sim$1 Torr.[14] The Na atoms are collimated into a beam by a collimator, not shown in

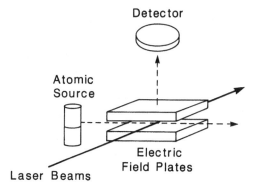

Side View

Fig. 14.7 Experimental apparatus configured with the electric field perpendicular to the collisional velocity (from ref. 14).

Fig. 14.7, and pass into the interaction region, between a pair of electric field plates, at which point the beam density can be up to $10^9 \, \text{cm}^{-3}$. The Na atoms are excited to an ns state by two 5 ns pulsed dye lasers tuned to the Na 3s → 3p and 3p → ns transitions at 5890 Å and ~4150 Å respectively. The ns atoms are allowed to collide with each other for a period of ~1 μs during which time the slow atoms in the beam are overtaken by the fast atoms. Thus the collision velocity is approximately the width of the velocity distribution of the beam. After the 1 μs period during which the atoms collide a high voltage pulse is applied to the lower plate of the interaction region. The amplitude of the pulse is set to ionize atoms in the np state, the higher lying final state of the collision, but not atoms in the ns state, the initial state of the collision. Ions produced by field ionization are expelled from the region between the plates and impinge upon a particle multiplier above the interaction region. The signal from the particle multiplier is recorded with a gated integrator. A static tuning voltage is applied to the lower plate to tune the atomic levels into resonance. The collisional resonances are observed by monitoring the field ionization signal from atoms in the np state as the tuning voltage is slowly swept through the collisional resonances. A typical example is shown in Fig. 14.8, a recording of the field ionization signal from the Na 17p state when the 17s state is populated by laser excitation.[12]

n scaling laws

One of the most striking aspects of resonant collisional energy transfer between Rydberg atoms is the magnitude of the cross sections. Accordingly, the first

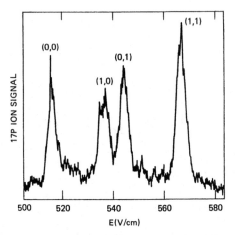

Fig. 14.8 The observed Na 17p ion signal after population of the 17s state vs dc electric field, showing the sharp collisional resonances. The resonances are labeled by the $|m|$ values of the lower and upper p states (from ref. 12).

experiments were focused on determining the magnitude of the cross sections and verifying the n scaling laws for the Na $ns + ns \rightarrow np + (n-1)$p cross sections and collision times.

The cross sections were measured by taking advantage of the fact that the number of atoms, N_p, which are in the np state after the atoms have been allowed to collide for a time interval T is given by[12]

$$N_p = \frac{N_s^2 \sigma \bar{v} T}{V},$$ (14.28)

where N_s is the number of atoms in the ns state, V is the volume of the sample of excited atoms, $\bar{v}$ is the average collision velocity, and σ is the cross section. Eq. (14.28) is valid only if the population of the ns state is not depleted significantly. If significant depletion of the population in the ns state occurs, corrections to Eq. (14.28) must be introduced. The number of atoms in the np state, N_p, is related to the signal associated with np state, N_p', by the conversion efficiency of the detector, Γ. Specifically $N_p' = \Gamma N_p$, and $N_s' = \Gamma N_s$. Using N_p' and N_s' we can rewrite Eq. (14.28) as

$$\sigma = \frac{N_p'}{(N_s')^2} \left[\frac{\bar{v} T \Gamma}{V} \right].$$ (14.29)

If the geometry of the experiment, timing gates, oven temperature, and detector gain are fixed, then the quantity in the square brackets of Eq. (14.29) is constant, and the relative cross section as a function of n is easily obtained by measuring the signals N_s' and N_p' as a function of n. The relative cross sections were put on an absolute basis by measuring the quantities in the brackets of Eq. (14.29), with the result shown in Fig. 14.9. The magnitudes and the n scaling of the

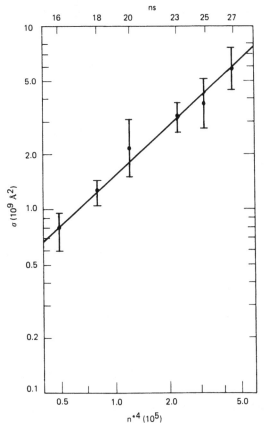

Fig. 14.9 The observed cross sections with their relative error bars (●), and the fit curve (—) (from ref. 12).

cross section as $n^{*3.7(5)}$ are in good agreement with Eq. (14.7).[12] Here n^* is the effective quantum number of the ns state, defined by $W_{ns} = -1/2n^{*2}$. The collisions resulting in the cross sections shown in Fig. 14.9 were most likely not between atoms of different velocities in a directed beam but between atoms moving in random directions. Both the excited atom sample volume and collision velocity were probably a factor of 2 larger than thought at the time, so the errors compensated to a large extent. In any event, from Fig. 14.9 it is apparent that the cross section exhibits the expected n^4 scaling and is of the correct magnitude to match the expected $\sigma = n^4/v$ behavior.

Verification that the collision time τ increases as n^2, or that the linewidth of the collisional resonances decreases as n^{-2}, is simply a matter of measuring the widths of the collisional resonances. In Fig. 14.10 we show a plot of the width of the (0,0) resonance as a function of n. The observed widths exhibit magnitudes and an $n^{*-1.95(20)}$ dependence in agreement with Eq. (14.8).

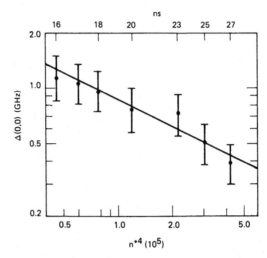

Fig. 14.10 The widths of the (0,0) resonances (●), and the fit curve (—) (from ref. 12).

Orientation dependence of v and E and intracollisional interference

One of the more interesting aspects of the collision process is the calculated dependence on the orientation of **v** relative to **E**. In particular, do the lineshapes for collisions with $\mathbf{v} \parallel \mathbf{E}$ and $\mathbf{v} \perp \mathbf{E}$ differ as dramatically as shown in Fig. 14.6? To answer this question unambiguously required two improvements upon the initial measurements. First, the Na atoms must be in a well defined beam, otherwise **v** is not well defined. This requirement is easily met by enclosing the interaction region with a liquid N_2 cooled box, which ensures that the only Na atoms in the interaction region are those in the beam. Second, the field homogeneity must be adequate, ~1 part in 10^4, to resolve the intrinsic lineshape of the collisional resonances. A pair of Cu field plates 1.592(2) cm apart with 1 mm diameter holes in the top plate to allow the ions to be extracted is adequate to meet this requirement.

Making these two modifications, but keeping the basic geometry of Fig. 14.7, substantially improves the resolution so that what appear to be four collisional resonances in Fig. 14.8 are actually nine.[14] All the resonances but the (0,0) resonance are split by the spin orbit splitting of the np $|m| = 1$ states. In the (0,1) and (1,0) resonances the upper and lower p states, respectively, are split, leading to doublets, while in the (1,1) resonance both upper and lower p states are split, leading to a quartet. In view of the fact that only the (0,0) resonance is a single line, it was selected for a detailed investigation of the cross section with $\mathbf{v} \parallel \mathbf{E}$ and $\mathbf{v} \perp \mathbf{E}$. To measure the cross section with $\mathbf{v} \perp \mathbf{E}$ the geometry of Fig. 14.8, with the liquid N_2 cooled shield, is adequate. To measure the cross section with $\mathbf{v} \parallel \mathbf{E}$ an arrangement, shown in Fig. 14.11, is used in which the atomic beam passes

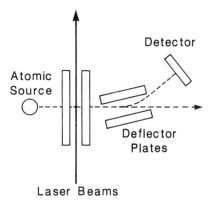

Top View

Fig. 14.11 Experimental apparatus configured so that the electric field and collisional velocity are parallel (from ref. 14).

through the field plates. In both cases the collision velocity is the average of the difference in velocities of different atoms in the beam.

The observed cross sections for the 18s (0,0) collisional resonance with $\mathbf{v} \parallel \mathbf{E}$ and $\mathbf{v} \perp \mathbf{E}$ are shown in Fig. 14.12. The approximately Lorentzian shape for $\mathbf{v} \perp \mathbf{E}$ and the double peaked shape for $\mathbf{v} \parallel \mathbf{E}$ are quite evident. Given the existence of two experimental effects, field inhomogeneties and collision velocities not parallel to the field, both of which obscure the predicted zero in the $\mathbf{v} \parallel \mathbf{E}$ cross section, the observation of a clear dip in the center of the observed $\mathbf{v} \parallel \mathbf{E}$ cross section supports the theoretical description of intracollisional interference given earlier. It is also interesting to note that the observed $\mathbf{v} \parallel \mathbf{E}$ cross section of Fig. 14.12(a) is clearly asymmetric, in agreement with the transition probability calculated with the permanent electric dipole moments taken into account, as shown by Fig. 14.6.

Velocity dependence of the collisional resonances

One of the potentially most interesting aspects of the resonant collisions is that, in theory, the collision time increases and the linewidth narrows as the collision velocity is decreased. According to Eqs. (14.6) and (14.8) the collision time is proportional to $1/v^{3/2}$. Collisions between thermal atoms with temperatures of ~ 500 K lead to linewidths of the collisional resonances that are a few hundred MHz at $n = 20$. In principle, substantially smaller linewidths can be observed if the collision velocity is reduced.

The most straightforward way of reducing the collision velocity is to velocity select the atoms in an atomic beam,[17] and a natural method of velocity selection is

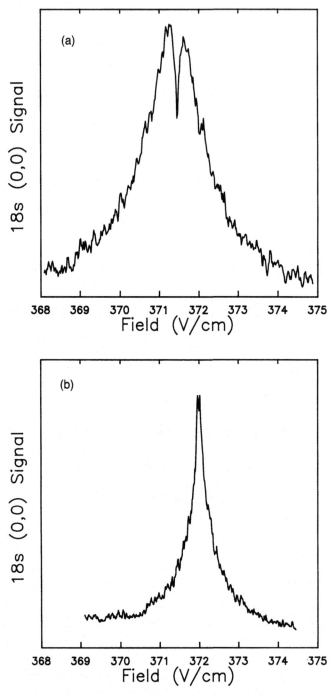

Fig. 14.12 Experimental measurements of the Na 18s (0,0) resonance for (a) $\mathbf{v} \| \mathbf{E}$ and (b) $\mathbf{v} \perp \mathbf{E}$. Note the definite dip near the center of the resonance in (a), as well as the slight asymmetry, both of which agree with the numerical predictions (from ref. 14).

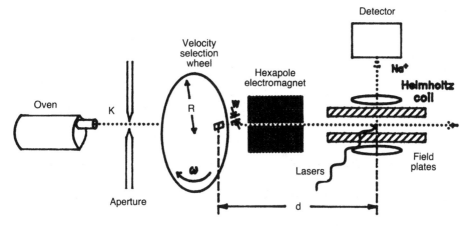

Fig. 14.13 Experimental arrangement for velocity selecting and focusing the atomic beam. The rotating slotted disc and the pulsed laser beam select atoms in a velocity group, and the hexapole magnet focuses them where they cross the laser beam (from ref. 18).

the one shown in Fig. 14.13.[18] Atoms effusing from the source pass through a chopper, a rotating 9.6 cm diameter disc which has radial slits as narrow as 1.5 mm near the outer edge. The disc rotates at frequencies as high as 200 Hz, at which frequency a 1.5 mm slit lets through a 25 μs wide pulse of atoms. The laser, which has a 5 ns pulse width, excites the atoms 250 μs later at the interaction region, which is 10 cm from the chopper. With a 1.5 mm wide slit the velocity distribution is ~10% of the velocity, which is usually chosen to be the velocity of the peak of the velocity distribution of the beam. Since a single velocity group of atoms is selected, it is possible to focus them spatially using the hexapole focusing magnet, substantially increasing the number density at the interaction region.[18,19]

The first and most obvious question is whether or not a narrower velocity distribution leads to narrower collisional resonances. In Fig. 14.14 we show the Na 26s + Na 26s → Na 26p + Na 25p resonances obtained under three different experimental conditions.[20] In Fig. 14.14(a) the atoms are in a thermal 670 K beam. In Figs. 14.14(b) and (c) the beam is velocity selected using the approach shown in Fig. 14.13 to collision velocities of 7.5×10^3 and 3.8×10^3 cm/s, respectively. The dramatic reduction in the linewidths of the collisional resonances is evident. The calculated linewidths are 400, 28, and 10 MHz, and the widths of the collisional resonances shown in Figs. 14.14(a)–(c) are 350, 40, and 23 MHz respectively. The widths decrease approximately as $1/v^{3/2}$ until Fig. 14.14(c), at which point the inhomogeneities of the electric field mask the intrinsic linewidth of the collisional resonance.

In the Na collision process of Eq. (14.1) the inhomogeneities in the static tuning fields required preclude the observation of very narrow collisional resonances. In

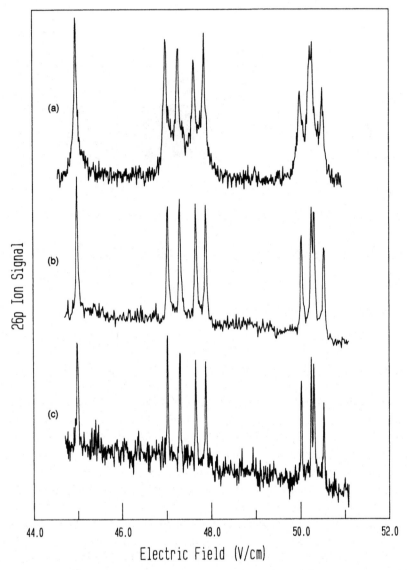

Fig. 14.14 Na 26s + Na 26s → Na 26p + Na 27p collisional resonances observed with (a) no velocity selection, collision velocity 4.6×10^4 cm/s, (b) velocity selection to a collision velocity 7.5×10^3 cm/s, (c) velocity selection to a collision velocity 3.8×10^3 cm/s (from ref. 20).

the K atom, however, dipole–dipole resonances occur near zero electric field,[18] where the homogeneity of the static field is less important. A K collision process which has been studied extensively is[17,18,21]

$$K\,29s + K\,27d \rightarrow K\,29p + K\,28p, \qquad (14.30)$$

which occurs in relatively small fields as shown in the energy level diagram of Fig. 14.15. As shown by Fig. 14.15 and Eq. (14.30), the 29s and 27d states must both be

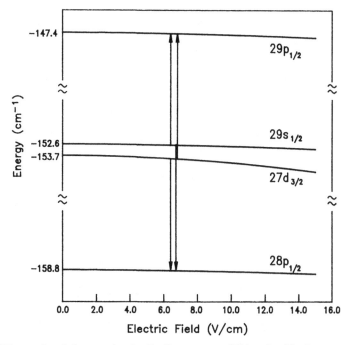

Fig. 14.15 Energy level diagram for the Rydberg states of K involved in the resonant energy transfer collisions being studied. The two transitions shown have been separated for clarity; they are degenerate in electric field (from ref. 21).

populated, and the fact that a resonant collision has occurred is ascertained by detecting the population in the higher lying 29p state by selective field ionization.

As shown by Fig. 14.15, the resonances occur near zero field, and it is easy to calculate the small Stark shifts with an accuracy greater than the linewidths of the collisional resonances. As a result it is straightforward to use the locations of the collisional resonances to determine the zero field energies of the p states relative to the energies of the s and d states. Since the energies of the ns and nd states have been measured by Doppler free, two photon spectroscopy,[22] these resonant collision measurements for $n = 27$, 28, and 29 allow the same precision to be transferred to the np states. If we write the quantum defect δ_p of the K np states as

$$\delta_p = \delta_p^0 + \delta_p^1(n^*)^{-2} + \delta_p^2(n^*)^{-4}, \tag{14.31}$$

these measurements have allowed a new determination of $\delta_p^{\ 0}$ with a factor of 5 smaller uncertainty than the previous value.[23] Explicitly, they yield $\delta_p^{\ 0} = 1.711925(3)$, in which the major uncertainly is from the Doppler free laser spectroscopy, not from the resonant collision measurements.[17]

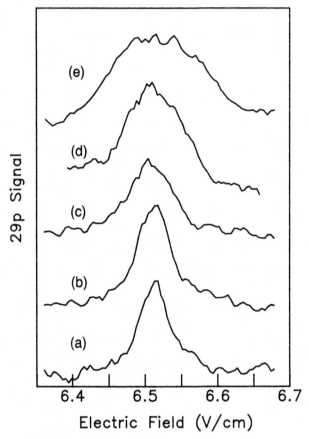

Fig. 14.16 Observed 29s resonance in K for several values of maximum-allowed collision time: (a) $\tau = 3.0\,\mu s$, FWHM = 1.4 MHz, (b), $\tau = 2.0\,\mu s$, FWHM = 1.8 MHz, (c) $\tau = 1.0$ μs, FWHM = 2.2 MHz, (d), $\tau = 0.4\,\mu s$, FWHM = 3.1 MHz, and (e) $\tau = 0.2\,\mu s$, FWHM = 5.2 MHz (from ref. 21).

Transform limited collisions

While collisional resonances 5 MHz wide are interesting for spectroscopic purposes, what makes them most interesting is that the 5 MHz linewidth implies that the collision lasts at least 200 *ns*, a time not much less than the 1 μs period allowed for the collisions to occur. If the collision linewidths can be reduced to the inverse of the time allowed for the collisions to occur, the collisional resonances become transform limited, and we know when each collision begins and ends.

Thomson *et al.* reached the transform limit by incorporating two improvements.[21] First, Helmholtz coils were placed around the interaction region to cancel the earth's magnetic field. Second, the time interval allowed for the collisions had previously been defined by the laser pulse and the field ionization pulse, which rises slowly at its onset. In these experiments the interval was terminated by the application of a rapidly rising, ~50 *n*s, low voltage detuning pulse which switches the field away from the field value of the collisional resonance at a well defined time.[24] These improvements allowed the observation of collisional resonances as narrow as 1 MHz. These narrow resonances were then transform broadened by reducing the time the atoms are allowed to collide to below 1 μs. In Fig. 14.16 we show collisional resonances observed as the time between the laser pulse and the detuning pulse is shortened from 1 μs to 0.4 μs, showing clearly the transform broadening of the collisional resonances as the time interval is shortened. If the collisional resonances are transform limited, we know when individual collisions begin and end, and we can perturb the colliding atoms at well defined times during the collision, for example, when the atoms are at their point of closest approach.

References

1. C. K. N. Patel, in *Lasers*, Vol. 2, ed. A. K. Levine (Marcel Dekker, New York, 1968).
2. A. Javan, W. R. Bennet, Jr., and D. R. Herriot, *Phys. Rev. Lett.* **8**, 470 (1962).
3. A. D. White and J. D. Rigden, *Proc. I.R.E.* **50**, 9167 (1962).
4. C. K. N. Patel, *Phys. Rev. Lett.* **13**, 617 (1964).
5. T. F. Gallagher, *Phys. Rept.* **210**, 319 (1992).
6. P. L. Houston, in *Advances in Chemical Physics*, Vol. 47, eds. I. Prigogine and S.A. Rice (Wiley, New York, 1981).
7. T. Oka, in *Advances in Atomic and Molecular Physics*, Vol. 9, eds. D. R. Bates and I. Esterman (Academic, New York, 1973).
8. P. W. Anderson, *Phys. Rev.* **76**, 647 (1949).
9. K. A. Smith, F. G. Kellert, R. D. Rundel, F. B. Dunning, and R. F. Stebbings, *Phys. Rev. Lett.* **40**, 1362 (1978).
10. T. F. Gallagher, G. A. Ruff, and K. A. Safinya, *Phys. Rev. A* **22**, 843 (1980).
11. K. A. Safinya, J. F. Delpech, F. Gounand, W. Sandner, and T. F. Gallagher, *Phys. Rev. Lett.* **47**, 405 (1981).
12. T. F. Gallagher, K. A. Safinya, F. Gounand, J. F. Delpech, W. Sandner, and R. Kachru, *Phys. Rev. A* **25**, 1905 (1982).
13. E. Fiordilino, G. Ferrante, and B. M. Smirnov, *Phys. Rev. A* **35**, 3674 (1987).
14. D. S. Thomson, R. C. Stoneman, and T. F. Gallagher, *Phys. Rev. A* **39**, 2914 (1989).
15. N. F. Ramsey, *Molecular Beams* (Oxford University Press, London, 1956).
16. D. S. Thomson, PhD Thesis, University of Virginia (1990).
17. R. C. Stoneman, M. D. Adams, and T. F. Gallagher, *Phys. Rev. Lett.* **58**, 1324 (1987).
18. M. J. Renn and T. F. Gallagher, *Phys. Rev. Lett.* **67**, 2287 (1991).
19. A. Lemonick, F. M. Pipkin, and D. R. Hamilton, *Rev. Sci. Instr.* **26**, 1112 (1955).
20. M. J. Renn, R. Anderson, and Q. Sun, private communication (1992).
21. D. S. Thomson, M. J. Renn, and T. F. Gallagher, *Phys. Rev. Lett.* **65**, 3273 (1990).
22. D. C. Thompson, M. S. O'Sullivan, B. P. Stoicheff, and Gen-Xing Xu, *Can. J. Phys.* **61**, 949 (1983).
23. P. Risberg, *Ark. Fys.* **10**, 583 (1956).
24. T. H. Jeys, K. A. Smith, F. B. Dunning, and R. F. Stebbings, *Phys. Rev. A* **23**, 3065 (1981).

15

Radiative collisions

A radiative collision is a resonant energy transfer collision in which two atoms absorb or emit photons during the collision.[1] Alternatively, a radiative collision is the emission or absorption of a photon from a transient molecule, and, as shown by Gallagher and Holstein,[2] radiative collisions can also be described in terms of line broadening. In a line broadening experiment there are typically many atoms and weak radiation fields, and in a radiative collision experiment there are few atoms and intense radiation fields. The only real difference is whether there are many atoms or many photons. Due to the short collision times, $\sim 10^{-12}$ s, simply observing radiative collisions between low lying states requires high optical powers, and entering the regime where the optical field is no longer a minor perturbation seems unlikely. Due to their long collision times and large dipole moments, Rydberg atoms provide the ideal system in which to study radiative collisions in a quantitative fashion. As we shall see, it is straightforward to enter the strong field regime in which the radiation field, a microwave or rf field to be precise, is no longer a minor perturbation. Ironically, while the experiments are radiative collision experiments, with few atoms and many photons, the description of the strong field regime is given in terms of dressed molecular states, which is more similar to a line broadening description.[3]

Using a simple, three level model we can develop a feeling for the microwave powers required to observe radiatively assisted collisional energy transfer between Rydberg atoms.[3] Consider the dipole–dipole atomic system shown in Fig. 15.1(a). In the Na $ns + ns \rightarrow np + (n-1)p$ resonant collisions described in the previous chapter the ns state corresponds to both s and s' of Fig. 15.1(a) and the $n-1$ and np states correspond to p and p' of Fig. 15.1(a), respectively. The collisions occurs via the interaction

$$V = \frac{\mu_1 \mu_2}{r^3}. \tag{15.1}$$

In Fig. 15.1(b) we show the level system of a collision radiatively assisted by the absorption of one photon of frequency ω. In this case the interaction leading to the production of the p and d states is given by

$$V = \frac{\mu_1 \mu_2 \mu_3 \, E}{r^3 \Delta}, \tag{15.2}$$

where Δ is the detuning between the real and virtual intermediate p' states and E is

(a)

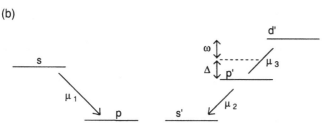

(b)

Fig. 15.1 (a) Energy levels and dipole matrix elements for the resonant collision of two atoms in the s and s' states resulting in the production of two atoms in the p and p' states. (b) Energy levels and matrix elements for the radiative collision in which an s and an s' atom collide to produce atoms in the p and d' states. The production of the d' state is via the virtual p' state which is detuned from the real p' state by an energy Δ.

the amplitude of the microwave field in which the collision occurs. The expressions of Eqs. (15.1) and (15.2) differ by the factor $\mu_3 E/\Delta$. For the radiatively assisted collision shown in Fig. 15.1(b) to be ~10% as probable as the resonant collision of Fig. 15.1a, we simply require that

$$\frac{\mu_3 E}{\Delta} = \frac{1}{3}.$$ (15.3)

If we assume typical values of $\mu_3 \simeq n^2$ and $\Delta \sim 0.1n^{-3}$, the required field is given by

$$E = \frac{1}{30n^5}.$$ (15.4)

Evaluating Eq. (15.4) at $n = 20$ leads to a field of 50 V/cm, or a power of ~6 W/cm^2, a power roughly six orders of magnitude lower than those used in optical radiative collision experiments with low lying atomic states.[4-7] These powers are not only readily obtained, but are easily exceeded by orders of magnitude, so that

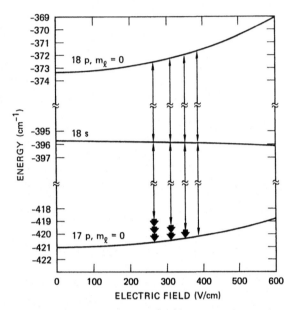

Fig. 15.2 Stark energy level diagram of the Na $|m_\ell| = 0$ states, relevant to the multiphoton-assisted collisions. The vertical lines indicate the collisional transfer and are drawn at the fields where they occur. The thick arrows correspond to the emitted photons (from ref. 8).

it is a straightforward matter to reach the strong field regime, in which the effect of the field is no longer a minor perturbation.[8]

The Na ns + Na $ns \rightarrow$ Na np + Na $(n-1)p$ collisions in combined static and microwave fields were the first Rydberg atom radiative collisions studied. Specifically, the process

$$\text{Na } 18s + \text{Na } 18s \rightarrow \text{Na } 18p + \text{Na } 17p + m\omega \qquad (15.5)$$

has been examined in detail. Here m is the number of photons emitted in the collision. When $m > 0$ there is emission, and when $m < 0$ there is absorption. The energy levels of the Na 18s, 17p, and 18p states are shown as a function of static field in Fig. 15.2.[8] For simplicity we have only shown the $m_\ell = 0$ levels in Fig. 15.2. The (0,0) resonant collision, with no microwaves, occurs at a field of 390 V/cm, and the radiative collisions accompanied by the stimulated emission of one, two, and three photons occur at lower static fields, as shown. We follow the convention of labelling the collisional resonances by the m_ℓ values of the lower and upper final p states of the collision, i.e. by (m_l, m_u), where m_l and m_u are the m_ℓ values of the lower and upper p states. The number of photons emitted is shown by the number of short bold arrows. In Fig. 15.2 the field displacement of a radiative collision resonance from the normal collisional resonance is proportional to the microwave frequency and inversely proportional to the sum of the differential Stark shifts of the p states. As shown in Fig. 15.2, the 18s state has a very small Stark shift, which we ignore. In principle, it is equally possible to observe collisions in which there is absorption, as opposed to emission, of several microwave photons. However, in the Na ns + Na $ns \rightarrow$ Na np + Na $(n-1)$p system these radiatively assisted collisions occur at higher static fields, where there are many collisional resonances due to the Stark states which are not shown in Fig. 15.2.[9] The $n = 17$ and $n=16$ Stark manifolds lie just below the 18p and 17p states, and at fields ~10% higher than the fields at which the resonant collisions of Fig. 15.2 occur, resonant collisions involving the Stark states appear. The presence of these collisional resonances precludes the unambiguous observation of ns + $ns \rightarrow np$ + $(n-1)p$ collisional resonances in which microwave photons are absorbed.

Initial experimental study of radiative collisions

The initial experimental study of radiative collisions between Rydberg atoms was done by simply introducing a microwave field into the region between the two field plates of an apparatus such as the one shown in Fig. 14.7.[10] The microwave field was introduced using a horn outside the plates, and several microwave frequencies between 12 and 18 GHz which propagated well between the plates were used. The maximum intensities were estimated to be ~3 W/cm^2. The data are accumulated in the same fashion as in the resonant collision experiment, the only difference being the presence of a microwave field. The atoms are excited by two pulsed lasers, and collisions are allowed to occur for a 1 μs period, after which a high voltage pulse is applied to the lower field plate to field ionize atoms in the np state. The signal from the np state atoms is recorded as the static field is scanned. On each scan the microwave field amplitude is held constant. The dependence of the radiative collision cross section on the microwave power is determined by repeating the static field scan at different microwave powers. In these experiments single photon radiative collisions were observed, at static fields slightly below the fields at which the resonant collisions were observed, as expected from Fig. 15.2. A typical example is shown in Fig. 15.3. The field displacement of the radiative collision signals from the resonant collision signal is equal to the field required to produce a Stark shift equal to the microwave frequency. Two points about this experiment are worth noting. First, the radiatively assisted collision cross sections are within an order of magnitude of the resonant collision cross sections for incident microwave intensities of a few W/cm^2, confirming the estimate of Eq. (15.4). Second, it is interesting to observe that at low microwave power the radiatively assisted collision signal is linear in the microwave power, but at higher powers the radiatively assisted collision signal rises less rapidly, suggesting the entrance into a nonperturbative regime.

Replacing the field plates of Fig. 14.7 with the resonant microwave cavity shown in Fig. 15.4 allowed an increase in the circulating microwave power by Q, the quality factor, of the cavity.[8] The cavity is a piece of WR-90 (X band) waveguide 20 cm long which is closed at both ends. The inside dimensions of the cavity are

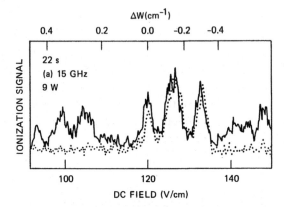

Fig. 15.3 The observed Na 22p ion signal after the population of 22s state vs the dc field in the presence of a 15 GHz microwave field (solid trace). The dotted trace was observed with no microwave field, and the sharp resonances in the center (for both solid and dotted traces) result from the resonant collisions, while the displaced resonances on the side (solid trace) are due to microwave assisted collisions. The scale on the top of the figure shows the detuning from the (0,0) resonance (from ref. 10).

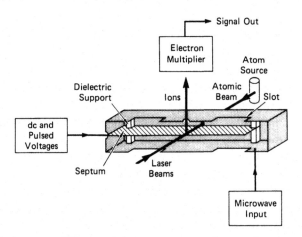

Fig. 15.4 Main features of the atomic beam apparatus, the atomic source, the microwave cavity, and the electron multiplier. The microwave cavity is shown sliced in half. The Cu septum bisects the height of the cavity. Two holes of diameter 1.3 mm are drilled in the sidewalls to admit the collinear laser and Na atomic beams, and a 1 mm diameter hole in the top of the cavity allows Na^+ resulting from field ionization of Na to be extracted. Note the slots for pumping (from ref. 8).

20.32 cm long, 1.02 cm high, and 2.28 cm wide. The cavity contains a septum, as shown in Fig. 15.4, which allows the application of a static tuning field and a pulsed field in the vertical direction. The presence of the septum does not affect the TE_{10n} cavity modes since they have only vertical electric fields.

The cavity is operated on modes of odd n, which have an electric field antinode at the center of the cavity. Specifically the $n = 15, 17$, and 19 modes, with resonant frequencies of 12.8, 14.4, and 15.5 GHz have been used. These modes have $Q \sim 1300$, and when excited with a power of 20 W, the circulating microwave intensity is $\sim 10\,\mathrm{kW/cm^2}$, and the microwave field is ~ 500 V/cm. As shown in Fig. 15.4, the atomic and laser beams enter the cavity through opposing holes in the sidewalls of the cavity ~ 1 mm above the septum. Thus a cylindrical volume of Rydberg atoms is produced. There is a small, ~ 1 mm diameter, hole in the center of the top of the cavity through which ions produced by field ionization of atoms directly beneath the hole escape from the cavity en route to the detector. Since the atoms only travel ~ 1 mm during the time between laser excitation and field ionization, the location of the hole in the top ensures that the signals observed are due to atoms at an antinode of the microwave electric field.

Data are accumulated in the same way as used to acquire those shown in Fig. 15.3. Not surprisingly, the radiative collisions which had previously been observed with 10 W of microwave power can be observed with milliwatts of power, due to the high Q of the cavity. The most interesting aspect of the data, though, is what happens in strong microwave fields. In Fig. 15.5 we show the development of the resonances observed for the radiative collision process of Eq. (15.5) as the 15.5 GHz microwave power incident on the cavity is increased.[8] With no microwave power only the four collisional resonances corresponding to the different $|m|$ values of the 17p and 18p states are observed. When the microwave power is raised this pattern of four resonances is repeated three times at lower static tuning fields, corresponding to the stimulated emission of one, two, and three microwave photons during the collision. A straightforward way of labeling the resonances is simply to extend our previous notation to account for the number of photons emitted. Explicitly, we label the collisional resonances as $(m_1,m_\mathrm{u})^m$ where m_1 and m_u are the m_ℓ values of the upper and lower p states and m is the number of photons emitted. In Fig. 15.5 the $(0,0)^m$ resonances are indicated by m with an arrow.

It is hardly surprising that, as the microwave power is raised, higher order multiphoton processes are observed. On the other hand, it may be surprising that for each $m \neq 0$ the cross sections first increase then decrease with microwave power. For example, the $m = 1$ cross section is clearly zero in the lowest trace. Similarly, the $m = 0$ cross section vanishes in the trace one above the lowest but reappears in the lowest trace. Such behavior, typical of the strong field regime, is not predicted by perturbation theory. Close inspection of Fig. 15.5 reveals that the positions of the collisional resonances shift to lower static field as the microwave power is raised. Finally, in contrast to the usual observation of broadening with increased power, the $(0,0)^m$ resonances, which are well isolated from other resonances, develop from broad asymmetric resonances to narrow symmetric ones as the microwave power is raised.

That the observed resonances for $m \neq 0$ become narrower and more symmetric as the power is increased can be understood in the following way, using the one-photon process as an example. The coupling matrix element of Eq. (15.2) has two

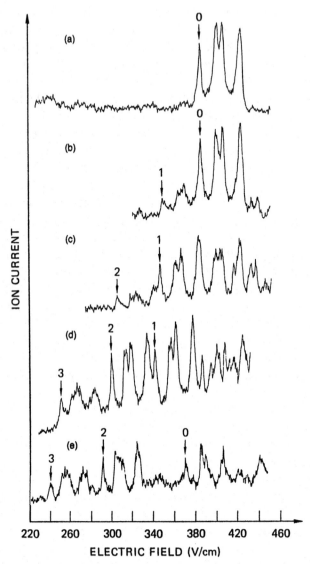

Fig. 15.5 Observed Na 18p ion signal after the population of the 18s level vs the static field with a 15.4 GHz microwave field. Trace (a) corresponds to no microwave power input to the cavity and shows the set of four zero-photon collisional resonances. Traces (b), (c), (d), and (e) correspond, respectively, to 13.5, 50, 105, and 165 V/cm microwave field amplitudes inside the cavity and show additional sets of four collisional resonances corresponding to one, two, and three-photon radiatively assisted collisions. The peaks labelled 0, 1, 2, and 3 correspond to the lowest field member of the set of four resonances corresponding to zero-, one-, two-, and three-photon assisted collisions, $(0,0)^0$, $(0,0)^1$, $(0,0)^2$, $(0,0)^3$ (from ref. 8).

factors, a dipole–dipole term, and a dipole–field term. For the collision to be observable, the time integral of the matrix element over the collision must exceed some minimum value ~1. Thus, if the microwave field is smaller, the radiative collision must occur with a smaller impact parameter to compensate. A small impact parameter has two effects. First, it implies a short collision time and a wide resonance. Second, as the atoms come closer together the molecular energy levels deviate from their $r = \infty$ values, producing the asymmetry. As the microwave field is raised, the impact parameter can become larger, and the resonances become narrower and more symmetric.

Theoretical description

The experimental data shown in Figs. 15.3 and 15.5 were obtained with a microwave frequency $\omega/2\pi > 1/\tau$ where τ is the time or duration of the collision and $1/\tau$ is the linewidth. In this case the resonances corresponding to the absorption or emission of different numbers of photons are resolved. Here we describe radiative collisions starting from the high frequency regime, $\omega/2\pi > 1/\tau$ and progressing to the low frequency regime, $\omega/2\pi < 1/\tau$.

The strong field, high frequency regime

To describe the shifts and intensities of the m-photon assisted collisional resonances with the microwave field Pillet *et al.* developed a picture based on dressed molecular states,[3] and we follow that development here. As in the previous chapter, we break the Hamiltonian into an unperturbed Hamiltonian H_0, and a perturbation V. The difference from our previous treatment of resonant collisions is that now H_0 describes the isolated, noninteracting, atoms in both static and microwave fields. Each of the two atoms is described by a dressed atomic state, and we construct the dressed molecular state as a direct product of the two atomic states. The dipole–dipole interaction V is still given by Eq. (14.12), and using it we can calculate the transition probabilities and cross sections for the radiatively assisted collisions.

We begin by developing the wavefunction for one atom in combined static and microwave fields, both of which are in the same direction. If the nominal $n\ell$ state has an energy $W_{n\ell}(E)$ in a static field E, we can describe the energy in the vicinity of the static field E_S by

$$W_{n\ell}(E_S + \Delta E) = W'_{n\ell} + k_{n\ell}\Delta E + \frac{1}{2}\alpha_{n\ell}(\Delta E)^2 \tag{15.6}$$

where

$$W'_{n\ell} = W_{n\ell}(E_S), \quad k_{n\ell} = d\frac{W_{n\ell}}{dE}\bigg|_{E_S}, \quad \text{and} \quad a_{n\ell} = \frac{d^2 W_{n\ell}}{dE^2}\bigg|_{E_S}$$

If we now imagine that ΔE varies in time, but slowly, its only effect is to cause a time variation of the energy of the $n\ell$ state. We assume that the spatial wavefunction is unaffected by ΔE and that no transitions occur. This approximation is the adiabatic approximation of Autler and Townes.[11] Now let us consider the time variation of ΔE of particular interest to us, $E_{mw} \cos \omega t$. If the assumptions stated above are valid, we can use the energy of Eq. (15.6) as the unperturbed Hamiltonian H_0 in the Schroedinger equation. Explicitly,

$$W(t)\,\psi_{n\ell}(t) = \left(W'_{n\ell} + k_{n\ell}\,E_{mw} \cos \omega t + \frac{a_{n\ell}}{2}(E_{mw} \cos \omega t)^2\right)\psi_{n\ell}(t) = \frac{i\partial\psi_{n\ell}(t)}{\partial t},$$

(15.7)

which has the solution

$$\psi_{n\ell}(t) = \psi_{n\ell}\,e^{-i\int_{t_0}^{t} W(t')\,dt'}.$$

(15.8)

Carrying out the integrations and discarding the phase factors from t_0 yields

$$\psi_{n\ell}(t) = \psi_{n\ell}\,e^{-i\left(W' + \frac{a_{n\ell}}{4}\right)t}\,e^{-i\frac{k_{n\ell}E_{mw}}{\omega}\sin \omega t}\,e^{-i\frac{a_{n\ell}E_{mw}^2}{8\omega}\sin 2\omega t}.$$

(15.9)

The exponentials with sinusoidal arguments may be expanded as Bessel functions using[12]

$$e^{ix \sin \omega t} = \sum_{-\infty}^{\infty} J_k(x)\,e^{ik\omega t}.$$

(15.10)

With the Bessel function expansion of Eq. (15.10), Eq. (15.9) becomes

$$\psi_{n\ell}(t) = \psi_{n\ell}\,e^{-i\left(W'_{n\ell} + \frac{a_{n\ell}E_{mw}^2}{4}\right)t}\left(\sum_{-\infty}^{\infty} J_k\left(\frac{k_{n\ell}'E_{mw}}{\omega}\right)e^{-ik\omega t}\right)\left(\sum_{-\infty}^{\infty} J_{k'}\left(\frac{a_{n\ell}E_{mw}^2}{8\omega}\right)e^{-i2k'\omega t}\right).$$

(15.11)

If $a_{n\ell}E_{mw}^2 \ll \omega$, as is usually the case for the experiments described here, in the last expansion of Eq. (15.11) only the $k' = 0$ term contributes and the summation is equal to one. In this case the wavefunction is written as

$$\psi_{n\ell}(t) = \psi_{n\ell}\,e^{-i\left(W'_{n\ell} + \frac{a_{n\ell}E_{mw}^2}{4}\right)t}\sum_{-\infty}^{\infty} J_k\left(\frac{k_{n\ell}E_{mw}}{\omega}\right)e^{-ik\omega t}.$$

(15.12)

We note that for collisions involving states of good parity, $k_{n\ell} = 0$, and the first summation of Eq. (15.11) collapses to 1. In this case we may not disregard the second summation of Eq. (15.11).

As discussed in Chapter 10, we can think of the effect of the microwave field on the $n\ell$ state as modulating its energy. Just as modulating the frequency of a radio wave produces sidebands, the microwave field can be thought of as modulating the energy of the $n\ell$ state and breaking it into carrier and sideband states.

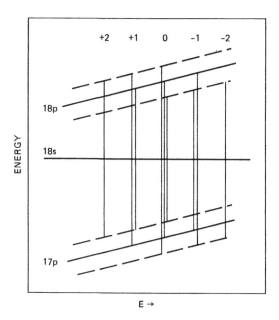

Fig. 15.6 Energy levels of the Na 17p, 18s, and 18p states showing the first upper and lower sideband states of the p states. The numbers $+2,\ldots,-2$ refer to the net number of photons emitted in a radiative collision at these tunings of the field. Note that there are several processes which lead to the net emission of, for example, zero photons (from ref. 3).

In Eq. (15.12) $\psi_{n\ell}$ is the spatial wavefunction of the $n\ell$ state at the static field E_S, $W'_{n\ell}$ is the energy in the static field E_S, $\alpha_{n\ell}E_{mw}^2/4$ is the ac Stark shift produced by the microwave field, and $W'_{n\ell} + \alpha_{n\ell}E_{mw}^2/4$ is the carrier energy. The Bessel function $J_k(kE_{mw}/\omega)$ gives the amplitude in the kth sideband state, which is displaced in energy from the carrier state by $k\omega$, k microwave photons. For the np states k_{np} is large, but for the ns states k_{ns} is very small, and the amplitudes of the s sidebands are negligible.

A graphic presentation of the effect of the microwave fields is shown in Fig. 15.6, a plot of the energies of the 18s, 17p and 18p states showing the carrier energies of these states as well as the energies of the ± 1 sideband states of the p states.[3] As mentioned above, and shown explicitly in Eq. (15.12), the amplitude of each sideband depends on the strength of the microwave field, and while this point is not made explicitly in Fig. 15.6, it is implicit in that no sidebands of the s state are shown and second and higher sidebands of the p states are not shown. Thus Fig. 15.6 corresponds to a microwave field too weak to have observable amplitudes for the second sidebands of the p states. With no microwaves all sidebands would have zero amplitude and the carriers amplitude one. A further interesting point to observe in Fig. 15.6 is that there are two processes corresponding to the net emission of ± 1 photon, and three corresponding to the net emission of zero photons. These processes correspond to the transitions between the

energetically allowed combinations of sideband states, raising the question of how
to account for these different processes, all of which occur at the same static field
and are therefore experimentally indistinguishable.

Using the dressed atomic states we construct the dressed molecular states as
direct products. Explicitly, we construct the two states $\psi_A(t)$ and $\psi_B(t)$ given by[3]

$$\psi_A(t) = \psi_{1ns}(t) \otimes \psi_{2ns}(t) \tag{15.13a}$$

and

$$\psi_B(t) = \psi_{1np}(t) \otimes \psi_{2(n-1)p}(t) \tag{15.13b}$$

in which each of the individual wavefunctions is given by Eq. (15.12). If we write
out the expression for ψ_B explicitly we find

$$\psi_B(t) = \psi_{1np}\psi_{2(n-1)p} \, e^{-iW_Bt} \sum_{kk'} J_k\left(\frac{k_{np}E_{mw}}{\omega}\right) J_{k'}\left(\frac{k_{(n-1)p}E_{mw}}{\omega}\right) e^{-i(k+k')\omega t}, \tag{15.14}$$

where $W_B = W'_{np} + W'_{(n-1)p} + (a_{np} + a_{(n-1)p})E_{mw}^2/4$, the sum of the carrier
energies of the np and $(n-1)p$ states. With the Bessel function identity,[13]

$$J_N(x \pm y) = \sum_{-\infty}^{\infty} J_{N \mp k}(x) J_k(y), \tag{15.15}$$

the double summation of Eq. (15.14) may be condensed to a single sum, yielding

$$\psi_B(t) = \psi_{1np}\psi_{2(n-1)p} \, e^{-iW_Bt} \sum_k J_k\left(\frac{k_B E_{mw}}{\omega}\right) e^{-ik\omega t}, \tag{15.16}$$

where $k_B = k_{np} + k_{(n-1)p}$. Similarly we can write the wavefunction for $\psi_A(t)$ as

$$\psi_A(t) = \psi_{1ns}\psi_{2ns} \, e^{-iW_At} \sum_{k'} J_{k'}\left(\frac{k_A E_{mw}}{\omega}\right) e^{-ik'\omega t}, \tag{15.17}$$

where $W_A = 2W'_{ns}$ and $k_A = 2k_{ns}$

Both $\psi_A(t)$ and $\psi_B(t)$ are solutions to H_0, and the total wavefunction is, in
general, a linear combination of the two. Explicitly, it is given by

$$\psi(t) = C_A(t)\psi_A(t) + C_B(t)\psi_B(t), \tag{15.18}$$

where $C_A(t)$ and $C_B(t)$ are time dependent coefficients analogous to those in Eq.
(14.15). If we insert the wavefunction of Eq. (15.18) into the time dependent
Schroedinger equation,

$$H\Psi(t) = i\partial\Psi/\partial t, \tag{15.19}$$

and use the fact that $\psi_A(t)$ and $\psi_B(t)$ are solutions of the unperturbed Schroedinger equation, we find the equation

$$VC_A(t)\psi_A(t) + VC_B(t)\psi_B(t) = i\,\dot{C}_A(t)\psi_A(t) + i\dot{C}_B(t)\psi_B(t). \tag{15.20}$$

Assuming that the diagonal matrix elements of V vanish, multiplying Eq. (15.20)
in turn by $\psi_A^*(t)$ and $\psi_B^*(t)$ and carrying out the spatial integrations yields

$$\langle\psi_B(t)|V|\psi_A(t)\rangle C_A(t) = i\dot{C}_B(t) \tag{15.21a}$$

and

$$\langle \psi_A(t)|V|\psi_B(t)\rangle \, C_B(t) = i\dot{C}_A(t). \tag{15.21b}$$

We note that Eqs. (15.21) are similar to Eqs. (14.16) but for the time dependence of $\psi_A(t)$ and $\psi_B(t)$. We recall that V is always time dependent since it depends on the internuclear separation of the atoms. However, this time dependence is slow compared to those of $\psi_A(t)$ and $\psi_B(t)$

Our interest is in computing the transition probability at resonance, a condition which is met when $W_A = W_B + m\omega$ where m is an integer. First, we explicitly write the matrix element $\langle \psi_B(t)|V|\psi_A(t)\rangle$ using the wavefunctions of Eqs. (15.16) and (15.17) and the Bessel function relation of Eq. (15.15). It is given by

$$\langle \psi_B(t)|V|\psi_A(t)\rangle = \langle \psi_B|V|\psi_A\rangle \sum_k J_k\left(\frac{(k_A - k_B)\,E_{mw}}{\omega}\right) e^{-i[W_A - W_B + k\omega]t}. \tag{15.22}$$

In this form it is apparent that the matrix element $\langle \psi_B(t)|V|\psi_A(t)\rangle$ is a product of the spatial matrix element, which varies slowly with time as the atoms move past each other, and the Bessel function sum, which contains an enormous range of frequencies. At resonance one term of the sum is constant in time and the others oscillate at multiples of ω. The oscillating terms are ineffective in causing transitions in collisions of duration much longer than the microwave period, since they average to zero. Thus at the resonance corresponding to the radiative collision in which m photons are emitted, i.e. when $W_A = W_B + m\omega$, the constant and significant part of the matrix element of Eq. (15.22) is given by

$$\langle \psi_B(t)|V|\psi_A(t)\rangle = \langle \psi_B|V|\psi_A\rangle J_m\left(\frac{K\,E_{mw}}{\omega}\right) \tag{15.23}$$

where $K = k_A - k_B$. The matrix element of Eq. (15.23) is identical to the matrix element for a normal resonant collision multiplied by a constant factor, the Bessel function, and as a result the transition probability can be computed using the same approaches as in Chapter 14.

All radiative collision experiments with Rydberg atoms have been done with the collision velocity perpendicular to the static and microwave fields. Thus the results obtained are an average over spatial orientations, and it makes little sense to use a detailed model of the interaction. Accordingly, we assume that

$$|\langle \psi_B|V|\psi_A\rangle| = \begin{cases} \dfrac{\chi}{b^3} & r < \dfrac{\sqrt{5}b}{2} \\[3mm] 0 & r > \dfrac{\sqrt{5}b}{2} \end{cases} \tag{15.24}$$

where b is the impact parameter and $\chi = \mu_1\mu_2$. With this assumption, the atoms interact for a time b/v. Using matrix elements of the form of Eq. (15.24) leads to sinusoidally oscillating solutions of $C_A(t)$ and $C_B(t)$ during the collision . If we choose as initial conditions $C_A(-\infty) = 1$ and $C_B(-\infty) = 0$, then after the collision,

$$C_A^2(t) = \cos^2\left[\frac{\chi}{b^2 v} J_m\left(\frac{KE_{mw}}{\omega}\right)\right], \tag{15.25a}$$

and the transition probability at impact parameter b, $P_m(b)$, is given by $C_B^2(t)$. Explicitly,

$$P_m(b) = C_B^2(t) = \sin^2\left[\frac{\chi}{b^2 v} J_m\left(\frac{KE_{mw}}{\omega}\right)\right]. \tag{15.25b}$$

The transition probability of Eq. (15.25b) is of the same form as the transition probability of Eq. (20.25), with $D = \chi|J_m(KE_{mw}/\omega)|$. Accordingly, if we define an impact parameter $b_m(E_{mw})$ such that

$$\frac{\chi}{b_m^2(E_{mw}) v}\left|J_m\left(\frac{KE_{mw}}{\omega}\right)\right| = \frac{\pi}{2}, \tag{15.26}$$

$P_m(b)$ oscillates between 0 and 1 as b increases from zero to $b_m(E_{mw})$, at which point $P_m(b) = 1$ for the last time. As b increases from $b_m(E_{mw})$ to infinity $P_m(b)$ decreases monotonically from 1 to 0.

To obtain the cross section for the m-photon assisted collision in the presence of a microwave field E_{mw} we integrate the transition probability over impact parameter. Explicitly

$$\sigma_m(E_{mw}) = \int_0^\infty 2\pi b \sin^2\frac{\chi}{b^2 v}\left|J_m\left(\frac{KE_{mw}}{\omega}\right)\right| db. \tag{15.27}$$

Carrying out the integral explicitly yields

$$\sigma_m(E_{mw}) = \frac{\pi^3}{4} b_m^2(E_{mw}) = \frac{\pi^2}{2}\frac{\chi}{v}\left|J_m\left(\frac{KE_{mw}}{\omega}\right)\right|. \tag{15.28}$$

A radiative collision with the emission of no photons in the absence of a microwave field is simply a resonant collision. Since $J_0(0) = 1$, we can express the radiative collision cross section of Eq. (15.28) as

$$\sigma_m(E_{mw}) = \sigma_R\left|J_m\left(\frac{KE_{mw}}{\omega}\right)\right|, \tag{15.29}$$

i.e. as the product of the resonant collision cross section σ_R and the magnitude of the $J_m(KE_{mw}/\omega)$ Bessel function.

Eq. (15.29) is only valid in the strong field regime. It is clear that if $J_m(KE_{mw}/\omega)$ is small, using Eq. (15.26) we compute a small value of $b_m(E_{mw})$. When this value approaches the size of the atom, Eq. (15.29) is no longer valid, for the integral of Eq. (15.27) is dominated by contributions for $b \sim b_m(E_{mw})$, where neither the assumption of the energies being independent of r nor the dipole approximation are valid. Under such circumstances a collision with a small impact parameter does not lead to as large a transition probability as indicated by Eq. (15.25b).

Connection to the weak field regime

In the experiments with Rydberg atoms it is very difficult to observe radiatively assisted collisions with cross sections more than a factor of 10 smaller than the resonant collision cross sections, so the deviations from Eq. (15.29) are not apparent. However, in other contexts, such as laser assisted collisions, this limitation does not apply, and it is interesting to consider how the above description passes over into the weak field regime, in which $J_m(KE_{mw}/\omega)$ is small. If we restrict the integration in Eq. (15.27) to the large r region of space, in which the approximations we have used are valid, we can rewrite Eq. (15.27) as

$$\sigma_m = \int_B^\infty \frac{\chi^2}{b^4 v^2} J_m^2 \left(\frac{KE_{mw}}{\omega}\right) 2\pi b \, db$$

$$= \frac{\pi \chi^2}{B^2 v^2} J_m^2 \left(\frac{KE_{mw}}{\omega}\right), \tag{15.30}$$

in which an obvious lower limit for B is the geometrical radius of the atom. Acccordingly, $B = n^2 a_0$ for a Rydberg atom of principal quantum number n. An important fact to note about Eq. (15.30) is that $\sigma_m(E_{mw})$ is proportional to $J_m^2(KE_{mw}/\omega)$. Using the fact that for small arguments $J_m(x) = (x/2)^m$,[12] at low microwave fields

$$\sigma_m \propto E_{mw}^{2m}$$

$$\propto I_{mw}^m, \tag{15.31}$$

i.e. the cross section for an m-photon assisted collision is proportional to the microwave intensity to the mth power, as expected on the basis of time dependent perturbation theory.

The low frequency regime

In the previous discussion of radiatively assisted collisions we have assumed that there are many cycles of the radiation field during the time of one collision, i.e. $\omega/2\pi \gg 1/\tau$. It is interesting to consider the opposite extreme, $\omega/2\pi \ll 1/\tau$. We consider collisions in the total field

$$E = E_s + E_{rf} \cos(\omega t + \phi), \tag{15.32}$$

and we assume that the collisional resonances are observed by scanning the static field. If, in the absence of the rf field, the resonance is found at $E_S = E_R$, it is found at

$$E_S = E_R - E_{rf} \cos(\omega t + \phi) \qquad (15.33)$$

with the added rf field of Eq. (15.32). If we can control the phase of the rf field at which the collision occurs the resonance is simply shifted as shown by Eq. (15.33). If we cannot control the phase, either because the exciting laser is not synchronized with the rf field, or because collisions can occur randomly over several rf cycles, then the probability of observing the resonant collision signal at any given static field is proportional to the likelihood of the total field's being equal to E_R at that static field. If the resonant collision cross sections is given as a function of field by $\sigma_R(E)$, then in the field $E_S + E_{rf} \cos \omega t$ the cross section observed at E_S is given by the convolution of $\sigma_R(E)$ with the function

$$f(E) = \begin{cases} \dfrac{1}{\pi E_{rf}[1 - (E - E_S/E_{rf})^2]^{1/2}} & |E - E_S| \leq E_{rf} \\ 0 & |E - E_S| > E_{rf} \end{cases} \qquad (15.34)$$

In other words $\sigma(E_S)$ is given by the time weighted average of the cross section in the static and oscillating fields. Since the oscillating field spends most of its time at its turning points, the observed cross section can be expected to exhibit the behavior shown in Fig. 15.7.[3]

The intermediate frequency regime

We have considered the high frequency regime, which is independent of the rf phase, and the low frequency regime which is critically dependent on the rf phase. We now consider the intermediate frequency regime, $\omega/2\pi \sim 1/\tau$, approaching it from the low frequency side. To date the only way of calculating the lineshape of the collisional resonances has been explicit numerical integration. As we shall see, when comparisons with experimental results are made, the calculations predict multiple peaks in the cross section. The separation between peaks does not, however, correspond to the rf frequency. We can use a simple picture to understand the origin of the peaks and their locations, although the picture does not give lineshapes.[14]

For concreteness we assume that the collision occurs over the time interval $-T/2 < t < T/2$ in the presence of a rf field of well defined phase so that the total field is given by Eq. (15.32). We consider the transitions between the two molecular states A and B which are degenerate (at $r = \infty$) at the static field E_R. We may assume with no loss of generality that, in the vicinity of E_R, B has no Stark shift and A has a linear Stark shift. Ignoring the dipole–dipole coupling, the energy,

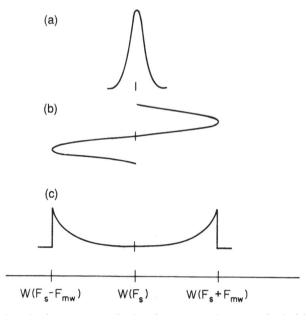

(a)

(b)

(c)

$W(F_s-F_{mw})$ $W(F_s)$ $W(F_s+F_{mw})$

Fig. 15.7 Collisions in the presence of a low frequency microwave field: (a) the resonant collision profile without the microwaves; (b) one period of microwave field showing both the amplitude and the time weighting of the extreme field values; (c) typical time weighted radiative collision lineshape when the microwave period becomes long compared to the duration of one collision (from ref. 3).

W_B, of molecular state B is a constant, and the energy of molecular state A is given by

$$W_A = W_B + k_A(E - E_R), \tag{15.35}$$

where E is given by Eq. (15.32). Inspecting Eq. (15.35) we see that W_A follows the rf field. Imagine that we choose the rf phase $\phi = 0$ so that W_A reaches a maximum in the center of the allowed collision time. In Figs. 15.8(a) and (b), we show the energies W_A and W_B as functions of time for two choices of the phase of a 0.75 MHz rf field, $\phi = 0$ and $\pi/2$, respectively. The allowed collision time $T = 0.7\ \mu s$. The dipole–dipole interaction lifts the degeneracy at resonance, leading to the avoided level crossings shown by the insets of Fig. 15.8. In Fig. 15.8(a) the energies are shown for two values of the static field E_S. For $E_S < E_R - E_{rf}$, the two levels never come into resonance. When $E_S = E_R - E_{rf}$ the two levels are resonant at $t = 0$, as shown by the broken line, and a broad resonance can be expected. For larger E_S the two levels come into resonance twice, as shown by the solid line, and the transition amplitude is the coherent sum of the amplitudes acquired at the two times the levels become resonant, at $t = \pm t'$.

As shown by Fig. 15.8(a), if the system is initially in state A it traverses two avoided level crossings to reach the final state B, and there are two possible paths to go from A to B. Whether the transition amplitudes for these two paths add

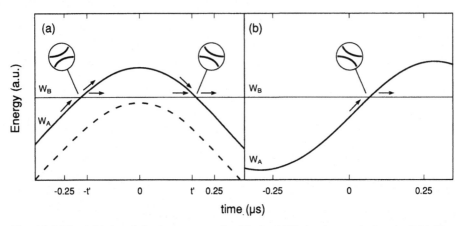

Fig. 15.8 The initial and final state energies W_A and W_B in the total electric field $E = E_S + E_{rf} \cos(\omega t + \phi)$ where the frequency $\omega/2\pi = 0.75$ MHz field and the allowed collision time $T = 0.7 \mu s$. W_B is field independent and W_A varies linearly with the field. (a) The phase $\phi = 0$. No resonance occurs for $E_S < E_R - E_{rf}$. As the static field is increased, when $E_S = E_R - E_{rf}$ (---) the two levels come into resonance at $t = 0$. When $E_S > E_R - E_{rf}$ (—) the two levels come into resonance twice, and there are two resonant interaction periods at $t = \pm t'$. The levels do not actually cross but have avoided crossings as shown by the insets. There are two paths by which atoms initially in state A can reach state B, and they interfere constructively when Φ_B, the phase difference accumulated between $-t'$ and t', is $2\pi N$. In (b) where $\phi = \pi/2$ and $E_S \approx E_R$, only one interaction period occurs for all values of E_S between $E_R - E_{rf}$ and $E_R + E_{rf}$ (after ref. 14).

contructively or destructively depends on the phase difference Φ between the two paths. Φ is simply the area between the two curves in Fig. 15.8(a), i.e.,

$$\Phi = \int_{-t'}^{t'} (W_A - W_B) \, dt. \tag{15.36}$$

If $\Phi = 2\pi N$, where N is an integer the interference is constructive, whereas if $\Phi = 2\pi(N + 1/2)$ it is destructive. As E_S is raised Φ passes through even and odd multiples of π leading to constructive and destructive interference in the cross section. Inspecting Fig. 15.8(a) we can see that the fringes should become closer together as E_S is increased. This description of the origin of the oscillations in the cross section makes it apparent that they are a form of Stückelberg oscillations observed in differential scattering.[15]

The two resonant interactions at $t = \pm t'$ can also be thought of as being analogous to the two separated oscillating field regions in a Ramsey magnetic resonance experiement, in which case the oscillations in the cross section can be thought of as Ramsey fringes.[16]

For $\phi = 0$, as shown in Fig. 15.8(a) the oscillations in the cross sections should extend up to the static field

$$E_S = E_R - E_{rf} \cos(\omega T/2). \tag{15.37}$$

With the rf phase $\phi = \pi$, the collisional resonance is reversed with respect to E_R. With $\phi = \pi/2$, as shown in Fig. 15.8(b) for $E_S \approx E_R$, the result is qualitatively different. The two levels come into resonance at most once for any value of E_S, and as a result there is one broad collisional resonance extending from $E_S = E_R - E_{rf}\sin(\omega T/2)$ to $E_S = E_R + E_{rf}\sin(\omega T/2)$.

The above description is obviously an extension of the low frequency description, but it can also be connected to the high frequency limit. Consider the case in which $\omega \gg 1/T$, i.e., there are many cycles of the field, and $|E_S - E_R| \ll E_{rf}$, the static tuning field is near resonance compared to the rf field amplitude. For a large transition probability the transition amplitudes of successive rf cycles must add constructively. This only happens when the phase difference Φ over an rf cycle satisfies $\Phi = 2\pi N$ with N an integer. Using the energies of Eq. (15.35),

$$\Phi = \int_{-\pi/\omega}^{\pi/\omega} k_A(E_S + E_{rf}\cos\omega t - E_R)\, dt$$

$$= \frac{2\pi k_A(E_S - E_R)}{\omega}. \tag{15.38}$$

Equivalently,

$$k_A(E_S - E_R) = N\omega. \tag{15.39}$$

In other words the resonances are detuned from E_R by the static field shift equivalent to N photons, the same result as obtained in the dressed state picture.

Eq. (15.39) tells us where the collisional resonances occur, but it does not tell us how strong they are. The strength is determined by the phase contributions of the two parts of the rf cycle, when $W_A < W_B$ and $W_A > W_B$. When the phases accumulated in these two half cycles are both integral multiples of 2π the collisional resonances are strong. This requirement leads to an oscillation in the intensity of the N photon assisted collisional resonance proportional to $\cos(k_A E_{rf}/\omega + \gamma)$, where γ is a small constant, which has the same period as the Bessel function expression of Eq. (15.29).

While the experiments with Na atoms have validated many aspects of the theory of radiatively assisted collisions in the strong field, high frequency regime, experiments with velocity selected K atoms have extended the measurements both to higher order processes and through the regime in which the collision time matches the frequency of the oscillating field.[14,17,18] Specifically, studying the process

$$K(n + 2)s + K\, nd \rightarrow K\, np + K\, (n - 1)p + m\omega \tag{15.40}$$

with velocity selected atoms has several attractive features. First, since the collisional resonances are so narrow, it is straightforward to satisfy the condition $\omega/2\pi \gg 1/\tau$ yet still observe collisions radiatively assisted by many photons without exceeding the available tuning range. Second, since the collision times are long, it is possible to observe the predicted phase dependence of the collisions in the transition region $\tau \sim 2\pi/\omega$.

The techniques employed are the same as those used to study the K resonant collisions described in the previous chapter. The apparatus shown in Fig. 14.13 is used, the only difference from the previous experiment being the application of a rf voltage to the top plate of the interaction region. The rf field, at frequencies from 0.5 MHz to 4.0 MHz is either left free running or phase locked to the laser excitation.

Most of the data are acquired in the same way as described above. K atoms are excited to the 29s and 27d states by the laser excitation, 4s → 4p → 29s, 27d. The atoms are allowed to collide for 1 μs, after which a rapidly rising detuning pulse is applied, followed by the more slowly rising field ionization pulse. Atoms which have made the transition to the 29p state are selectively ionized by the field ionization pulse and detected. This signal is monitored as the small static tuning field is scanned. The amplitude and phase of the rf field are changed as parameters.

Using such low collision frequencies it is possible to study radiative collision processes of high order, a point made explicitly by Fig. 15.9. In Fig. 15.9 we show the development of the radiative collision signal for collisions in the presence of 4 MHz rf fields of amplitudes from 0 to 0.76 V/cm.[17] Unlike the Na case described above, there is only a single collisional resonance when there is no rf field, as shown by Fig. 15.9(a). As the rf field amplitude is raised, progressively higher order resonant collision processes are observed, out to the seven photon assisted collisions. It is interesting to note that both $m > 0$ and $m < 0$ radiatively assisted collisions are observed, approximately symmetrically about the position of the collisional resonance at 6.44 V/cm. A slight asymmetry is expected at the extreme static tuning fields due to the second order Stark shift.

Close inspection of Fig. 15.9 suggests that the strength of a particular m-photon assisted cross section oscillates with the rf field as implied by the Bessel function dependence of Eq. (15.29). To demonstrate this point explicitly, the static field was fixed at the fields corresponding to several m photon assisted resonances and the amplitude of the rf field swept. Following this procedure for $m = 0, +1, -2$, and $+3$ leads to the results shown in Fig. 15.10, which shows quite clearly several zeroes in each of the σ_m cross sections. Furthermore, the experimental results agree almost perfectly with the expression of Eq. (15.28) and the dependence obtained by simply numerically integrating Eq. (14.16) using time dependent energies W_A and W_B. Close examination of Fig. 15.10 reveals that the observed cross sections at the highest rf field amplitudes are smaller than the calculated cross sections. This discrepancy is attributed to the fact that the static fields at which the collisional resonances occur shift slightly due to the second order ac Stark shift as the rf field is increased.

A final aspect of these radiation collisions, shown experimentally by Thomson et al.,[17] is that the cross section integrated over all the collisional resonances increases with the rf field. Since the mth collisional resonance has a cross section of $\sigma = \sigma_R |J_m(KE_{mw}/\omega)|$ and $\Sigma_m J_m^2(x) = 1$, in general, $\Sigma \sigma_m > \sigma_R$.[17]

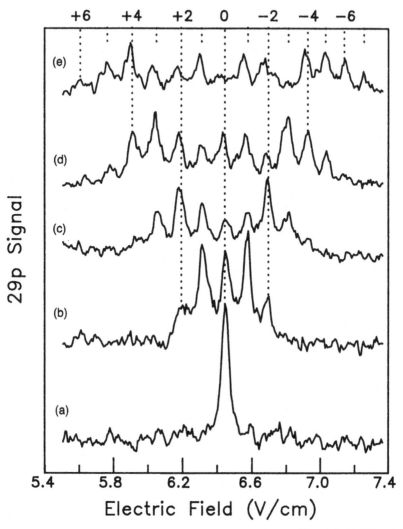

+6   +4   +2   0   −2   −4   −6

29p Signal

(e)

(d)

(c)

(b)

(a)

5.4   5.8   6.2   6.6   7.0   7.4

Electric Field (V/cm)

Fig. 15.9 The $n = 29$ resonance of K in the presence of a 4 MHz rf field. The rf field amplitudes for the five sets of data are: (a) 0 V/cm, (b) 0.19 V/cm, (c) 0.38 V/cm, (d) 0.57 V/cm, and (e) 0.76 V/cm. The top axis labels the sideband resonances according to the number of rf photons emitted by the atomic system (from ref. 17).

As we pointed out above, an attractive feature of the velocity selected K collisions is that we can examine the region $\omega/2\pi < 1/\tau$. We consider first the case in which there is no control over the phase of the field at which the collisions occur. This case is exemplified by Figs. 15.11(b)–(d), which show the effect of adding progressively larger amplitude 1 MHz rf fields of uncontrolled phase. Since the laser fires at an uncontrolled phase of the rf field, the observed cross section can be calculated using Eq. (15.34). As shown by Fig. 15.11 the cross section is broadened in approximately the manner shown in Fig. 15.7. As shown by a

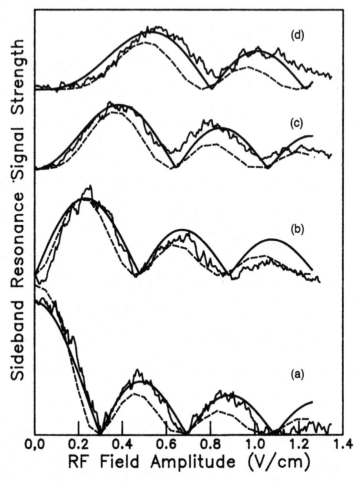

Fig. 15.10 Cross sections as a function of rf field strength for the first four orders of sideband resonances of the K 29s + K 27d radiative collisions in a 4 MHz rf field. (a) The zero-photon resonant collision cross section, (b) the +1 sideband resonance, (c) the −2 sideband resonance, and (d) the +3 sideband resonance. The solid line shows the experimental data, the bold line indicates the prediction the Floquet theory, and the dashed line is the result of numerical integration of the transition probability (from ref. 17).

comparison of Figs. 15.11(d) and 15.11(e), obtained with 1 and 0.5 MHz 0.2 V/cm rf fields, respectively, the broadening of the collisional resonance is independent of the frequency of the applied field, depending only on its amplitude, as predicted by Eq. (15.34).[18]

When the phase of a low frequency rf field is controlled the observed collisional resonances change dramatically. With the field of Eq. (15.32), when the atoms are allowed to collide during the interval $-0.5 \, \mu s < t < 0.5 \, \mu s$ with $\phi = 0$ or π, we observe the collisional resonances shown in Figs. 15.12(b) and (c) respectively, in agreement with Eq. (15.33). With no rf field the resonance occurs at $E_S = 6.44$

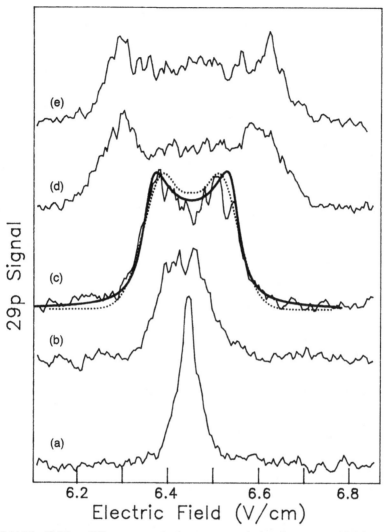

Fig. 15.11 The K 29s + 27d resonance in the presence of a low frequency rf field. In zero rf field (a), the FWHM is 1.6 MHz. In (b)–(d), a 1.0 MHz field of strength 0.05 V/cm, 0.1 V/cm, and 0.2 V/cm respectively is present. The solid line in (b) is a numerical integration of the transition probability, and the bold line is the convolution of a Lorentzian lineshape with a sinusoidal shift from resonance. In (e), the rf frequency is 0.5 MHz and its strength is 0.2 V/cm. For these low frequencies, the features are no long frequency dependent but rather are field strength dependent (from ref. 18).

V/cm, as shown by Fig. 15.12(a). When $\phi = 0$ the collision occurs at the maximum of the rf field, and the resonance occurs at a lower static field, as shown in Fig. 15.12(b). Similarly when the collision occurs at the minimum of the rf field the resonance occurs at a higher static field, as shown in Fig. 15.12(c).

Finally we consider the case in which $\omega/2\pi \approx 1/\tau$. In Fig. 15.13 we show the resonances obtained with an allowed collision time $T = 0.8\,\mu$s (the collisions are

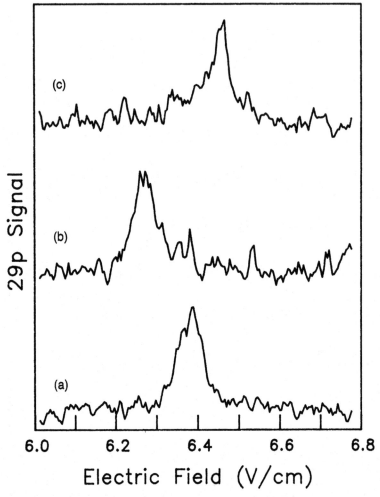

Fig. 15.12 (a) The K 29s + K 27d resonance with a FWHM of 2.0 MHz. In (b) and (c), a 0.5 MHz, 0.1 V/cm rf field is present with a phase of 0 and π respectively. A phase of 0 means the cosine wave is at maximum during the temporal center of the collision (from ref. 18).

allowed to occur for $-0.4\,\mu s < t < 0.4\,\mu s$ and two choices of the phase of an rf field, $E_{rf} \cos(\omega t + \phi)$, of amplitude $E_{rf} = 0.21$ V/cm and frequency 0.75 MHz.[14] In Fig. 15.13(a) we show the resonance with no rf field at 6.44 V/cm. In Fig. 15.13(b), in which $\phi = 0$, the experimental curve exhibits a broad peak at $E_S = 6.27$ V/cm, a subsidiary peak at 6.35 V/cm, and a possible third subsidiary peak at 6.40 V/cm. The dotted lines are numerical calculcations and the locations of the peaks for $N = 0$–2 from Eq. (15.36) are shown as well. As can be seen, the locations of the experimental peaks agree well with both the numerical calculation and the interference picture of Fig. 15.8(a) which leads to Eq. (15.36). When $\phi = \pi/2$, as shown in Fig. 15.13(c), a single broad resonance is observed, with no apparent

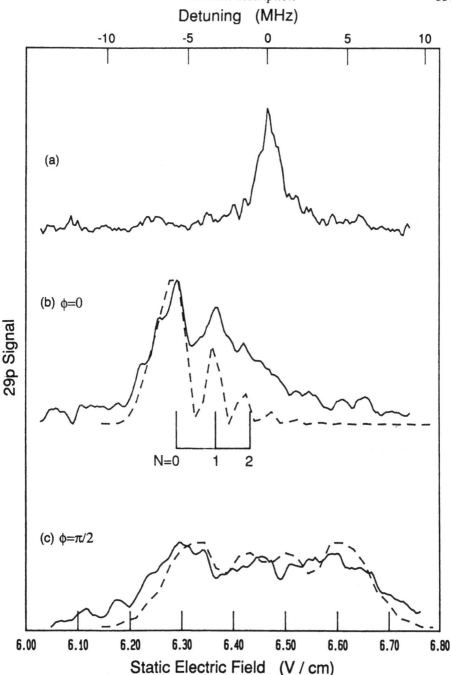

Fig. 15.13 The experimental (—) and calculated (---) lineshapes of the $n=29$ radiatively assisted resonance of K as functions of the static field E_s. (a) The unperturbed resonance with no rf field present, FWHM = 1.4 MHz. (b), (c) The resonance in the presence of a 0.75 MHz, 0.21 V/cm rf electric field and a phase of (b) $\phi = 0$ and (c) $\phi = \pi/2$. Indicated below the spectra are the calculated positions of the peaks using the $\Phi = 2\pi N$ condition (from ref. 14).

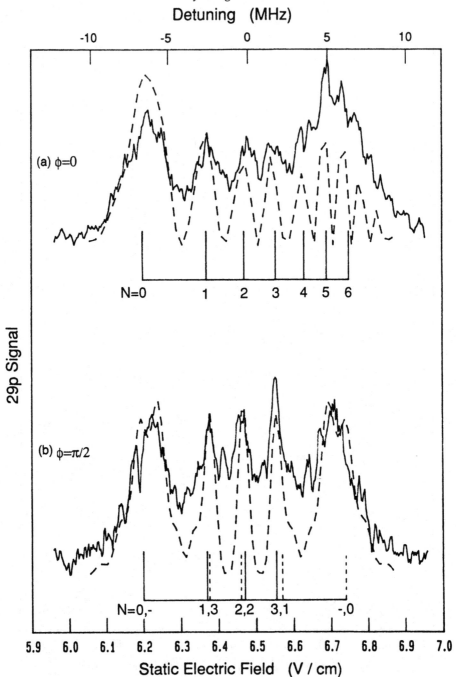

Fig. 15.14 The experimental (——) and calculated (– – –) lineshapes as functions of the static field E_S in a 0.31 V/cm, 1.48 MHz, rf field of phases (a) $\phi = 0$ and (b) $\phi = \pi/2$. The allowed collision time is 0.8 μs. The calculated fringe positions using $\Phi = 2\pi N$ are indicated below the spectra. In (b) the fringe positions due to interference between the first and second (solid vertical marker) and the second and third (dashed marker) interactions are shown (from ref. 14).

subsidiary peaks. As shown, the numerical calculations agree with the experimental result. Since the levels are only degenerate once, there is no possibility of interference, as shown by Fig. 15.8(b).

As the rf frequency is raised, interference becomes possible for all phases, a point made by Figs. 15.14(a) and (b), showing the results for a 1.48 MHz field of amplitude 0.31 V/cm and phases $\phi = 0$ and $\pi/2$, respectively. The result for $\phi = \pi$ is the mirror image of the $\phi = 0$ result. As shown, the experimental data match the numerical calculation and the prediction of Eq. (15.36).

In addition to the fact that the $\phi = \pi/2$ trace shows interference peaks, we note that they occur at approximately the same static field as the $\phi = 0$ peaks, although they are narrower. As the rf frequency is raised, the phase dependence slowly disappears, until at 4 MHz it is undetectable, and we are in the high frequency regime described by the dressed state picture.

References

1. L. I. Gudzenko and S. S. Yakovlenko, *Zh. Eksp. Teor. Fiz.* **62**, 1686 (1972) (Sov. Phys. JETP **35**, 877 (1972)).
2. A. Gallagher and T. Holstein, *Phys. Rev. A* **16**, 2413 (1977).
3. P. Pillet, R. Kachru, N. H. Tran, W. W. Smith, and T. F. Gallagher, *Phys. Rev. A* **36**, 1132 (1987).
4. R. W. Falcone, W. R. Green, J. C. White, J. F. Young, and S. E. Harris, *Phys. Rev. A* **15**, 1333 (1977).
5. P. Cahuzac and P. E. Toschek, *Phys. Rev. Lett.* **40**, 1087.
6. A. V. Hellfeld, J. Caddick, and J. Weiner, *Phys. Rev. Lett.* **40**, 1369 (1978).
7. W. R. Green, M. D. Wright, J. Lukasik, J. F. Young, and S. E. Harris, *Opt. Lett.* **4**, 265 (1979).
8. P. Pillet, R. Kachru, N. H. Tran, W. W. Smith, and T. F. Gallagher, *Phys. Rev. Lett.* **50**, 1763 (1983).
9. R. Kachru, T. F. Gallagher, F. Gounand, P. Pillet, and N. H. Tran, *Phys. Rev. A* **28**, 2676 (1983).
10. R. Kachru, N. H. Tran, and T. F. Gallagher, *Phys. Rev. Lett.* **49**, 191 (1982).
11. S. H. Autler and C. H. Townes, *Phys. Rev.* **100**, 703 (1955).
12. M. Abramowitz and I. A. Stegun, *Handbook of Mathematical Functions* National Bureau of Standards Applied Mathematics Series No. 55 (US GPO, Washington, DC, 1964).
13. F. Bowman, *Introduction to Bessel Functions* (Dover, New York, 1958).
14. M.J. Renn and T.F. Gallagher, *Phys. Rev. Lett.* **67**, 2287 (1991).
15. D. Coffey Jr., D.C. Lorents, and F.T. Smith, *Phys. Rev.* **187**, 201 (1969).
16. N. F. Ramsey, *Molecular Beams* (Oxford University Press, London, 1956).
17. D. S. Thomson, M. J. Renn, and T. F. Gallagher, *Phys. Rev. A* **45**, 358 (1992).
18. M. M. Renn, D. S. Thomson, and T. F. Gallagher, *Phys. Rev. A* **49**, 409 (1944).

16

Spectroscopy of alkali Rydberg states

In the first two chapters we have seen that the Na atom, for example, differs from the H atom because the valence electron orbits about a finite sized Na$^+$ core, not the point charge of the proton. As a result of the finite size of the Na$^+$ core the Rydberg electron can both penetrate and polarize it. The most obvious manifestation of these two phenomena occurs in the lowest ℓ states, which are substantially depressed in energy below the hydrogenic levels by core penetration. Core penetration is a short range phenomenon which is well described by quantum defect theory, as outlined in Chapter 2.

In the higher ℓ states the Rydberg electron is classically excluded from the core by the centrifugal potential $\ell(\ell + 1)/2r^2$, and, as a result, core penetration does not occur in high ℓ states, but core polarization does. Since it is not a short range effect, it cannot be described in terms of a phase shift in the wave function due to a small r deviation from the coulomb potential. However, the polarization energies of each series of $n\ell$ states exhibit an n^{-3} dependence, so the series can be assigned a quantum defect. Unlike the low ℓ states, in which the valence electron penetrates the core, measurements of the $\Delta\ell$ intervals of a few high ℓ states enable us to describe all the quantum defects of the high ℓ states in terms of the polarizability of the ion core. Irrespective of whether the quantum defect is due to core polarization or core penetration, it is important to have good values for these quantum defects as their values are important in determining the Rydberg state wavefunctions and, as a result, the Rydberg atom properties.

Although the fine structure intervals are invisible on the relatively coarse energy scale of Fig. 2.2, the spin orbit splittings of the alkali Rydberg levels differ from the hydrogenic intervals in a systematic fashion. Not surprisingly the fine structure intervals for high ℓ states are hydrogenic, but those of the low ℓ levels differ radically from H.

We are interested in measuring alkali Rydberg energy levels to extract quantum defects and fine structure intervals. There are two ways of going about these measurements. The first is to make purely optical measurements from low lying states to determine directly the quantum defects and fine structure intervals of the optically accessible Rydberg states. The second is to measure the intervals between the Rydberg states by rf resonance or related techniques. For example, measuring the intervals between several high ℓ states allows us to fit the observed intervals to a core polarization model yielding the quantum defects.

The data can be represented either as quantum defects for each fine structure series or as a quantum defect for the center of gravity of the ℓ level and a fine structure splitting. For the moment we shall use the latter convention, although it is by no means universal. Explicitly, we represent the energy of an $n\ell j$ state, where $\mathbf{j} = \boldsymbol{\ell} + \mathbf{s}$ and s is the electron spin, as

$$W_{n\ell j} = -\frac{1}{2(n - \delta_\ell)^2} + S\,\boldsymbol{\ell} \cdot \mathbf{s},$$

(16.1)

where S is an empirical constant. The center of gravity of the $n\ell$ state has the energy

$$W_{n\ell} = -\frac{1}{2(n - \delta_\ell)^2}.$$

(16.2)

Optical measurements

Historically the energies of the low ℓ Rydberg states were measured by optical absorption or emission spectroscopy. Today optical spectroscopy is still a valuable tool for acquiring spectra, the most useful modern forms being Doppler free techniques. One example is the single photon excitation of atoms in a collimated beam by a single frequency laser beam which intersects the atomic beam at a right angle. While the Doppler broadening can never be eliminated due to the divergence of the atomic beam, it can be substantially suppressed by this approach. For example, Fredriksson *et al.*[1] excited Cs atoms in a collimated beam from the ground $6s_{1/2}$ state to the $7p_{3/2}$ state using a 4555 Å laser, and 11% and 1% of the atoms in the $7p_{3/2}$ state decayed to the $5d_{5/2}$ and $5d_{3/2}$ states respectively. From the $5d_j$ states the atoms were excited to the $nf_{5/2}$ and $nf_{7/2}$ states by a single mode dye laser beam crossing the atomic beam at a right angle. The production of nf atoms was monitored by observing the nf–$5d$ fluorescence. They observed 10 MHz wide lines, yielding excellent values for the fine structure intervals of the Cs nf states.

The second Doppler free optical technique is two photon spectroscopy of atoms moving randomly in a cell. The basic notion of two photon spectroscopy is straightforward. A laser beam is retroreflected upon itself, usually so as to produce coincident foci of the beams propagating in both directions in the center of the cell. In its own reference frame an atom moving to the right with velocity v sees photons coming from the right as having frequency $v(1 + v/c)$ and photons coming from the left as having frequency $v(1 - v/c)$, where v is the laser frequency and c is the speed of light. The sum of the two frequencies is $2v$, irrespective of the velocity of the atom. Thus all the atoms, irrespective of velocity, are equally likely to absorb one photon from each beam when $2v$ equals an allowed two photon

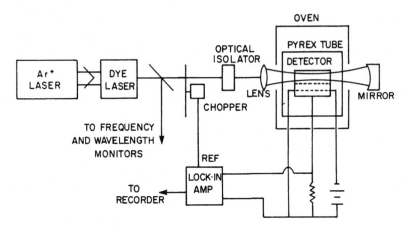

Fig. 16.1 Schematic diagram of the experimental arrangement for two photon spectroscopy of Rb Rydberg states (from ref. 5).

interval. The narrow two photon absorption is accompanied by a much weaker Doppler broadened absorption due to two photons from either beam.

This method has been used quite successfully to measure the two photon absorption spectra of K[2–4], Rb[5,6] and Cs[7,8]. The most often used detection approach is shown in Fig. 16.1. The atoms are contained in a glass cell, and the two photon absorption is detected by collisional ionization of the Rydberg atoms. The cell shown in Fig. 16.1 is usually called a space charge limited diode.[9] The basic principle of the diode is that electrons are thermionically emitted from a hot wire, but the bias voltage on the wire is so low that the current is limited by the electron space charge near the wire. If an atom is excited to a Rydberg state in a fairly dense vapor, it is collisionally ionized, and the ion formed is attracted to the wire emitting the electrons. Near the wire the ion partially neutralizes the electron space charge allowing a larger electron current to flow. Since the ion moves very slowly relative to an electron, many electrons escape through the hole in the space charge created by each ion reaching the wire. Gains of 10^6 have been reported for such diodes, and, not surprisingly, they have found wide application in optical spectroscopy of Rydberg states.

The optical data give the energies of the members of the Rydberg series relative to some lower state, not relative to the ionization limit, and it is convenient to use the term energy $T_{n\ell_j}$ of the $n\ell_j$ state. It is the energy above the energy of the ground state. Knowing the energy of the initial state of the optical transition, all the observed transition frequencies in an optical spectrum are readily converted to term energies. To extract the fine structure intervals is straightforward. They are simply the differences in the term energies of the same n and ℓ but different j.

To extract the quantum defects requires that we also determine the ionization limit. We define the ionization limit IP as the binding energy of the ground state, or equivalently, as the term energy of an $n = \infty$ state. A straightforward graphical approach can be used to determine both the ionization limit and the quantum defect of the observed Rydberg series. When the fine structure energies are removed, following Eqs. (16.1) and (16.2), the resulting observed term energies, $T_{n\ell}$, of the centers of gravity of the $n\ell$ levels are related to the quantum defect δ_ℓ and the ionization limit IP by

$$T_{n\ell} = IP + W_{n\ell} = IP - \frac{1}{2(n - \delta_\ell)^2}, \tag{16.3a}$$

or in conventional units by

$$T_{n\ell} = IP - \frac{Ry}{(n - \delta_\ell)^2}, \tag{16.3b}$$

where Ry is the Rydberg constant for the atom under study. Assuming that $\delta_\ell \ll n$ we expand the right hand side of Eq. (16.3b) in powers of δ_ℓ/n, which, when the higher order terms are ignored, leads to

$$T_{n\ell} + \frac{Ry}{n^2} = IP - \frac{2Ry\delta_\ell}{n^3}. \tag{16.4}$$

If we simply plot $T_{n\ell} + Ry/n^2$ vs $1/n^3$ we obtain a straight line the y intercept of which gives the ionization limit and the slope of which gives the quantum defect. In Fig. 16.2 this procedure is applied to the Li 5p to 15p term energies from Moore[10] to determine the ionization limit and quantum defect of the Li np series. The ionization limit from Fig. 16.2 is 43,486(1) cm^{-1}, and the np quantum defect is 0.046(1).

Radiofrequency resonance

The techniques described above allow accurate determination of the energy levels of optically accessible Rydberg series. To measure the intervals to other Rydberg series the primary technique has been microwave resonance. The basic notions are identical to those used in molecular beam resonance experiments. In the alkali experiments a tunable laser is used to excite an optically accessible Rydberg state, which is then driven to another Rydberg state by a microwave field which is slowly swept through resonance with the atomic transition. To detect that the transition has taken place requires a method of selectively detecting the final state of the microwave transition. Since initial and final states of the microwave transition are necessarily energetically close to each other, discriminating between them is not always easy.

Why microwave resonance techniques are so attractive for the spectroscopy of Rydberg states becomes clear when we estimate how much microwave power is required to drive the transitions. To drive the electric dipole $nd \rightarrow nf$ transition A

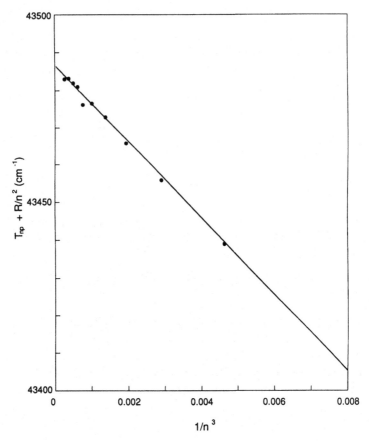

Fig. 16.2 Use of the term energies of the Li np Rydberg series to determine the ionization limit and the quantum defect. $T_{np} + Ry/n^2$ for the 5p to 15p states is plotted vs $1/n^3$. The y intercept is the ionization limit and the slope is $-2Ry\delta_p$.

$\rightarrow$ B in a time τ requires that the Rabi frequency be equal to the inverse of the time τ, or that

$$\mu E = \pi/\tau, \qquad (16.5)$$

where μ is the dipole transition moment and E is the microwave electric field. In a Rydberg atom transition in which $\Delta n = 0$ and $\Delta \ell = 1$, $\mu \sim n^2$, and the exposure time of the atoms to the microwaves can be several microseconds ($\sim 10^{11}$ au.). Thus for $n \sim 20$, an electric field of $\sim 10^{-13}$ is sufficient to drive the transition. This is a field of 500 μV/cm, corresponding to a power density of 10^{-11} W/cm^2.

The fact that the single photon transitions require so little power prompts us to consider two photon transitions. Consider the Na 16d $\rightarrow$ 16g transition via the virtual intermediate 16f$'$ state which is detuned from the real intermediate 16f state as shown by the inset of Fig. 16.3. If the detuning between the real and virtual intermediate states is Δ and the matrix elements between the real states are μ_1 and μ_2, the expression analogous to Eq. (16.5) for a two photon transition is

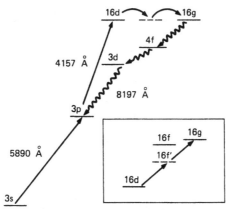

Fig. 16.3 Relevant energy levels for the observation of the two photon, d–g resonance in the $n = 16$ state of Na. The straight arrows are the two laser pumping steps; the wavy arrows down indicate the most probable fluorescent decay of the 16g state. The 3d–3p fluorescence at 8197 Å is observed. The inset shows the location of the d, f, and g states, as well as the virtual f′ state through which the two photon process proceeds (from ref. 12).

$$\frac{\mu_1 \mu_2 E^2}{\Delta} = \frac{\pi}{\tau}. \tag{16.6}$$

While Eq. (16.6) depends on the precise detuning, it has not been uncommon to have detunings of $10^{-2}n^{-3}$. If we assume that $\mu_1 = \mu_2 = n^2$, $\Delta = 10^{-2}n^{-3}$, and $\tau = 10^{11}$, when we evaluate Eq. (16.6) for $n = 20$ we find $E \sim 10^{-11}$, or 50 mV/cm, corresponding to $1\ \mu\text{W/cm}^2$.

A very simple method of observing the microwave transitions is to detect the fluorescence from only the final state. Gallagher *et al.*[11,12] used this technique to detect one, two, and three photon Na transitions from the optically accessible nd states to the $3 \le \ell \le 5$ states. The scheme used to detect the Na 16d→16g transitions is shown in Fig. 16.3. Two pulsed dye lasers at wavelengths of 5890 Å and ~ 4150 Å are required to excite the Na nd states, which decay primarily by emitting a 4150 Å photon. All of the $\ell \ge 3$ states are likely to decay via the 3d state which yields an 8200 Å photon when it decays to the 3p state. Monitoring the fluorescence at 8200 Å with a color filter and a photomultiplier allowed the microwave transitions from the d states to $\ell \ge 3$ states to be detected easily. A typical resonance is the Na $16\text{d}_{3/2}$–$16\text{g}_{7/2}$ two photon resonance shown in Fig. 16.4.

There are several considerations to bear in mind when using fluorescence detection. First, the approach is most useful when the photons to be detected have a vastly different wavelength than the exciting light and the most probable decay of the optically excited state, which need not be the same. Second, the branching ratio for the detected transition should be favorable. Third, the lifetimes of the initial and final state of the microwave transitions must be taken into account. If the microwaves are always on, at resonance, radiative decay occurs from the coupled pair of states. If the initial state of the microwave transition has a much

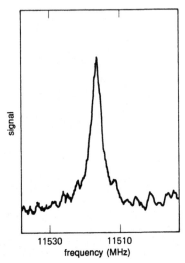

Fig. 16.4 Two photon Na $16d_{3/2}$–$16g_{7/2}$ resonance on a sweep of increasing microwave frequency (from ref. 12).

faster radiative decay rate, few of the atoms will decay from the final state, and the resonance signal will be small. If the laser excitation is pulsed and only a pulse of microwaves is used, a difference in the radiative lifetimes is unimportant. Finally, in a case such as the one shown in Fig. 16.4, there is the possibility of a cascade decay from the nd state, via lower np and nf states, to the 3d state. The cascade decay leads to a background signal tending to obscure partially the resonance signal.

A way of doing microwave spectroscopy peculiar to the study of Rydberg atoms is to use selective field ionization to discriminate between the initial and final states of the microwave transition. An example of the application of this technique is the measurement of millimeter wave intervals between Na Rydberg states by Fabre *et al.*[13] using the arrangement shown in Fig. 16.5.

Using two pulsed tunable dye lasers, Na atoms in a beam are excited to an optically accessible ns or nd state as they pass between two parallel plates. Subsequent to laser excitation the atoms are exposed to millimeter wave radiation from a backward wave oscillator for 2–5 μs, after which a high voltage ramp is applied to the lower plate to ionize selectively the initial and final states of the microwave transition. For example, if state A is optically excited and the microwaves induce the transition to the higher lying state B, atoms in B will ionize earlier in the field ramp, as shown in Fig. 16.5. The A–B resonance is observed by monitoring the field ionization signal from state B at t_B of Fig. 16.5 as the microwave frequency is swept.

In the method shown in Fig. 16.5, the excitation, microwave transition and selective ionization all occur in essentially the same place, since the thermal atoms do not move more than a few millimeters in a few microseconds. It is also possible

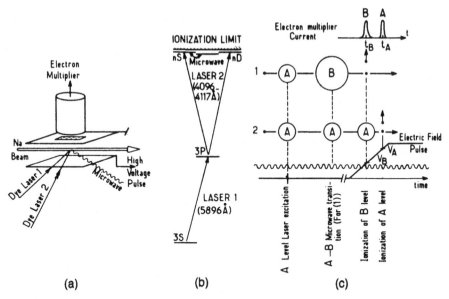

Fig. 16.5 (a) Schematic diagram of the experimental setup. (b) Energy diagram of the Na atom showing the levels populated by the stepwise laser excitation and the microwave transitions. (c) Sketch of the sequence of events experienced by the Na atoms (from ref. 13).

to separate the three parts of the experiment spatially, and an example is the measurement of the two photon Na 39s → 40s transitions by Goy *et al.*[14] They used two pulsed lasers to excite Na atoms from the 3s to the 3p to the 39s state as they entered a Fabry–Perot microwave cavity. In the cavity, which is fed by a phase locked backward wave oscillator, the microwave field drives two photon transitions to the 40s state. After the atoms pass out of the microwave cavity they pass between two parallel plates. A voltage, ramped in time, is applied to selectively field ionize the Rydberg atoms which have made the transition to the 40s state. Separating the microwave interaction region from the field ionization region and allowing the atoms to interact with the microwaves for a long time enabled Goy *et al.* to obtain resonance linewidths of ~10 kHz. In the discussion of fine structure measurements we give another slightly different example of the use of selective field ionization to detect transitions between fine structure levels. Due to its high efficiency field ionization is a very attractive technique, which is limited mainly by the separation in the ionization fields of the levels in question. This issue is discussed in Chapter 7.

A third resonance method which has been used to measure the intervals of alkali and Ba Rydberg atoms is delayed field ionization which takes advantage of the increase of the lifetime with ℓ. The method used by Safinya *et al.*[15] in the study of Cs nf → nh and nf → ni transitions is typical. Atoms are excited to the nf states in a manner similar to the one used by Fredriksson *et al.*,[1] except that pulsed rather than cw lasers are used. The atoms are then exposed to a 1 μs pulse of microwaves. After ~15 μs, about three lifetimes of the initial nf state, but less than one lifetime of the nh state, all the Rydberg atoms are ionized. Atoms in the optically excited nf initial state have largely decayed, but those which have undergone the $nf \rightarrow nh$ transition have not, and as a result, when the microwaves are tuned to a resonance the ionization signal increases. The bigger the difference in lifetimes the better this method works. The major limitation is that black body radiation redistributes population from the initial state to other final states, producing a substantial nonresonant background signal.[15]

Other methods which have been used to detect resonant microwave transitions between Rydberg states are selective photoionization,[16] and the detection of final states reached by black body radiation from only one of the two states involved in the microwave transition.[17] A surprisingly accurate way of measuring the $\Delta\ell$ intervals is the resonant collision approach used by Stoneman *et al.*[18] to measure the separations between the K ns, np, and nd states in small static electric fields. Using the readily calculable Stark shifts of all the levels it is straightforward to determine the zero field energy levels of the p states relative to the well known s and d energies, which have been measured by Doppler free two–photon spectroscopy.[2,3]

Core polarization

The data from the microwave resonance experiments give the differences in energies of the levels studied. For the high ℓ states, in which the difference in energy is due only to core polarization, we can calculate the quantum defects themselves using a core polarization model first proposed by Mayer and Mayer.[19] The model assumes that the Rydberg electron is slowly moving in a hydrogenic orbit around an ion core, which is polarized by the electrostatic interaction with the electron. Slowly moving in this context means that the excited states of the ion core are far removed in energy from the ionic ground state compared to the intervals between the Rydberg states. The polarization energy W_{pol} is given by

$$W_{pol} = \frac{-\alpha_d}{2}\langle r^{-4}\rangle - \frac{\alpha_q}{2}\langle r^{-6}\rangle, \tag{16.7}$$

where α_d and α_q are the dipole and quadrupole polarizabilities of the ionic core. We can express the energy of the $n\ell$ state as

$$W_{n\ell} = -\frac{1}{2n^2} + W_{pol}$$

$$\approx -\frac{1}{2n^2} - \frac{\delta_\ell}{n^3}. \tag{16.8}$$

In Eq. (16.8) we follow the convention of Eqs. (16.1) and (16.2) of separating the quantum defect from the fine structure.

Values of $\langle r^{-4} \rangle$ and $\langle r^{-6} \rangle$ for hydrogenic wavefunctions have been calculated analytically, and the expressions are given in Table 2.3. Examining the forms of Table 2.3, it is apparent that both $\langle r^{-4} \rangle$ and $\langle r^{-6} \rangle$ exhibit n^{-3} scalings, due to the normalization of the radial wavefunction. However, they exhibit different ℓ dependences; $\langle r^{-4} \rangle$ scales as ℓ^{-5}, and $\langle r^{-6} \rangle$ scales as ℓ^{-8}. As a result of the very different ℓ scalings the contributions of the dipole and quadrupole polarizabilities to the quantum defect are easily separated by measurements of the intervals between several ℓ series. Furthermore, for high ℓ, $\langle r^{-4} \rangle \gg \langle r^{-6} \rangle$, and as a result, for high ℓ,

$$W_{\text{pol}} \simeq \frac{-3\alpha_d}{4n^3 \ell^5}, \tag{16.9}$$

or, in terms of the quantum defect,

$$\delta_\ell \simeq \frac{3\alpha_d}{4\ell^5}. \tag{16.10}$$

It is a straightforward matter to apply the description of Eqs. (16.7) and (16.8) to the observed energy level intervals. In doing so we shall follow the convention of Edlen[20] and express the energies in wavenumbers using the Rydberg constant, Ry, corrected for the nuclear mass of the atom under study. From Eq. (16.8) we can see that for the ℓ and ℓ' states of the same n the energy difference is

$$\Delta W = W_{\text{pol}_{n\ell'}} - W_{\text{pol}_{n\ell}}, \tag{16.11}$$

which is implicitly a function of n, ℓ, and ℓ'. We may write ΔW as

$$\Delta W = \alpha_d \left(P_{n\ell} - P_{n\ell'} \right) + \alpha_q \left(P_{n\ell} Q_{n\ell} - P_{n\ell'} Q_{n\ell'} \right), \tag{16.12}$$

where

$$P_{n\ell} = Ry \langle r^{-4} \rangle_{n\ell}, \tag{16.13}$$

and

$$Q_{n\ell} = \frac{\langle r^{-6} \rangle_{n\ell}}{\langle r^{-4} \rangle_{n\ell}}. \tag{16.14}$$

In Eq. (16.12) ΔW is expressed in cm^{-1}, $P_{n\ell}$ in cm^{-1}/a_0^2, and $Q_{n\ell}$ in a_0^2. Following Safinya *et al.*[15] we can write Eq. (16.12) more compactly as

$$\Delta W = \alpha_d \, \Delta P + \alpha_q \, \Delta PQ, \tag{16.15}$$

where

$$\Delta P = P_{n\ell} - P_{n\ell'}, \tag{16.16}$$

and

$$\Delta PQ = P_{n\ell} Q_{n\ell} - P_{n\ell'} Q_{n\ell'}. \tag{16.17}$$

In Eqs. (16.15)–(16.17) ΔP and ΔPQ are also implicit functions of n and ℓ.

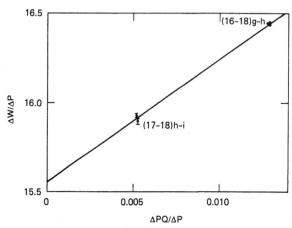

Fig. 16.6 Plot of $\Delta W/\Delta P$ vs $\Delta PQ/\Delta P$ used to extract Cs^+ effective dipole and quadrupole polarizabilities, $\alpha_d{}'$ and $\alpha_q{}'$. α'_d is the y intercept of the line through the data points while α'_q is its slope (from ref. 15).

Dividing Eq. (16.15) by ΔP removes the n^{-3} scaling as well as the contribution of the dipole polarizability to the observed $\Delta\ell$ intervals. Eq. (16.15) then becomes

$$\frac{\Delta W}{\Delta P} = \alpha_d + \alpha_q \frac{\Delta PQ}{\Delta P}. \tag{16.18}$$

Eq. (16.18) consists of the measured frequency intervals, ΔW, variables ΔP and ΔPQ which are readily calculated from the hydrogenic expressions in Table 2.3, the Rydberg constant, and the unknown core polarizabilities α_d and α_q. Dividing the observed frequencies ΔW by ΔP and plotting them vs $\Delta PQ/\Delta P$ gives a straight line such as the one shown in Fig. 16.6 for the Cs ng–nh–ni intervals. The y intercept gives the Cs^+ dipole polarizability, $\alpha_d = 15.544(30)a_o{}^3$, and the slope gives the Cs^+ quadrupole polarizability $\alpha_q = 70.7(20)a_o{}^5$. If the Cs nf–ng intervals are added to Fig. 16.6 they are found to be 20% too large to fit on the line, indicating that core penetration contributes ~20% to the quantum defect of the Cs nf states. The $\ell \geq 2$ states of Na and Li have also been studied by microwave spectroscopy, and the Na d states clearly exhibit core polarization while the Li d states do not.

In Table 16.1 we list the experimentally measured[15,21,22] core polarizabilities of the alkali ions Li^+, Na^+, and Cs^+ as well as the theoretically calculated values.[23–27] We have followed the convention of Freeman and Kleppner[28] of expressing the experimentally determined polarizabilities using primes, i.e. as α'_d and α'_q. The primes serve to remind us that we have assumed that the outer electron is very slowly moving. As discussed by Freeman and Kleppner, the fact that the outer electron is moving necessarily leads to a deviation of α'_d and α'_q from α_d and α_q. However, as shown by Table 16.1, the agreement between the measured values α'_d and α'_q and the calculated values α_d and α_q is quite good, and

Table 16.1. *Dipole and quadrupole polarizabilities.*

	measured		Calculated	
	α_d' (a_0^3)	α_q' (a_0^5)	α_d (a_0^3)	α_q (a_0^5)
Li^+	0.1884 (20)[a]	0.046 (7)[a]	0.189[b]	0.1122[c]
Na^+	0.9980 (33)[d]	0.35 (8)[d]	0.9459[e]	1.53[e]
Cs^+	15.544 (30)[f]	70.7 (29)[f]	19.03[g]	118.26[h]

[a] (see ref. 21)
[b] (see ref. 23)
[c] (see ref. 24)
[d] (see ref. 22)
[e] (see ref. 25)
[f] (see ref. 15)
[g] (see ref. 26)
[h] (see ref. 27)

the core polarization approach is a good way to describe the energies of high ℓ states. This point is made graphically by Fig. 16.7, a plot of alkali quantum defects vs. ℓ^5.[29] The linearity of the plot for high ℓ shows that high ℓ states have quantum defects which are largely due to the dipole core polarizability and that the quantum defects are well represented by Eq. (16.10). Since the quadrupole contribution to the core polarization is small, the deviation of the measured quantum defects from an ℓ^{-5} dependence is generally indicative of the onset of core penetration. From Fig. 16.7 it is apparent that the Li np, Na nd, and Cs nf states are the highest ℓ states to have appreciable core penetration. As we shall see, the presence or absence of core penetration plays an important role in the fine structure of the alkali atoms. The quantum defects of alkali atoms have been measured far more accurately than can be shown on a graph such as Fig. 16.7.[2–8,10–15,21,22,30–35] Accurate measurements reveal an energy dependence of the quantum defects, which is conveniently parameterized with a modified Rydberg–Ritz expression.[34] The quantum defects are slightly n dependent, and they are often specified for the resolved j states, not their center of gravity, as in Eq. (16.2). Explicitly, the quantum defect of an $n\ell j$ state is given by

$$\delta_{n\ell j} = \delta_0 + \frac{\delta_2}{(n - \delta_0)^2} + \frac{\delta_4}{(n - \delta_0)^4} + \frac{\delta_6}{(n - \delta_0)^6} + \frac{\delta_8}{(n - \delta_0)^8}, \cdots, \quad (16.19)$$

The parameters $\delta_0, \delta_2 \ldots$ are given in Table 16.2. For high n states the first two terms of Eq. (16.19) are often sufficient.

The term energies of these levels, $T_{n\ell j}$ are readily computed using

$$T_{n\ell j} = IP - \frac{Ry_{alk}}{(n - \delta_{n\ell j})^2}, \quad (16.20)$$

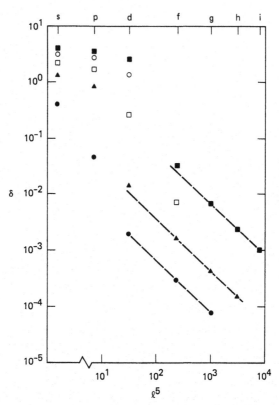

Fig. 16.7 Plot of the quantum defects δ of the alkali atoms vs ℓ^5: Li ($\bullet$), Na ($\blacktriangle$), K ($\square$), Rb ($\bigcirc$), and Cs ($\blacksquare$). Note the ℓ^{-5} dependence for high ℓ where the dipole polarizability dominates (from ref. 29).

where Ry_{alk} is the Rydberg constant for the alkali atom. Table 16.3 is a compilation of the ionization limits from the centers of gravity of the ground hyperfine levels and the values of Ry_{alk} for the most common isotopes of the alkali atoms.

Fine structure intervals

Measuring the $\Delta\ell$ intervals by microwave resonance techniques generally yields the fine structure intervals as well. However, $\Delta\ell = 0$ transitions between the fine structure levels can also be examined by several other techniques. The first of these is rf resonance. Since the transition involves no change in ℓ it is not an electric dipole transition but rather a magnetic dipole transition, and a straight-forward approach is magnetic resonance, which has been used by Farley and Gupta[36] to measure the 6f and 7f fine structure intervals in Rb. Their approach is

Table 16.2. *Alkali quantum defect parameters for Eq. (16.19).*[a]

Series		δ_0	δ_2	δ_4	δ_6	δ_8
[7]Li	$ns_{1/2}$	0.399468	0.030233	−0.0028	0.0115	
	$np_{1/2,3/2}$	0.47263	−0.02613	0.0221	−0.0683	
	$nd_{3/2,5/2}$	0.002129	−0.01491	0.1759	−0.8507	
	$nf_{5/2,7/2}$[b]	0.0003055(40)	−0.00126(5)			
[23]Na	$ns_{1/2}$[c]	1.3479692(4)	0.06137(10)			
	$np_{1/2}$[c]	0.855424(6)	0.1222(2)			
	$np_{3/2}$[c]	0.854608	0.1220(2)			
	$nd_{3/2,5/2}$	0.015543	−0.08535	0.7958	−4.0513	
	$nf_{5/2,7/2}$[d]	0.001663(60)	−0.0098(3)			
[39]K	$ns_{1/2}$	2.180197(15)	0.136(3)	0.0759	0.117	−0.206
	$np_{1/2}$	1.713892(30)	0.2332(50)	0.16137	0.5345	−0.234
	$np_{3/2}$	1.710848(30)	0.2354(60)	0.11551	1.105	−2.0356
	$nd_{3/2}$	0.276970(6)	−1.0249(10)	−0.709174	11.839	−26.689
	$nd_{5/2}$	0.277158(6)	−1.0256(20)	−0.59201	10.0053	−19.0244
	$nf_{5/2,7/2}$	0.010098	−0.100224	1.56334	−12.6851	
[85]Rb	$ns_{1/2}$	3.13109(2)	0.204(8)	−1.8		
	$np_{1/2}$	2.65456(15)	0.388(60)	−7.904	116.437	−405.907
	$np_{3/2}$	2.64145(20)	0.33(18)	−0.97495	14,6001	−44,7265
	$nd_{3/2,5/2}$	1.347157(80)	−0.59553	−1.50517	−2.4206	19.736
	$nf_{5/2,7/2}$	0.016312	−0.064007	−0.36005	3.2390	
[133]Cs	$ns_{1/2}$[e]	4.049325(15)	0.246(5)			
	$np_{1/2}$[e]	3.591556(30)	0.3714(40)			
	$np_{3/2}$[e]	3.559058(30)	0.374(40)			
	$nd_{3/2}$[e]	2.475365(20)	0.5554(60)			
	$nd_{5/2}$[e]	2.466210(15)	0.0167(5)			
	$nf_{5/2}$[e]	0.033392(50)	−0.191(30)			
	$nf_{7/2}$[e]	0.033537(28)	−0.191(20)			

[a] (see ref. 34 unless stated otherwise))
[b] (see ref. 21)
[c] (see refs. 35)
[d] (see ref. 22)
[e] (see ref. 33)

shown schematically in Fig. 16.8. Rb atoms contained in a cell are placed in a static magnetic field and excited to the $5p_{3/2}$ state by an unpolarized resonance lamp. Then they are excited to the $nd_{5/2}$ state by a circularly polarized cw laser beam. Some of the atoms decay from the $nd_{5/2}$ state to the 6f and 7f states, and there is enough polarization left in the f states that the detected nf–4d fluorescence has a circular polarization. There is, in addition to the static magnetic field, an oscillating magnetic field of frequency ~400 MHz perpendicular to the static field. The magnetic field is swept, and when it brings a $\Delta m_j = \pm 1$ transition between the fine structure levels into resonance with the rf magnetic field it reduces the polarization of the emitted fluorescence, allowing the detection of the transition.

Table 16.3. *Alkali ionization potentials and Rydberg constant.*[a]

atom	Ry^{alk}	IP
	(cm^{-1})	(cm^{-1})
^{7}Li	109728.64	43487.15(3)
^{23}Na	109734.69	41449.44(3)
^{39}K	109735.774	35009.814(1)
^{85}Rb	109736.605	33690.798(2)
^{133}Cs	109736.86	31406.471(1)

[a] (see ref. 34)

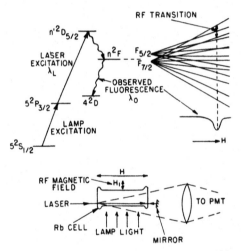

Fig. 16.8 Schematic illustration of the magnetic resonance technique used to measure the Rb nf fine structure intervals. The Rb atoms in the $n^{2}F$ states are populated by spontaneous decay of the $n'D_{5/2}$ states, which are populated by stepwise excitation of the ground state atoms. The rf transitions, induced among the magnetic sublevels of the $n^{2}F$ states, are detected as a change in the intensity of the polarized $n^{2}F \rightarrow 4^{2}D$ fluorescence. The lower part of the figure shows a sketch of the experimental arrangement (from ref. 36).

Both $\Delta j = 0$ and $\Delta j = 1$ transitions can be detected, the $\Delta j = 1$ intervals being more sensitive to the fine structure interval.

While the fine structure transitions are inherently magnetic dipole transitions, it is in fact easier to take advantage of the large $\Delta \ell = 1$ electric dipole matrix elements and drive the transitions by the electric resonance technique, commonly used to study transitions in polar molecules.[37] In the presence of a small static field of ~1 V/cm in the z direction the Na nd_j fine structure states acquire a small amount of nf character, and it is possible to drive electric dipole transitions between them at a Rabi frequency of 1 MHz with an additional rf field of ~1 V/cm.

A good example of the use of the electric resonance technique is the measurement of the Na nd fine structure intervals and tensor polarizabilities.[38] These transitions were observed using selective field ionization, although they appear to be unlikely prospects for field ionization detection because of the small separations of the levels, $\sim$20 MHz. The $nd_{3/2}$ states were selectively excited from the $3p_{1/2}$ state in a small static electric field and the $\Delta m_j = 0$ transitions to the $nd_{5/2}$ states induced by a rf field parallel to the static field. When a slowly rising field ionization pulse is applied to atoms in the Na $nd_{3/2}$ and $nd_{5/2}$ states of the same m_j, the $nd_{5/2}$ state atoms pass adiabatically to a high field state in which $|m| = |m_j| - 1/2$, while the $nd_{3/2}$ state atoms evolve to a state with $|m| = |m_j| + 1/2$ as shown by the correlation diagram of Fig. 7.10. The lower m states are more easily ionized, and the peak of the ionizing field pulse is set so as to ionize the $nd_{5/2}$ state but not the $nd_{3/2}$ state of the same m_j. When the radio frequency is scanned through resonance with the fine structure interval a sharp increase in the field ionization current is observed. Due to the fact that the ionization fields depend more strongly on m than on ℓ, it is easier to observe these 20 MHz transitions between the nd fine structure states than it is to observe the 20 GHz d $\rightarrow$ f transitions. The technique has also been used to measure Na np and K nd fine structure intervals.[38,39]

Quantum beats and level crossing

When the fine structure frequencies fall below $\sim$100 MHz they can also be measured by quantum beat spectroscopy. The basic principle of quantum beat spectroscopy is straightforward. Using a polarized pulsed laser, a coherent superposition of the two fine structure states is excited in a time short compared to the inverse of the fine structure interval. After excitation, the wavefunctions of the two fine structure levels evolve at different rates due to their different energies. For example if the $nd_{3/2}$ and $nd_{5/2}$ $m_j = 3/2$ states are coherently excited from the $3p_{3/2}$ state at time $t = 0$, the nd wavefunction at a later time t can be written as[40]

$$|nd_{\frac{3}{2}}\rangle = \{a|\tfrac{5}{2}\tfrac{3}{2}\rangle\,e^{-iW_{FS}t/2} + b|\tfrac{3}{2}\tfrac{3}{2}\rangle\,e^{iW_{FS}t/2}\}\,e^{(-i\overline{W}_n - \Gamma/2)t}. \qquad (16.21)$$

Here $\overline{W}_n$ is the average energy of the two fine structure levels. The real numerical coefficients a and b, where $a^2 + b^2 = 1$, depend on the polarizations used in the excitation scheme, but are constant in time. Thus the relative amounts of $d_{5/2}$ and $d_{3/2}$ states do not change with time but simply decay together at the radiative decay rate Γ. However, the relative amounts of m character oscillate at the fine structure frequency, and this oscillation is manifested in any property which depends upon m, such as the fluorescence polarized in a particular direction, or the field ionization signal due to a particular value of m. This fact becomes more apparent when we write the $|jm_j\rangle$ levels of Eq. (16.21) in terms of the uncoupled $|\ell m\rangle|sm_s\rangle$ states. Explicitly,

$$|\tfrac{5}{2}\tfrac{3}{2}\rangle = \sqrt{\tfrac{1}{5}}|22\rangle\,|\tfrac{1}{2} - \tfrac{1}{2}\rangle + \sqrt{\tfrac{4}{5}}|21\rangle\,|\tfrac{1}{2}\tfrac{1}{2}\rangle \qquad (16.22a)$$

and

$$|\tfrac{3}{2}\tfrac{3}{2}\rangle = \sqrt{\tfrac{4}{5}}|22\rangle\,|\tfrac{1}{2} - \tfrac{1}{2}\rangle - \sqrt{\tfrac{1}{5}}|21\rangle\,|\tfrac{1}{2}\tfrac{1}{2}\rangle. \qquad (16.22b)$$

Written in terms of the uncoupled states, Eq. (16.21) becomes

$$|\,n\mathrm{d}_{j\tfrac{3}{2}}\rangle = \{A|22\rangle\,|\tfrac{1}{2} - \tfrac{1}{2}\rangle + B|21\rangle\,|\tfrac{1}{2}\tfrac{1}{2}\rangle\,|\tfrac{1}{2}\tfrac{1}{2}\rangle\}\,e^{-\Gamma t/2}, \qquad (16.23)$$

where

$$|A|^2 = \tfrac{1}{5}\,(a^2 + 4b^2 + 4ab\cos W_{\mathrm{FS}}\,t)$$

and

$$|B|^2 = \tfrac{1}{5}\,(4a^2 + b^2 - 4ab\cos W_{\mathrm{FS}}\,t). \qquad (16.24)$$

As shown by Eq. (16.24), the amount of $m = 2$ and $m = 1$ character oscillates at the fine structure frequency.

The first measurements of Na nd fine structure intervals using quantum beats were the measurements of Haroche *et al*,[41] in which they detected the polarized time resolved nd–3p fluorescence subsequent to polarized laser excitation for n=9 and 10. Specifically, they excited Na atoms in a glass cell with two counterpropagating dye laser beams tuned to the $3s_{1/2} \rightarrow 3p_{3/2}$ and $3p_{3/2} \rightarrow n\mathrm{d}_j$ transitions. The two laser beams had orthogonal linear polarization vectors e_1 and e_2 as shown in Fig. 16.9.

The $n\mathrm{d}_j$–3p$_{3/2}$ fluorescence with polarization vector e_d emitted at a right angle to the directions of propagation of the two laser beams was detected. With $e_1 \perp e_d$ and $e_2 \perp e_d$ the phase of the beat signal is reversed, as shown in Figs. 16.9(a) and 16.9(b). Subtracting the trace of Fig. 16.9(b) from that of Fig. 16.9(a) leads to the beat signal of Fig. 16.9(c). Fig. 16.9 shows the beat signal for the Na 9d state. The 4 ns pulse duration of the exciting laser and the 2 ns risetime of the phototube set an upper limit of 150 MHz on the frequency of the beat signals they could detect. Fabre *et al*.[42] used the same technique to extend the measurements up to n=16.

An alternative method for detecting the beats is field ionization, which is particularly useful for higher n states. If the field ionization pulse is applied rapidly compared to the fine structure interval, the passage from the $\ell s j m_j$ fine structure states to the uncoupled $\ell m s m_s$ states is diabatic and the fine structure states are projected onto the uncoupled $\ell m s m_s$ states, as required for the detection of a beat signal. Such experiments were first carried out by Leuchs and Walther[43] who applied a rapidly rising field ionization pulse which projected Na nd atoms from the coupled $\ell s j m_j$ states onto the uncoupled $\ell m s m_s$ states and only ionized the atoms in $m = 0$ and 1 levels but not those in $m = 2$ levels. Their observed beat signals are shown in Fig. 16.10. Using this technique they measured the Na nd fine structure intervals over the range $21 \le n \le 31$.

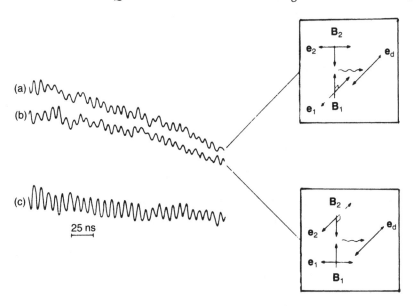

Fig. 16.9 Recording of fine structure beats in the Na 9d level (averaging of 1000 runs) obtained by time resolved fluorescence detection. Trace (a): signal obtained with configuration of polarizers $e_1 \perp e_2$, $e_d \| e_2$ as shown in the upper inset; trace (b): signal obtained with $e_1 \perp e_2$, $e_d \| e_2$ as shown in the inset; trace (c): result of subtracting trace (b) from trace (a). e_1, e_2, and e_d are the electric polarization vectors of laser beams 1 and 2 and the detected fluorescence, and B_1 and B_2 are the propagation vectors of the two beams (from ref. 41).

Jeys et al.[40] extended the measurements of Leuchs and Walther to higher n by a taking advantage of the fact that the entire field ionization pulse need not be rapid, only the initial rise to a field large enough to uncouple the spin and orbital motions. In fact, the requirements are quite modest. To project the fine structure states onto the uncoupled states a rather small field, ~1 V/cm, is required. Jeys et al. applied a rapidly rising "freezing pulse" of ~1 V/cm at a variable time after laser excitation to project the zero field fine structure states onto the uncoupled states. The freezing pulse was followed by a slowly rising ionizing pulse which allowed them to separate easily the ionization of the diabatically ionizing $m = 2$ levels from the adiabatically ionizing $m = 0$ and 1 levels. Taking the ratio of the $m = 2$ signal to the total signal allowed them to record clear beat signals for n up to 40.

Level crossing spectroscopy has been used by Fredriksson and Svanberg[44] to measure the fine structure intervals of several alkali atoms. Level crossing spectroscopy, the Hanle effect, and quantum beat spectroscopy are intimately related. In the above description of quantum beat spectroscopy we implicitly assumed the beat frequency to be high compared to the radiative decay rate Γ. We show schematically in Fig. 16.11(a) the fluorescent beat signals obtained by

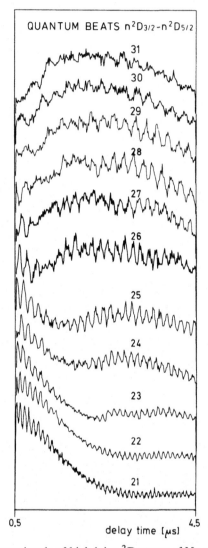

Fig. 16.10 Quantum beat signals of high lying ^{2}D states of Na obtained by time resolved selective field ionization. The variation of the beat frequency with principal quantum number is shown. Several quantum beat frequencies appear due to a Zeeman splitting of the fine structure levels in the earth's magnetic field (from ref. 43).

detecting the linearly polarized fluorescence with two orthogonal polarizations. When $W_{FS} < \Gamma$, the two polarized fluorescence signals might be as shown in Fig. 16.11(b). Examining Figs. 16.11(a) and (b) it is apparent that in Fig. 16.11(a), while there is a phase difference between the intensities of the two polarizations, the temporally integrated intensities are independent of W_{FS} for $W_{FS} \gg \Gamma$. On the other hand, in Fig. 16.11(b) it is evident that when $W_{FS} < \Gamma$ not only are the detailed time dependences different but also the temporally integrated intensities. The integrated intensity for vertical polarization increases as W_{FS} decreases and the integrated intensity for horizontal polarization decreases.

We have discussed the fluorescence intensity plots as if they were produced by a short pulse of light. In fact they also represent the probability of observing fluorescence subsequent to excitation by a single photon. In a level crossing experiment there is usually a weak but continuous stream of exciting photons, and we have no way of knowing when any given photon excited the atom, so all we can hope to do is to measure the time integrated fluorescence. As we have already seen, the integrated polarized fluorescence changes when $W_{FS} < \Gamma$, and if we could simply sweep W_{FS} from $W_{FS} < -\Gamma$ to $W_{FS} > \Gamma$ we would see a pronounced change in the integrated polarized fluorescence.

In a level crossing experiment the field dependence of two magnetic levels which cross in a magnetic field enables us to vary their energy spacing. In Fig. 16.8 we show the energy levels of the Rb nf states in a magnetic field.[44] The $nf_{5/2}$ and $nf_{7/2}$ levels of different m_j cross, and with light polarized perpendicular to the field, σ polarization, levels of m_j differing by 2 may be coherently excited. Away from the avoided crossings there is no difference in the time integrated fluorescence polarized parallel or perpendicular to the field. However, at the level crossing there is a readily detectable increase in the ratio of the parallel to perpendicular polarized fluorescence. From the field at which the level crossing signal is observed and the known orbital and spin gf factors it is straightforward to extract the fine structure intervals. Fredriksson and Svanberg used a Na resonance lamp and a tunable cw dye laser to excite Na atoms in a thermal beam to nd states via the 3p states. They scanned the static magnetic field to observe the level crossing signals for the $n = 5$ to $n = 9$ nd states, most of which have fine structure intervals too large to be measured by quantum beat spectroscopy, and their signals for the lowest field level crossings for $n = 5$ to $n=9$ are shown in Fig. 16.12.

To put the alkali fine structure intervals in perspective it is useful to compare them to the hydrogenic intervals. For H the energy of Eq. (16.4) is valid if[45]

$$\delta_\ell = 0 \text{ for all } \ell,$$

and

$$S = \frac{\alpha^2}{[\ell(2\ell + 1)(\ell + 1)]n^3}, \qquad (16.25)$$

where α is the fine structure constant. Using Eqs. (16.4) and (16.25) we can express the hydrogenic fine structure intervals, the difference in energy between the $j = \ell + 1/2$ state and the $j = \ell - 1/2$ state as

$$W_{FS} = \alpha^2/2\ell(\ell + 1)n^3, \qquad (16.26)$$

i.e. the interval scales as n^{-3}. In the alkali atoms the interval scales only

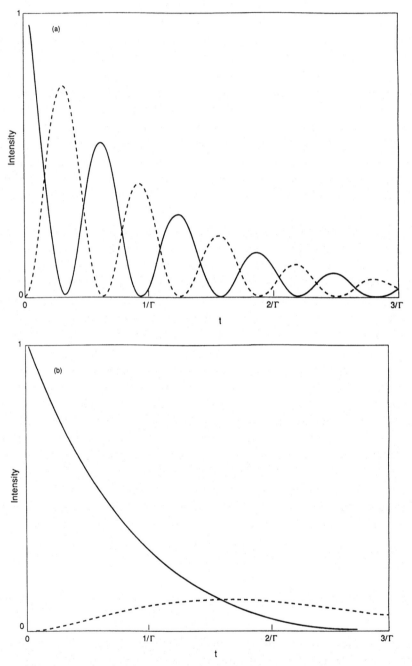

Fig. 16.11 Intensity of x(——) and y (– – –) polarized fluorescence as a function of time after excitation with x polarized light for different ratios of the fine structure interval W_{FS} to the radiative decay fluorescence Γ: (a) $W_{FS} = 10\Gamma$, (b) $W_{FS} = \Gamma$. In (a) there is only a phase difference in the intensity of the emitted fluorescence. In (b) there is a clear difference in the temporally integrated fluorescence as well.

approximately as n^{-3}. A better representation of the intervals using the effective quantum number is

$$W_{FS} = \frac{A}{(n-\delta)^3} + \frac{B}{(n-\delta)^5} + \frac{C}{(n-\delta)^7}. \qquad (16.27)$$

In Table 16.4 we tabulate the reported values of the parameters A, B, and C for H and the alkali atoms.[46–49]

From Table 16.4 several phenomena are evident. First, the Li intervals and the nonpenetrating f states of Na have very nearly hydrogenic intervals. Second the

Table 16.4. *Fine structure parameters.*

	A	B	C
H np^a	87.65 GHz		
H nd^a	29.22 GHz		
H nf^a	14.61 GHz		
Li np^b	74.53 GHz		
Li nd^c	29.096 GHz		
Na np^d	5.376(5) Thz		
Na nd^e	−97.8 (10) GHz	520 (20) GHz	
Na nf^f	14.2(1) GHz		
K np^g	20.0847 (12) THz	−6.78 (75) THz	13.05 (40) THz
K nd^h	−1.131 (18) THz	−15.53 (66) THz	90.4 (30) THz
Rb np^i	85.865 THz		
Rb nd^j	10.800 (15) THz	−84.87 (10) THz	
Rb nf^k	−152 GHz	+1.82 THz	
Cs np^l	213.925 (20) THz	−56 (4) THz	390 (10) THz
Cs nd^m	60.2183 (60) THz	−58 (8) THz	
Cs nf^n	−979.6 GHz	12.22 THz	−33.76 THz

[a] (see ref. 45)
[b] (see ref. 10)
[c] (see refs. 21 and 46)
[d] (see ref. 35)
[e] (see ref. 47)
[f] (see ref. 48)
[g] (see ref. 18)
[h] (see ref. 3)
[i] (see ref. 49)
[j] (see ref. 5)
[k] (see ref. 36)
[l] (see ref. 33)
[m] (see ref. 8)
[n] (see ref. 1)

State   Position (G)

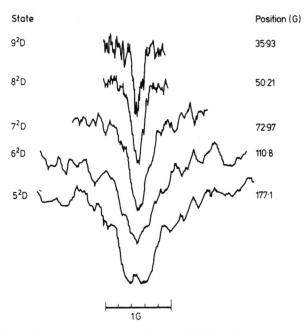

State	Position (G)
9^2D	35·93
8^2D	50·21
7^2D	72·97
6^2D	110·8
5^2D	177·1

1G

Fig. 16.12 Experimental level crossing curves for the Na 5d to 9d states. In each case the recording is made at the lowest field $\Delta m_j = 2$ level crossing. Each curve has been sampled for 1–2 h (from ref. 44).

low ℓ states for all the alkalis but Li have much larger fine structure intervals than does H. Finally, the Na and K nd states and the Rb and Cs nf states are inverted. These are the highest ℓ states of these atoms to exhibit core penetration.

Two general methods have been used to address the fine structure inversion theoretically. One is to reduce the relativistic operators to a nonrelativistic form and use them to calculate perturbations to the nonrelativistic solution. This general approach has been used by Holmgren *et al.*[50] and Sternheimer *et al.*[51] The other approach, used by Luc-Koenig,[52] is to solve the Dirac equation in the relativistic potential of the ionic core. Both approaches give similar numerical results, and it has recently been shown by Lindgren and Martensson[53] that the approaches are in some cases equivalent up to terms of order α^2. From the point of view of a non relativistic calculation, the inversion is due to polarization of the core by the valence electron. As pointed out by Sternheimer *et al.*, for this effect to be strong, it is imperative that the core have electrons with nonzero angular momenta. As a result, while the Na nd states are clearly inverted it is not surprising that the Li nd fine structure intervals deviate from the hydrogenic values by less than 1%.

References

1. K. Fredriksson, H. Lundberg, and S. Svanberg, *Phys. Rev. A* **21**, 241 (1980).
2. C. J. Lorenzen, K. Niemax, and L.R. Pendrill, *Opt. Comm.* **39**, 370 (1980).
3. D.C. Thompson, M.S. O'Sullivan, B.P. Stoicheff, and G-X Xu, *Can. J. Phys.* **61**, 949 (1983).
4. C.D. Harper and M.D. Levenson, *Phys. Lett.* **36A**, 361 (1976).
5. K.C. Harvey and B.P. Stoicheff, *Phys. Rev. Lett.* **38**, 537 (1978).
6. B.P. Stoicheff and E. Weinberger, *Can. J. Phys.* **57**, 2143 (1979).
7. C. J. Lorenzen, K.H. Weber, and K. Niemax, *Opt. Comm.* **33**, 271 (1980).
8. M.S. O'Sullivan and B.P. Stoicheff, *Can. J. Phys.* **61**, 940 (1983).
9. D. Popescu, I. Popescu, and J. Richter, *Z. Phys.* **226**, 160 (1969).
10. C.E. Moore, *Atomic Energy Levels*, NBS Circular 467 (U.S. Government Printing Office, Washington, 1949).
11. T.F. Gallagher, R.M. Hill, and S.A. Edelstein, *Phys. Rev. A* **13**, 1448 (1976).
12. T.F. Gallagher, R.M. Hill, and S.A. Edelstein, *Phys. Rev. A* **14**, 744 (1976).
13. C. Fabre, S. Haroche, and P. Goy, *Phys. Rev. A* **18**, 229 (1978).
14. P. Goy, C. Fabre, M. Gross, and S. Haroche, *J. Phys. B* **13**, L83 (1980).
15. K. A. Safinya, T.F. Gallagher, and W. Sandner, *Phys. Rev A* **22**, 2672 (1981).
16. W.E. Cooke and T.F. Gallagher, *Opt. Lett.* **4**, 173 (1979).
17. T.F. Gallagher, in *Atoms in Intense Laser Fields*, ed. M. Gavrila (Academic Press, Cambridge, 1992).
18. R.C. Stoneman, M.D. Adams, and T.F. Gallagher, *Phys. Rev. Lett.* **58**, 1324 (1987).
19. J.E. Mayer and M.G. Mayer, *Phys. Rev.* **43**, 605 (1933).
20. B. Edlen in *Handbuch der Physik*, ed. S. Flugge (Springer-Verlag, Berlin, 1964).
21. W.E. Cooke, T.F. Gallagher, R.M. Hill, and S.A. Edelstein, *Phys. Rev. A* **16**, 1141 (1977).
22. L. G. Gray, X. Sun, and K.B. MacAdam, *Phys. Rev. A* **38**, 4985 (1988).
23. H.D. Cohen, *J. Chem. Phys.* **43**, 3558 (1966).
24. J. Lahiri and A. Mukherji, *Phys. Rev.* **141**, 428 (1966).
25. J. Lahiri and A. Mukherji, *Phys. Rev.* **153**, 386 (1967).
26. J. Heinrichs, *J. Chem. Phys.* **52**, 6316 (1970).
27. R. M. Sternheimer, *Phys. Rev. A* **1**, 321 (1970).
28. R.R. Freeman and D. Kleppner, *Phys. Rev. A* **14**, 1614 (1976).
29. T.F. Gallagher, in *Progress in Atomic Spectroscopy*, eds. H.J. Beyer and H. Kleinpoppen, (Plenum, New York, 1987)
30. C. Corliss and J. Sugar, *J. Phys. Chem. Ref. Data* **8**, 1109 (1979).
31. C.J. Lorenzen and K. Niemax, J. Quant, *Spectrosc. Radiative Transfer* **22**, 247 (1979).
32. K.B.S. Eriksson and I. Wenaker, *Phys. Scr.* **1**, 21 (1970).
33. P. Goy, J.M. Raimond, G. Vitrant, and S. Haroche, *Phys. Rev. A* **26**, 2733 (1982).
34. C. J. Lorenzen and K. Niemax, *Phys. Scr.* **27**, 300 (1983).
35. C. Fabre, S. Haroche, and P. Goy, *Phys. Rev. A* **22**, 778 (1980).
36. J. Farley and R. Gupta, *Phys. Rev. A* **15**, 1952 (1977).
37. H.K. Hughes, *Phys. Rev.* **72**, 614 (1947).
38. T.F. Gallagher, L.M. Humphrey, R.M. Hill, W.E. Cooke, and S.A. Edelstein, *Phys. Rev. A* **15**, 1937 (1977).
39. T.F. Gallagher and W.E. Cooke, *Phys. Rev. A* **18**, 2510 (1978).
40. T.H. Jeys, K.A. Smith, F.B. Dunning, and R.F. Stebbings, *Phys. Rev. A* **23**, 3065 (1981).
41. S. Haroche, M. Gross, and M.P. Silverman, *Phys. Rev. Lett.* **33**, 1063 (1974).
42. C. Fabre, M. Gross, and S. Haroche, *Opt. Comm.* **13**, 393 (1975).
43. G. Leuchs and H. Walther, *Z. Physik A* **293**, 93 (1979).
44. K. Fredricksson and S. Svanberg, *J. Phys. B* **9**, 1237 (1976).
45. H.A. Bethe and E.A. Salpeter, *Quantum Mechanics of One and Two Electron Atoms*, (Academic Press, New York, 1957).
46. J. Wangler, L. Henke, W. Wittman, H.J. Plöhn, and H.J. Andrä, *Z. Phys. A.* **299**, 23 (1981).
47. C. Fabre and S. Haroche, in *Rydberg States of Atoms and Molecules*, eds. R.F. Stebbings and F.B. Dunning (Cambridge University Press, Cambridge, 1983).

48. N.H. Tran, H.B. van Linden van den Heuvell, R. Kachru, and T.F. Gallagher, *Phys. Rev. A* **30**, 2097 (1984).
49. S. Liberman and J. Pinard, *Phys. Rev. A* **20**, 507 (1979).
50. L. Holmgren, I. Lindgren, J. Morrison, and A.M. Martensson, *Z. Physik A* **276**, 179 (1976).
51. R.M. Sternheimer, J.E. Rodgers, T. Lee, and T.P. Das, *Phys. Rev. A* **14**, 1595 (1976).
52. E. Luc-Koenig, *Phys. Rev. A* **13**, 2114 (1976).
53. I. Lindgren and A.M. Martensson, *Phys. Rev. A* **26**, 3249 (1982).

17

Rf spectroscopy of alkaline earth atoms

The spectra of bound alkaline earth atoms contain two features which are absent in the spectra of the alkali atoms and He. First, there are perturbations in the regularity of the bound Rydberg series caused by the presence of doubly excited states converging to higher ionization limits. Second, in the higher angular momentum states there is clear evidence for the breakdown of the adiabatic core polarization model which we used in Chapter 16 to describe the alkali atoms.[1] Although it may not be apparent, these two features are slightly different manifestations of the same phenomena. The perturbations of the spectra have been studied by both rf and optical spectroscopy, but core polarization can only be studied by rf spectroscopy.

Theoretical description of series perturbation and nonadiabatic effects in core polarization

For simplicity we shall consider the Ba atom, although precisely the same procedure could be carried out for any alkaline earth atom. We shall assume that Ba consists of an inert spherical closed shell Ba^{2+} core which has two electrons.[3] The basic notion of the treatment of Van Vleck and Whitelaw[2] is readily understood by looking at the energy level diagram of Fig. 17.1. Each one electron

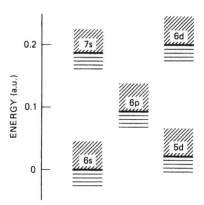

Fig. 17.1 Low lying energy levels of Ba^+ (——), Rydberg series of Ba (——) converging to them, and the continua (///) above each of them (from ref. 3).

state of Ba$^+$ supports an entire system of bound and continuum states of the outer electron. The bound 6s$n\ell$ Rydberg states are the lowest lying system of bound states. As a first approximation we can construct wavefunctions for the two electron states ignoring the interaction between the two electrons. However, when the interaction is introduced as a perturbation, there is a second order correction to the energies of the bound 6s$n\ell$ states which in most cases depresses their energies. This description reduces, in the adiabatic limit, to the core polarization model of Mayer and Mayer.[1]

The Hamiltonian for the outer two Ba electrons is given by

$$H = -\left[\frac{\nabla_1^2}{2} + \frac{\nabla_2^2}{2} + f(r_1) + f(r_2) - \frac{1}{r_{12}}\right], \tag{17.1}$$

where $-f(r)$ is the potential of an electron at a distance r from the Ba^{2+} core and r_{12} is the distance between the two electrons. As $r \rightarrow \infty$, $f(r) \rightarrow 2/r$. Separating H into

$$H = H_0 + H_1 \tag{17.2}$$

leads to

$$H_0 = -\left[\frac{\nabla_1^2}{2} + \frac{\nabla_2^2}{2} + f(r_1) + \frac{1}{r_2}\right] \tag{17.3a}$$

and

$$H_1 = -f(r_2) + 1/r_2 + 1/r_{12}. \tag{17.3b}$$

Using H_0, we can write the time independent Schroedinger equation as

$$-\left[\frac{\nabla_1^2}{2} + f(r_1) + \frac{\nabla_2^2}{2} + \frac{1}{r_2}\right]\Psi(\mathbf{r}_1,\mathbf{r}_2) = W\Psi(\mathbf{r}_1,\mathbf{r}_2), \tag{17.4}$$

where $\mathbf{r}_1$ and $\mathbf{r}_2$ are the vectors from the Ba^{2+} core to electrons 1 and 2. Eq. (17.4) may be separated into two independent equations,

$$-\left[\frac{\nabla_1^2}{2} + f(r_1)\right]\psi_1(\mathbf{r}_1) = W_1\psi_1(\mathbf{r}_1) \tag{17.5a}$$

and

$$-\left(\frac{\nabla_2^2}{2} + \frac{1}{r_2}\right)\psi_2(\mathbf{r}_2) = W_2\psi_2(\mathbf{r}_2), \tag{17.5b}$$

where

$$\Psi(\mathbf{r}_1,\mathbf{r}_2) = \psi_1(\mathbf{r}_1)\psi_1(\mathbf{r}_2) \tag{17.6a}$$

and

$$W = W_1 + W_2. \tag{17.6b}$$

The solutions to Eq. (17.5a) are the Ba^+ wavefunctions $\phi_{n'\ell'm'}(\mathbf{r}_1)$, and the solutions to Eq. (17.5b) are H wavefunctions $u_{n\ell m}(\mathbf{r}_2)$. Explicitly, Eq. (17.6a) becomes

$$\Psi(\mathbf{r}_1,\mathbf{r}_2) = \phi_{n'\ell'm'}(\mathbf{r}_1)u_{n\ell m}(\mathbf{r}_2), \tag{17.7}$$

and Eq. (17.6b) becomes

$$W = W_{n'\ell'} - 1/2n^2, \tag{17.8}$$

where $W_{n'\ell'}$ is the energy of the Ba^+ $n'\ell'$ state relative to the Ba^{2+} ionization limit. The wavefunction of Eq. (17.7) is an excellent representation of the high ℓ Rydberg states even though it is not antisymmetrized and takes no account of exchange effects.

From Eq. (17.8) it is apparent that all states of the same n, ℓ, and m are degenerate. This degeneracy is removed by including H_1 as a perturbation. For high ℓ states in which $r_2 > r_1$, the potential $f(r_2)$ can be written as $f(r_2) = 2/r_2$. Replacing $f(r_2)$ by $2/r_2$ in Eq. (17.3b) and expanding $1/r_{12}$ in terms of r_1 and r_2 yields

$$H_1 = \frac{-2}{r_2} + \frac{1}{r_2} + \left[\frac{1}{r_2} + \frac{r_1}{r_2}P_1(\cos\theta_{12}) + \frac{r_1^2}{r_2^3}P_2(\cos\theta_{12})\ldots\right]$$

$$= \frac{r_1}{r_2^2}P_1(\cos\theta_{12}) + \frac{r_1^2}{r_2^3}P_2(\cos\theta_{12})\ldots, \tag{17.9}$$

where θ_{12} is the angle between $\mathbf{r}_1$ and $\mathbf{r}_2$. It is useful to rewrite the Legendre polynomials in terms of spherical harmonics. Explicitly,[4]

$$P_k(\cos\theta_{12}) = \left(\frac{4\pi}{2k+1}\right)\sum_m Y_{km}^*(\theta_1,\phi_1)Y_{km}(\theta_2,\phi_2) \tag{17.10}$$

where θ_1, ϕ_1, and θ_2, ϕ_2 are the angular coordinates of electrons 1 and 2 relative to an axis through the Ba^{2+} core. Since the angular eigenfunctions of Eq. (17.7) are spherical harmonics and the ground state of Ba^+ is an s state, the diagonal matrix elements of H_1 vanish for the Ba $6sn\ell$ states, and there is no first order correction to the energy from H_1. The second order corrections to the energies of the Ba $6sn\ell$ states can be expressed to adequate accuracy using the dipole and quadrupole terms of Eq. (17.9). Explicitly, we write the second order energy W_2 as

$$W_2 = W_d + W_q, \tag{17.11}$$

where

$$W_d = \sum_{n'\ell'n''\ell''}\left[\frac{\left|\left\langle 6sn\ell\left|\frac{r_1}{r_2^2}P_1(\cos\theta_{12})\right|n'\ell'n''\ell''\right\rangle\left\langle n'\ell'n''\ell''\left|\frac{r_1}{r_2^2}P_1(\cos\theta_{12})\right|6sn\ell\right\rangle\right|}{W_{6sn\ell} - W_{n'\ell'n''\ell''}}\right],$$

$$\tag{17.12}$$

and

$$W_q = \sum_{n'\ell'n''\ell''} \left[\frac{\left\langle 6sn\ell \left| \frac{r_1^2}{r_2^3} P_2 (\cos \theta_{12}) \right| n'\ell'n''\ell'' \right\rangle \left\langle n'\ell'n''\ell'' \left| \frac{r_1^2}{r_2^3} P_2 (\cos \theta_{12}) \right| 6sn\ell \right\rangle}{W_{6sn\ell} - W_{n'\ell'n''\ell''}} \right].$$

(17.13)

In Eqs. (17.12) and (17.13) summations over n' and n'' implicitly include the continua.

The second order energy shifts given by Eqs. (17.12) and (17.13) lead to both the series perturbation and core polarization. We consider first the perturbation of the $6sn\ell$ Rydberg series which occurs when a low lying Ba $5dn''\ell''$ state lies in the middle of the $6sn\ell$ Rydberg series. The 5d7d state is a good example of such a perturber. Due to its proximity to the $6sn\ell$ Rydberg states the $5dn''\ell''$ state dominates the sum of Eq. (17.13) and the energy shift of a $6sn\ell$ level is given by

$$\Delta W = W_q = \frac{\left| \left\langle 6sn\ell \left| \frac{r_1^2}{r_2^3} P_2(\cos \theta_{12}) \right| 5dn''\ell'' \right\rangle \right|^2}{W_{6sn\ell} - W_{5dn''\ell''}}.$$

(17.14)

The $6sn\ell$ Rydberg states lying above the perturbing $5dn''\ell''$ state are shifted up in energy, while those lying below the perturbing $5dn''\ell''$ state are shifted down in energy. The Rydberg states lying closest to the perturber are shifted the most. When there is more than one perturbing level it is often easier to use the quantum defect theory than to apply Eq. (17.14).

While the perturbation of the $6sn\ell$ Rydberg series typically comes from the interaction with a single $5dn''\ell''$ state, core polarization comes from the interaction

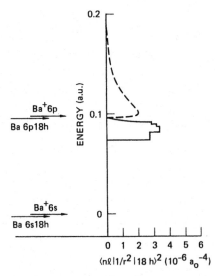

Fig. 17.2 The squared matrix elements $\langle ng|1/r^2|18h\rangle^2$ (——) and $\langle ni|1/r^2|18h\rangle^2$ (– – –) plotted vs energy to show the energy distribution of the squared matrix elements (from ref. 3).

with entire Rydberg series converging to excited states of Ba^+, i.e. many terms of the sums of Eqs. (17.12) and (17.13) are important. The sums do not extend over all possible values of n', ℓ', n'', and ℓ''. With the help of Eq. (17.10) we can see that in the expression for W_d, Eq. (17.12), $\ell' = 1$ $\ell'' = \ell \pm 1$, and that in the expression for W_q, Eq. (17.13), $\ell' = 2$ and $\ell'' = \ell$, $\ell \pm 2$. In Ba, the 6s–6p dipole matrix element and 6s-5d quadrupole matrix elements are by far the largest matrix elements in the summations of Eqs. (17.12) and (17.13), and these terms also have the smallest energy denominators. Thus we can, to an excellent approximation, restrict the sum over n' to one term in each case.

The most straightforward way of evaluating the angular matrix elements of Eqs. (17.12) and (17.13) is to use the method of Edmonds.[5] The matrix elements are evaluated as the scalar products of tensor operators operating on the wavefunctions of electrons 1 and 2. Using this approach we can write W_d as

$$W_d = \frac{|\langle 6s|r_1|6p\rangle|^2}{3} \left[\sum_{n''} \frac{\ell \left|\langle n''\ell - 1 \left| \frac{1}{r_2^2} \right| n\ell \rangle\right|^2}{(2\ell + 1)(W_{6sn\ell} - W_{6pn''\ell-1})} \right.$$

$$\left. + \sum_{n''} \frac{(\ell + 1)\left|\langle n''\ell + 1 \left| \frac{1}{r_2^2} \right| n\ell \rangle\right|^2}{(2\ell + 1)(W_{6sn\ell} - W_{6pn''\ell+1})} \right], \tag{17.15}$$

and W_q as

$$W_q = |\langle 6s|r_1^2|5d\rangle|^2 \left[\frac{3}{10(4\ell^2 - 1)(2\ell + 3)}\right]$$

$$\times \left[(2\ell - 1)(\ell + 1)(\ell + 2)\sum_{n''} \frac{\left|\langle n''\ell + 2 \left| \frac{1}{r_2^3} \right| n\ell \rangle\right|^2}{W_{6sn\ell} - W_{5dn''\ell + 2}} + \frac{2(\ell^2 + \ell)(2\ell + 1)}{3}\right.$$

$$\times \left. \sum_{n''} \frac{\left|\langle n''\ell \left| \frac{1}{r_2^3} \right| n\ell \rangle\right|^2}{W_{6sn\ell} - W_{5dn''\ell}} + (2\ell + 3)(\ell^2 - \ell)\sum_{n''} \frac{\left|\langle n''\ell - 2 \left| \frac{1}{r_2^3} \right| n\ell \rangle\right|^2}{W_{6sn\ell} - W_{5dn''\ell - 2}}\right]. \tag{17.16}$$

In Eqs. (17.15) and (17.16) the n'' summations extend over the continua as well. The ranges of n'', or equivalently, of energies relative to the 6p and 5d states of Ba^+, which contribute to the sums are different. This point is shown explicitly by Figs. 17.2 and 17.3. In these figures we show the squared matrix elements for the initial bound 6s18h state. To show the matrix elements between the 18h state and bound $n''\ell''$ and continuum $\varepsilon''\ell''$ virtual intermediate states in a consistent way we have multiplied each of the squared bound matrix elements $|\langle n''\ell''|1/r^k|18h\rangle|^2$ by n''^3 and assigned it the energy width $1/n''^3$ centered at an energy $1/2n''^2$ below the Ba^+5d or 6p state. In this way the sum over the bound $n''\ell''$ states becomes an integral over energy of the $n''\ell''$ series, which smoothly continues above the series limit. As shown by Fig. 17.2, the $\langle n''g|1/r^2|18h\rangle|^2$ matrix elements from n'' slightly lower than $n'' = 18$ provide the dominant contribution to the $n''g$ sum, while for the $n''i$ sum it is the continuum above the Ba$^+$ 6p limit which provides the largest contribution. In Fig. 17.3 the contribution to the $n''f$ sum comes from very low lying states, even below the bound 6s18h state, while the contributions to the $n''h$ sum come from the region of the Ba$^+$ 5d limit, and the contributions to the $n''k$ sum come from well above the 5d limit. From Figs. 17.2 and 17.3 it is apparent that the energy range of contributing states to the dipole sums is $\sim \pm 20\%$ of the Ba$^+$ 6s–6p energy interval, while for the quadrupole sums the range of contributing energies exceeds the Ba$^+$ 6s–5d energy interval. It is the energy range of the contributing terms relative to the ionic energy spacings which leads to the nonadiabatic effects in core polarization. To show this point explicitly we shall now ignore the obvious energy variation of Figs. 17.2 and 17.3 and assume that $W_{6sn\ell} - W_{6pn''\ell''} = W_{6s} - W_{6p}$ and that $W_{6sn\ell} - W_{5dn''\ell''} = W_{6s} - W_{5d}$. This approximation is poor for Ba, but excellent for He and alkali atoms. With these approximations we can remove the energy denominators from the summations of Eq. (17.15) and reexpress W_d as

$$W_d = \frac{|\langle 6s|r_1|6p\rangle|^2}{3(W_{6s} - W_{6p})}\left[\frac{\ell}{2\ell + 1}\sum_{n''}\left|\left\langle n''\ell - 1\left|\frac{1}{r_2^2}\right|n\ell\right\rangle\right|^2\right.$$

$$\left. + \frac{\ell + 1}{2\ell + 1}\sum_{n''}\left|\left\langle n''\ell + 1\left|\frac{1}{r_2^2}\right|n\ell\right\rangle\right|^2\right]. \tag{17.17}$$

If we use the fact that the set of $n''\ell''$ functions of arbitrary n'', but constant ℓ'', constitutes a complete set of radial functions, we can write

$$\sum_{n''}\left|\left\langle n''\ell''\left|\frac{1}{r_2^2}\right|n\ell\right\rangle\right|^2 = \sum_{n''}\left\langle n\ell\left|\frac{1}{r_2^2}\right|n''\ell''\right\rangle\left\langle n''\ell''\left|\frac{1}{r_2^2}\right|n\ell\right\rangle$$

$$= \sum_{n''}\left\langle n\ell\left|\frac{1}{r_2^2}\cdot 1\cdot\frac{1}{r_2^2}\right|n\ell\right\rangle$$

$$= \left\langle n\ell\left|\frac{1}{r_2^4}\right|n\ell\right\rangle. \tag{17.18}$$

Using Eqs. (17.17) and (17.18) we can write the dipole energy, W_d, as

$$W_d = \frac{|\langle 6s|r_1|6p\rangle|^2}{3(W_{6s} - W_{6p})}\left\langle n\ell\left|\frac{1}{r_2^4}\right|n\ell\right\rangle \tag{17.19}$$

$$\cong \frac{-\alpha_d}{2}\left\langle n\ell\left|\frac{1}{r_2^4}\right|n\ell\right\rangle. \tag{17.20}$$

Similarly, we can write the quadrupole energy W_q as

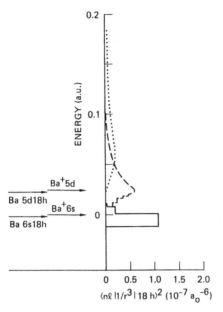

Fig. 17.3 The squared matrix elements $\langle nf|1/r^3|18h\rangle^2$ (———), $\langle nh|1/r^3|18h\rangle$ (– – –), and $\langle nk|1/r^3|18h\rangle^2$ ($\cdots$) plotted vs energy to show the energy distribution of the squared matrix elements. Note that most of the $\langle nf|1/r^3|18h\rangle^2$ matrix element lies below the 6s18h state (from ref. 3).

$$W_q = \frac{|\langle 6s|r_1^2|5d\rangle|^2}{5(W_{6s} - W_{5d})} \left\langle n\ell \left| \frac{1}{r_2^6} \right| n\ell \right\rangle \tag{17.21}$$

$$\cong \frac{-\alpha_q}{2} \left\langle n\ell \left| \frac{1}{r_2^6} \right| n\ell \right\rangle. \tag{17.22}$$

In Eqs. (17.20) and (17.22) the equalities are only approximate since we have not used all of the n'p and n'd states in computing the dipole and quadrupole polarizabilities. Aside from this approximation, the above development shows that the method of Van Vleck and Whitelaw[2] reduces to the adiabatic core polarization model[1] when the ionic energy separation is large compared to the energy ranges spanned by the contributing matrix elements. However, when the ionic energy spacings are relatively small, as shown by Figs. 17.2 and 17.3, the nonadiabatic effects become apparent. When they become smaller yet, series pertubations become important.

It is useful to introduce the factors k_d and k_q to show explicitly the departure from the adiabatic model.[3] Specifically we may write the dipole and quadrupole energies of Eqs. (17.15) and (17.16) in a form equivalent to the adiabatic forms. Explicitly

$$W_d = \frac{-\alpha_d k_d}{2} \left\langle n\ell \left| \frac{1}{r_2^4} \right| n\ell \right\rangle, \tag{17.23}$$

Table 17.1. *Values of* k_d *and* k_q *calculated for the Ba* $6sn\ell$ *states.*[a]

	n	$\ell = 4$	$\ell = 5$	$\ell = 6$	$\ell = 7$
k_d	all	0.945	0.953	0.965	0.975
k_q	18		−0.430	1.67	1.11
	23		−0.355	1.89	1.14

[a] (from ref. 3)

and

$$W_q = \frac{-a_q k_q}{2} \left\langle n\ell \left| \frac{1}{r_2^6} \right| n\ell \right\rangle. \tag{17.24}$$

For the adiabatic case $k_q = k_d = 1$. The factors k_d and k_q, defined implicitly by Eqs. (17.23) and (17.24), are given by

$$k_d = \frac{(W_{6p} - W_{6s})}{\left\langle n\ell \left| \frac{1}{r_2^4} \right| n\ell \right\rangle} \left[\frac{\ell}{2\ell + 1} \sum_{n''} \frac{\left| \left\langle n''\ell - 1 \left| \frac{1}{r_2^2} \right| n\ell \right\rangle \right|^2}{W_{6pn''\ell-1} - W_{6sn\ell}} \right.$$

$$\left. + \frac{\ell + 1}{2\ell + 1} \sum_{n''} \frac{\left| \left\langle n''\ell - 1 \left| \frac{1}{r_2^2} \right| n\ell \right\rangle \right|^2}{W_{6pn''\ell+1} - W_{6sn\ell}} \right], \tag{17.25}$$

and

$$k_q = \frac{W_{5d} - W_{6s}}{\left\langle n\ell \left| \frac{1}{r_2^6} \right| n\ell \right\rangle} \left(\frac{3}{10(4\ell^2 - 1)(2\ell + 3)} \right)$$

$$\times \left[(2\ell - 1)(\ell + 1)(\ell + 2) \sum_{n''} \frac{\left| \left\langle n''\ell + 2 \left| \frac{1}{r_2^3} \right| n\ell \right\rangle \right|^2}{W_{5dn''\ell+2} - W_{6sn\ell}} \right.$$

$$+ \frac{2(\ell^2 + \ell)(2\ell + 1)}{3}$$

$$\times \left. \sum_{n''} \frac{\left| \left\langle n''\ell \left| \frac{1}{r_2^3} \right| n\ell \right\rangle \right|^2}{W_{5dn''\ell} - W_{6sn\ell}} + (2\ell + 3)(\ell^2 - \ell) \sum_{n''} \frac{\left| \left\langle n''\ell - 2 \left| \frac{1}{r_2^3} \right| n\ell \right\rangle \right|^2}{W_{3dn''\ell-2} - W_{6sn\ell}} \right]. \tag{17.26}$$

In Table 17.1 we give the values of k_d and k_q calculated for the Ba $6sn\ell$ states. The values of k_d are n independent, mainly because the Ba$^+$ 6p state lies well above the 6s state, and even the lowest lying $6pn\ell$ state contributing to the k_d sum lies far from the $6sn\ell$ states. For k_q there is a smooth but noticeable n dependence since the lowest lying $5dn\ell''$ states, which contribute substantially to k_q, lie relatively near the $6sn\ell$ states, and, as a result, the precise energy of the $6sn\ell$ Rydberg state is important. It is also interesting to note that k_q is negative for the $6snh$ states since the dominant contribution is from the 5d4f state which lies below the $6snh$ states of $n \approx 20$. Since it is almost impossible to calculate k_q accurately for $6sng$ states using Eq. (17.26), it is not given in Table 17.1.

As shown by Table 17.1, the nonadiabatic corrections become smaller, i.e. $k_d \rightarrow 1$ and $k_q \rightarrow 1$ as ℓ increases. This notion can be appreciated qualitatively by inspecting Figs. 17.2 and 17.3. As ℓ increases the lowest member of the contributing $n''\ell''$ sums of Eqs. (17.15) and (17.16) moves to higher energies since the lowest value of n'' increases with ℓ''.

As mentioned earlier, Eqs. (17.20) and (17.22) are approximate since we have neglected higher Ba^+ states in Eqs. (17.15) and (17.16). In a more accurate calculation one would apply the approach we have outlined to each contributing state of the Ba^+ core, a straightforward procedure. It is also important to note that the contributions from higher lying states of Ba^+ will be more adiabatic than those from lower lying states.

Experimental methods

Three different methods have been used to make microwave resonance measurements of intervals between alkaline earth Rydberg states. In all of these measurements state selective laser excitation of alkaline earth atoms in a beam was combined with one of three forms of state selective ionization of the final states.

The first detection technique, delayed field ionization, used by Vaidyanathan *et al.*[6] to measure $4snf$ $^1F_3$–$4sng$ $^1G_4$ intervals in Ca, is illustrated by the timing diagram of Fig. 17.4. They optically excited a beam of Ca atoms in two steps, $4s \rightarrow 4p$, $4p \rightarrow nf$. The second step required the presence of a small field of 20 V/cm during the excitation. About 200 ns after the laser excitation the field was turned off, and 1 μs afterwards the atoms were exposed to a 1.6 μs pulse of microwaves. Finally, about 25 μs later all the Rydberg atoms present were ionized by a field ionization pulse. This method is based upon the fact that the lifetimes of the $4snf$ $^1F_3$ levels are much shorter than 25 μs, while the lifetimes of the $4sng$ $^1G_4$ states are approximately 25 μs. For example, the lifetimes of the 24f and 24g states are 2.5(5) and 25 μs respectively.[6] Only if atoms are driven from the nf to the ng state by the microwaves will there be Rydberg atoms left after 25 μs. The microwaves must be turned off, otherwise all the atoms will decay from the nf state, albeit with half the usual rate. The experiment of Vaidyanathan *et al.* was done at a temperature of 77 K. In similar experiments with Ba and Cs done at

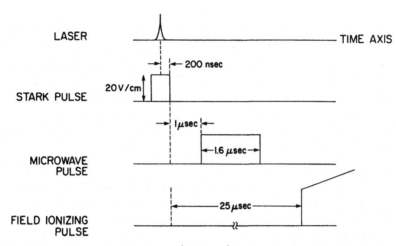

Fig. 17.4 Timing diagram for the Ca $4snf\ ^1F_3$–$4sng\ ^1G_4$ experiment. A Stark pulse is used to allow excitation of the F state. The microwave pulse drives the F → G transition, and the field ionizing pulse is to detect the G state (from ref. 6).

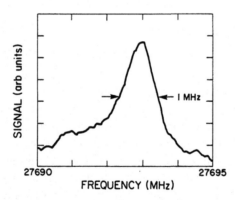

Fig. 17.5 Field ionization signal for the $4s25f^1F_3 \rightarrow 4s25g^1G_4$ transition in Ca (from ref. 6).

room temperature, the black body radiation induced transitions from the initially populated states to longer lived states produced a significant background signal.[3,7] In Fig. 17.5 we show the $4s25f\ ^1F_3$–$4s25g\ ^1G_4$ resonance observed by Vaidya-nathan *et al.*[6] As shown, it has a good signal to noise ratio and it is ~1 MHz wide, close to the 0.6 MHz linewidth expected for the 1.6 μs microwave pulse.

The second technique, selective field ionization, was used by Gentile *et al.*[8] to measure many intervals between low ℓ states of Ca. In their experiments the laser excitation, microwave transition, and detection were separated spatially as well as temporally, an approach similar to the one of Goy *et al.*[9] described in Chapter 16.

Finally, a method peculiar to alkaline earth atoms was used by Cooke and Gallagher.[10] They took advantage of the fact that the optical transitions from the Sr 5snd states to the autoionizing 5pnd states occur at different wavelengths than do the transitions from the 5snf states to the 5pnf states. The reason for this dependence of the spectra on ℓ quantum defect differences is discussed in Chapter 21. In their experiment two pulsed dye lasers were used to excite Sr atoms in a beam from the ground 5s5s state to the 5snd state via the intermediate 5s5p state. The Sr 5snd atoms were exposed first to microwaves for 1 μs and then to a third laser pulse tuned to the 5snf $\rightarrow$ 5pnf transition. After the third laser pulse the atoms were exposed to a small field pulse which extracted the ions from the autoionizing 5pnf states, but did not field ionize the 5snd atoms. Using this method they measured the 5s(n+2)d $^1D_2 \rightarrow$ 5snf $^1F_3$ transitions for $36 \le n \le 38$.

Low angular momentum states and series perturbations

Gentile *et al.*[8] have measured the 26–80 GHz intervals between the low angular momentum states of Ca. From these intervals they have determined the quantum defects of the observed series and fit them to two forms. The first is

$$\delta(n) = \delta_0 + \frac{\delta_1}{(n - \delta_0)^2} + \frac{\delta_2}{(n - \delta_0)^4}, \qquad (17.27)$$

which is a form often used to take into account the slow variation of the quantum defect with energy. In the alkaline earth series if there is a perturber in the bound series the energy perturbation of Eq. (17.14) can be represented in the quantum defect as[8]

$$\delta(n) = \delta_0' + \frac{\delta_1'}{(n - \delta_0')^2} \cdots + \frac{\alpha}{(n - \delta_0')^2 - 2W_p}, \qquad (17.28)$$

where W_p is the energy of the perturber.

The quantum defects of the Ca 4sns $^1S_0$, 4snp $^1P_1$, 4sns $^3S_1$ and 4snp $^3P_1$ states are all slowly varying enough to be fit to Eq. (17.27), while the 4snd $^1D_2$ and $^3D_2$ states are much better fit by Eq. (17.28). The values of the parameters extracted from the microwave measurements are given in Table 17.2.

A measure of the extent to which the quantum defects of the 4snd series are dominated by perturbers is that, while the quantum defects of the 4snd $^1D_2$ states are ~1.2, the value of δ_0' given in Table 17.2 is 0.88.

Table 17.2. *Quantum defects of the s, p,*
and d states of Ca.[a]

$4sns\ {}^1S_0$	$\delta_0 = 2.33793016\ (30)$
	$\delta_1 = -0.1143\ (30)$
$4sns\ {}^3S_1$	$\delta_0 = 2.44095574\ (30)$
	$\delta_1 = 0.3495\ (30)$
$4snp\ {}^1P_1$	$\delta_0 = 1.88558437\ (30)$
	$\delta_1 = -3.2409\ (50)$
	$\delta_2 = -23.8\ (25)$
$4snp\ {}^3P_1$	$\delta_0 = 1.96470945\ (30)$
	$\delta_1 = 0.2276\ (30)$
$4snd\ {}^1D_2$	$\delta_0' = 0.88585\ (50)$
	$\delta_1' = 0.126\ (40)$
	$\alpha = 9.075\ (90) \times 10^{-4}$
$4snd\ {}^3D_2$	$\delta_0' = 0.88334\ (50)$
	$\delta_1' = -0.025\ (40)$
	$\alpha = 8.511\ (90) \times 10^{-4}$

[a](from ref. 8)

Using the Sr $5snd\ {}^1D_2$ quantum defects extrapolated from the calculations of Esherick[11] and their observed $5s(n + 2)d\ {}^1D_2$–$5snf\ {}^1F_3$ intervals, Cooke and Gallagher[10] determined the $5snf\ {}^1F_3$ quantum defects to be 0.0853.

Nonadiabatic core polarization

Perhaps the most interesting aspect of microwave spectroscopy of alkaline earth Rydberg states is that it affords easy access to the nonadiabatic effects in core polarization, and such experiments have been done with both Ca[6] and Ba.[3]

Vaidyanathan *et al.*[6] measured the Ca $4snf\ {}^1F_3$–$4sng\ {}^1G_4$ intervals given in Table 17.2. They compared their measurements to adiabatic and nonadiabatic theories of core polarization in two ways. First, they compared their measured $4snf$–$4sng$ intervals to those calculated[12] assuming adiabatic and nonadiabatic core polarization models. Vaidyanathan and Shorer[12] used the nonadiabatic model of Eissa and Öpik[13] rather than the one of Van Vleck and Whitelaw.[2] Not surprisingly, the departures from the adiabatic model are parametrized with factors identical to k_d and k_q, although there are small higher order corrections as well. For both the adiabatic and nonadiabatic core polarization models they used the Ca^+ dipole and quadrupole polarizabilities, $\alpha_d = 89(9)a_0^3$ and $\alpha_q = 987(90)a_0^5$, calculated using relativistic Hartree–Fock wavefunctions.[12]

The adiabatic model predicts $4snf$–$4sng$ intervals a factor of 2 too high and the nonadiabatic model predicts intervals 10% lower than the measured frequencies.

Table 17.3. *Ba$^+$ Dipole and quadrupole polarizabilities, α_d and α_q, determined from an analysis of measured 6snℓ intervals and calculated using coulomb wavefunctions.*[a]

	α_d (a_0^3)	α_q (a_0^5)
measured	125.5 (10)	2050 (100)
calculated	122.6	2589

[a](from ref. 3)

The 10% difference is probably due to core penetration in the 4snf states, which is not taken into account, and, according to the theoretical work of Vaidyanathan and Shorer[12] should be of the same approximate size as the discrepancy. In any case, it is clear that the nonadiabatic core polarization model reproduces the observed intervals quite well, while the adiabatic model is substantially in error.

The second approach used by Vaidyanathan *et al.* in analyzing their data was not to assume anything about the 4snf states. Rather the 4snf $^1F_3$ quantum defect, determined from optical measurements, was used with the measured 4snf–4sng intervals to determine the quantum defects of the 4sng states. The 4snf quantum defect from the optical measurements of Borgström and Rubbmark[14] is $9.61(5) \times 10^{-2}$. Together with the microwave resonance intervals of Table 17.3 this value leads to a quantum defect of $3.10(8) \times 10^{-2}$ for the Ca 4sng states of $23 \le n \le 25$.[6] Since the quantum defects of the $23 \le n \le 25$ states are the same, to the experimental accuracy, it is not possible to determine both the dipole and quadrupole polarizabilities from the 4sng quantum defects alone. The dominant contribution, however, is from the dipole polarizability. Accordingly it is reasonable to use the computed Ca$^+$ quadrupole polarizability as a small correction and extract a good value of the dipole polarizability from the 4sng quantum defect. Using the nonadiabatic model this process leads to $\alpha_d = 87(3)a_0^3$, in good agreement with the calculated value of $89a_0^3$. Using the adiabatic model the value of α_d extracted is $63a_0^3$, far from the calculated value.

An alternative method of investigating nonadiabatic effects was used by Gallagher *et al.*,[3] who measured the Ba 6sng–6snh–6sni–6snk intervals. Starting from the optically excited 6sng $^1G_4$ states of $18 \le n < 23$ they observed the resonant microwave transitions to the higher ℓ states. They observed doublets in the one-, two-, and three-photon transitions to the 6snℓ states of $ℓ > 4$. They did not assign the doublets but simply used the average of the two observed lines to determine the energies of the 6snℓ, $ℓ > 4$ states. The doublet splittings are about two orders of magnitude larger than the He magnetic intervals, and they decrease rather slowly with ℓ, unlike an exchange effect.

Since more than one rf interval has been measured, it is, in principle, possible to determine both the quadrupole and dipole polarizabilities from the microwave resonance data alone. The procedure is identical to that used earlier for alkali atoms,[15] except that the factors k_d and k_q must be introduced. If we define

$$P' = k_d Ry \left\langle n\ell \left| \frac{1}{r_2^4} \right| n\ell \right\rangle \tag{17.29a}$$

and

$$Q' = \frac{k_q \left\langle n\ell \left| \frac{1}{r_2^6} \right| n\ell \right\rangle}{k_d \left\langle n\ell \left| \frac{1}{r_2^4} \right| n\ell \right\rangle}, \tag{17.29b}$$

where Ry is the Rydberg constant, in cm^{-1}, for Ba, then the polarization energy of the $n\ell$ state (in cm^{-1}) is given by

$$W_{pol_{n\ell}} = -\alpha_d P' - \alpha_q P' Q', \tag{17.30}$$

and the interval between the $n\ell$ and $n\ell'$ states is given by

$$\Delta W = W_{pol_{n\ell'}} - W_{pol_{n\ell'}}$$

$$= \alpha_d \Delta P' + \alpha_q \Delta P' Q', \tag{17.31}$$

where $\Delta P' = P'_{n\ell} - P'_{n\ell}$ and $\Delta P' Q' = P'_{n\ell} Q'_{n\ell} - P'_{n\ell'} Q'_{n\ell'}$. Dividing Eq. (17.31) by $\Delta P'$ to remove the variation in the dipole polarization energy, W_d, with n and ℓ yields

$$\frac{\Delta W}{\Delta P'} = \alpha_d + \alpha_q \frac{\Delta P' Q'}{\Delta P'}. \tag{17.32}$$

If we plot the observed intervals by plotting $\Delta W/\Delta P'$ vs $\Delta P' Q'/\Delta P'$ we obtain a linear plot, the intercept giving α_d and the slope α_q.

In Fig. 17.6 we show a plot of $\Delta W/\Delta P'$ vs $\Delta P' Q'/\Delta P'$. Although the 6sng–6snh–6sni–6snk intervals were all measured, the 6sng-6snh intervals are not shown in Fig. 17.6 since the 6sng series is perturbed by the 5d7d state at $n = 24$. Originally Gallagher *et al.* removed the perturbation from the experimental 6sng–6snh intervals by using an expression similar to Eq. (17.16), and then removed the approximately corresponding term from the k_q sum. Specifically the term from the 5dnd state just below the Ba^+ 6s limit was removed from the k_q summation of Eq. (17.26). The fact that the data modified in this way appeared reasonable in a plot such as Fig. 17.6 was probably fortuitous.

From the 6snh–6sni and 6sni–6snk intervals shown in Fig. 17.6 reasonable values of α_d and α_q may be obtained, as shown by Table 17.3. The experimental uncertainties given in Table 17.3 are those from fitting the data to Eq. (17.32). We also give in Table 17.3 the values calculated using coulomb wave functions for

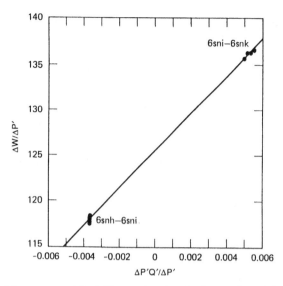

Fig. 17.6 A plot of the measured Ba $\Delta\ell$ intervals using the factors k_d and k_q to correct for the nonadiabatic effects (from ref. 3).

Ba[+].[16] In calculating the dipole and quadrupole polarizabilities only the 6p and 5d states were used, leading to values of α_d and α_q which are too low by an estimated 1% and 10% respectively. The agreement of the measured and calculated dipole polarizabilities is good, but between the measured and calculated quadrupole polarizabilities it is at best fair. Either the calculation of the polarizability is wrong or the k_q correction factors are wrong. In fact, it is almost certain that the k_q factors of the 6s*n*h states are incorrect due to our having ignored the spins of the electrons. For the 6s*n*g and 6s*n*h states this omission leads to significant errors. The reason is shown in Fig. 17.3, the plot of the squared matrix elements contributing to k_q for the 6s18h state. As shown, the spinless 5d4f state lying 1750 cm^{-1} below the Ba[+] 6s limit has a large matrix element, and, due to its proximity to the bound 6s*n*h states, it makes the largest contribution to k_q. Consequently, the precise location of this state is important to the calculation of k_q. It is not well known for two reasons. First, in reality, the Ba[+] 5d state is split into the 5d$_{3/2}$ and 5d$_{5/2}$ states, which are 800 cm^{-1} apart, and since there are 4f$_{5/2}$ and 4f$_{7/2}$ states converging to both of the Ba[+] 5d limits, it is reasonable to expect the real Ba 5d4f states to fall in a range ~800 cm^{-1} wide around the energy of the spinless 5d4f state. An 800 cm^{-1} range is hardly small compared to the 1500 cm^{-1} energy difference between the 6s18h state and the spinless 5d4f state and is potentially a significant source of error. Second, the quantum defects of the 5d*n*f states are not zero,[17] but ~0.05. For both of the above reasons we cannot expect the values of k_q calculated for the 6s*n*f states using spinless wavefunctions to be particularly accurate. For the 6s*n*i and higher angular momentum states the terms contributing to k_q all lie ~1000 cm^{-1} above the 6s limit and are thus less susceptible to errors than the 6snh states. For the 6sng states, on the other hand, k_q can be calculated with even less confidence than for the 6snh states.

While the agreement of the measured and calculated Ba$^+$ quadrupole polarizabilities is not very good, compared to an analysis based on adiabatic core polarization the agreement in Table 17.3 is superb. The adiabatic core polarization model leads to $\alpha_d = 146a_0^3$ and $\alpha_q = -5800a_0^5$. The ground state of Ba$^+$ cannot have a negative quadrupole polarizability. Taken together, the Ca and Ba experiments show clearly that the nonadiabatic effects in core polarization in the alkaline earth atoms are important and may be calculated with some accuracy.

References

1. J. E. Mayer and M. G. Mayer, *Phys. Rev.* **43**, 605 (1933).
2. J. H. Van Vleck and N. G. Whitelaw, *Phys. Rev.* **44**, 551 (1933).
3. T. F. Gallagher, R. Kachru, and N. H. Tran, *Phys. Rev. A* **26**, 2611 (1982).
4. H. A. Bethe and E. E. Salpeter, *Quantum Mechanics of One- and Two-Electron Atoms* (Springer, Berlin, 1957).
5. A. R. Edmonds, *Angular Momentum in Quantum Mechanics* (Princeton University Press, Princeton, 1960).
6. A. G. Vaidyanathan, W. P. Spencer, J. R. Rubbmark, H. Kuiper, C. Fabre, D. Kleppner, and T. W. Ducas, *Phys. Rev. A* **26**, 3346 (1982).
7. K. A. Safinya, T. F. Gallagher, and W. Sandner, *Phys. Rev. A* **22**, 6 (1980).
8. T. R. Gentile, B. J. Hughey, D. Kleppner, and T. W. Ducas, *Phys. Rev. A* **42**, 440 (1990).
9. P. Goy, C. Fabre, M. Gross, and S. Haroche, *J. Phys.* **B13**, L83 (1980).
10. W. E. Cooke and T. F. Gallagher, *Opt. Lett.* **4**, 173 (1979).
11. P. Esherick, *Phys. Rev.* **A15**, 1920 (1977).
12. A. G. Vaidyanathan and P. Shorer, *Phys. Rev. A* **25**, 3108 (1982).
13. H. Eissa and U. Öpik, *Proc. Phys. Soc. London* **92**, 556 (1960).
14. S. A. Borgström and J. R. Rubbmark, *J. Phys. B* **10**, 18 (1977).
15. B. Edlen, in *Handbuch der Physik*, ed. S. Flugge (Springer-Verlag, Berlin, 1964).
16. M. L. Zimmerman, M. G. Littman, M. M. Kash, and D. Kleppner, *Phys. Rev. A* **20**, 2251 (1979).
17. E. A. J. M. Bente and W. Hogervorst, *J. Phys. B* **24**, 3565 (1989).

The He atom differs from the atoms we have discussed in the last two chapters in that many of the intervals between high lying states can be calculated with ease to an accuracy better than 1%.[1,2] Consequently, the challenge is to calculate the intervals more accurately and to make measurements to test these sophisticated calculations. While the sophistication of the calculations is substantially beyond the scope of this book, the measurements of the intervals between the He $1sn\ell$ states constitute a significant fraction of the measurements of Rydberg energy level spacings. Accordingly, the focus of this chapter is on the experimental measurements which have been made.

Theoretical description

The Hamiltonian for two spinless electrons at points $\mathbf{r}_1$ and $\mathbf{r}_2$ relative to a He^{2+} core is given by[3]

$$H = -\left(\frac{\nabla_1^2}{2} + \frac{\nabla_2^2}{2} + \frac{2}{r_1} + \frac{2}{r_2} - \frac{1}{r_{12}}\right), \tag{18.1}$$

where $r_{12} = |\mathbf{r}_2 - \mathbf{r}_1|$, and $\mathbf{r}_1$ and $\mathbf{r}_2$ are the positions of the two electrons relative to the He^{2+} ion.

To describe an atom with one electron in a Rydberg state a reasonable approach, termed Heisenberg's method by Bethe and Salpeter,[3] is to separate the Hamiltonian of Eq (18.1) into two parts. Explicitly,

$$H = H_0 + H_1, \tag{18.2a}$$

where

$$H_0 = -\left(\frac{\nabla_1^2}{2} + \frac{\nabla_2^2}{2} + \frac{2}{r_1} + \frac{1}{r_2}\right) \tag{18.2b}$$

and

$$H_1 = \frac{-1}{r_2} + \frac{1}{r_{12}}. \tag{18.2c}$$

Using the Hamiltonian H_0 of Eq. (18.2b) in the time independent Schroedinger equation the problem is separable into independent solutions for the two electrons, and the solutions can be written as

$$\Psi = \phi_{n_1 \ell_1 m_1} (\mathbf{r}_1) \psi_{n_2 \ell_2 m_2}(\mathbf{r}_2), \tag{18.3}$$

where $\phi_{n_1 \ell_1 m_1}$ and $\psi_{n_1 \ell_1 m_1}$ are the wavefunctions for He^+ and H. Correspondingly, the energies are given by

$$W_{n_1 n_2} = -\frac{2}{n_1^2} - \frac{1}{2 n_2^2} \tag{18.4}$$

relative to the double ionization limit. We are, at the moment, only interested in the bound Rydberg states of He in which electron 1 is in the 1s state, so we shall specify the wavefunction of Eq. (18.3) as

$$\Psi_{n\ell m} = \phi_{1s} (\mathbf{r}_1) \, \psi_{n\ell m} (\mathbf{r}_2). \tag{18.5}$$

For consistency with other bound Rydberg states we shall measure the energies relative to the ground state of He^+. Expressed in this way, the energy of the state given in Eq. (18.5) is

$$W_{n\ell m} = -1/2n^2. \tag{18.6}$$

Following the Pauli exclusion principle we must antisymmetrize the wavefunctions when we include the spins of the two electrons. The wavefunction is given by a product of spatial and spin wave functions, i.e.

$$\psi_{\pm n\ell m} = \frac{1}{\sqrt{2}} [\phi_{1s} (\mathbf{r}_1) \, \psi_{n\ell m} (\mathbf{r}_2) \pm \phi_{1s} (\mathbf{r}_2) \, \psi_{n\ell m} (\mathbf{r}_1)] \phi_{\pm}. \tag{18.7}$$

Here $\phi_{\pm}$ is the spin wavefunction, and the $+$ and $-$ signs refer to the orthogonal singlet, $S = 0$ and triplet, $S = 1$ states, where S is the total electron spin. Using the wavefunctions of Eq. (18.7) we can calculate the first order corrections to Eq. (18.6) using the Hamiltonian of Eq. (18.2c). Explicitly,

$$W_1 = \int \psi_{\pm n\ell m}^* H_1 \, \psi_{\pm n\ell m} \, d\mathbf{r}_1 \, d\mathbf{r}_2. \tag{18.8}$$

Assuming that ϕ_{1s} and $\psi_{n\ell m}$ are orthogonal we can express the integrals implied by Eq. (18.8) in terms of the direct and exchange integrals, J and K. The J integral compensates for the incomplete screening of the outer electron from the doubly charged core. Equivalently, the J integral reflects the penetration of the He^+ core by the outer $n\ell$ electron. From either point of view, it is apparent that $J < 0$. The exchange integral K can be viewed as the rate at which the two electrons exchange their states. Explicitly, we can express the penetration and exchange energies as W_{pen} and W_{ex} as

$$W_{\text{pen}} = J = \int | u_1 (\mathbf{r}_1) |^2 | u_{n\ell m} (\mathbf{r}_2)|^2 \left(\frac{1}{r_{12}} - \frac{1}{r_2} \right) d\mathbf{r}_1 \, d\mathbf{r}_2 \tag{18.9}$$

and

$$W_{ex} = K = \int u_1^*(\mathbf{r}_1) u_{n\ell m}^*(\mathbf{r}_1) u_1(\mathbf{r}_2) u_{n\ell m}(\mathbf{r}_2) \frac{1}{r_{12}} \, d\mathbf{r}_1 \, d\mathbf{r}_2.$$

$$(18.10)$$

Since both J and K integrals depend upon the spatial overlap between the ground state He^+ wavefunction and the Rydberg $n\ell m$ wavefunction, they decrease very rapidly with increasing ℓ and can be expressed as rapidly converging series.[3]

In addition to the first order corrections to the energy given by Eq. (18.11), we must account for the polarization of the He^+ core by the field from the outer $n\ell$ electron.[4,5] As we saw in Chapter 17, this correction is really a second order correction,[6] but if written in terms of the polarizabilities of He^+ it appears as a first order correction. Core polarization is relatively most important in $\ell \geq 2$ Rydberg states, for which the inner classical turning point occurs at $r_2 \geq 6a_0$, while the classical outer turning point of the He^+ 1s electron is at $1a_0$. Thus, it is an adequate approximation to assume that $r_2 > r_1$, ignore exchange effects, and use the unsymmetrized wavefunction of Eq. (18.5) to calculate the polarization energy. In this approximation the polarization energy, W_{pol}, is given by[4,5]

$$W_{pol} = -\frac{\alpha_d}{2} \langle r_2^{-4} \rangle_{n\ell m} - \frac{\alpha_q}{2} \langle r_2^{-6} \rangle_{n\ell m} ,$$

$$(18.11)$$

where $\langle r^{-k} \rangle_{n\ell m}$ is the expectation value of r^{-k} for a hydrogenic $n\ell m$ state. Explicit expressions for these expectation values are given in Chapter 2, and the He^+ polarizabilities are given by $\alpha_d = 9a_0^3/32$ and $\alpha_q = 15a_0^5/64$.[5]

When the exchange, penetration, and polarization energies are included, the energies of the singlet and triplet $n\ell m$ states are given by

$$W_{\pm n\ell m} = \frac{-1}{2n^2} + W_{pen} \pm W_{ex}.$$

$$(18.12)$$

The penetration and exchange energies are typically of the same size and decrease rapidly with increasing ℓ.[3] The polarization energy decreases more slowly with ℓ, and for $\ell > 2$ it is the dominant contribution to the deviation from the hydrogenic energy in Eq. (18.12). For $\ell = 2$ the three energies are comparable, and for $\ell < 2$ the exchange and penetration energies exceed the polarization energies. W_{pol} and W_{pen} are both negative, and W_{ex} is positive.

Introducing spin adds significant magnetic interactions, which may be accounted for with the Breit–Bethe theory.[3] The theory is based on the unsymmetrized wavefunctions used to calculate the polarization energy. There are two terms which make significant contributions, the spin–orbit interaction and the spin–spin interactions. These operators have the matrix elements[3]

$$\langle \psi_{n\ell SJ} | H_{SO} | \psi_{n\ell S'J} \rangle = A M_{SO} (\ell S S' J)$$

$$(18.13)$$

and

$$\langle \psi_{n\ell SJ} | H_{SS} | \psi_{n\ell S'J} \rangle = A M_{SS}(\ell S S' J),$$

$$(18.14)$$

where

$$A = \frac{\alpha^2}{4} \langle r_2^{-3} \rangle,$$

$$(18.15)$$

Table 18.1. *Matrix elements of the Breit–Bethe spin–orbit and spin–spin operators, Eqs.* (18.13) *and* (18.14).

S, S', J	$M_{SO}(\ell SS'J)^{a,b}$	$M_{SS}(\ell, SS'J)^{a}$
$0, 0, \ell$	0	0
$1, 1, \ell + 1$	$-\ell$	$2\ell/(2\ell + 3)$
$1, 1, \ell$	-1	-2
$1, 1, \ell - 1$	$\ell + 1$	$(2\ell+2)/(2\ell - 1)$
$1, 0, \ell$	$-3\sqrt{\ell(\ell + 1)}$	0

[a] $(S = S'$ matrix elements from ref. 3) (see ref. 10)
[b] $(S = 0, S' = 1$ element from ref. 7)

$J = \ell + S$, and α is the fine structure constant, $1/137.07$. The values of $M_{SO}(\ell SS'J)$ and $M_{SS}(\ell SS'J)$ are given in Table 18.1. The diagonal elements, $S = S' = 1$, are given by Bethe and Salpeter,[3] and the off diagonal element is given by Miller *et al.*[7] Due to the $\langle r_2^{-3} \rangle$ dependence, the magnetic fine structure decreases slowly with increasing ℓ.

For low ℓ the polarization and exchange energies are both far larger than the magnetic fine structure energies. The singlet states are well above the triplet states, and in computing the magnetic fine structure the off diagonal spin orbit matrix element of Eq. (17.13) can be ignored. On the other hand, for high ℓ states the polarization energy is largest, followed by the magnetic fine structure energies, followed by the penetration and exchange energies. In this case the states are no longer all good singlets and triplets. The $J = \ell \pm 1$ states are triplets, but the $J = \ell$ states are roughly 50–50 mixtures of singlet and triplet. In the limit of the zero exchange energy the energies of the two $J = \ell$ states are symmetrically displaced from the polarization energy by $\pm AM_{SO}(\ell, 1, 0, \ell)$.

Experimental approaches

Most of the high precision spectroscopy of He Rydberg states has been done by microwave resonance, which is probably the best way of obtaining the zero field energies. Wing *et al.*[8–12] used a 30–1000 μA/cm^2 electron beam to bombard He gas at 10^{-5}–10^{-2} Torr. As electron bombardment favors the production of low ℓ states, it is possible to detect $\Delta \ell$ transitions driven by microwaves. The microwave power was square wave modulated at $\sim$40 Hz, and the optical emission from a specific Rydberg state was monitored. When microwave transitions occurred to or

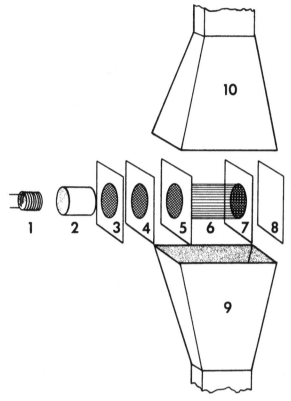

Fig. 18.1 The resonance module of MacAdam and Wing: 1, heater; 2, dispenser cathode; 3, accelerating grid; 4, grid; 5, grid; 6, the "cage," consisting of wires joining the perimeters of grids 5 and 6; 7, grid; 8, collector; 9. and 10, X-band waveguide horns. The grid supports are nonmagnetic stainless-steel plates, 2.54 cm square. The cage is 2.54 cm long by 1/6 cm diameter (from ref. 11).

from the state monitored, an increase or decrease in the detected emission was observed in phase with the modulation of the microwave power.

In Fig. 18.1 we show the apparatus used by MacAdam and Wing.[11] Electrons from a dispenser cathode were accelerated and passed down the axis of a cage formed by W wires, where they excited He atoms to Rydberg states. The cage, which was periodically heated to incandescence, ensured that there were no stray electric fields due to long term buildup of oil or other charge trapping insulators near the Rydberg atoms. The cage was positioned between two X band microwave horns, which brought microwave radiation to the Rydberg atoms. Microwave power was fed into the cage from one or both horns allowing the simultaneous use of two frequencies. Light from the Rydberg atoms was collected by an ellipsoidal light pipe with one focus at the center of the cage of Fig. 18.1 and the other at the entrance slit of a 1/4 m monochromator. A typical resonance signal is shown in Fig. 18.2, for the two photon 9d $^1D_2 \rightarrow$ 9d $^1G_4$, and

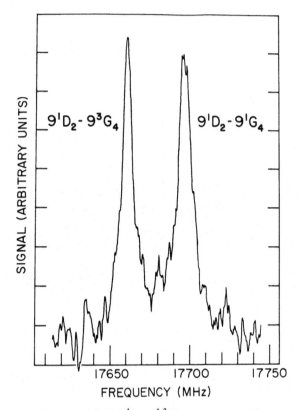

Fig. 18.2 Chart recorder trace of the 9^1D_2–$9^{1,3}G_4$ resonances. The resonances are power broadened by a factor of 2–3 for purposes of display. Sweep rate 0.4 MHz/s, lock-in time constant 2 s (from ref. 11).

9d $^1D_2 \rightarrow$ 9d$^3G_4$ transitions.[11] The resonances of Fig. 18.2 were obtained by monitoring the 9d $^1D_2$–2p $^1P_1$ fluorescence at 3872 Å. The transitions shown in Fig. 18.2 are power broadened by a factor of 2–3 above the instrumental linewidth. They are also exhibit an ac Stark shift, linear in the microwave power, and to extract the correct intervals the measurements must be made at several powers and the measured resonance frequencies extrapolated to zero power. For reference, the zero power intervals corresponding to Fig. 18.2 are $^1D_2$–$^1G_4$, 17697.065(41) MHz, and $^1D_2$–$^3G_4$, 17661.096(22) MHz.

Perhaps the most precise measurements are those made using fast He beams. The first experiments were those of Cok and Lundeen.[13] A duoplasmatron ion source produced a 5–20 keV beam of He$^+$, usually with an energy of 13 keV. The beam passed through a charge exchange cell containing Ar at a pressure of 100 μTorr. Unlike electron bombardment, charge exchange produces useful populations in higher ℓ states. For example, Cok and Lundeen estimated that there was a flux of 10^8 He 8f atoms/s in their beam. The He beam passed coaxially down a piece of circular waveguide. The waveguide had microwaves travelling either

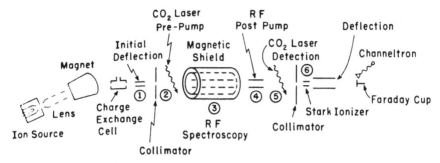

Fig. 18.3 Fast He beam apparatus used to measure the radio frequency intervals between the n=10 g, h, i and k intervals (from ref. 14).

parallel or antiparallel to the He beam, and by making measurements with the microwaves travelling in both directions it was possible to cancel the rather large Doppler shifts, 0.1% of the microwave frequency. Downstream from the waveguide either the 4d $^1D_2$–2p $^1P_1$, or 4d $^3D_2$–2p $^3P_1$ emission was detected, depending upon whether ^1F or ^3F Rydberg states were being studied. In either case, it was the second step of the 8f $\rightarrow$ 4d $\rightarrow$ 2p cascade. There was minimal emission from the 4d state atoms populated directly by charge exchange since they decayed before reaching the detection region. Using this technique Cok and Lundeen measured $n = 7$ and 8 f–g, f–h, and f–i intervals. The minimum resonance linewidth was limited to about 4 MHz by the transit time of atoms through the waveguide.[13]

Palfrey and Lundeen modified the previous approach by using a different detection scheme.[14] They used a Doppler tuned CO_2 laser to drive transitions from specific ℓ states of $n = 10$ to $n = 27$ and subsequently field ionizing the $n = 27$ He atoms. A diagram of their apparatus is shown in Fig. 18.3. The fast He beam initially contains all 10ℓ states, but one of the $\ell \geq 4$ states is depleted by driving the transition to the $n=27$ state with the CO_2 laser. The depleted beam passes through the resonance region where a rf field, when tuned to a resonance, drives the $10 (\ell + 1) \rightarrow 10\ell$ transition, repopulating the empty 10ℓ state. The increase in population of the 10ℓ state at the 10ℓ–$10(\ell+1)$ resonance is detected by using the CO_2 laser beam a second time to drive the 10ℓ atoms to the $n = 27$ state and field ionizing them. This approach is much more sensitive to transitions between high ℓ states than one based on optical detection, and they used it to make very precise measurements of 10g–10h–10i–10k intervals.[14] This approach has been continuously refined, and, using it, Hessels *et al.* have been able to measure transition frequencies to better than 1 kHz.[15]

A completely different approach is anticrossing spectroscopy,[16] employed by Miller *et al.*,[7,17] Beyer and Kollath,[18–21] and Derouard *et al.*[22,23] This method is, in principle, a direct way of measuring the mixing of the singlet and triplet states and the singlet–triplet $^1L_\ell$–$^3L_\ell$ separation and has been primarily applied to the $^1D_2$ and $^3D_2$ states. To understand the method we begin with the magnetic field

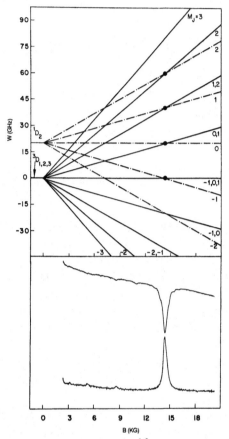

Fig. 18.4. Energy level diagram (top) of the $n = 6^{1,3}$D states as a function of magnetic field and actual spectra (bottom) of the $n = 6^{1,3}$D anticrossing. Each of the spectral traces was obtained in approximately 10 min running time. Top trace shows the decrease in light intensity of the 6^1D emission line at 4144 Å, while the lower trace shows the corresponding increase in the 6^3D emission at 3819 Å (from ref. 17).

dependence of the energies. In Fig. 18.4, we show a plot of the He 6d energy levels as a function of magnetic field.[17] To the zero field Hamiltonian we add

$$H_B = \mu_B \mathbf{B} \cdot (g_s \mathbf{S} + g_\ell \mathbf{L}), \tag{18.16}$$

where μ_B is the Bohr magneton and g_ℓ and g_s are the g factors of the orbital and spin magnetic moments, 1 and 2, respectively. It is useful to define the average energy of the singlet and triplet states, W_{av}, which is given by

$$W_{av} = -\frac{1}{2n^2} + W_{pen} + W_{pol}. \tag{18.17}$$

The spin–orbit energies are very small compared to the other energies, so we can treat the spin and orbital angular momenta as being uncoupled. If we ignore the offdiagonal spin–orbit matrix element of Eq (18.13), an excellent approximation for the He nd states, the energies W_+ and W_- of the $^1D_2$ and $^3D_2$ states, respectively, are given by

$$W_+ - W_{av} = W_{ex} + \mu B g_\ell m_{\ell+} \tag{18.18a}$$

$$W_- - W_{av} = -W_{ex} - A + \mu B(g_s m_{s-} + g_\ell m_{\ell-}) \tag{18.18b}$$

where $m_{\ell+}$ is the m_ℓ value of the $^1D_2$ state and m_{s-} and $m_{\ell-}$ are the m_s and m_ℓ values of the $^3D_2$ state. In the magnetic field the total azimuthal angular momentum, $m_j = m_s + m_\ell$, is a good quantum number, and, as shown by Fig. 18.4, 6d $^{1,3}D_2$ states of the same m_j, and $m_{\ell-} = m_{\ell+} - 1$ intersect at a field of 15 kG.[17] Equating the $^1D_2$ and $^3D_2$ energies of Eq. (18.18) for $m_{\ell+} = m_{\ell-}+1$ and $m_{s-} = 1$ yields

$$W_+ - W_- = \mu B_c, \tag{18.19}$$

where B_c is the magnetic field at the intersection of the singlet and triplet levels.

If there were no coupling between the singlet and triplet levels, they would cross at B_c. However, the off diagonal spin–orbit matrix element of Eq. (18.13), which we ignored in computing the energies of Eq. (18.18), couples them, and there is an avoided crossing, or anticrossing, at B_c. At the anticrossing the two levels are separated by twice the off diagonal matrix element of Eq (18.13), i.e. by $6A\sqrt{5}$, and the eigenstates are not the $^1D_2$ and $^3D_2$ states but 50–50 mixtures of the two states.

In the experiment, electron beam excitation from the ground $^1S_0$ state produces primarily singlet Rydberg states. If the 6d $^1D_2$–2p $^1P_1$ fluorescence at 4144 Å is observed, as the magnetic field is swept through the anticrossing a drop in the fluorescence is observed at the anticrossing, as shown in Fig. 18.4.[17] The origin of the decrease in the fluorescence can be understood in the following way. The same number of Rydberg atoms is excited by the electron beam irrespective of the value of the magnetic field. Away from the avoided crossing only the $^1D_2$ state is excited, but at the avoided crossing half of the excitation is into each of the two mixed states. Away from the avoided crossing the $^1D_2$ atoms excited by the electron beam can only decay to lower lying singlet states, but at the avoided crossing the two mixed states can decay to lower lying triplet states as well. To the extent that decay to the lower lying triplet state occurs, the singlet fluorescence is diminished. If the decay rates to lower lying singlet and triplet states are equal, at the anticrossing the singlet emission drops by 50%.

On the other hand, if the 6d$^3D_2$–$^3P_1$ radiation at 3819 Å is observed, normally very little light is detected, and a sharp increase is seen at the anticrossing, as shown in Fig. 18.4. According to Eq (18.10) the field of the anticrossing gives the singlet triplet splitting. The width of the anticrossing gives the spin–orbit matrix element.

In practice there are several factors which complicate anticrossing spectroscopy. The four anticrossings of Fig 18.4 apparently occur at the same field, given

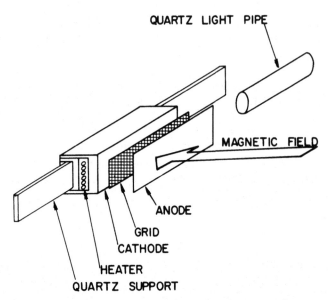

QUARTZ LIGHT PIPE

MAGNETIC FIELD

ANODE

GRID

CATHODE

HEATER

QUARTZ SUPPORT

Fig. 18.5. Apparatus used to observe anticrossing emission from the 7^1D state (from ref. 7).

by Eq. (18.19). However, we have ignored corrections such as the quadratic Zeeman effect and the motional Stark effect, which split the four anticrossings, but not enough to be well resolved.[5] Unfortunately, their relative strengths are not well known since the populations produced by electron impact excitation are not known.[17] In most of the He $^1D_2$-$^3D_2$ anticrossing measurements the fact that there are four overlapping anticrossing signals leads to unavoidable ambiguities in the interpretation of the observed signals. Beyer and Kollath[20] removed some of the ambiguities by applying an electric field perpendicular to the magnetic field to split the unresolved anticrossings into well resolved anticrossings. Using this technique they were able to make the most precise anticrossing measurements of $^1D_2$-$^3D_2$ splittings. Applying electric fields also enabled them to observe antic-rossings of the nd states with states of higher ℓ, and in this way they made the first determinations of the energies of the $\ell > 3$ states.[21]

A typical experimental arrangement for anticrossing spectroscopy is shown in Fig. 18.5.[7] A Cu vacuum chamber, which fits between the pole faces of a magnet, encloses the assembly of Fig. 18.5 and contains flowing He at a pressure of 1 mTorr. The cathode produces ~100 μA of electrons which are accelerated by applying 30–50 V between the cathode and grid and a comparable voltage between the grid and the anode. The light, leaving in the direction perpendicular to the field, is collected by a light pipe, and the desired wavelength selected by an interference filter before the light is detected by a photomultiplier tube.

A final technique which has been used by Panock *et al.*[24,25] is laser magnetic resonance. They excited He atoms in a 20–140 kG magnetic field by electron impact. They used a line tunable CO_2 laser to drive atoms in the $7s$ $^1S_0$ state to the

$n = 9 \ \ell \geq 2$ levels, which contained 9p $^1P_1$ character due to the strong magnetic field. By changing lines, the laser frequency can be varied in steps of ~ 1.3 cm^{-1}, and the magnetic field is continuously tunable. Together these two tuning methods give continuous tuning over a broad frequency range. The 7s $\rightarrow$ 9ℓ transitions were observed by monitoring the 7s $^1S_0$–2p $^1P_1$ radiation as the magnetic field was slowly scanned through the resonances.

To observe a 7s $\rightarrow$ 9ℓ transition requires that there be a 9p admixture in the 9ℓ state. For odd ℓ this admixture is provided by the diamagnetic interaction alone, which couples states of ℓ and $\ell \pm 2$, as described in Chapter 9. For even ℓ states the diamagnetic coupling spreads the 9p state to all the odd 9ℓ states and the motional Stark effect mixes states of even and odd ℓ. Due to the random velocities of the He atoms, the motional Stark effect and the Doppler effect also broaden the transitions. Together these two effects produce asymmetric lines for the transitions to the odd 9ℓ states, and double peaked lines for the transitions to even 9ℓ states. The difference between the lineshapes of transitions to the even and odd 9ℓ states comes from the fact that the motional Stark shift enters the transitions to the odd 9ℓ states once, in the frequency shift. However, it enters the transitions to the even 9ℓ states twice, once in the frequency shift and once in the transition matrix element. Although peculiar, the line shapes of the observed transitions can be analyzed well enough to determine the energies of the 9ℓ states of $\ell \geq 2$ quite accurately.[25]

Quantum defects and fine structure

The quantum defects of the $n = 10$ states of He are given in Table 18.2. The quantum defects for the s and p states are derived from optical measurements, specifically from the term energies given by Martin.[26] The quantum defects of the $\ell = 8$ and 9 (10ℓ and 10m) states are calculated using Eq. (18.2), and the quantum defects of the $2 \leq \ell \leq 7$ states are obtained from the calculated $\ell = 8$ quantum defect and the measured intervals reported by Hessels *et al.*[15] The quantum defects given by Table 18.2 are by no means indicative of the possible precision of the measurements. For example, Hessels *et al.* have reported measurements of $n = 10$ intervals which have precisions of parts per million.[15]

One of the more interesting aspects of the He atom is the evolution of the $J = \ell$ states from well defined singlet and triplet states at low ℓ to states which are mixtures of singlet and triplet at high ℓ. This evolution is shown in Fig. 18.6 for the 9d–9g levels. The intervals are taken from Farley *et al.*[12] In the 9d state the exchange splitting is orders of magnitude larger than the spin–orbit and spin–spin splitting, and the 9d states are good singlet and triplet states. The amplitude of $^1D_2$ wavefunction in the $^3D_2$ states is $\sim 10^{-2}$ and vice versa.[9] In the 9f state the exchange energy is the same order of magnitude as the magnetic energies. As a

Table 18.2. *Quantum defects of the He states of* $n \approx 10$

$^1S_0{}^a$	0.14
$^3S_1{}^a$	0.30
$^1P_1{}^a$	−0.012
$^3P_1{}^a$	0.063
$^1D_2{}^b$	2.08×10^{-3}
$^3D_2{}^b$	2.82×10^{-3}
f^b	4.26×10^{-4}
g^b	1.17×10^{-4}
h^b	4.29×10^{-5}
i^b	1.90×10^{-5}
k^c	7.79×10^{-6}
ℓ^c	3.68×10^{-6}
m^c	1.93×10^{-6}

[a] (from ref. 25)
[b] (from ref. 15)
[c] (from Eq. (18.11)

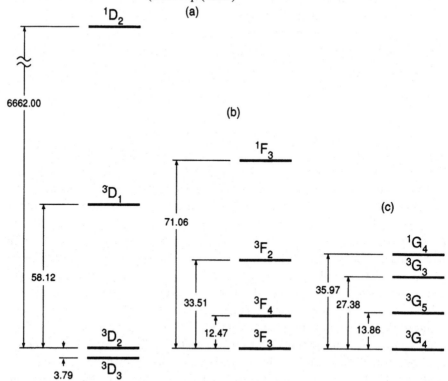

Fig. 18.6. Exchange and magnetic energies of (a) the He 9d states, (b) the He 9f states, and (c) the He 9g states. In (a) the 9d states, the singlet triplet splitting is not shown to scale. The 9d states are good singlet and triplet states. The 9f states are approximately singlets and triplets, but the $^3F_3$ state is below the $^3F_4$ state. The 9g states do not display singlet–triplet structure. All intervals are given in MHz.

result, while the $^1F_3$ state is still clearly removed from the triplet states, the off diagonal spin–orbit matrix element depresses the energy of the $^3F_3$ level to below that of the $^3F_4$ level. Correspondingly, the nf $^{1,3}F_3$ states are not good singlets and triplets; there is a singlet amplitude of 0.55 in the triplet state and vice versa.[10] In the 9g state the exchange energy is negligible, and only the $J = 3$ and $J = 5$ states are good triplets. The $J = 4$ states are 50–50 mixtures of singlet and triplet. For the 9g state the intervals shown in Fig. 18.6 can be calculated within 1 MHz using Eqs. (18.4) and (18.5). However, for the 9d and 9f states a more sophisticated approach is required. Using many body perturbation theory Chang and Poe[27] have calculated $^1D_2$–$^3D_2$ and $^1F_3$–$^3F_3$ intervals of 6615 and 69.09 MHz, in good agreement with the experimental results shown in Fig. 18.6. However, using an extended adiabatic model Cok and Lundeen[28] were able to reproduce all six 9d and 9f intervals to within less than 1 MHz.

References

1. R.J. Drachman, *Phys. Rev. A* **31**, 1253 (1985).
2. G.W.F. Drake, *Phys. Rev. Lett.* **65**, 2769 (1990).
3. H.A. Bethe and E.E. Salpeter, *Quantum Mechanics of One- and Two-Electron Atoms* (Springer, Berlin, 1957).
4. J.E. Mayer and M.G. Mayer, *Phys. Rev.* **43**, 605 (1933).
5. C. Deutsch, *Phys. Rev. A* **2**, 43 (1970).
6. J.H. van Vleck and N.G. Whitelaw, *Phys. Rev.* **44**, 551 (1933).
7. T.A. Miller, R.S. Freund, F. Tsai, T.J. Cook, and B.R. Zegarski, *Phys. Rev. A* **9**, 2474 (1974).
8. W.H. Wing and K.B. MacAdam, in *Progress in Atomic Spectroscopy*, eds. W. Hanle and H. Kleinpoppen (Plenum, New York, 1978).
9. K.B. MacAdam and W.H. Wing, *Phys. Rev. A* **12**, 1464 (1975).
10. K.B. MacAdam and WH. Wing, *Phys. Rev. A* **13**, 2163 (1976).
11. K.B. MacAdam and W.H. Wing, *Phys. Rev. A* **15**, 678 (1977).
12. J.W. Farley, K.B. MacAdam, and W.H. Wing, *Phys. Rev. A* **20**, 1754 (1979).
13. D.R. Cok and S.R. Lundeen, *Phys. Rev. A* **23**, 2488 (1981).
14. S.L. Palfrey and S.R. Lundeen, *Phys. Rev. Lett.* **53**, 1141 (1984).
15. E.A. Hessels, P.W. Arcuni, F.J. Deck, and S.R. Lundeen, *Phys. Rev. A* **46**, 2622 (1992).
16. T.G. Eck, L.L. Foldy, and H. Weider, *Phys. Rev. Lett* **10**, 239 (1963).
17. T.A. Miller, R.S. Freund, and B.R. Zegarski, *Phys. Rev. A* **11**, 753 (1975).
18. H.J. Beyer and K.J. Kollath, *J. Phys. B* **8**, L326 (1975).
19. H.J. Beyer and K.J. Kollath, *J. Phys. B* **9**, L185 (1976).
20. H.J. Beyer and K.J. Kollath, *J. Phys. B* **10**, L5 (1977).
21. H.J. Beyer and K.J. Kollath, *J. Phys. B* **11**, 979 (1978).
22. J. Derouard, R. Jost, M. Lombardi, T.A. Miller, and R.S. Freund, *Phys. Rev. A* **14**, 1025 (1976).

23. J. Derouard, M. Lombardi, and R. Jost, *J . Phys. (Paris)* **41**, 819 (1980).
24. R. Panock, M. Rosenbluth, B. Lax, and T.A. Miller, *Phys. Rev. A* **22**, 1050 (1980).
25. R. Panock, M. Rosenbluth, B. Lax, and T.A. Miller, *Phys. Rev. A* **22**, 1041 (1980).
26. W.C. Martin, *J. Phys. Chem. Ref. Data* **2**, 257 (1973).
27. T.N. Chang and R.T. Poe, *Phys. Rev. A* **10**, 1981 (1974).
28. D.R. Cok and S.R. Lundeen, *Phys. Rev. A* **19**, 1830 (1979).

19

Autoionizing Rydberg states

Autoionizing states are those atomic states in which there are at least two excited electrons which together have enough energy that one can escape from the atom. We shall consider only states in which there are two excited electrons, one of which is in a Rydberg state.[1] From a spectroscopic point of view an autoionizing state is one which is coupled to a continuum, and from a collision point of view it is a long lived scattering resonance. In other words, autoionization is located at the intersection of collision physics and spectroscopy, and the theory commonly used to describe autoionizing states, quantum defect theory, is based on scattering theory.

Basic notions of autoionizing Rydberg states

It is straightforward to extend the picture of the spinless Ba atom given in Chapter 17 to autoionizing states. We again use the Hamiltonian of Eq. (17–1), which separates into

$$H = H_0 + H_1, \tag{19.1}$$

where

$$H_0 = -\left[\frac{\nabla_1^2}{2} + \frac{\nabla_2^2}{2} + f(r_1) + \frac{1}{r_2} \right] \tag{19.2}$$

and

$$H_1 = -f(r_2) + \frac{1}{r_2} + \frac{1}{r_{12}}. \tag{19.3}$$

Here $\mathbf{r}_1$ and $\mathbf{r}_2$ are the locations of the two electrons relative to the Ba^{2+} ion, and $r_{12} = |\mathbf{r}_1 - \mathbf{r}_2|$. In Eqs. (19.2) and (19.3) $f(r)$ is the potential experienced by an electron at a distance r from the Ba^{2+} core. As $r \to \infty$, $f(r) \to 2/r$.

Using H_0 only, the Schroedinger equation

$$H_0\Psi(\mathbf{r}_1,\mathbf{r}_2) = W\Psi(\mathbf{r}_1,\mathbf{r}_2) \tag{19.4}$$

is separable into the two equations

$$-\left[\frac{\nabla_1^2}{2} + f(r_1)\right]\phi(\mathbf{r}_1) = W_1\phi(\mathbf{r}_1), \tag{19.5a}$$

and

$$-\left(\frac{\nabla_2^2}{2} + \frac{1}{r_2}\right)\psi(\mathbf{r}_2) = W_2\,\psi\,(\mathbf{r}_2), \tag{19.5b}$$

in which

$$\Psi(\mathbf{r}_1,\mathbf{r}_2) = \phi_{n'\ell'm'}(\mathbf{r}_1)\psi_{n\ell m}(\mathbf{r}_2), \tag{19.6a}$$

and

$$W = W_1 + W_2. \tag{19.6b}$$

Here ϕ is a Ba^+ wavefunction and ψ is a hydrogenic wavefunction. The wavefunction of Eq. (19.6a) is a good zero order representation of the Ba atom as long as the inner turning point of the outer $n\ell$ electron is at a larger radius than the outer turning point of the Ba^+ $n'\ell'$ electron.

Since the two wavefunctions are independent, the energies do not depend on m', m, or ℓ, and we can write the energy of the $n'\ell'n\ell$ state relative to the ground state of Ba^{2+} as

$$W_{n'\ell'n\ell} = W_{n'\ell'} - \frac{1}{2n^2}. \tag{19.7}$$

Physically, the wavefunction of Eq. (19.6a) corresponds to adding one electron to Ba^{2+} to make the Ba^+ $n'\ell'$ states and then adding the second $n\ell$ electron to form the neutral Ba $n'\ell'n\ell$ states. On each state of Ba^+ there is built an entire system of Rydberg states and continua, as shown in Fig. 19.1 for the three lowest lying states of Ba^+. Note that the Rydberg states converging to the 5d and 6p states of Ba^+ are above the Ba^+ 6s state. With the independent electron Hamiltonian H_0 these states can only decay radiatively. They are not coupled to the degenerate continua above the Ba^+ 6s state.

Adding H_1 to the Hamiltonian introduces the couplings shown by the arrows in Fig. 19.1. First, and most important, it couples the doubly excited states above the lowest limit to the degenerate continua, introducing autoionization. Second, it leads to couplings between nominally bound series converging to different Ba^+ limits. The perturbations of the regularities of the energies of the bound Rydberg series are the result of this interaction between nearly degenerate states converging to different limits, while core polarization is the result of the coupling between energetically well separated states. These two phenomena were discussed in Chapter 17. One of the most interesting examples of interseries interaction is that between autoionizing Rydberg series. It is often the case that two interacting states overlap, due to their finite autoionization widths, leading to striking spectral profiles. We shall, in the following chapter, discuss these states in detail,

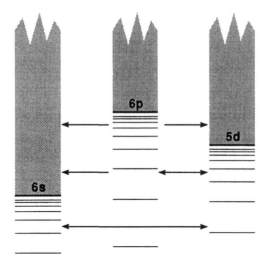

Fig. 19.1 The three lowest lying energy levels of Ba^+; 6s, 6p, and 5d (——); and the neutral Ba Rydberg energy levels (—) and continua (///) obtained by adding the second valence electron to these Ba^+ levels. The horizontal arrows show the possible interactions between channels associated with different ion levels. Interactions with other bound states lead to series perturbations while interactions with continua lead to autoionization.

but for the moment we focus on isolated autoionizing states coupled only to the degenerate continua, ignoring any interseries interaction between autoionizing series converging to different limits. Finally, although it is not shown in Fig. 19.1, the quadrupole term of H_1 lifts the degeneracy of $n'\ell'n\ell$ states of ℓ', $\ell \geq 1$; i.e. states of different total angular momentum do not have the same energy due to the differing orientations of ℓ' relative to ℓ.

The autoionization rate Γ of a Ba $6pn\ell$ state converging to the Ba^+ 6p state is given by the product of the squared coupling matrix element and the density of final continuum states. Explicitly,[2]

$$\Gamma = 2\pi \left(\sum_{\ell'=\ell\pm 1} |\langle 6pn\ell|H_1|6s\varepsilon\ell'\rangle|^2 + \sum_{\ell''=\ell\pm 1} |\langle 6pn\ell|H_1|5d\varepsilon\ell''\rangle|^2 \right), \quad (19.8)$$

where $6s\varepsilon\ell'$ and $5d\varepsilon\ell''$ are the continua associated with the 6s and 5d states of Ba^+. To account properly for the density of final states Eq. (19.8) requires that the $6s\varepsilon\ell'$ and $5d\varepsilon\ell''$ wavefunctions be normalized per unit energy.

A $6pn\ell$ state with an autoionization rate Γ is broadened, and its spectral density is described by a Lorentzian of width Γ centered at its energy W_0. A reasonable first approximation to the wavefunction is obtained by multiplying the zero order wavefunction of Eq. (19.6a) by the square root of a Lorentzian of width Γ;

$$\Psi^1_{6pn\ell} = \phi_{6p}(\mathbf{r}_1)\psi_{n\ell}(\mathbf{r}_2) \sqrt{\frac{\Gamma}{(W - W_0)^2 - (\Gamma/2)^2}}. \quad (19.9)$$

If we assume that ℓ is large enough that we may approximate $f(r_2)$ by $2/r_2$, then

$$H_1 = \frac{r_1}{r_2^2} P_1(\cos \theta_{12}) + \frac{r_1^2}{r_2^3} P_2(\cos \theta_{12}) \dots , \qquad (19.10)$$

and the leading term in the matrix element of Eq. (19.8) is the dipole term. Explicitly,

$$\Gamma = 2\pi \left[\sum_{\ell'=\ell\pm1} |\langle 6pn\ell| \frac{r_1 P_1(\cos \theta_{12})}{r_2^2} |6s\varepsilon\ell'\rangle|^2 \right.$$

$$\left. + \sum_{\ell''=\ell\pm1} |\langle 6pn\ell| \frac{r_1 P_1(\cos \theta_{12})}{r_2^2} |5d\varepsilon\ell''\rangle|^2 \right]. \qquad (19.11)$$

The quadrupole term of H_1 cannot couple the $6pn\ell$ states to the 6s and 5d continua, and we ignore the octopole term which couples the $6pn\ell$ state to the $5d\varepsilon\ell''$ continua.

The matrix elements of Eq. (19.11) can be broken into angular and radial parts. The angular parts can be evaluated as the scalar products of two tensor operators using the methods of Edmonds.[3] The radial parts consist of the Ba^+ $\langle 6p|r_1|6s\rangle$ or $\langle 6p|r_1|5d\rangle$ radial matrix element and the hydrogenic $\langle n\ell \mid 1/r_2^2|\varepsilon\ell'\rangle$ radial matrix elements. Since some of the angular matrix elements are always of order one, others being smaller, and the Ba^+ radial matrix elements do not depend on n or ℓ, all of the dependence on the quantum numbers n and ℓ enters via the radial matrix element coupling the Rydberg $n\ell$ state to the continuum. The radial matrix element of $1/r_2^2$ is most sensitive to the small r part of the $n\ell$ Rydberg wavefunction and the $\varepsilon\ell'$ continuum wavefunction, and the only significant n dependence is through the $1/n^{3/2}$ normalization factor of the bound Rydberg wavefunction, assuming $n \gg \ell$. When the matrix elements are squared to give the autoionization rate, we find the autoionization rate has a $1/n^3$ dependence. Explicitly,

$$\Gamma \propto 1/n^3. \qquad (19.12)$$

While it is straightforward to show that the autoionization rates scale as $1/n^3$, it is not possible to make an equally simple argument as to how they scale with ℓ. Consider a typical matrix element of interest, $\langle n\ell|1/r_2^2|\varepsilon(\ell + 1)\rangle$, which depends on the small r part of the wavefunctions. How does the matrix element change if n is kept fixed and ℓ is varied from $\ell \ll n$ to $\ell = n - 1$? Increasing ℓ increases the classical inner turning point r_i of the outer $n\ell$ electron's wavefunction according to

$$r_i \cong \frac{\ell(\ell + 1)}{2}, \qquad (19.13)$$

and as a result both the bound $n\ell$ and continuum $\varepsilon(\ell + 1)$ wavefunctions are found at increasing r_2 as ℓ is increased. Accordingly, the matrix element of $1/r_2^2$

decreases rapidly with increasing ℓ. Exactly how the matrix element decreases with ℓ depends on the energy ε, especially as $\ell \to n$.

In the example described above the autoionization of Ba $6pn\ell$ states to the 6s and 5d states of Ba^+ is by means of a dipole coupling. The Ba $5dn\ell$ states, however, autoionize by a quadrupole coupling. The quadrupole processes are often as strong as dipole processes, counter to the intuition developed from optical transitions. Inspecting Eq. (19.10) we can see that the factor by which multipole processes of adjacent orders differ is r_1/r_2. This factor is approximately the radius of the ion divided by the inner turning point of the Rydberg electron's orbit. For low ℓ states this ratio is of order one. For an optical transition the analogous factor is r/λ, where λ is the wavelength of the light, and for visible radiation $r/\lambda \sim 10^{-3}$.

Autoionization can be viewed as a scattering process, and a simple classical picture leads to the same n and ℓ scaling as the quantum mechanical treatment given above. The basic mechanism responsible for autoionization is superelastic scattering of the Rydberg electron from the excited Ba^+ ion to produce a ground state Ba^+ ion and a more energetic electron. Consider a single passage of the electron past a Ba^+ ion in the 6p state. For the electron to induce the transition to the Ba^+ 6s state it must come close enough that its field components are strong enough and high enough in frequency to induce the Ba^+ 6p $\to$ 6s transition. How close the electron comes to the Ba^+ and how rapidly it goes past the Ba^+ depend mostly on the ℓ of the electron relative to the Ba^+. Increasing ℓ decreases the field from the electron at the Ba^+ ion and lowers the frequency components of the field, thus reducing the probability of superelastic scattering, and of autoionization.

We have thus far considered the probability of superelastic scattering on a single orbit. To obtain the scattering rate, or autoionization rate, we simply multiply this probability by the orbital frequency, $1/n^3$.[4] Once again we find that $\Gamma \propto 1/n^3$ and that Γ decreases with increasing ℓ. The scattering description we have just given is a two channel description. This picture, when many channels are present, forms the basis of multichannel quantum defect theory.[5]

Experimental methods

The classic method of studying autoionizing states is vacuum ultraviolet (vuv) absorption spectroscopy using the continuum radiation from either a lamp, or more recently, a synchrotron in conjunction with a spectrometer.[6] This approach is straightforward in principle and generally applicable, but it is limited to the observation of single photon transitions. Usually the sample of absorbing atoms is composed of ground state atoms. However, it is possible to use a laser to produce substantial populations in excited states and observe the absorption from them.[7] An alternative approach, employed by Garton *et al.* was to measure the absorption spectrum of Ba heated to high temperatures in a shock tube.[8] In addition to producing excited states the shock heating also produced ions and electrons leading to the initial observation of forced autoionization.[8]

While vuv absorption spectroscopy is relatively straightforward, one of its inherent properties is that both the doubly excited autoionizing state and the degenerate continuum have nonzero excitation amplitudes from the ground state, and these amplitudes can interfere constructively or destructively. Typically, if the interference is constructive on the high energy side of the autoionizing state it is destructive on the low energy side and vice versa. The interference term between the two excitation amplitudes changes sign across the autoionizing resonance due to the continuum phase shift of π on going through the resonance.[2] The resulting asymmetric line profiles are termed Beutler–Fano profiles, and we shall discuss them more quantitatively in the next chapter. For the moment we note that while the line profiles of autoionizing states are not symmetric, it is still possible to extract the energies and widths of the autoionizing states from the vuv spectra.

As a representative example, we show in Fig. 19.2 the spectrum of Ba obtained by vuv absorption spectroscopy.[9] The light from a microwave excited Kr or Xe continuum lamp is focused on the slits of a spectrograph by a spherical mirror. Between the mirror and the spectrograph the light passes through a King furnace containing Ba vapor. In essence the King furnace consists of a tube which is heated to 900–1100°C, over the central 120 cm of its length.[10] In the central region the Ba vapor pressure is 2–20 Torr, and 1–10 Torr of flowing Ar is present to confine the Ba vapor to the heated zone, keeping it away from both the mirror and the spectrograph slits.

Several features typical of spectra in the autoionizing region are apparent in Fig. 19.2. First, there is substantial continuum absorption. Second, the autoionizing Rydberg series converging to both the Ba^+ $6p_{1/2}$ and $6p_{3/2}$ limits are clearly visible. Since the spectrum is from the ground, $6s^2\,{}^1S_0$, state the autoionizing states of Fig. 19.2 are odd parity $6p_j\,ns_{1/2}$ and $6p_j nd_j$ $J = 1$ states. Third, the lineshapes are very asymmetric. Each autoionizing state appears not as an additional absorption peak added to the continuum absorption but as something more similar to a dispersion lineshape. In spite of the dispersion like lineshapes, which are Beutler–Fano profiles, the regularity of the Rydberg series converging to the $6p_{3/2}$ limit is apparent, and it is possible to determine the quantum defects and widths of the autoionizing states converging to the $6p_{3/2}$ limit. On the other hand, below the $6p_{1/2}$ limit, where series converging to both the $6p_{1/2}$ and $6p_{3/2}$ limits are present, the spectrum is irregular due to the interseries interaction. Such spectra are particularly difficult to analyze because both the bound–continuum excitation interference and the interseries interactions lead to peculiar lineshapes, and it is difficult to disentangle the two effects present in the spectra.

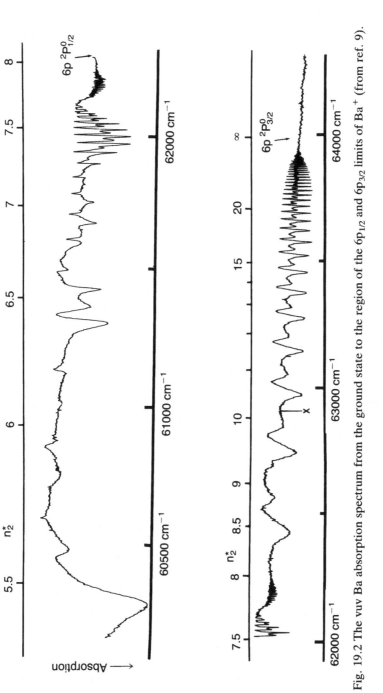

Fig. 19.2 The vuv Ba absorption spectrum from the ground state to the region of the $6p_{1/2}$ and $6p_{3/2}$ limits of Ba$^+$ (from ref. 9).

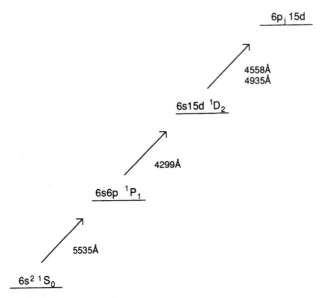

Fig. 19.3 Ba energy levels showing the excitation of the Ba $6p_j15d$ states. The first two lasers excite the outer electron and the third laser excites the inner electron to the $6p_{1/2}$ or $6p_{3/2}$ state when it is tuned to 4935 Å or 4558 Å.

A laser method which has been used to study autoionizing states of alkaline earth atoms is the isolated core excitation (ICE) method first used by Cooke *et al.* to study the autoionizing $5pn\ell$ states of Sr.[11] It has since been used to study autoionizing states of Mg, Ca, and Ba as well.[12–14]

The idea of the method is illustrated by the study of the Ba $6pnd$ states. Using two pulsed lasers tuned to the Ba $6s^2$ $^1S_0 \rightarrow 6s6p$ $^1P_1$ and $6s6p$ $^1P_1 \rightarrow 6s15d$ $^1D_2$ transitions Ba atoms in a thermal beam are excited to the long lived, bound $6snd$ $^1D_2$ state, as shown by Fig. 19.3. The autoionizing $6p_{1/2}15d$ or $6p_{3/2}$ $15d$ states are observed by scanning the wavelength of the third laser, at 4935 Å or 4558 Å, across the $6s15d \rightarrow 6p_j15d$ transition while monitoring the ions or electrons resulting from the decay of the autoionizing $6p_j15d$ state.

An example of a typical spectrum is shown in Fig. 19.4, the spectrum of the excitation from the Ba $6s15d$ $^1D_2$ state to the Ba $6p_{3/2}15d_j$ $J = 3$ state, obtained with all three lasers circularly polarized in the same sense.[15] As shown by Fig. 19.4 the lineshape is approximately Lorentzian, as expected from Eq. (19.9). The width may be measured directly, and the energy of the state is obtained by adding the laser frequency at the center of the resonance to the known energy of the $6s15d$ $^1D_2$ state. In most cases the width of the autoionizing levels is due predominantly to autoionization. There is, however, a radiative contribution to the width. For the Ba $6pn\ell$ Rydberg states the dominant source of radiative width is the radiative decay of the inner electron from $6p$ to $6s$, which has a width of 26 MHz.[16,17]

As shown by Fig. 19.4 the observed profile directly reflects the spectral profile of the autoionizing state, as implied by Eq. (19.9), and there is evidently negligible

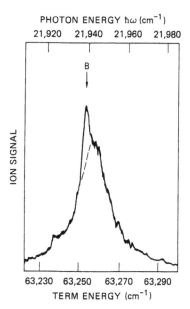

PHOTON ENERGY ℏω (cm⁻¹)

Fig. 19.4 Low power scan of the third laser across the Ba 6s15d $^1D_2 \rightarrow 6p_{3/2}15d\, J = 3$ transition taken with all three lasers circularly polarized in the same sense. This scan yields the position and width of the $6p_{3/2}15d\, J = 3$ state. The narrow peak denoted by B is a coincidence with the 6s6p $^1P_1 \rightarrow$ 6s10d $^1D_2$ transition. The term energy of the 6s15d $^1D_2$ state is 41315.5 cm⁻¹ (from ref. 15).

excitation of the degenerate continuum; only the doubly excited state is excited. In the 6s15d → 6p15d transition the inner electron undergoes the Ba⁺ 6s → 6p transition, the Ba⁺ resonance line transition, and the outer 15d electron remains a spectator, making minor readjustments in its orbit to account for the difference in the quantum defects of the 6snd and 6pnd states. This process is shown schematically in Fig. 19.5, which shows the orbits of the two valence electrons in the 6s², 6s15d, and 6p15d states.[18] In the excitation from the Ba 6snd to the 6pnd states the quantum defects of the bound and autoionizing nd states differ by ~0.1, in which case only a single strong peak is observed in the spectrum.[11,15] However, when the quantum defect of the autoionizing series differs from that of the bound state by 1/2, two strong transitions from the bound $n\ell$ state to the $n\ell$ and $(n + 1)\ell$ states are observed.[11,15]

Why the continuum excitation is negligible is apparent if we consider the two alternative ways the 6s15d atom can absorb the visible photon required to reach the energy of the 6p15d state. The first way is the inner electron 6s → 6p transition with the outer electron remaining a spectator. The Ba⁺ 6s → 6p transition has an oscillator strength of one[17] spread over the 10 cm⁻¹ width of the 6p15d state, yielding df/dW ~ 10⁴. In general df/dW for this excitation scales as n^3, due to the widths of the autoionizing states. The second excitation possibility is for the Rydberg electron to be directly photoionized by the visible photon. The Rydberg

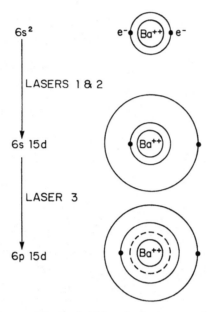

Fig. 19.5 Schematic picture of the Ba 6p15d excitation process. Lasers 1 and 2 excite one of the 6s electrons to the 15d state. Laser 3 drives the remaining 6s electron to the 6p state, while the 15d electron is a spectator (from ref. 18).

electron is most likely to be found at its outer turning point, far from the ionic core, where the spatial variation in the wavefunction is slow. In contrast, at the energy of the 6pnd state the continuum wavefunction exhibits rapid spatial variation far from the ionic core, and the large r parts of the wavefunctions contribute little to the photoionization matrix element. Only the small r part of the Rydberg wavefunction contributes to the matrix element. Accordingly, df/dW for direct photoionization of the 6snd states scales as $1/n^3$, due to the normalization of the Rydberg wavefunction near the core. Thus the ratio of the two oscillator strengths per unit energy for the inner and outer electron absorptions scales as n^6. For the direct photoionization of the 15d state $df/dW \sim 10^{-4}$ so the ratio of the two oscillator strengths is $\sim10^8$. Even the interference cross term is a factor of 10^4 smaller than the peak cross section.

While the isolated core excitation method makes the analysis of single isolated autoionizing states trivial, its real strength lies in unravelling the spectra of interacting series. One example of the interaction between different series of autoionizing states is the interaction among the Ba $J = 1$ odd parity states converging to the $6p_{1/2}$ and $6p_{3/2}$ limits. In the vuv spectrum of Fig. 19.2, the region below the $6p_{1/2}$ limit exhibits strong interseries interaction, but due to the added complication of attendant continuum excitation it is difficult to separate the effect of interseries interaction from the bound–continuum interference in the excitation. As we shall show in the following chapter, using the ICE method it is straightforward to examine the effects of interseries interaction with negligible

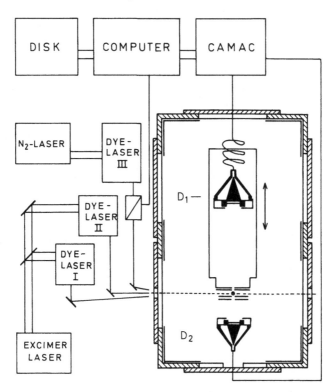

Fig. 19.6 A schematic view of an apparatus for measuring photoexcitation cross sections and photoelectron energy and angular distributions. The atom beam comes out of the page, and D_1 and D_2 are the electron and ion detector, respectively (from ref. 25).

interferences in the excitation amplitudes.[4,19] It has also proven useful for experiments involving multiphoton excitation of the inner electron.[20–24]

Other strengths of the ICE method derive from the fact that the atoms are in a low density beam, not a high pressure absorption cell. As shown in Fig. 19.1, an atom in a Ba $6pn\ell$ state can autoionize into the degenerate continua above lower lying states of Ba^+. By analyzing the energy and angular distributions of the ejected electrons it is possible to determine the branching ratios for autoionization to the possible final states of Ba^+ and the ejected electron. Such measurements are far more stringent tests of calculations than measurements of total autoionization rates. The apparatus used to make the electron spectroscopy measurements is not much more complicated than that required to make the total photoexcitation cross section measurements. In Fig. 19.6 we show the apparatus used by Sandner *et al.*[25] to measure the energy and angular distributions of electrons ejected from the Ba $6pns$ states. The electron spectrometer of Fig. 19.6 is a time of flight spectrometer, which has a 50 meV resolution at electron energies of 1 eV.

We have described the excitation from the bound Ba $6snd$ to autoionizing $6pnd$ states, and it is possible by purely optical means to excite the $6sn\ell$ states of $\ell \leq 4$.

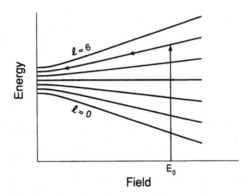

Fig. 19.7 Schematic diagram of the Stark switching technique applied to n = 7 Rydberg states. The arrow shows laser excitation of the Stark state in the field E_0. The field is then reduced to zero adiabatically to produce the $\ell = 5$ state. The zero field separations of the ℓ states are exaggerated for clarity.

It is possible to take advantage of the long lifetimes and large dipole moments of the bound $6sn\ell$ Rydberg states to excite bound $6sn\ell$ Rydberg states of arbitrary ℓ and observe the transitions to their autoionizing $6pn\ell$ analogues. To produce higher ℓ states a Stark switching technique, originally proposed by Freeman and Kleppner,[26] can be used. It was first used by Cooke *et al.*[11] to populate Sr $5sn\ell$ states of ℓ up to 7. The basic idea is shown in Fig. 19.7, a schematic energy level diagram of the $n = 7, m = 0$ Stark manifold as a function of static electric field. For clarity the zero field separations of the ℓ states are exaggerated. At the field E_0 the Stark levels are well resolved and a single Stark state can be excited with a laser. The field is then reduced to zero adiabatically, and the initially excited Stark level evolves into a single zero field angular momentum state. In Fig. 19.7 we show the laser excitation of the Stark state which evolves adiabatically into the zero field $\ell = 5$ state as the field is turned off. The ℓ state produced is determined by which Stark state is initially excited with the laser.

The first application of the Stark switching technique to a wide range of ℓ states was made by Jones and Gallagher who produced the Ba $6sn\ell$ states of $11 \le n \le 13$ and $\ell \ge 4$.[16] They initially excited the Ba atoms from the ground $6s^2\ {}^1S_0$ state to the $6s6p\ {}^1P_1$ state and then to a $6snk$ Stark state in a field of 3 kV/cm. This field is strong enough that the Stark levels are ~2 cm^{-1} apart and there is sufficient $6snd$ character in each $6snk$ Stark level for excitation. The field was reduced to zero in 2–3 μs. These times are determined by two limits. The zero field separation between the highest ℓ states is 200 MHz, setting a lower limit on the switching time. On the other hand, if the field is switched too slowly, black body radiation drives population to other states, the atoms in $\ell \sim 4$ states decay radiatively, and the atoms pass out of the interaction region.[16] This method has also been used by Pruvost *et al.*[27] to study Ba $6pn\ell$ and $6dn\ell$ states of high ℓ and by Eichmann *et al.*[28] to study Sr $n'gn\ell$ states of high ℓ.

An alternative technique, originally proposed by Delande and Gay,[29] and first used by Hare *et al.*[30] has been used by Roussel *et al.*[31] to produce atoms in circular Ba ($\ell = m = n - 1$) $6sn\ell$ states for subsequent excitation to autoionizing $6pn\ell$ states. The atoms are excited in a strong electric field to the highest energy $6snk$ Stark state and pass adiabatically into a region where there is only a modest magnetic field perpendicular to the direction of the electric field. The highest energy Stark state is one which is originally a Stark state of $m = 0$ in the electric field but which evolves into the $m = n - 1$ state in the magnetic field.

Measurement of the autoionization rates by sweeping the laser wavelength across the bound $\rightarrow$ autoionizing state transition is a method which works well for autoionizing states of widths large compared to the laser linewidth. For high ℓ autoionizing states this condition is not satisfied, and it is necessary to use the depletion broadening approach first used by Cooke *et al.*[32] The method takes advantage of two facts. First, the laser line usually has an approximately Gaussian profile, which has line wings which drop more rapidly than those of a Lorentzian, while the autoionizing state has a profile which is a Lorentzian. Thus, at a detuning from line center much larger than the laser linewidth the observed signal is determined by the Lorenztian tail of the cross section at the detuned laser wavelength, not by the light in the wing of the laser line at the maximum of the cross section. In this case, when the laser is tuned far from line center, the width of the laser can be ignored. The second important fact is that since the optical cross sections are so large, it is possible to observe strong signals even when the detuning is such that the cross section is two or three orders of magnitude below the peak cross section.

For the optical transition from the bound $6sn\ell$ state to the autoionizing $6pn\ell$ state the optical cross section is given by the Lorentzian form

$$\sigma_{n\ell}(\Delta\omega) = \frac{\Omega\Gamma_{n\ell}}{(\Delta\omega)^2 + (\Gamma_{n\ell}/2)^2} \qquad (19.14)$$

where $\Delta\omega$ is the detuning from line center, $\Gamma_{n\ell}$ is the FWHM due to autoionization and radiative decay and Ω is proportional to the ionic $6s$–$6p$ dipole matrix element squared. Now consider the $6sn'\ell' \rightarrow 6pn'\ell'$ transition for $\ell' \neq \ell$ or $n' \neq n$. The cross section $\sigma_{n'\ell'}(\Delta\omega)$ is obtained from $\sigma_{n\ell}(\Delta\omega)$ by replacing $\Gamma_{n\ell}$ by $\Gamma_{n'\ell'}$.

If the sample of N_0 bound $6sn\ell$ Rydberg atoms is exposed to an integrated photon flux Φ, then the number of atoms excited to the autoionizing states is given by

$$N_{n\ell} = N_0\sigma_{n\ell}\Phi \qquad (19.15)$$

if $\sigma_{n\ell}\Phi \ll 1$. However, with available laser powers the regime $\sigma_{n\ell}\Phi \gg 1$ is easily reached, in which case

$$N_{n\ell} = N_0(1 - e^{-\sigma_{n\ell}(\Delta\omega)\Phi}). \tag{19.16}$$

If the flux of the laser is high enough that $\sigma_{n\ell}(0) \, \Phi \gg 1$, the observed signal is broadened to a width substantially in excess of the width of the cross section of Eq. (19.14).

We assume that we can measure directly the width of the $6sn\ell \rightarrow 6pn\ell$ transition, i.e. that Γ_ℓ exceeds the laser linewidth. We also assume that we can use a high enough laser power so that both the $6sn\ell \rightarrow 6pn\ell$ and the $6sn'\ell' \rightarrow 6pn'\ell'$ transitions are broadened to a width substantially in excess of the laser linewidth. In this case, if we compare the detunings $\Delta\omega_{n\ell}$ and $\Delta\omega_{n'\ell'}$ at which the signals are half their maximum values, from Eq. (19.16) we can see that

$$\sigma_{n\ell}(\Delta\omega_{n\ell})\Phi = \sigma_{n'\ell'}(\Delta\omega_{n'\ell'})\Phi = \ln 2. \tag{19.17}$$

Dividing by the common laser flux of Φ yields

$$\sigma_{n\ell}(\Delta\omega_{n\ell}) = \sigma_{n'\ell'}(\Delta\omega_{n'\ell'}). \tag{19.18}$$

Using the form of the cross sections given by Eq. (19.14), Eq. (19.18) may be written as

$$\Gamma_{n'\ell'} = \Gamma_{n\ell}\left[\frac{(\Delta\omega_{n'\ell'})^2 + (\Gamma_{n'\ell'}/2)^2}{(\Delta\omega_{n\ell})^2 + (\Gamma_{n\ell}/2)^2}\right]. \tag{19.19}$$

Since the detuning must be large compared to the laser linewidth, if we are measuring a width $\Gamma_{n'\ell'}$ less than the laser linewidth, Eq. (19.9) reduces to

$$\Gamma_{n'\ell'} \cong \frac{\Gamma_{n\ell}(\Delta\omega_{n'\ell'})^2}{(\Delta\omega_{n\ell})^2 + (\Gamma_{n\ell}/2)^2}. \tag{19.20}$$

It is worth noting that there is nothing special about the point $N_{n\ell}/N_0 = 1/2$, where the signal is equal to half the peak signal. We could equally well have chosen $N_{n\ell}/N_0 = 1/3$ or any other value, but at these values of $N_{n\ell}/N_0$ it is harder to determine $\Delta\omega$ experimentally.

Experimental observations of autoionization rates

Earlier in this chapter we presented arguments to show that the autoionization rates of, for example, the Ba $6pn\ell$ states should scale as $1/n^3$, provided that $n \gg \ell$. This point is shown rather clearly in Fig. 19.8, a plot of the observed widths of the three Ba $6p_j nd$, $J = 3$ series.[33] The two $6p_{3/2} nd\, J = 3$ series were labelled $+$ and $-$ by Mullins *et al.*[34] since they are reached from the bound $6snd\, {}^1D_2$ and ${}^3D_2$ states, respectively, using isolated core excitation. The $1/n^3$ dependence is quite evident for all three series, as is the deviation of the $6p_{1/2}nd_{5/2}$ widths at $n^* = 8$ and 17. At these values of n^* the interaction with approximately degenerate $6p_{3/2}nd$ states alters the $1/n^3$ dependence of the widths. There are, in fact, many examples of the $1/n^3$ scaling of the autoionization rates, but it is important to keep in mind that this scaling law is only valid for $n \gg \ell$. When $n \sim \ell$ the autoionization rates of an ℓ

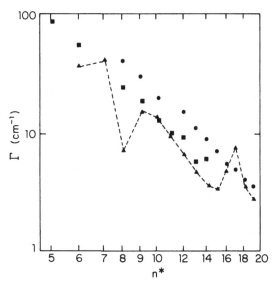

Fig. 19.8 Autoionization rates of the three $6pnd$ $J = 3$ series $6p_{1/2}nd_{5/2}$ ($\triangle$), $6p_{3/2}nd^+$ ($\bullet$), and $6p_{3/2}nd^-$ ($\blacksquare$). The $6p_{3/2}nd^+$ and $6p_{3/2}nd^-$ states are excited from the $6snd$ $^1D_2$ and $6snd$ $^3D_2$ states respectively. The perturbations in the $6p_{1/2}nd_{5/2}$ series are quite evident (from ref. 33).

series can increase with n. This phenomenon is observed in the Ba $4fng$ series. The $4f6g$ state is broader than the $4f5g$ state.[35] Part of the reason is that for $n = \ell + 1$ the $n\ell$ wavefunction has only one lobe in the minimum of the combined coulomb–centrifugal potential. For $n = \ell + 2$ the inner turning point moves to noticeably smaller radius. For example, the inner turning point of the $5g$ state is at $13.8a_0$, while it is at $12.0a_0$ for the $6g$ state. As n is further increased, the inner turning point continues to move to smaller radius, but very slowly, and the effect of the $1/n^{3/2}$ normalization factor of the ng wavefunction quickly becomes more important. Another possible contributing factor is wavefunction collapse. In La the $4f$ wavefunction is collapsed into an inner well in the radial potential, while in Ba, with a smaller nuclear charge, the $4f$ electron is just on the verge of collapsing.[36,37] It may be that the presence of the $6g$ electron at a smaller orbital radius screens the $4f$ electron from the nuclear charge more efficiently than does the $5g$ electron. Thus collapse of the $4f$ electron is more likely in the $4f5g$ state than in the $4f6g$ state. The partial collapse of the $4f$ wavefunction in the $4f5g$ state might lead to a smaller autoionization rate for the $4f5g$ state than for the $4f6g$ state. Connerade has pointed out that it may be possible to control externally the extent to which a wavefunction collapses.[38] These observations may support his suggestion, but careful calculations need to be carried out to verify this point.

Autoionization rates also decrease rapidly with ℓ, a point first shown experimentally by Cooke *et al.*[11] who measured the autoionization rates of the Sr $5p15\ell$ states of $\ell = 2$–7. The $\ell > 2$ states were populated using the Stark switching

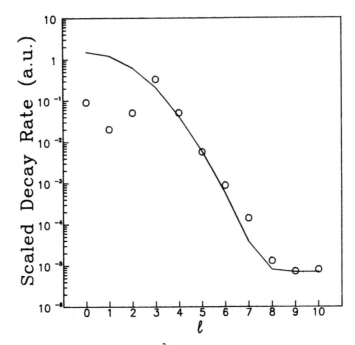

Fig. 19.9 Plot of scaled total decay rates $n^3\Gamma$ of Ba $6p_{1/2}n\ell$ $J = \ell + 1$ autoionizing states in atomic units vs ℓ. For $\ell = 0$–4 the measured rates (○) shown are the average rates from many n values. The data for the rates for $\ell > 4$ are for $n = 12$. The solid line is a simple theoretical calculation based on the dipole scattering of a hydrogenic Rydberg electron from the 6p core electron. Note that the core penetration of the lower ℓ states reduces the actual rate from the one calculated using the dipole scattering model. The constant total decay rate for $\ell > 8$ is the spontaneous decay rate of the Ba$^+$ 6p state (from ref. 39).

technique described earlier. Their measurements show a dramatic decrease of the observed autoionization width from 15 cm^{-1} at $\ell = 2$ to 1 cm^{-1} at $\ell = 5$. The measurements for $\ell > 5$ were limited by the linewidth of the laser. The widths of the Ba $6p_{1/2}n\ell$ states of $11 \leq n \leq 13$ and $4 \leq \ell \leq n - 1$ were measured by Jones and Gallagher[16] using both the Stark switching technique[26] and the depletion broadening technique to measure decay rates below the laser linewidth.[32] Their results showed clearly the rapid decrease of the autoionization rate with ℓ, out to $\ell \sim 9$ where the autoionization rate falls below the radiative decay rate of the Ba$^+$ $6p_{1/2}$ ion, and the total decay rate is dominated by the radiative decay rate, which is independent of n and ℓ.

In Ba, the autoionization rates of many $6pn\ell$ states have been measured,[4,16,19,27,34,39–42] and it is interesting to combine the autoionization rates of the high ℓ states,[16] with those measured for low ℓ states. In Fig. 19.9 we show the measured scaled autoionization rates, $n^3\Gamma$, for the $6p_{1/2}n\ell$ states of $J = \ell+1$.[39] The measured values are compared to the values calculated using Ba$^+$ coulomb wavefunctions and hydrogenic wavefunctions, i.e. the approach outlined earlier. The calculations show uniformly decreasing autoionization rates with increasing

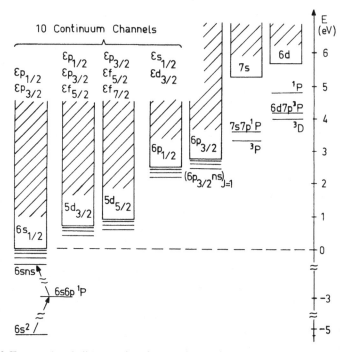

Fig. 19.10 Energy level diagram for the autoionization of Ba $6p_{3/2}ns_{1/2}$ $J = 1$ states. As shown, above the $6p_{1/2}$ limit there are ten continuum channels. Below the $6p_{1/2}$ limit there are only eight continuum channels. The two step laser excitation of the spherically symmetric $6sns$ $^1S_0$ state is also shown. Not shown is the laser excitation from the $6sns$ state to the $6p_{3/2}ns_{1/2}$ state (from ref. 25).

ℓ. For $\ell \geq 3$ the calculated and observed rates agree. For $\ell < 3$ they do not. The disagreement is not surprising since the $\ell < 3$ states have significant core penetration, and the calculation is almost certain to be wrong for $\ell < 3$. A more surprising observation is that the measured $6p_{1/2}nf$ autoionization rate is the highest, substantially higher than the autoionization rates of the $\ell < 3$ states, which we might intuitively expect to be larger. A possible reason why this is so is that the $\ell < 3$ states are all distinctly core penetrating, while the nf states are not. In other words, the inner turning point of the nf wavefunction approximately coincides with the outer turning point of the inner $6p$ electron's wavefunction. In this case the two wavefunctions overlap where both electrons are moving slowly, and therefore interacting strongly.

Electron spectroscopy

If we replace the schematic energy level diagram of Fig. 19.1 by Fig. 19.10, which more accurately represents the Ba atom, we can see that the Ba $6p_{3/2}ns_{1/2}$ $J = 1$

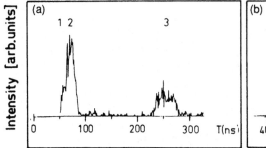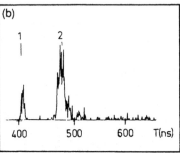

Fig. 19.11 (a) Time of flight spectra of Ba $6p_{3/2}ns$ $J = 1$ autoionization electrons. Peak 1, decay into the Ba$^+$ 6s continuum; peak 2, into the Ba$^+$ $5d_{3/2,5/2}$ continua; peak 3, into the Ba$^+$ $6p_{1/2}$ continuum. The drift length was approximately 10 cm. The spectra were recorded with a gate width of 14 ns. (b) The same as (a) but with a drift length of approximately 45 cm and a gate width of 6 ns (from ref. 25).

atoms, for example, can autoionize into ten continua of different ℓ above the $6s_{1/2}$, $5d_j$, and $6p_{1/2}$ ionization limits.[25] Below the $6p_{1/2}$ limit the Ba $6p_j$ $ns_{1/2}$ $J = 1$ states can only ionize into the eight continua above the $6s_{1/2}$ and $5d_j$ ionization limits. The allowed final states have been probed by measuring the energy and angular distributions of electrons ejected from the Ba $6p_j$ $ns_{1/2}$ $J = 1$ states. There are two quantities of interest for each autoionizing state, the branching ratios to the final ion states, and the anisotropy parameters of the angular distributions of the ejected electrons in the autoionization to specific ionic states. When the autoionizing Ba $6p_j$ $ns_{1/2}$ $J = 1$ states are excited from the spherically symmetric $6sns$ $^1S_0$ states by electric dipole excitation, the angular distribution of the electrons ejected at any energy takes the form[43]

$$I(\Theta) = I_0[1 + \beta P_2(\cos \Theta)] \tag{19.21}$$

where Θ is the angle between the **E** vector of the exciting laser and the momentum vector of the escaping electron and β is the anisotropy parameter. An interesting point about the ICE method is that since there is no excitation of the continuum, which has a varying phase, the β parameter and the branching ratio are constant across the autoionizing state's profile.

Conservation of momentum requires that virtually all the kinetic energy of autoionization appears in the electron. Thus the final state of the ion is implied by the energy of the ejected electron, and measuring the energy and angular distributions of the ejected electrons yields the branching ratios for autoionization into the possible final states of the ion.

In Fig. 19.11 we show the time of flight spectra, obtained with $\Theta = 0$, for electrons from a $6p_{3/2}$ $ns_{1/2}$ state lying above the $6p_{1/2}$ ionization limit.[25] With a 10 cm flight path for the electrons, the electrons ejected in autoionization to the Ba$^+$ $6s_{1/2}$ and $5d_j$ states are barely resolved, while with the higher resolution afforded by a 45 cm flight path the electrons from autoionization to the $6s_{1/2}$ and $5d_j$ states are well resolved, showing little autoionization to the Ba$^+$ $5d_{5/2}$ state. To convert the time of flight spectra, such as those shown in Fig. 19.11, to branching ratios, they must be either measured at all values of the angle Θ and integrated over Θ, or measured at the magic angle, $\Theta = 54.7°$, where $P_2(\cos \Theta) = 0$. Rotating the polarization of the laser driving the 6sns $^1S_0 \rightarrow 6p_j ns_{1/2}$ $J = 1$ transition yields the angular distributions of the ejected electrons, and they have been measured by two groups.[25,44,45] Angular distributions from the autoionizing states of Sr, Ca, and Mg have also been reported.[46–48]

An interesting point to note in Fig. 19.11(b) is that a significant number of the observed electrons are from autoionization to the Ba^+ $6p_{1/2}$ state. In fact, 40% of the Ba $6p_{3/2}ns_{1/2}$ atoms autoionize to the $6p_{1/2}$ state of Ba^+, more than to any other single ion state,[25] in spite of the fact that it is a quadrupole process, whereas autoionization to either the Ba^+ $6s_{1/2}$ or $5d_j$ states occurs by a dipole process. The preferential population of the Ba^+ $6p_{1/2}$ state when Ba $6p_{3/2}np$ states of $n > 12$ are populated by a laser has been used by Bokor *et al.* to make a laser on the Ba^+ $6p_{1/2} \rightarrow 5d_{3/2}$ transition at 600 nm and the Ba^+ $6p_{1/2} \rightarrow 6s_{1/2}$ transition at 493 nm.[49]

References

1. W. Sandner, *Comm. At. Mol. Phys.* **20**, 171 (1987).
2. U. Fano, *Phys. Rev.* **124**, 1866 (1961).
3. A. R. Edmonds, *Angular Momentum in Quantum Mechanics*, (Princeton University Press, Princeton, 1960).
4. W. E. Cooke and C. L. Cromer, *Phys. Rev. A* **32**, 2725 (1985).
5. M. J. Seaton, *Rep. Prog. Phys.* **46**, 167 (1983).
6. J. Berkowitz, *Photoabsorption, Photoionization, and Photoelectron Spectroscopy* (Academic Press, New York, 1979).
7. J. L. Carlsten, T. J. McIlrath, and W. H. Parkinson, *J. Phys. B* **8**, 38 (1962).
8. W. R. S. Garton, W. H. Parkinson, and E. M. Reeves, *Proc. Phys. Soc. London* **80**, 860 (1962).
9. C. M. Brown and M. L. Ginter, *J. Opt. Soc. Am.* **68**, 817 (1978).
10. C. M. Brown, R. H. Naber, S. G. Tilford, and M. L. Ginter, *Appl. Opt.* **12**, 1858 (1973).
11. W. E. Cooke, T. F. Gallagher, S. A. Edelstein, and R. M. Hill, *Phys. Rev. Lett.* **40**, 178 (1978).
12. G. W. Schinn, C. J. Dai, and T. F. Gallagher, *Phys. Rev. A* **43**, 2316 (1991).
13. V. Lange, V. Eichmann, and W. Sandner, *J. Phys. B* **22**, L245 (1989).
14. L. D. von Woerkem and W. E. Cooke, *Phys. Rev. Lett.* **57**, 1711 (1986).
15. N. H. Tran, P. Pillet, R. Kachru, and T. F. Gallagher, *Phys. Rev. A* **29**, 2640 (1984).
16. R. R. Jones and T. F. Gallagher, *Phys. Rev. A* **38**, 2946 (1988).
17. A. Lindgard and S. E. Nielson, *At. Data Nucl. Data Tables* **19**, 613 (1977).
18. T. F. Gallagher, *J. Opt. Soc. Am. B* **4** 794 (1987).
19. F. Gounand, T. F. Gallagher, W. Sandner, K. A. Safinya, and R. Kachru, *Phys. Rev. A* **27**, 1925 (1983).
20. L. A. Bloomfield, R. R. Freeman, W. E. Cooke, and J. Bokor, *Phys. Rev. Lett.* **53**, 2234 (1984).
21. P. Camus, P. Pillet, and J. Boulmer, *J. Phys. B* **18**, L481 (1987).
22. N. Morita, T. Suzuki, and K. Sato, *Phys. Rev. A* **38**, 551 (1988).
23. U. Eichmann, V. Langé, and W. Sandner, *Phys. Rev. Lett.* **64**, 274 (1990).
24. R. R. Jones and T. F. Gallagher, *Phys. Rev. A* **42**, 2655 (1990).
25. W. Sandner, U. Eichman, V. Lange, and M. Völkel, *J. Phys. B* **19**, 51 (1986).

26. R. R. Freeman and D. Kleppner, *Phys. Rev. A* **14**, 1614 (1976).
27. L. Pruvost, P. Camus, J.-M. Lecompte, C. R. Mahon, and P. Pillet, *J. Phys. B* **24**, 4723 (1991).
28. U. Eichmann, V. Lange, and W. Sandner, *Phys. Rev. Lett.* **68**, 21 (1992).
29. D. Delande and J. C. Gay, *Europhys. Lett.* **5**, 303 (1988).
30. J. Hare, M. Gross, and P. Goy, *Phys. Rev. Lett.* **61**, 1938 (1988).
31. F. Roussel, M. Cheret, L. Chen, T. Bolzinger, G. Spiess, J. Hare, and M. Gross, *Phys. Rev. Lett.* **65**, 3112 (1990).
32. W. E. Cooke, S. A. Bhatti, and C. L. Cromer, *Opt. Lett.* **7**, 69 (1982).
33. T.F. Gallagher, in *Electronic and Atomic Collisions*, eds D.C. Lorents, W. E. Meyerhof, and J. R. Peterson, (Elsevier, Amsterdam, 1986).
34. O. C. Mullins, Y. Zhu, E. Y. Xu, and T. F. Gallagher, *Phys. Rev. A* **32**, 2234 (1985).
35. R. R. Jones, P. Fu, and T. F. Gallagher, *Phys. Rev. A* **44**, 4260 (1991).
36. M. G. Mayer, *Phys. Rev.* **60**, 184 (1941).
37. D. C. Griffin, K. L. Andrew, and R. D. Cowan, *Phys. Rev.* **177**, 62 (1969).
38. J. P. Connerade, *J. Phys. B* **11**, L381 (1978).
39. R. R. Jones, C. J. Dai, and T. F. Gallagher, *Phys. Rev. A* **41**, 316 (1990).
40. J. G. Story and W. E. Cooke, *Phys. Rev. A* **39**, 5127 (1989).
41. X. Wang, J. G. Story, and W. E. Cooke, *Phys. Rev. A* **43**, 3535 (1991).
42. B. Carre, P.d'Oliveira, P. R. Fournier, F. Gounand, and M. Aymar, *Phys. Rev. A* **42** (6545) 1990.
43. C. N. Yang, *Phys. Rev.* **74**, 764 (1948).
44. W. Sandner, R. Kachru, K. A. Safinya, F. Gounand, W. E. Cooke, and T. F. Gallagher, *Phys. Rev. A* **27**, 1717 (1983).
45. R. Kachru, N. H. Tran, P. Pillet, and T. F. Gallagher, *Phys. Rev. A* **31**, 218 (1985).
46. Y. Zhu, E. Y. Xu, and T. F. Gallagher, *Phys. Rev. A* **36**, 3751 (1987).
47. V. Lange, U. Eichmann, and W. Sandner, *J. Phys. B* **22**, L245 (1989).
48. M. D. Lindsay, L.-T. Cai, G. W. Schinn, C.-J. Dai, and T. F. Gallagher, *Phys. Rev. A* **45**, 231 (1992).
49. J. Bokor, R. R. Freeman, and W. E. Cooke, *Phys. Rev. Lett.* **48**, 1242 (1982).

20

Quantum defect theory

In the previous chapter we considered isolated autoionizing states coupled to degenerate continua. The perturbative treatment presented there is perfectly adequate to describe isolated autoionizing states but is at best awkward for the description of an entire series of interacting autoionizing series. A straightforward way of describing these phenomena is to use quantum defect theory (QDT), a multichannel scattering approach first developed by Seaton.[1]

Depending on the choice of basis functions the QDT equations can take many forms.[2] Two choices are commonly used. The first is based on the separated ion and electron and uses reactance, or $\underline{\mathbf{R}}$, matrices.[1–7] The second is based on the normal modes of the short range electron–ion scattering.[8–10] In this chapter we briefly describe the essential ideas of QDT, following the approach of Cooke and Cromer.[3] This approach has the advantage of showing the connection between the two formulations. The essential notions presented here are adequate to understand many of the subtle features of the spectra of autoionizing states, but for a more complete description the reader is referred to the original articles.[1–10]

Quantum defect theory (QDT)

In the second chapter we described a Rydberg electron outside a spherically symmetric ground state ion core using single channel QDT. Only a single channel is necessary since the ionic core cannot exchange energy or angular momentum with the Rydberg electron. An implicit assumption of single channel QDT is that all excited states of the ion are so far removed in energy that they may be ignored.

In a Rydberg state of any atom but H, when the distance, r, of the Rydberg electron from the ion core is greater than a core radius, r_c, the potential is a coulomb potential, but for $r < r_c$ the potential is usually deeper than a coulomb potential. The effect of the deeper potential is that if an incoming coulomb wave scatters from the ion core, the reflected wave has a phase shift of $\pi\mu$ compared to what it would have if it scattered from a proton. In other words the standing wavefunction for all $r > r_c$ is given by

$$\Psi = \frac{1}{r}[f(W, \ell, r) \cos \pi\mu + g(W, \ell, r) \sin \pi\mu] \tag{20.1}$$

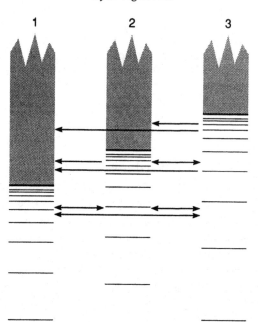

Fig. 20.1 A three channel, three limit problem. Three levels of the ion, shown by the bold lines, have Rydberg series converging to them from below and continua above them. The interactions between the series are shown by the horizontal arrows. Bound-continuum interactions are shown by single headed arrows while bound interactions are shown by double headed arrows.

where f and g are the regular and irregular coulomb wavefunctions. For scattering from a proton the g wave would be absent. The utility of QDT is that μ is nearly energy independent, a property which derives from the fact that if r_c is small, $\sim 1a_0$, the kinetic energy of the electron when it collides with the ion core at $r < r_c$ is large, ~ 10 eV and over an energy range of <1 eV the phase shift $\pi\mu$ is constant, as shown in Chapter 2.

An alternative way of expressing the wavefunction of Eq. (20.1) is

$$\Psi = \frac{\cos \pi\mu}{r} [f (W, \ell, r) + \tan (\pi\mu) g (W, \ell, r)], \qquad (20.2)$$

which resembles a scattering wavefunction.

If the ion does not have only a single spherically symmetric state it is no longer true that the energy and angular momentum of the Rydberg electron and the ion are separately conserved, and multichannel QDT must be used to account for the fact that several combinations of ion and electron energy, angular momentum, and parity may conserve the total atomic energy, angular momentum and parity, given by W, J, and Π, respectively. Consider the case shown in Fig. 20.1, in which there are three ionic states of different energies. In Fig. 20.1 we show the bound states converging to these ionic limits and the continua above them. Each set of bound Rydberg states and associated continuum, both having the same angular momentum and spin, constitutes a channel, and Fig. 20.1 depicts three such channels. In an isolated atom for the channels to be interacting they must have the same total angular momentum and parity. It is easy to describe the channels when the Rydberg electron is far from the ionic core, i.e. as $r \to \infty$. For example, as $r \to \infty$, the Ba $J = 1$ odd parity channels are well described as products of a specific Ba$^+$ ionic state and an electron in a well defined state. Examples are $6s_{1/2}\varepsilon p_{1/2}$, $6s_{1/2}\varepsilon p_{3/2}$, $5d_{3/2}\varepsilon p_{1/2}$, $5d_{3/2}\varepsilon p_{3/2}$, $6p_{1/2}ns_{1/2} \cdots$. In each case the relative orientation of the ionic, electron, and spin angular momenta is specified by the total angular momentum J. These channels are usually called the dissociation or collision channels, and we shall describe them using the wavefunctions ϕ_i. In the collision channels we can unambiguously partition the total energy into an ionic and a Rydberg electron part according to

$$W = W_i + I_i, \tag{20.3}$$

where W_i is the Rydberg electron's energy and I_i is the ionic core's energy. A channel is open if $W_i > 0$ and closed if $W_i < 0$. In these two cases the Rydberg electron is in a continuum and a bound state respectively. Far from the ionic core the logical wavefunctions with which to describe the collision channels are the ϕ_i wavefunctions given by

$$\phi_i = \frac{1}{r}[\chi_i f(W_i, \ell, r) \cos \pi v_i + \chi_i g(W_i, \ell, r) \sin \pi v_i], \tag{20.4}$$

where χ_i is the product of the total ionic wavefunction and the angular part of the Rydberg electron wavefunction, including spin. In Eq. (20.4) it is apparent that v_i specifies the fraction of the regular and irregular coulomb functions. Equivalently, πv_i is the phase shift from the regular hydrogenic f function, as in Eq. (20.1).

Although the properties of the f and g functions are outlined in chapter 2, it is worth summarizing their properties here.[8] The f and g coulomb functions are termed regular and irregular since as $r \to 0$, $f \propto r^{\ell+1}$ and $g \propto r^{-\ell}$. Due to the $r = 0$ behavior of the g function, in H only the f wave exists. As $r \to \infty$ for $W_i > 0$ the f and g waves are sine and cosine functions, and if $W_i > 0$, πv_i simply specifies the phase of the wavefunction relative to the hydrogenic f wave. If $W_i < 0$ the f and g waves both have exponentially increasing and decreasing parts, and, as we have seen in Chapter 2, only if

$$W_i = \frac{-1}{2v_i^2} \tag{20.5}$$

does the wavefunction of Eq. (20.4) vanish as $r \to \infty$. Unless Eq. (20.5) is satisfied the wavefunction diverges exponentially as $r \to \infty$ if $W_i < 0$.

Although as $r \to \infty$ the wavefunctions of different collision channels are very different, at small r they are similar, and it is possible to find normal modes of the scattering from the ionic core. Over an energy range ΔW we can find a radius R, $R > r_c$, such that for $r < R$ the wavefunctions of Eq. (20.4) are energy independent, aside from the normalization factors of the closed channels. If we disregard the $r = \infty$ boundary condition for the closed channels, ν_i can take any value, not just $\nu_i = 1/\sqrt{-2W_i}$. If we in addition apply the continuum normalization per unit energy to these wavefunctions, then in the region $r_c < r < R$ we can think of all the ϕ_i wavefunctions as energy independent scattering, or continuum, wavefunctions. Our goal is to find the normal modes of the scattering from the ionic core.

The wavefunctions of the normal scattering modes are the standing waves produced by a linear combination of incoming coulomb wavefunctions which is reflected from the ionic core with only a phase shift. The composition of the linear combination is not altered by scattering from the ionic core. These normal modes are usually called the α channels, and have wavefunctions in the region $r_c < r < R$ given by

$$\Psi_\alpha = \frac{1}{r}\left[\sum_i U_{i\alpha}\chi_i f(W_i, \ell, r) \cos \pi\mu_\alpha - \sum_i U_{i\alpha}\chi_i g(W_i, \ell, r) \sin \pi\mu_\alpha\right], \quad (20.6)$$

where $\pi\mu_\alpha$ is the phase shift analogous to μ in Eq. (20.1) and $U_{i\alpha}$ is a unitary transformation. Often μ_α is termed an eigen quantum defect. If we compare Eq. (20.1) to Eq. (20.6) we can see that in the former case a single incident incoming hydrogenic wave results in a phase shifted outgoing wave, whereas in the latter case a linear combination of incoming hydrogenic waves results in the same linear combination of outgoing waves, only phase shifted. Since the ϕ_i wavefunctions are energy independent for $r < R$, the μ_α quantum defects are also. More generally, if the wavefunctions are slowly varying with energy, the values of μ_α are also.

The normal modes with the wavefunctions Ψ_α defined by Eq. (20.6) are in essence the eigenfunctions which match the boundary condition at r_c, but they do not match the $r \to \infty$ boundary condition if there are any closed channels. In contrast the Ψ_i wavefunctions match the $r \to \infty$ boundary condition but not the $r = r_c$ boundary condition. In general for $r > r_c$ the wavefunction Ψ can be written as a linear combination of either the Ψ_i or Ψ_α wavefunctions. Explicitly

$$\Psi = \sum_i A_i\phi_i = \sum_\alpha B_\alpha\Psi_\alpha. \quad (20.7)$$

Writing ϕ_i and Ψ_α explicitly using Eqs. (20.4) and (20.6) and equating the coefficients of $\chi_i f(W_i,\ell,r)$ and $\chi_i g(W_i,\ell,r)$ yields

$$A_i \cos \pi\nu_i = \sum_\alpha U_{i\alpha}B_\alpha \cos \pi\mu_\alpha \quad (20.8a)$$

$$A_i \sin \pi\nu_i = -\sum_\alpha U_{i\alpha}B_\alpha \sin \pi\mu_\alpha \quad (20.8b)$$

Multiplying Eqs. (20.8a) and (20.8b) by sin πv_i and cos πv_i respectively and adding and subtracting them yields

$$A_i = \sum_a U_{ia} \cos \pi(v_i + \mu_a)B_a, \qquad (20.9a)$$

$$0 = \sum_a U_{ia} \sin \pi(v_i + \mu_a)B_a. \qquad (20.9b)$$

Using the fact that U_{ia} is a unitary matrix for which $\underline{U}^T = \underline{U}^{-1}$, Eqs. (20.8a) and (20.8b) can be written as

$$\sum_i U_{ia}A_i \cos \pi v_i = B_a \cos \pi \mu_a, \qquad (20.10a)$$

$$\sum_i U_{ia}A_i \sin \pi v_i = -B_a \sin \pi \mu_a. \qquad (20.10b)$$

Multiplying Eqs. (20.10a) and (20.10b) by cos $\pi\mu_a$ and -sin $\pi\mu_a$ respectively and adding and subtracting yields

$$\sum_i U_{ia} \cos \pi(v_i + \mu_a)A_i = B_a, \qquad (20.11a)$$

$$\sum_i U_{ia} \sin \pi(v_i + \mu_a)A_i = 0. \qquad (20.11b)$$

Eqs. (20.10a) and (20.10b) appear in the Fano formulation of QDT.[8] Following Cooke and Cromer,[3] we shall use Eqs. (20.11a) and (20.11b). Irrespective of which set is more convenient, nontrivial solutions can only be found if

$$\det |U_{ia} \sin \pi(v_i + \mu_a)| = 0. \qquad (20.12)$$

Eq. (20.12) defines the quantum defect surface. For bound states, all channels are closed, $W_i < 0$ for all i, and the dimensionality of the surface defined by Eq. (20.12) is one less than the number of channels. For a two channel problem Eq. (20.12) defines a line while for a three channel problem it defines a two dimensional surface. From Eq. (20.12) it is apparent that replacing v_1 by $v_1 + n$, where n is an integer, either leaves the matrix unchanged or reverses the sign of an entire row. In either case Eq. (20.12) is satisfied for the same value of v_1. The same reasoning may be applied to v_2, so that the quantum defect surface defined by Eq. (20.12) repeats modulo 1 in v_i.

Let us for a moment consider the three channel problem depicted in Fig. 20.1. We have just discussed the case in which $W_i < 0$ for all i, i.e. the region below the lowest limit. The quantum defect surface defined by Eq. (20.12) is a two dimensional surface inscribed in a cube of length $\Delta v_i = 1$ on a side. Now let us consider the region between the first and second ionization limits, where channel 1 is open. Since ϕ_1 is a continuum wave the $r \to \infty$ boundary condition does not

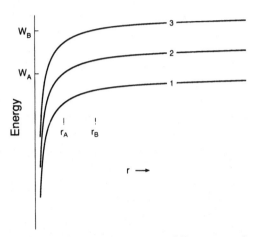

Fig. 20.2 Radial potentials due to three ionic states at different energies. At energy W_A, for $r > r_A$ only channel 1 is classically allowed and only ϕ_1 exists in this region. At energy W_B, for $r > r_B$ channels 1 and 2 are classically allowed and the wavefunction is an energy dependent linear combination of ϕ_1 and ϕ_2.

constrain the possible values of $\pi\nu_1$. However, due to the interchannel couplings at small r, only the values of $\pi\nu_1$ specified by Eq. (20.12) satisfy both the $r = r_c$ and $r \to \infty$ boundary conditions. In this case the quantum defect surface is identical to the surface below the first limit. Now however $\pi\nu_1$ is only the continuum phase shift, sometimes labeled $\pi\tau$.

It is also useful to think of the problem in scattering terms. In Fig. 20.2 we have drawn the radial potentials for channels 1, 2, and 3. At an energy W_A, above the first limit, but below the second limit for $r > r_A$ the wavefunction is composed entirely of the channel 1 wavefunction ϕ_1, given by Eq. (20.1). We can imagine putting a spherical box of radius r_A around the ionic core and asking, "What is the normal scattering mode for the scattering of an electron from the contents of the box?" Since there is only one continuum wave, the only issue is its phase shift $\pi\nu_1$, and depending upon the proximity of the energy to the energies of the bound states of channels 2 and 3 the phase shift $\pi\nu_1$, corresponding to the normal mode, has different values.

If the energy is raised above the second limit there are two open channels. In Fig. 20.2 at an energy W_B for $r > r_B$ the wavefunction is composed of a linear combination of ϕ_1 and ϕ_2. If we put a radial box of radius r_B around the ionic core we can again ask, "What are the normal modes for electron scattering from the contents of the box?" In other words, what linear combinations of incoming coulomb wavefunctions will suffer at most a phase shift when scattering from the contents of the box? There are two wavefunctions, labelled by $\rho = 1,2$. They are linear combinations of ϕ_1 and ϕ_2, given by

$$\Psi_\rho = \sum_{i=1,2} A_{i\rho}\chi_i f(W_i, \ell, r)\cos \pi\tau_\rho + \sum_{i=1,2} A_{i\rho}\chi_i g(W_i, \ell, r)\sin \pi\tau_\rho. \quad (20.13)$$

In Eq. (20.13) the eigen phase shift $\pi\tau_\rho$ plays the same role μ_α plays in Eq. (20.6). However, the values of τ_ρ are not energy independent due to the bound states of channel 3. The phase shifts τ_ρ depend on the proximity of the energy to the energies of the states of channel 3. Similarly, the composition of the normal modes of scattering from the box of radius r_B is not energy independent, and the $A_{i\rho}$ values depend strongly on the energy.

To find the values of the eigen phase shift τ_ρ we simply replace $\pi\nu_i$ by $\pi\tau$ for all the open channels in the determinant of Eq. (20.12) and solve Eq. (20.12) for the two possible values of τ, which are τ_ρ, $\rho = 1,2$. If there are P open channels there are P values of τ_ρ. While it is not transparent from the discussion up to this point that this procedure is reasonable, as we shall see, these values of τ_ρ lead to scattering and reactance matrices which are diagonal, and continuum wavefunctions which are the normal scattering modes.

When there are multiple values of τ, the dimensionality of the quantum defect surface is reduced. In the region below the second limit the quantum defect surface defined by Eq. (20.12) is a two dimensional surface. Above the second limit, for each value of ν_3 there are two values of τ_ρ, and the quantum defect surface is two lines.

Finally, if we consider the energy region above the third limit of Fig. 20.2, all three channels are open for all values of r. In this case we can put a box of any $r > r_c$ about the ionic core and inquire as to the normal modes for electron scattering from the contents of the box. We have in fact already solved this problem. The normal modes are the Ψ_α wavefunctions. Since the energy is above all the ionization limits we no longer need to ignore the $r \to \infty$ boundary conditions; they play no role.

Geometrical interpretation of the quantum defect surface

In any orbit of the Rydberg electron, most of the time is spent when the electron is far from the nucleus, where the wavefunction is most reasonably characterized in terms of A_i, the coefficients of the ϕ_i wavefunctions. Correspondingly, many of the properties depend upon the values of A_i in a very direct way. As shown by Cooke and Cromer,[3] a particularly attractive feature of QDT is that the values of A_i^2, i.e. the composition of the wavefunction in terms of the collision channels, can be determined by inspecting the quantum defect surface. If we define the cofactor matrix $C_{i\alpha}$ of the matrix of Eq. (20.12) by

$$C_{i\alpha} = \text{Cofactor} \, |U_{i\alpha} \sin \pi(\nu_i + \mu_\alpha)|, \qquad (20.14)$$

then we can rewrite the determinant of Eq. (20.12) as

$$\sum_a U_{ia} \sin \pi(\nu_i + \mu_a) C_{ia} = 0 \tag{20.15a}$$

or

$$\sum_i U_{ia} \sin \pi(\nu_i + \mu_a) C_{ia} = 0. \tag{20.15b}$$

Comparing Eqs. (20.15a) and (20.15b) to Eqs. (20.10a) and (20.11a), we see that

$$C_{ia} = G A_i B_a, \tag{20.16}$$

where G is a constant.

Consider now the function

$$f(\nu_1 \nu_2 \cdots \nu_i) = \det \left(U_{ia} \sin \pi(\nu_i + \mu_a) \right), \tag{20.17}$$

the same determinant set equal to zero in Eq. (20.12) to define the allowed values of ν_i. If we construct the gradient ∇f we find the normal to the surface of Eq. (20.12). Writing out the components,[3]

$$\frac{\partial f}{\partial \nu_j} = \frac{\partial}{\partial \nu_j} \det |U_{ia} \sin \pi(\nu_i + \mu_a)|$$

$$= \frac{\partial}{\partial \nu_j} \sum_a U_{ja} \sin \pi(\nu_j + \mu_a) C_{ja}$$

$$= \pi \sum_a U_{ja} \cos \pi(\nu_j + \mu_a) G A_i B_a$$

$$= \pi G A_j^2. \tag{20.18}$$

If the gradient defined by Eq. (20.18) is represented by a vector perpendicular to the quantum defect surface, its projections on the ν_i axes are proportional to the values of A_i^2. In a two limit problem Eq. (20.18) reduces to the simple form[8]

$$\frac{A_2^2}{A_1^2} = \frac{-\partial \nu_1}{\partial \nu_2}. \tag{20.19}$$

Normalization

The scattering f and g wavefunctions we have used are normalized per unit energy, and now we consider how to normalize the wavefunctions based on them in different energy regions. First we consider the bound states. We require a bound state wavefunction to satisfy

$$\int \Psi^* \Psi \, d^3 r = 1. \tag{20.20}$$

To do this we must change the ϕ_i wavefunctions from normalization per unit energy to normalization per state, i.e. Eq. (20.20). Using the derivative $dW_i/dv_i = 1/v_i^3$ we may convert the squared wavefunctions $|\phi_i|^2$ from energy to state normalization by multiplying by $1/v_i^3$. Equivalently, a bound ϕ_i wavefunction which is normalized per unit energy has a normalization integral of v_1^3. Since the wavefunction $\Psi = \Sigma A_i \phi_i$ is composed of bound wavefunctions normalized per unit energy, its normalization integral, N^2, is given by

$$N^2 = \sum A_i^2 v_i^3, \tag{20.21}$$

in which the higher v_i states carry proportionally more weight than do the lower v_i states. As pointed out by Cooke and Cromer,[3] this weighting arises from the fact that the A_i values reflect the fraction of orbits the Rydberg electron spends in the various Ψ_i wavefunctions of the state Ψ, and the factor v_i^3 reflects the time duration of an orbit of effective quantum number v_i. Using the normalization integral of Eq. (21), the properly normalized bound wavefunction is given by

$$\Psi = \frac{1}{N} \sum_i A_i \Psi_i, \tag{20.22}$$

in which the Ψ_i wavefunctions are still normalized per unit energy. If we wished to express Eq. (20.22) in terms of conventional bound wavefunctions normalized per state, Ψ_i^B, using $\Psi_i = v_1^{3/2} \Psi_i^B$, Eq. (20.21) becomes

$$\Psi = \frac{1}{N} \sum A_i v_i^{3/2} \Psi_i^B. \tag{20.23}$$

If one or more of the channels is open, the wavefunction is a continuum wavefunction, since it extends to $r = \infty$, and it must be normalized per unit energy. Each of the Ψ_i continuum wavefunctions is separately normalized per unit energy, so we simply require for each ρ solution

$$\sum_i A_{i\rho}^2 = 1, \tag{20.24}$$

where the sum extends over the open channels.

Using the normalization relations of Eqs. (20.21) and (20.24) and the geometric relation between the A_i values we are able to construct properly normalized wavefunctions at any energy.

Energy constraints

As pointed out early in this chapter, πv_i is really the phase of the i channel wavefunction as $r \to \infty$. For each bound channel, we have already introduced the constraint $W_i = -1/2v_i^2$ which sets the phase πv_i for any energy W_i. If we consider, for example, the region below the first ionization limit of Fig. 20.1, all three channels are closed, and the states have discrete energies and discrete values of v_1, v_2, and v_3. Yet the quantum defect surface, defined by Eq. (20.12), is continuous over v_1, v_2, and v_3. More values of the phase shifts πv_i satisfy Eq. (20.12) than are physically observed. What constrains the continuous variation in v_i allowed by Eq. (20.12) is the energy relation

$$W = W_i + I_i, \tag{20.25}$$

which ties together the v_i values of all closed channels. Below the lowest ionization limit of Fig. 20.1, Eq. (20.25) is given by

$$W = \frac{-1}{2v_1^2} + I_1 = \frac{-1}{2v_2^2} + I_2 = \frac{-1}{2v_3^2} + I_3, \tag{20.26}$$

a line, and its intersection with the quantum defect surface defined by Eq. (20.12) yields a set of points corresponding to the bound atomic states. Between the first and second limits only channels 2 and 3 are closed, and Eq. (20.25) becomes

$$W = \frac{-1}{2v_2^2} + I_2 = \frac{-1}{2v_3^2} + I_3, \tag{20.27}$$

which defines a surface independent of v_1. The intersection of this surface with the quantum defect surface defines a line, and in this case all values of v_2 and v_3 are sampled, although not all pairs v_2 and v_3. The freedom of the continuum phase to take any value at any energy allows all values of v_2 and v_3 to be sampled.

Alternative $\underline{R}$ matrix form of QDT

The use of QDT which has drawn our attention to it is to represent autoionizing states. Consider the simplest case of a series of autoionizing states degenerate with a single continuum, a two channel QDT problem. If we use the ICE method to observe the autoionizing states we observe their positions and widths, which may be characterized by two parameters, a quantum defect δ and a scaled width $n^3\Gamma$. The relation of these quantities to the $U_{i\alpha}$ and μ_α parameters of QDT is not obvious. In fact it is not even unique. In the two channel problem we are considering there are two measured parameters, but there are three parameters required to specify the two channel QDT, μ_1, μ_2, and a rotation angle θ to specify $U_{i\alpha}$. The continuum phase is a superfluous piece of information for the interpretation of an ICE experiment. If in our two channel problem the autoionizing states are channel 2 and the continuum is channel 1, the wavefunction is given by

$$\Psi = A_1\phi_1 + A_2\phi_2, \tag{20.28}$$

where $A_1 = 1$. The spectral density of the autoionizing state is given by A_2^2, which according to Eq. (20.19) is given by the derivative dv_1/dv_2 since $A_1^2 = 1$. The position and widths of the autoionizing states of channel 2 are determined by the derivative of the continuum phase but do not depend at all upon the absolute continuum phase. Thus if we used a different set of $U_{i\alpha}$ and μ_α parameters with the same value of $d\nu_1/\partial\nu_2$ at each value of ν_2, these parameters would represent the series of autoionizing states equally well, if only the positions and widths are observed.

To arrive at a more unique way of characterizing autoionizing states in the absence of information about the continuum phase we can recast the QDT equations into an **R** matrix form which is at the same time similar to the original development of QDT from multichannel scattering theory.[2] The relation between the different forms of scattering matrix is discussed by Mott and Massey,[11] Seaton,[2] and Fano and Rau.[12]

If we rewrite Eq. (20.12) using $\sin(A + B) = \sin A \cos B + \cos A \sin B$, it is given by

$$\cos(\pi\mu_\alpha) \sum_i [U_{i\alpha} \tan \pi\nu_i + U_{i\alpha} \tan \pi\mu_\alpha] \cos(\pi\nu_i)A_i = 0. \qquad (20.29)$$

Since $\cos \pi\mu_\alpha$ multiplies the entire left-hand side of Eq. (20.29) we can ignore it and write the remaining equation as a matrix equation. Explicitly

$$[\mathbf{U}^\mathrm{T} \tan \rho\nu + \tan \rho\mu \, \mathbf{U}^\mathrm{T}] \cos \rho\nu \, \mathbf{A} = 0 \qquad (20.30)$$

where $\tan \pi\nu$, $\tan \pi\mu$, and $\cos \pi\nu$ are diagonal matrices, and **A** is a vector. If we define the matrix **R** by

$$\mathbf{R} = \mathbf{U} \tan \mathbf{U}^\mathrm{T}, \qquad (20.31)$$

and the vector **a** by

$$a_i = \cos \pi\nu_i A_i, \qquad (20.32)$$

then Eq. (20.30) can be rewritten as

$$(\mathbf{R} + \tan \pi\nu)\mathbf{a} = 0. \qquad (20.33)$$

The **R** matrix, or reactance matrix, is real and symmetric. If there is no coupling between the i channels, off diagonal elements of the **R** matrix vanish and the diagonal elements are given by $\tan\pi\delta_i$, where δ_i is the quantum defect of the ith channel. To a first approximation the interchannel couplings are given by the off diagonal elements of the **R** matrix. In the form of Eq. (20.33) the **R** matrix both accounts for the quantum defects of isolated channels and the interchannel couplings. As pointed out by Cooke and Cromer,[3] these are two rather different functions. The quantum defect reflects only the phase shift of the nonhydrogenic coulomb wave, and is a spherical, single channel effect. The energy, angular momentum, and parity of the ionic core and Rydberg electron are thus separately conserved. The interchannel couplings, on the other hand, destroy the separate conservation of the above three quantities. We may remove the quantum defects of the i channels from the **R** matrix by using phase shifted coulomb waves as basis functions instead of the regular and irregular f and g waves. We express the collision channel wavefunctions of Eq. (20.3) as[2-4]

$$\phi_i = \frac{1}{r} [\chi_i f'(W_i, \ell, r) \cos \pi \nu_i + \chi_i g'(W_i, \ell, r)], \tag{20.34}$$

where f' and g' are phase shifted coulomb functions given by

$$f'(W_i, \ell, r) = f(W_i, \ell, r) \cos \pi \delta_i - g(W_i, \ell, r) \sin \pi \delta_i \tag{20.35a}$$

and

$$g'(W_i, \ell, r) = f(W_i, \ell, r) \sin \pi \delta_i + g(W_i, \ell, r) \cos \pi \delta_i \tag{20.35b}$$

in which

$$\nu_i' = \nu_i + \delta_i . \tag{20.36}$$

If we define

$$a_i' = A_i \cos \pi \nu_{i'} \tag{20.37}$$

we can derive from Eq. (20.29) an expression analogous to Eq. (20.33). Specifically,

$$(\underline{\mathbf{R}}' + \tan \pi \nu_i')\mathbf{a}' = 0, \tag{20.38}$$

where

$$\underline{\mathbf{R}}' = [\cos \pi \delta + \mathbf{R} \sin \pi \delta]^{-1} [\mathbf{R} \cos \pi \delta - \sin \pi \delta]. \tag{20.39}$$

If there are no interchannel couplings, $\underline{\mathbf{R}}' = 0$, corresponding to the fact that the i channels have quantum defects of δ_i and occur when $\nu_i' = 0$. In general the $\underline{\mathbf{R}}'$ matrix is a real symmetric matrix with zeroes on the diagonal.

Eqs. (20.33) and (20.38) are precisely analogous to Eq. (20.12), and we can find the quantum defect surface from Eqs. (20.33) and (20.38), for example, by setting

$$\det [\mathbf{R} + \tan \pi \nu] = 0 \tag{20.40a}$$

or

$$\det [\underline{\mathbf{R}}' + \tan \pi \nu'] = 0. \tag{20.40b}$$

Another useful property of the $\mathbf{R}$ matrix formulation is that we can write it in blocks corresponding to the bound (closed) and continuum (open) channels. Explicitly, we can write Eq. (20.33) as[3]

$$\begin{bmatrix} \underline{\mathbf{R}}_{bb} + \tan \pi \nu_b & \underline{\mathbf{R}}_{bc} \\ \underline{\mathbf{R}}_{cb} & \underline{\mathbf{R}}_{cc} + \tan \pi \tau \end{bmatrix} \begin{matrix} \mathbf{a}_b \\ \mathbf{a}_c \end{matrix} = 0. \tag{20.41}$$

Eq. (20.38) may be written in exactly the same form. In Eq. (20.41) we have replaced all the continuum phases by $\pi \tau$ to reflect the fact that we are searching for the normal scattering modes which satisfy Eq. (20.40a). Following Cooke and Cromer[3] we can write Eq. (20.41) in separate bound and continuum matrix equations. Explicitly

$$-[\underline{\mathbf{R}} + \tan \pi\nu]_{bb}^{-1}\mathbf{R}_{bc}\mathbf{a}_c = \mathbf{a}_b \qquad (20.42a)$$

and

$$[\mathbf{R}_{cb}[\mathbf{R} + \tan \pi\nu]_{bb}^{-1}\mathbf{R}_{bc} - \mathbf{R}_{cc}]\mathbf{a}_c = [\tan \pi\tau_\rho]\mathbf{a}_c \qquad (20.42b)$$

where τ_ρ is one of the allowed values of τ. From Eq. (20.42b) it is apparent that the eigenvector $\mathbf{a}_c$, i.e. that linear combination of i channels, which satisfies Eq. (20.42b) has a diagonal continuum–continuum $\mathbf{R}$ matrix given by the left-hand side of Eq. (20.42b). In other words, it is the normal scattering mode. When we solved Eq. (20.12) with some open channels we simply replaced $\pi\nu$ by $\pi\tau$ for all the open channels. Eq. (20.42b) justifies this procedure.

The role of QDT

QDT provides a framework which relates a few energy independent parameters to a wealth of spectroscopic data. It is used both as an efficient way to parametrize data and as a way of comparing theoretical results to experimental data. Which of the several parametrizations to use is usually unimportant for comparing theoretical results to experimental observations, since all the parametrizations are equivalent. On the other hand, if a set of data is to be represented by QDT parameters, it is useful to use the set of parameters which allows the experimental data to be fit with the minimum number of free parameters. For example, to fit ICE data, the phase shifted $\mathbf{R}$ matrix approach is by far the most convenient, for the absolute continuum phase does not enter.

The QDT parameters can also be generated theoretically, by *ab initio*[13] or semiempirical methods.[14-18] The $\mathbf{R}$ matrix approach is a very successful example of the latter. In the application of the $\mathbf{R}$ matrix method to Sr, for example, a model spherical potential is constructed which correctly reproduces the energies of Sr^+. The Schroedinger equation for both valence electrons is then solved using this radial potential. The Schroedinger equation is not solved over all space but only in a spherical box of radius r_0 using a truncated basis set of trial wavefunctions. The size of the box is larger than the wavefunction of the highest energy state of Sr^+ considered in the calculation. This restriction sets an upper limit on the energies of the states which can be treated in this way, but makes solving the coupled equations a fairly straightforward matter. The restriction also ensures that the wavefunction at the edges of the box is a one electron wavefunction. The Schroedinger equation is solved for the normal scattering modes, and the logarithmic derivative of the solutions at the box is related to the scattering phase shift $\pi\mu_\alpha$. The results can be expressed as the phase shifts and the matrix which connects the wavefunctions of the normal modes to, for example, the LS coupled basis, or an $\underline{\mathbf{R}}$ matrix can be specified for the LS coupled basis.

References

1. M. J. Seaton, *Proc. Phys. Soc. London* **88**, 801 (1966).
2. M. J. Seaton, *Rep. Prog. Phys.* **46**, 167 (1983).
3. W. E. Cooke and C. L. Cromer, *Phys. Rev. A* **32**, 2725 (1985).
4. A. Giusti, *J. Phys. B* **13**, 3867 (1980).
5. A. Giusti-Suzor and U. Fano, *J. Phys. B* **17**, 215 (1984).
6. K. Ueda, *Phys. Rev. A* **35**, 2484 (1987).
7. J. P. Connerade, A. M. Lane, and M. A. Baig, *J. Phys. B* **18**, 3507 (1985).
8. U. Fano, *Phys. Rev. A* **2**, 353 (1970).
9. K. T. Lu and U. Fano, *Phys. Rev. A* **2**, 81 (1970).
10. C. M. Lee and K. T. Lu, *Phys. Rev. A* **8**, 1241 (1975).
11. N. F. Mott and H. S. W. Massey, *The Theory of Atomic Collisions* (Oxford University Press, New York, 1965).
12. U. Fano and A. R. P. Rau, *Atomic Collisions and Spectra* (Academic Press, Orlando, 1986).
13. W. R. Johnson, K. T. Cheng, K. N. Huang, and M. LeDourneuf, *Phys. Rev. A* **22**, 989 (1980).
14. C. H. Greene and L. Kim, *Phys. Rev. A* **36**, 2706 (1987).
15. C. H. Greene, *Phys. Rev. A* **28**, 2209 (1983).
16. M. Aymar, E. Lue-Koenig, and S. Watanabe, *J. Phys. B* **20**, 4325 (1987).
17. M. Aymar and J. M. Lecompte, *J. Phys. B* **22**, 223 (1989).
18. C. H. Greene and Ch. Jungen, in *Advances in Atomic and Molecular Physics*, Vol. 21, eds. D. Bates and B. Bederson (Academic Press, Orlando, 1985).

21

Optical spectra of autoionizing Rydberg states

QDT enables us to characterize series of autoionizing states in a consistent way and to describe how they are manifested in optical spectra. We shall first consider the simple case of a single channel of autoionizing states degenerate with a continuum. Of particular interest is the relation of the spectral density of the autoionizing states to how they are manifested in optical spectra from the ground state and from bound Rydberg states using isolated core excitation. We then consider the case in which there are two interacting series of autoionizing states, converging to two different limits, coupled to the same continuum.

First we consider the two channel problem shown in Fig. 21.1. Our present interest is in the region above limit 1, i.e. the autoionizing states of channel 2. Later we shall consider the similarity of the interactions above and below the limit. A typical quantum defect surface obtained from Eq. (20.12) or (20.40) for all energies below the second limit is shown in Fig. 21.2. The surface of Fig. 21.2 may be obtained with either of two sets of parameters, $\delta_1 = 0.56$, $\delta_2 = 0.53$, and $R'_{12} = 0.305$, $R'_{11} = R'_{22} = 0$ or $\mu_1 = 0.4$, $\mu_2 = 0.6$, and $U_{11} = U_{22} = \cos\theta$ and $U_{12} = -U_{21} = \sin\theta$, with $\theta = 0.6$ rad.[1,2] To conform to the usual convention, in Fig. 21.2 the ν_i axis is inverted. The wavefunction is given in terms of the collision channels by

$$\Psi = A_1 \phi_1 + A_2 \phi_2. \tag{21.1}$$

Between limit 1 and limit 2, since ϕ_1 is open, $A_1^2 = 1$. If we choose $A_1 = 1$, it is straight forward to show that

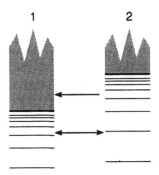

Fig. 21.1 Two atomic channels associated with two states of the ion (ionization limits), shown by bold lines. Above the first limit and below the second the nominally bound states of channel 2 autoionize into the continuum of channel 1.

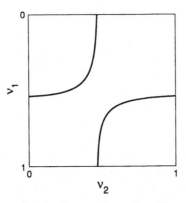

Fig. 21.2 Quantum defect surface for the two channel problem showing ν_1, or equivalently τ, the phase shift in the open channel, channel 1, divided by $-\pi$. The ν_1 axis is reversed to conform to convention.

$$A_2^2 = -R_{12}'\sqrt{\frac{1 + \tan^2 \pi\nu_2'}{(R_{12}')^4 + \tan^2 \pi\nu_2'}}.\tag{21.2}$$

The probability of finding an autoionizing state of channel 2 at any energy is given by A_2^2, which we shall term the spectral density. Squaring Eq. (21.2) we find

$$A_2^2 = (R_{12}')^2\left[\frac{1 + \tan^2 \pi\nu_2'}{(R_{12}')^4 + \tan^2 \pi\nu_2'}\right].\tag{21.3}$$

From Eq. (21.3) several points are apparent. First A_2^2 is periodic in ν_2'. Second the maximum value, $A_2^2 = 1/(R_{12}')^2$, occurs for $\nu_2' = 0 \pmod 1$, and the minimum value $A_2^2 = (R_{12}')^2$, occurs for $\nu_2' = 0.5 \pmod 1$. For $|R_{12}'| \ll 1$ Eq. (21.3) can be approximated by a Lorentzian, i.e.[1]

$$A_2^2 = (R_{12}')^2 \frac{1}{(R_{12}')^4 + (\pi\nu_2')^2}\tag{21.4}$$

from which it is apparent that the half maximum points of A_2^2 occur at

$$\nu_2' = \pm\frac{(R_{12}')^2}{\pi}.\tag{21.5}$$

Thus the FWHM is given by $2(R_{12}')^2/\pi$ or in terms of the energy width[1]

$$\Gamma = \frac{2(R_{12}')^2}{\pi\nu_2^3}.\tag{21.6}$$

Eq. (21.6) shows both the expected ν^{-3} scaling of the autoionization rates and the utility of the parametrization in terms of δ_1, δ_2, and R_{12}'.

It is useful to recall the geometric interpretation of the quantum defect surface of Fig. 21.2. As we have already discussed, we can draw the normal to the curve of Fig. 21.2, and its projections on the ν_1 and ν_2 axes give the relative values of A_1^2 and A_2^2. In Fig. 21.2 the normal points nearly vertically, i.e. in the ν_1 direction

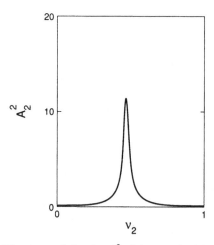

Fig. 21.3 The spectral density A_2^2 of the autoionizing state vs v_2.

except at $v_2 = 0.4$, where it has a substantial horizontal, or v_2, component as well. In other words, only at $v_2 \approx 0.4$ is there an appreciable value of A_2^2, and here is where the autoionizing state of channel two is located. One might ask why we bothered with the normal to the quantum defect surface, since for the two channel case Eq. (20.19) shows that A_2^2/A_1^2 is given simply by $-\partial v_1/\partial v_2$. The reason is that the normal can be generalized to three dimensions. In any case, we plot A_2^2 in Fig. 21.3, and it is clear that all the information contained in Eq. (21.3) is in the A_2^2 curve.

The autoionizing states of channel 2 are represented by Fig. 21.3, however they do not necessarily appear as in Fig. 21.3 in a photoionization spectrum. Let us first consider photoexciting the autoionizing states of channel 2 and the degenerate continuum of channel 1 from a compact initial state, g, such as the ground state. Since the initial state is spatially localized near the ionic core, only the part of the Rydberg wavefunction near the core plays an active role in the excitation. We can write the dipole matrix element for the excitation in either of two ways,[1-3]

$$\langle \Psi_g | \mu | \Psi \rangle = \sum_\alpha B_\alpha d_\alpha \tag{21.7a}$$

or

$$\langle \Psi_g | \mu | \Psi \rangle = \sum_i A_i d_i \cos \pi(v_i + \phi_i). \tag{21.7b}$$

In Eqs. (21.7) d_α and d_i are energy independent dipole matrix element constants defined by

$$d_\alpha = \langle \Psi_g | \mu | \Psi_\alpha \rangle \tag{21.8a}$$

and

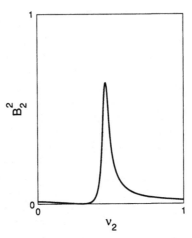

Fig. 21.4 The value of the squared α channel coefficient B_2^2. Apart from a factor of ω it is proportional to the photoionization cross section, if $d_{\alpha=1} = 0$ as assumed.

$$d_i \cos \pi(\nu_i + \phi_i) = \langle \Psi_g | \mu | \phi_i \rangle. \tag{21.8b}$$

Using the α channels takes advantage of the fact that near the origin the phase of the α channels is energy invariant. Furthermore, it is often the case that only one of the d_α is nonzero. For example, if the α channels are LS coupled and the initial state g is a singlet state, d_α for the triplet α channel vanishes. Writing the excitation matrix element in terms of the i channels, we must account for the phase variation of the ν_i channels with energy. Although such excitations can be described using the i channels, most often they are described using the α channels, because it is simpler to do so. In either case, the photoionization cross section is given by

$$\sigma = \frac{4\pi^2 \omega^2}{c} |\langle \Psi_g | \mu | \Psi \rangle|^2. \tag{21.9}$$

As a typical example, we compute $\langle \Psi_g | \mu | \Psi \rangle$ assuming that $d_\alpha = 0$ for $\alpha = 1$ and $d_\alpha = 1$ for $\alpha = 2$. Using Eq. (21.2) we calculate A_2, assuming $A_1 = 1$, and using Eq. (20.11a) we calculate B_2 (we do not need B_1 since $d_{\alpha=1} = 0$). In Fig. 21.4 we show the squared dipole moment, B_2^2, which exhibits the asymmetric Beutler–Fano lineshape characteristic of autoionizing states.[4] These lineshapes are often characterized by the Fano q parameter which is the ratio of the matrix elements per unit energy connecting the initial state to the discrete autoionizing state and to the continuum. When $q > 0$ the asymmetry is as shown in Fig. 21.4, and when $q < 0$ it is reversed; the zero in the cross section is on the high energy side of the autoionizing state. When $|q| \gg 1$ the continuum excitation is negligible and the Beutler–Fano profile becomes a symmetric Lorentzian, and when $q = 0$ the Beutler–Fano profile is a symmetric Lorentzian dip in the cross section. When $|q| \sim 1$ the lineshape is most asymmetric.

If we again consider Fig. 21.4, we can see that the cross section vanishes at $v_2 = 0.32$ and that the profile does not match the spectral density, A_2^2, of the autoionizing state. The Beutler–Fano profile of Fig. 21.4 is periodic in v_2 with period 1, so the spectrum from the ground state consists of a series of Beutler–Fano profiles. At higher values of v_2 the profiles become compressed in energy since $dW/dv_2 = 1/v_2^3$. Fig. 19.2 shows two regular series of Beutler–Fano profiles between the Ba^+ $6p_{1/2}$ and $6p_{3/2}$ limits. In this case the absorption never vanishes because there is more than one continuum.

Now let us consider the ICE of the autoionizing states of channel 2. For example, imagine that we are able to start from a bound Ba $6sn\ell$ Rydberg state and drive the transition to the autoionizing $6pn\ell$ state of channel 2. Channel 1 is the degenerate continuum. For the moment we ignore the spins of the electrons. The 6s electron makes the dipole transition to the 6p state and the outer electron is projected onto the autoionizing $n\ell$ state. For the reasons given in Chapter 19 we can ignore the amplitude to the continuum. For concreteness, we define

$$\phi_1 = \text{continuum},$$
$$\phi_2 = 6pn\ell, \qquad (21.10)$$
$$\Psi_b = 6sn\ell_b.$$

The wavefunction Ψ of the autoionizing state is given by Eq. (21.1), and to calculate the photoionization cross section we need the dipole matrix element $\langle \Psi_b | \mu | \Psi \rangle = \langle \Psi_b | \mu | A_2 \phi_2 \rangle$. We can write $\langle \Psi_b |$ and $| \phi_2 \rangle$ as product wave functions;

$$\langle \Psi_b | = \langle 6s | \langle v_b \ell^B | \langle \Omega_b | \qquad (21.11)$$

and

$$| \phi_2 \rangle = | \Omega_2 \rangle | 6p \rangle | v_2^B \ell_b \rangle, \qquad (21.12)$$

where $\langle 6s |$ and $| 6p \rangle$ are the ionic wavefunctions, $\langle v_b \ell^B |$ and $| v_2 \ell \rangle$ are radial wavefunctions for states of effective quantum numbers v_b and v_2, and $\langle \Omega_b |$ and $| \Omega_2 \rangle$ are angular wavefunctions. The superscript B in $| v_b \ell^B \rangle$ reflects the fact that it is normalized per unit state, while $| v_2 \ell \rangle$ is normalized per unit energy. Substituting the expressions of Eq. (21.12) into Eq. (21.11) and computing the squared matrix element leads to

$$\langle \Psi_b | \mu | \Psi \rangle |^2 = |\langle 6s | \mu | 6p \rangle|^2 \delta_{\ell \ell_b} A_2^2 |\langle v_b \ell_b^B | v_2 \ell \rangle|^2. \qquad (21.13)$$

In Eq. (21.13) $\langle 6s | \mu | 6p \rangle$ is just the Ba^+ 6s–6p dipole moment, and $\delta_{\ell \ell_b}$ is the Kronecker delta function. Since $\ell_b = \ell$ we shall simply write ℓ in place of ℓ_b.

All the energy dependence in the cross section arises from the factors A_2^2 and $|\langle v_b \ell^B | v_2 \ell \rangle|^2$. These are, respectively, the spectral density of the channel 2 autoionizing states, and the overlap integral between the bound and continuum $n\ell$ states with effective quantum numbers v_b and v_2. We have already seen that A_2^2 is simply given by $-\partial v_1 / \partial v_2$, the derivative of the quantum defect surface, and repeats modulo 1 in v_2. The overlap integral is given is closed form by[5]

$$\langle \nu_{\mathrm{b}} \ell^{\mathrm{B}} | \nu_2 \ell \rangle = \frac{\sin[\pi(\nu_{\mathrm{b}} - \nu_2)] 2(\nu_{\mathrm{b}} \nu_2)^{1/2} \nu_2^{3/2}}{\pi(\nu_{\mathrm{b}} - \nu_2)(\nu_{\mathrm{b}} + \nu_2)}. \tag{21.14}$$

This expression is valid for any ℓ state, as long as $\ell \ll n$. The factor $\nu_2^{3/2}$ accounts for the fact that $|\nu_2 \ell \rangle$ is normalized per unit energy. The overlap integral $\langle \nu_{\mathrm{b}} \ell^{\mathrm{B}} | \nu_2 \ell^{\mathrm{B}} \rangle$ between two bound state wavefunctions would not have the factor $\nu_2^{3/2}$ and would be equal to one for $\nu_2 = \nu_{\mathrm{b}}$ and zero for ν_2 different from ν_{b} by an integer, as expected. We can write the cross section as

$$\sigma = \frac{4\pi^2 \omega}{c} A_2^2 |\langle \nu_{\mathrm{b}} d^{\mathrm{B}} | \nu_2 d \rangle|^2. \tag{21.15}$$

In Fig. 21.5 we plot A_2^2, $|\langle \nu_{\mathrm{b}} d^{\mathrm{B}} | \nu_2 d \rangle|^2$, and the resulting cross section for the case in which the initial state has an effective quantum number $\nu_{\mathrm{b}} = 12.35$ and the autoionizing states of channel 2 are located at $\nu_2 = 0.28$ (mod 1) and have fractional widths $\nu_2^3 \Gamma = 0.1$.[6] In this case the central lobe of the overlap integral coincides with the autoionizing state peaked at $\nu_2 = 12.28$, and there is one strong peak in the cross section, at $\nu_2 = 12.28$. There are also much weaker subsidiary peaks but for the moment we ignore them, our present interest being to verify that we can calculate the basic features of the ICE spectra. Since the width of the central lobe of the overlap integral is much larger than A_2^2, the width of the cross section is effectively equal to A_2^2, as asserted in the discussion of ICE in Chapter 19.

It is interesting to consider what the spectrum would look like if the autoionizing states of the same width were located at $\nu_2 = 0.85$(mod 1) instead of 0.25(mod 1), i.e. if the quantum defects of the bound and autoionizing states differed by 1/2. In this case the spectrum is given by Fig. 21.6.[6] The central lobe of the overlap integral contains now the two autoionizing states at $\nu_2 = \nu_{\mathrm{b}} \pm 1/2$, so in this case we expect two strong peaks in the ICE spectrum with noticeable subsidiary shake up, $\Delta \nu \neq 0$, peaks. In this case the peaks of the cross section do not match the variation in A_2^2 so well since they occur on the sloping sides of the central lobe of the overlap integral. The severity of the distortion depends upon how wide the autoionizing states are. It is most often the case that the quantum defects of the bound and autoionizing states do not differ by more than 0.2, in which case the overlap integral is constant across A_2^2 and the ICE spectrum reflects A_2^2 accurately.

The shake up satellites

If we examine Eq. (21.14) and Fig. 21.5 it is apparent that apart from $\nu_2 = \nu_{\mathrm{b}}$, the cross section has zeroes when $\nu_2 = \nu_{\mathrm{b}}$(mod 1). Furthermore, the combination of these zeroes and the periodic variations in A_2^2 lead to satellite structure which is

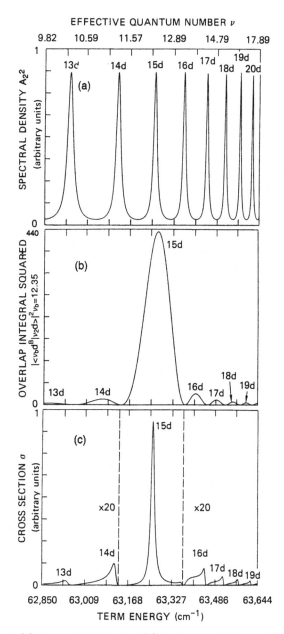

Fig. 21.5 Plots of (a) the spectral density, (b) the square of the overlap integral between initial and final states, and (c) their product which is proportional to the optical cross section σ. All are calculated for the third laser excitation from the 6s15d ($\nu_b = 12.35$) state to the $(6p_{3/2}nd)_{J=3}$ channel (from ref. 6).

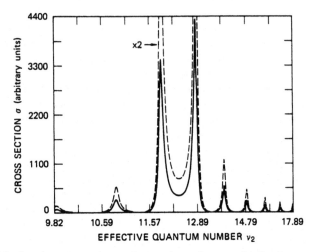

Fig. 21.6 Calculated cross section σ from a bound state of effective quantum number $\nu_b = 12.35$ to an autoionizing Rydberg series with width $\Gamma = 0.11\nu^{-3}$ and quantum defect $\delta_2 = 0.15$ so that the autoionizing states are found at $\nu_2 = 0.85 \pmod 1$ (from ref. 6).

very sensitive to the difference between δ_b and δ_2, as shown by Figs. 21.5 and 21.6. In particular it is very easy to see small changes in $\delta_2 - \delta_b$, if they are almost equal. Although the structure away from $\nu_2 = \nu_b$ is weak, especially in Fig. 21.5, it is possible to observe it with high laser power and compare it to the spectra calculated using Eq. (21.13). The high power saturates the center of the spectrum where $\nu_2 \approx \nu_b$, and to account for the saturation the ion signal must be expressed as[7,8]

$$I = I_0(1 - e^{-\sigma\Phi}) \tag{21.16}$$

where Φ is the time integrated photon flux per unit area.

Tran *et al.*[9] have excited Ba atoms in a thermal beam to the 6s15d $^1D_2$ state with two circularly polarized lasers. They then scanned the wavelength of a third circularly polarized laser in the vicinity of the Ba$^+$ 6s$_{1/2} \to$ 6p$_{3/2}$ transition and detected the ions resulting from the excitation of the 6p$_{3/2}$nd $J = 3$ states. Their observed 6s15d $^1D_2$–6p$_{3/2}$nd $J = 3$ spectrum is shown in Fig. 21.7. The two sharp peaks at 21850 and 22170 cm^{-1} are two photon resonances and should be disregarded. From the 6snd $^1D_2$ states only a single 6p$_{3/2}$nd $J = 3$ series is excited, so for all practical purposes this is a two channel problem, a single series of autoionizing states and the degenerate continuum. The presence of many continua is irrelevant as long as the excitation is insensitive to the continuum phase. In Fig. 21.7 the broken curve is calculated using Eqs. (21.15) and (21.16) with $\sigma\Phi \sim$ 100 at the peak of the cross section, and the agreement between the experimental and theoretical spectra is excellent. It is also interesting to note that the zeroes are still evident even though the signal at the peak cross section is saturated by a factor of 100.

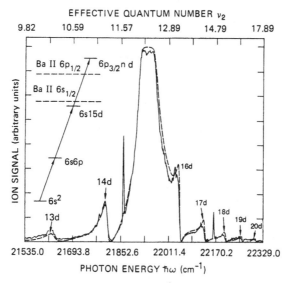

EFFECTIVE QUANTUM NUMBER ν_2

9.82   10.59   11.57   12.89   14.79   17.89

Ba II $6p_{1/2}$ / $6p_{3/2}$ n d

Ba II $6s_{1/2}$

6s15d

6s6p

$6s^2$

13d

14d

16d

17d

18d

19d

20d

ION SIGNAL (arbitrary units)

21535.0   21693.8   21852.6   22011.4   22170.2   22329.0

PHOTON ENERGY $\hbar\omega$ (cm^{-1})

Fig. 21.7 Photoexcitation spectrum of the $6s15d\,^1D_2 \rightarrow (6p_{3/2}nd)$ $J = 3$ transition in Ba as a function of the frequency of the third laser. All three lasers are circularly polarized in the same sense. The broken line is the spectrum calculated from Eqs. (21.15) and (21.16). The energy level inset is not to scale (from ref. 9).

We have so far considered the excitation to two channel systems, whereas in reality most ICE experiments have been done with autoionizing states converging to the lowest p states of alkaline earth ions, as shown schematically in Fig. 21.7. We now wish to consider exciting the Ba $6p_jnd$ state of high n and following the cross section across the $6p_j$ limit. To describe this cross section requires that we calculate the overlap integral to the continuum as well as to bound autoionizing states. In the explicit expression for $\langle\nu_b d^B|\nu_2 d\rangle$ of Eq. (21.14) we can see that the integral is zero when $\nu_2 = \nu_b \pmod 1$, by orthogonality, and maximum for $\nu_2 = \nu_b + 1/2 \pmod 1$. The same reasoning can be extended to the continuum. If the phase $\pi\nu_2$ is such that $\nu_2 = \nu_b$ the overlap integral vanishes by orthogonality, and if $\nu_2 = \nu_b + 1/2 \pmod 1$ it is a maximum. In fact, for any continuum phase $\pi\nu_2$ the overlap integral is the extrapolation of the bound $\langle\nu_b\ell^B|\nu_2\ell\rangle$ integral for the same value of $\nu_2 \pmod 1$. In Fig. 21.8 we show overlap integrals from the bound Ba $6s12d\,^1D_2$ and $6s23d\,^1D_2$ states to both the $6p_{1/2}nd$ and $6p_{3/2}nd$ channels.[6] The overlap integrals in the continuum are shown for the phases $\pi(\delta_b \pm 1/2)$ which give the maximum absolute values of the overlap integral.

The phase dependent continuum excitation has been observed by Tran *et al.*[6] and more clearly by Story and Cooke.[10] They excited the Ba $6s19d\,^1D_2$ states to the region of the $6p_{1/2}$ limit and observed the fluorescence from the excited $6p_{1/2}$ state of Ba$^+$. The resulting spectrum is shown in Fig. 21.9. Using this technique they do not detect excitation below the $6p_{1/2}$ limit at all efficiently, but are able to see the variations in the cross section above the $6p_{1/2}$ limit with remarkable clarity.

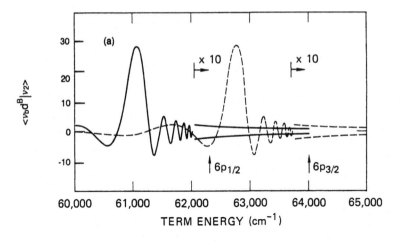

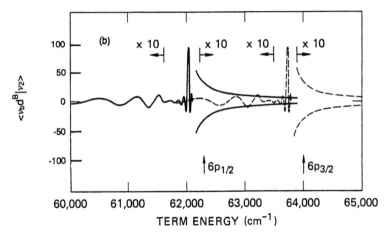

Fig. 21.8 Overlap integrals $\langle \nu_b d^B | \nu_2 d \rangle$ from (a) the Ba 6s12d state ($\nu_b = 9.36$) and (b) the 6s23d state ($\nu_b = 20.29$). In (a) the integrals from 6s12d to the $6p_{1/2}nd$ channel (——) and $6p_{3/2}nd$ channel (- - -) are multiplied by 10 at 62,000 and 63,800 cm^{-1} as indicated. Note that the integrals extend smoothly across the $6p_{1/2}$ and $6p_{3/2}$ limits. Note also that the overlap integrals peak at ~61,000 and ~62,800 cm^{-1} near the $6p_{1/2}12d$ and $6p_{3/2}12d$ states. The maximum value is $(9.36)^{3/2}$, not 1, because of the continuum normalization. In (b) the integrals from 6s23d to the $6p_{1/2}nd$ channel (——) and $6p_{3/2}nd$ channel (- - -) are each multiplied by 10 in the energy ranges far from their peak values at 62,000 and 63,800 cm^{-1}, respectively. In the continua above the $6p_{1/2}$ and $6p_{3/2}$ limits the maximum positive and negative values of the overlap integral are plotted (from ref. 6).

Since the features due to the $6p_{3/2}nd$ states decrease in intensity with increasing n, it is evident that most of the excitation is due to the $6p_{1/2}\varepsilon d$ continuum, not to the $6p_{3/2}nd$ states.

As shown by Fig. 21.8, in Ba the overlap integrals for the $6p_{1/2}nd$ and $6p_{3/2}nd$ channels overlap, admitting the possibility of interference in the excitation amplitudes to the $6p_{1/2}nd$ and $6p_{3/2}nd$ channels. However the fine structure

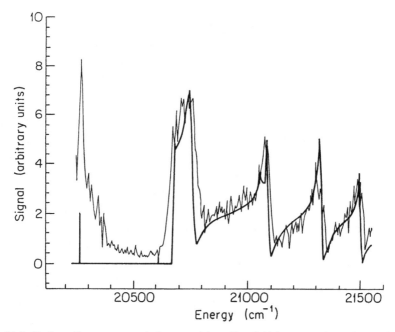

Fig. 21.9 Shake-off spectrum of the transition, Ba $6s19d \rightarrow$ Ba $6p_{1/2}\varepsilon d$, obtained by detecting the fluorescence from the excited Ba^+ $6p_{1/2}$ ions. The resonances are Ba $6p_{3/2}nd$ states with $11 \leq n \leq 14$ (from ref. 10).

splitting of the Ba $6p_j$ states is large enough that the interference is not unambiguously detectable. Below the $6p_{1/2}$ limit it is difficult to see small interference effects due to the $6p_{3/2}nd$ excitation amplitude in the presence of the enormous effects of the interseries interaction. Above the $6p_{1/2}$ limit the excitation amplitude to the $6p_{1/2}\varepsilon d$ continuum appears much the same as the excitation of the $6p_{3/2}nd$ states as shown in Fig. 21.9. The situation in Sr is very much the same. However, in the lighter alkaline earth atoms Ca and Mg the effect of interference in the two excitation amplitudes is very apparent and must be properly taken into account to reproduce the observed spectra.[11,12]

The presence of the overlap integral in the ICE cross section appears at first glance to be at best a minor inconvenience, certainly not something which is useful. However, the presence of the overlap integral variation has several uses. Sandner *et al.*[13] have used the overlap integral to determine the effective quantum number of the initial bound Ba $6s24d$ $^1D_2$ state to one part in 10^4. Combining the implied binding energy with the term energy of the initial Rydberg state provides a novel way of measuring the ionization limit. The overlap integral also provides a way of measuring small admixtures of different states. For example, Kachru *et al.* have determined that there is a 2% admixture of the $6p_{3/2}nd_j$ state in the $6p_{3/2}ns_{1/2}$ state and vice versa.[14] This admixture is so small that it would normally be quite difficult to detect, especially in an autoionizing state. However, the overlap integral variation allows it to be observed clearly.

Interacting autoionizing series

We would now like to consider the more complex case of two series of autoionizing states interacting with each other and with a degenerate continuum. The Ba $6pn\ell$ states just below the Ba$^+$ $6p_{1/2}$ limit constitute a good example. As shown by Fig. 19.2, in the vuv spectrum from the ground state, just below the $6p_{1/2}$ limit the spectrum has structure due to the series converging to both the $6p_{1/2}$ and $6p_{3/2}$ limits. How much of the structure is due to the interseries interaction and how much to Beutler–Fano interference profiles is not clear. Unfortunately, there is no purely experimental way to tell. In contrast, the ICE method allows us to differentiate between interference in the excitation amplitudes and the effects of interseries interactions. As a first example we would like to consider the Ba $6pnd$ $J = 3$ series just below the Ba$^+$ $6p_{1/2}$ limit, the region shown in Fig. 21.10. We shall treat this problem as a three channel quantum defect theory problem with the three channels

$$\left.\begin{aligned}
\phi_1 &= J = 3 \text{ continuum,} \\
\phi_2 &= 6p_{1/2}nd_{5/2} \quad J = 3, \\
\phi_3 &= 6p_{3/2}nd_j \qquad J = 3.
\end{aligned}\right\} \tag{21.17}$$

In Fig. 21.10 we show the $6pnd$ states as being degenerate with only a single continuum, above the $6s_{1/2}$ limit.[15] This simplifying assumption allows us to treat the problem with only three channels. It is important to recall that in recording an ICE spectrum we do not directly excite the continuum, it acts only as a sink for electrons. As a result, the fact that the continuum is not well characterized is not important.

Above the $6p_{1/2}$ limit the Ba $6p_{3/2}nd$ $J = 3$ states accessible from the bound $6snd$ $^1D_2$ states and the linear combination of continua into which they autoionize can be treated as a two channel problem. Below the $6p_{1/2}$ limit we must use the three channels of Eq. (21.17). In the region below the Ba$^+$ $6p_{1/2}$ limit the $J = 3$ wavefunction is given by

$$\Psi = A_1\phi_1 + A_2\phi_2 + A_3\phi_3, \tag{21.18}$$

where $A_1 = 1$. An interesting point to note about Fig. 21.10 is that the $6p_{3/2}10d$ state is degenerate with the $6p_{1/2}nd_{5/2}$ states of $n \sim 20$. If we observe the spectrum from the $6s20d$ $^1D_2$ state to the $6p_{1/2}20d_{5/2}$ state we observe a single Lorentzian line, as we expect for ICE. If, on the other hand, we examine the $6s10d \rightarrow 6p_{3/2}10d_j$ spectrum we find the spectrum of Fig. 21.11, a broad envelope containing sharp structure.[16] The structure evidently comes from the $6p_{1/2}nd_{5/2}$ states.

At first glance it seems evident that the observed spectrum of Fig. 21.11 is a sequence of Beutler–Fano interference profiles, which reverse in the Fano q parameter across the line. Although the solution may ultimately be expressed in the same mathematical form,[1] a simple consideration of the excitation amplitudes

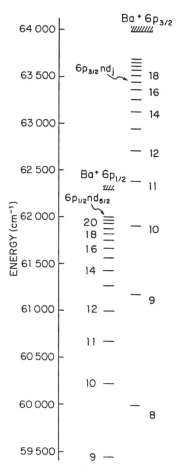

Fig. 21.10 The $6pnd$ $J = 3$ levels of Ba converging to the $6p_{1/2}$ and $6p_{3/2}$ limits. Note that the $6p_{3/2}$ $10d_j$ state is degenerate with the $6p_{1/2}20d_{5/2}$ state. The $6p_{3/2}11d_j$ and higher states are above the $6p_{1/2}$ limit. All the $6pnd$ states lie in the continuum above the Ba^+ $6s_{1/2}$ state (from ref. 15).

indicates that the structure is not due to interfering amplitudes. There are three energetically possible outcomes of the $6s10d$ state absorbing a photon; transitions to the $6p_{3/2}10d_j$, $6p_{1/2}20d_{5/2}$, and $6s_{1/2}\varepsilon f$ states.[15] If we represent the $6s10d$ $^1D_2$ state by the wavefunction Ψ_b and use the fact that it has an effective quantum number $\nu_b = 7.3$, we can write the dipole matrix elements from the initial $6s10d$ $^1D_2$ state to the three i channels of Eq. (21.17) as

$$\begin{aligned}
\langle\Psi_b|\mu|\phi_1\rangle &\cong \langle 6s|6s\rangle\langle\nu_b^B|\mu|\varepsilon f\rangle, \\
\langle\Psi_b|\mu|\phi_2\rangle &= \langle 6s|r|6p\rangle\Omega_2 A_2\langle\nu_b d^B|\nu_2 d\rangle, \\
\langle\Psi_b|\mu|\phi_3\rangle &= \langle 6s|r|6p\rangle\Omega_3 A_3\langle\nu_b d^B|\nu_3 d\rangle
\end{aligned} \qquad (21.19)$$

where $\langle 6s|r|6p\rangle$ is the Ba^+ radial matrix element, and Ω_2 and Ω_3 are angular factors of order one which account for the angular momenta and spins of the two

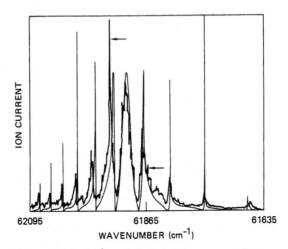

ION CURRENT

62095   61865   61635

WAVENUMBER (cm⁻¹)

Fig. 21.11 The experimental $6s10d\ ^{1}D_{2}-6p_{3/2}10d_{j}$ ICE spectrum is shown by the bold line
(——). The structure produced by interaction with the $6p_{1/2}nd_{5/2}$ states is quite evident. A
three channel, QDT fit is also shown by the light line (——) (from ref. 16).

electrons. In Eq. (21.19) the continuum excitation is vanishingly small, due to the
very different spatial variations of the ν_{b}^{B} and εf wavefunctions. The overlap
integral $\langle \nu_{b}d^{B}\ |\ \nu_{3}d\rangle$ is an overlap integral between two 10d wavefunctions, both
having $\nu \simeq 7.3$, and is equal to $\nu_{3}^{3/2}$ while the overlap integral $\langle \nu_{b}d^{B}\ |\ \nu_{2}d\rangle$ is an
overlap integral between 10d and 20d wavefunctions, and is thus much smaller
than the channel 3 overlap integral $\langle \nu_{b}d^{B}\ |\ \nu_{3}d\rangle$. In other words, only the amplitude
to the $6p_{3/2}10d_{j}$ state is important, and to a good approximation

$$
\left.\begin{aligned}
\sigma &= \frac{4\pi^{2}\omega}{c}\int \Psi_{b}|\mu|\Psi\rangle^{2} \\
&= \frac{4\pi^{2}\omega}{c}|\langle \Psi_{b}|\mu|A_{3}\phi_{3}\rangle|^{2} \\
&= \frac{4\pi^{2}\omega}{c}|\langle 6s\ |er|6p\rangle|\Omega_{3}^{2}A_{3}^{2}|\langle \nu_{b}d^{B}|\nu_{3}d\rangle|^{2}
\end{aligned}\right\} \qquad (21.20)
$$

In the last expression of Eq. (21.20) the squared overlap integral is approximately
constant and equal to ν_{3}^{3} for $\nu_{3} \approx \nu_{b}$, and $\langle 6s|r|6p\rangle$ and Ω_{3} are constant as well, so
the cross section is directly proportional to A_{3}^{2}. Knowing that the structure of Fig.
21.11 has no contribution from interference in the excitation amplitudes vastly
simplifies the task of interpreting the spectrum. In fact, knowing that all the
information appearing in the spectrum is in A_{3}^{2}, we know that it must be found in
the quantum defect surface.

In Fig. 21.12 we show the quantum defect surface for the three channel 6pnd
$J = 3$ problem.[16] This surface, inscribed in a cube, is valid for all energies below
the $6p_{1/2}$ limit. If there were no interchannel interactions the surface of Fig. 21.12
would consist of three intersecting planes at $\nu_{1} = 0.2$, $\nu_{2} = 0.2$, and $\nu_{3} = 0.3$. The

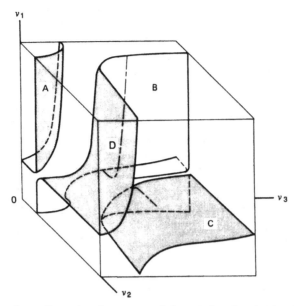

Fig. 21.12 The three dimensional, quantum defect surface for the 6pnd states below the 6p$_{1/2}$ limit. Here v_2 and v_3 are the effective quantum numbers relative to the 6p$_{1/2}$ and 6p$_{3/2}$ limits, respectively. πv_1 is the continuum phase. The direction of the normal to the surface at any point indicates the amounts of 6p$_{1/2}$$nd_{5/2}$, 6p$_{3/2}$$nd_j$, and continuum character in the wavefunction (from ref. 16).

interactions between the pairs of channels lead to the avoided crossings between the planes. For example, the avoided crossings between the $v_1 = 0.3$ and $v_2 = 0.2$ planes occurs because of the coupling between channels 1 and 2.

As we have described in two dimensions, constructing the normal to the surface at any point allows us determine the relative values of A_i^2 at that point; the projections of the normal in the i directions are proportional to A_i^2. For example, in the region around the letter C the normal points in the v_1 direction, so the wavefunction is predominantly channel 1, the continuum. More precisely, $A_2 = 1$, $A_2^2 \ll 1$ and $A_3^2 \ll 1$. At letters A and B the normal points in the v_2 direction so the wavefunction is predominantly channel 2, i.e. $A_2 = 1$, $A_2^2 \gg 1$, and $A_3^2 \ll A_2^2$. At letter D the normal points in the v_3 direction, and $A_2 = 1$, $A_2^2 \ll 1$, and $A_3^2 \gg 1$. In other words, simply inspecting the quantum defect surface of Fig. 21.12 allows us to determine visually the values of A_i^2 at any point on the surface.

Now we need to determine which points of the surface of Fig. 21.12 correspond to the energy scan of the spectrum of Fig. 21.11. The path along the quantum defect surface is determined by the energy constraint

$$I_2 - \frac{1}{2v_2^2} = I_3 - \frac{1}{2v_3^2}. \tag{21.21}$$

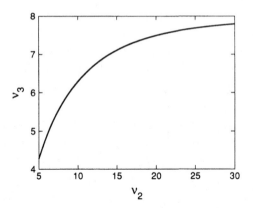

Fig. 21.13 A plot of ν_3 vs ν_2. As $\nu_3 \to 8.058$ $\nu_2 \to \infty$.

For the Ba$^+$ 6p$_j$ states $I_3 - I_2 = 1690$ cm^{-1}, or 7.70 x 10^{-3} au. Expressing ν_3 as a function of ν_2, Eq. (21.21) becomes

$$\nu_3 = \sqrt{2(I_3 - I_2) + \frac{1}{\nu_2^2}}, \tag{21.22}$$

Eq. (21.22) defines a ϕ independent two dimensional surface. In Fig. 21.13 we plot ν_3 as a function of ν_2, i.e. the projection of the surface onto the $\phi = 0$ plane, which shows that $\nu_3 \to 8.058$ as $\nu_2 \to \infty$, and that near the 6p$_{1/2}$ limit ν_2 increases much more rapidly than does ν_3. We can plot Fig. 21.13 modulo 1 in both ν_2 and ν_3 over the range of ν_2 and ν_3 which corresponds to the energy scan of Fig. 21.11. Over the energy range of Fig. 21.11 ν_2 varies from 12.89 to 23.33, and ν_3 varies from 6.82 to 7.61, and in Fig. 21.14 we show ν_3(mod 1) vs ν_2(mod 1) over this range. In each branch the curve reaches the right-hand side at $\nu_2 = 1$ and some value of ν_3. The intersection of the surface defined by Eq. (21.22) and shown in projection in Fig. 21.14 with the surface of Fig. 21.12 gives the path along the quantum defect surface corresponding to the spectrum of Fig. 21.11. The next higher branch of the curve starts at $\nu_2 = 0$ at the same value of ν_3. Equivalently, the projection of the path along the quantum defect surface on the $\nu_2\nu_3$ plane as ν_3 varies from 6.82 to 7.61, and ν_2 from 12.89 to 23.33, is shown by Fig. 21.14.

Before we see how the spectrum of Fig. 21.11 emerges from Fig. 21.12, let us try a simpler case, the 6s16d ^{1}D$_2 \to$ 6p$_{1/2}$16d$_{5/2}$ ICE transition, which exhibits a Lorentzian peak at $\nu_2 = 13.3$.[16] Since the peak falls at the center of the $\langle \nu_b d^B | \nu_2 d \rangle$ overlap integral the cross section is proportional to A_2^2, which should thus have a maximum at $\nu_2 = 13.3$. Inspecting Fig. 21.14 we can see that the 6p$_{1/2}$16d$_{5/2}$ state lies on the last full branch of the ν_2, ν_3 curve, from $\nu_3 = 6.85$ to $\nu_3 = 6.98$ and $\nu_2 = 13$ to $\nu_2 = 14$. This branch of the curve of Fig. 21.14 is almost parallel to the ν_2 axis and lies at $\nu_3 \sim 6.9$ where the normal of the quantum defect surface points in the ν_1 direction except at $\nu_2 \sim 13.3$ where it points predominantly in the ν_2 direction. In other words the normal to the quantum defect surface also tells us that A_2^2 is

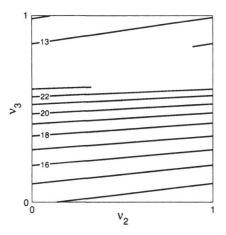

Fig. 21.14 A plot of v_3(mod 1) vs v_2(mod 1) over the range $6.82 \leq v_3 < 7.62$ and $12.89 < v_2 < 23.33$, corresponding to the energy scan of Fig. 21.12.

peaked at $v_2 = 13.3$, the location of the $6p_{1/2}16d_{5/2}$ state. The projection of the normal on the v_3 axis is minimal, so A_3^2 is very small.

Now we return to the $6p_{3/2}10d$ spectrum of Fig. 21.11. It extends from $v_3 = 6.82$ to $v_3 = 7.61$ and correspondingly $v_2 = 12.89$–23.33. From Fig. 21.14 we can see that the range $v_3 = 6.8$ to $v_3 = 7.6$ corresponds to ten branches of the $v_1 v_3$ curve of Fig. 21.14. Correspondingly, we follow the quantum defect surface above these branches in the v_2, v_3 plane. A_3^2 is the ratio of the projections of the normal in the v_3 and v_1 directions. Inspecting Fig. 21.12, it is apparent that for $v_3 \approx 7$ there is almost no component in the v_3 direction. As v_3 increases to 7.2 the v_3 component of the normal increases uniformly except when $v_2 \sim 0.2$(mod 1), the locations of the $6p_{1/2}nd_{5/2}$ states, and where the three planes of the quantum defect surface intersect, when it drops sharply in favor of the normal in the v_2 direction. In other words, A_3^2 has a broad maximum with holes from the $6p_{1/2}nd_{5/2}$ states. This description is in qualitative agreement with the observed spectrum of Fig. 21.11. In Fig. 21.11 the light line shows the spectrum calculated using the three channels of Eq. (21.17). Specifically, A_3^2 is shown. It was obtained from the expression of Eq. (20.16) for A_i with the requirement that $A_2^2 = 1$.

Using the phase shifted **R** matrix approach Cooke and Cromer[1] have derived an algebraic expression for the calculated spectrum of Fig. 21.11. In problems analogous to this one Connerade[17] has treated the broad state analogous to the $6p_{3/2}10d$ state as a continuum of finite bandwidth into which a series of states analogous to the $6p_{1/2}nd_{5/2}$ states autoionize, and Cooke and Cromer[1] have shown that the QDT expressions reduce to this form.

The three channel treatment leading to the calculated curve of Fig. 21.11 is too simple to reproduce the experimental spectrum with great accuracy. First, having only one continuum forces the theoretical spectrum to have zeroes which are not present in the experimental spectrum. Second, we have ignored one of the Ba

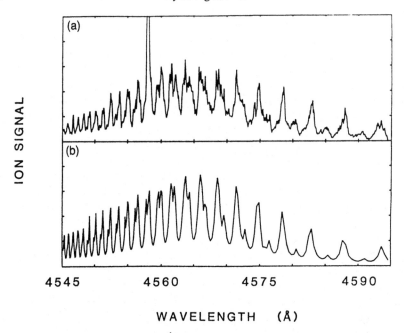

ION SIGNAL

WAVELENGTH   (Å)

Fig. 21.15 (a) Observed Ba 6s12s $^1S_0 \rightarrow$ 6p$_{3/2}$12s$_{1/2}$ spectrum showing the structure produced by the interacting 6p$_{1/2}$ns$_{1/2}$ and 6p$_{1/2}$nd$_{3/2}$ states. (b) Calculated spectrum using a six channel, QDT model (from ref. 1).

6p$_{3/2}$nd $J = 3$ series. While it is not visibly excited from the 6snd $^1D_2$ states, it is slightly coupled to the 6p$_{1/2}$nd$_{5/2}$ states, and thus cannot properly be ignored in an analysis of the spectrum. A more realistic model does a much better job of representing the observed spectra. An elegant example is the ICE spectrum from the bound 6s12s $^1S_0$ state to the 6pns $J = 1$ states just below the 6p$_{1/2}$ limit, the region containing the 6p$_{3/2}$12s$_{1/2}$ $J = 1$ state and 6p$_{1/2}$ns$_{1/2}$ and 6p$_{1/2}$nd$_{3/2}$ states of n ≈ 25. The experimental spectrum is shown in Fig. 21.15.[1] Two points are worth noting. First, this spectrum corresponds to the range 61,996 cm^{-1} to 62,236 cm^{-1} in Fig. 19.2, i.e. the uninterpretable region just below the Ba$^+$ 6p$_{1/2}$ limit. Second, applying the arguments used in connection with Eq. (21.20), we can see that the spectrum of Fig. 21.15 reflects A$_i^2$ of the 6p$_{3/2}$ns$_{1/2}$ $J = 1$ channel. In Fig. 21.15 we also show the calculated spectrum based on a six channel QDT model, the 6p$_{1/2}$ns$_{1/2}$, 6p$_{1/2}$nd$_{3/2}$, and 6p$_{3/2}$ns$_{1/2}$ $J = 1$ closed channels and three continua. As shown, the agreement is superb.

Thus far our discussion of the interacting series has focused on the spectrum of the lower ν state converging to the higher 6p$_{3/2}$ limit. There are, however, several interesting aspects to the 6p$_{1/2}$ series as well. At first glance the 6sns $\rightarrow$ 6p$_{1/2}$ns$_{1/2}$ spectra are unremarkable; they are simple Lorentzian peaks. However, if we return to our original consideration of interacting bound series, we would expect significant energy shifts of the 6p$_{1/2}$ns$_{1/2}$ states. In fact this is not the case, as is shown by Fig. 21.16, a plot of the quantum defects of the 6p$_{1/2}$ns$_{1/2}$ states, which

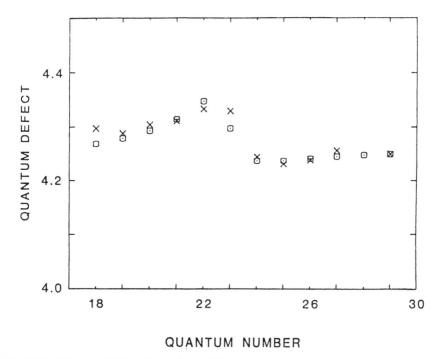

QUANTUM NUMBER

Fig. 21.16 Observed (X) and calculated (□) quantum defects of the Ba $6p_{1/2}ns_{1/2}$ states, which are degenerate with the $6p_{3/2}12s_{1/2}$ state. Note that there is little variation of the quantum defects, unlike what is observed for interacting bound series. The calculated values are from the same model used to produce Fig. 21.15(b) (from ref. 1).

show only a small variation as the $6p_{3/2}12s_{1/2}$ state is crossed. In a bound series this would produce one vertical cycle of the plot, as in Fig. 21.2, not the curve of Fig. 21.16.[1] As is shown by the excellent agreement between experiment and theory in Fig. 21.16, QDT predicts this phenomenon. As pointed out by Cooke and Cromer,[1] the lack of energy displacements can be understood in a simple way. A $6p_{1/2}ns_{1/2}$ state has a second order repulsion from the $6p_{3/2}12s_{1/2}$ state, which is distributed in energy. Thus the contributions from the parts of the $6p_{3/2}12s_{1/2}$ state above and below the $6p_{1/2}ns_{1/2}$ state are opposite in sign. The resulting cancellation of the energy shifts leads to the muted dispersion curve of Fig. 21.16.

A second interesting aspect of interacting series is shown in Fig. 21.17, a plot of the scaled or reduced widths, i.e. $\nu^3\Gamma$, of the $6p_{1/2}ns_{1/2}$ states.[1] They show a pronounced variation across the $6p_{3/2}12s_{1/2}$ state. Because of the coupling of the $6p_{1/2}ns_{1/2}$ and $6p_{3/2}12s_{1/2}$ states there are two autoionization amplitudes coupling the nominal $6p_{1/2}ns_{1/2}$ states to a continuum. Schematically, these are

$$\left.\begin{array}{l} 6p_{1/2}ns_{1/2} \rightarrow J = 1 \text{ continuum,} \\ 6p_{1/2}ns_{1/2} \rightarrow 6p_{3/2}ns_{1/2} \rightarrow J = 1 \text{ continuum.} \end{array}\right\} \quad (21.23)$$

Since the couplings are to the same continuum, they interfere. As a result of this, if there is only one continuum, at some point it is possible to have complete

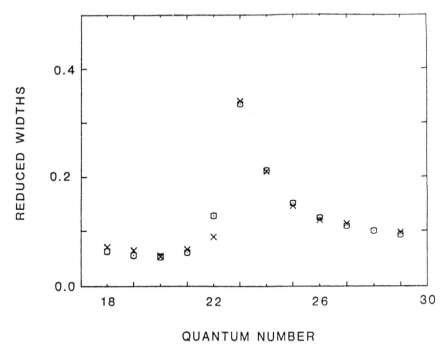

Fig. 21.17 Observed (**X**) and calculated (□) reduced, or scaled, widths of the Ba $6p_{1/2}\,ns_{1/2}$ states degenerate with $6p_{3/2}12s_{1/2}$ state. Note the enhancement at $n = 21$, which originates in the constructive interference of the two autoionization paths (from Ref. 1).

destructive interference, resulting in a vanishing autoionization rate. This phenomenon, first proposed by Cooke and Cromer,[1] is referred to as inhibited autoionization,[18] stabilization,[19] and bound states in the continuum.[20] In zero field it is unlikely that the energies will be such that complete cancellation of the autoionization rate will occur. Nonetheless, van Woerkom *et al.*[21] have observed a 190 ns lifetime of the Ba $5d_{3/2}\,26d_{5/2}\,J = 0$ state, which represents a decrease of five orders of magnitude in the autoionization rate relative to that expected for an unperturbed $5d_{3/2}\,nd_{5/2}\,J = 0$ series.

Comparisons between experimental spectra and **R** matrix calculations

In this chapter we have shown that complicated spectra can be understood using QDT. We have considered examples of fitting spectra to QDT. Now we would like to compare the results of **R** matrix calculations of spectra to the experimental spectra in cases in which fitting would have been hopeless or nearly so.

The first spectra which were synthesized using the **R** matrix approach were, not surprisingly, vuv spectra, and synthetic spectra of Mg, Ca, Sr, and Ba have all been calculated.[22–25] In Fig. 19.2 we show the Ba vuv spectrum. While the region above the Ba$^+$ $6p_{1/2}$ limit is understandable, the region below the limit exhibits

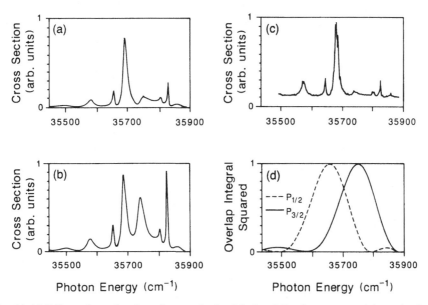

Fig. 21.18 Effect of overlap interference in the Mg 3pnd $J = 3$ spectrum: (a) synthesized spectrum including interference, (b) same as (a) but using only direct-excitation terms, (c) measured spectrum, and (d) corresponding overlap integral squared (from ref. 12).

clear evidence of interseries interaction, but defied further analysis until the R matrix calculation of Aymar was able to reproduce it almost perfectly.[25]

The **R** matrix method has also been applied with great success to ICE spectra.[12,26–29] An impressive example occurs in the spectra of the Mg 3pnd $J = 3$ states.[12,27] In Mg, because the Mg$^+$ 3p$_{1/2}$ and 3p$_{3/2}$ limits are so close together (92 cm^{-1}) the overlap integrals to the channels converging to these two limits overlap substantially for $n = 10$ to $n = 20$ Rydberg states. As a consequence, nonzero excitations to channels converging to both limits are the rule, and the spectra are almost impossible to fit using approaches which work well for Sr and Ba. However, the **R** matrix method works very well, even in this awkward case. In Fig. 21.18 we show the spectrum from the Mg 3s12d ^{1}D$_2$ state to the Mg 3pnd $J = 3$ states.[12] The experimental spectrum of Fig. 21.18 is obtained by exciting Mg atoms in a beam to the 3s12d ^{1}D$_2$ state with two lasers, and scanning the wavelength of a third laser across the region of the ionic 3s$_{1/2}$ → 3p$_j$ transitions while detecting the ions which result from the excitation of the autoionizing 3pnd states. With all three lasers circularly polarized in the same sense only $J = 3$ final states are produced. Inspecting Fig. 21.18, it is not obvious how to describe the experimental spectrum in terms of states converging to the 3p$_{1/2}$ and 3p$_{3/2}$ limits. Part of the problem derives from the fact that these states are not particularly good jj coupled states, and part from the fact that the overlap integrals to both ionic channels overlap, as shown in Fig. 21.18. It is not a coincidence that these problems occur together. The theoretical spectra are computed using a **R** matrix

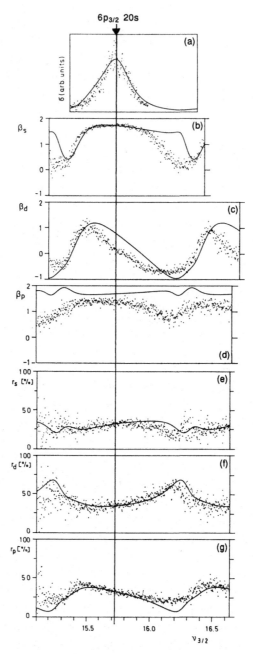

Fig. 21.19 The spectrum Ba $6p_{3/2}20s_{1/2}$ $J = 1$ state observed by ICE: (a) cross section σ; (b)–(d) Anisotropy parameters β for electron ejection to the continua above the 6s, 5d, and $6p_{1/2}$ limits. (b) $6s_{1/2}$ limit, β_s; (c) $5d_j$ limits, β_d; and (d) $6p_{1/2}$ limit, β_p; (e)–(g) Branching ratios r for ejection of electrons to the continua above the (e) $6s_{1/2}$ limit, r_s; (f) 5d limits, r_d; and (g) $6p_{1/2}$ limit, r_p. The vertical line is drawn through the maximum in the photoionization cross section. (from ref. 29).

computed by Greene,[27] both with and without the interference in the excitation amplitudes taken into account. (The $\underline{K}$ matrix used in ref. 12 is identical to the $\underline{R}$ matrix defined in Chapter 20. If the spectrum is computed without taking into account the interference in the excitation amplitudes, a spurious feature at 35,700 cm^{-1} appears, which is conspicuous by its absence in the experimental spectrum. When the interference in the excitation amplitudes is included the calculated and experimental spectra are nearly identical. In fact, Fig. 21.18 is typical of how well the $\underline{R}$ matrix method reproduces the ICE spectra of Mg and shows how the $\underline{R}$ matrix method allows the interpretation of spectra which would be hard to unscramble otherwise.

The Mg problem of Fig. 21.18 is not intrinsically complex; there are only two continua. In Sr$^+$ and Ba$^+$, on the other hand, there are low lying d states, and the analogous Sr $5pn\ell$ and Ba $6pn\ell$ states are degenerate with approximately ten continua. In spite of the increased complexity, the $\underline{R}$ matrix method still performs well.[28,29] An illustration of how well is afforded by the comparison of Lange *et al.*[29] between experimental observations of the ICE spectrum of the Ba $6sns$ $^1S_0 \rightarrow$ $6p_{3/2}20s_{1/2}$ $J=1$ transition and their counterparts calculated by $\underline{R}$ matrix techniques. As shown by Fig. 21.19, not only does the total photoionization cross section agree, but also the angular distributions and branching ratios of the electrons to their possible final states.

References

1. W.E. Cooke and C.L. Cromer, *Phys. Rev. A* **32**, 2725 (1985).
2. U. Fano, *Phys. Rev. A* **2**, 353 (1970).
3. C.J. Dai, S.M. Jaffe, and T.F. Gallagher, *J. Opt. Soc. Am. B* **6**, 1486 (1989).
4. U. Fano, *Phys. Rev.* **124**, 1866 (1961).
5. S.A. Bhatti, C.L. Cromer, and W.E. Cooke, *Phys. Rev. A* **24**, 161 (1981).
6. N.H. Tran, P. Pillet, R. Kachru, and T.F. Gallagher, *Phys. Rev. A* **29**, 2640 (1984).
7. W.E. Cooke, S.A. Bhatti, and C.L. Cromer, *Opt. Lett.* **7**, 69 (1982).
8. S.A. Bhatti and W.E. Cooke, *Phys. Rev. A* **28**, 756 (1983).
9. N.H. Tran, R. Kachru, and T.F. Gallagher, *Phys. Rev. A* **26**, 3016 (1982).
10. J.G. Story and W.E. Cooke, *Phys. Rev. A* **39**, 4610 (1989).
11. V. Lange, U. Eichmann, and W. Sandner, *J. Phys. B* **22**, L245 (1989).
12. C.J. Dai, G.W. Schinn, and T.F. Gallagher, *Phys. Rev. A* **42**, 223 (1990).
13. W. Sandner, G.A. Ruff, V. Lange, and U. Eichmann, *Phys. Rev. A* **32**, 3794 (1985).
14. R. Kachru, H. B. van Linden van den Heuvell, and T.F. Gallagher, *Phys. Rev. A* **31**, 700 (1985).
15. T.F. Gallagher, *J. Opt. Soc. Am. B* **4**, 794 (1987).
16. F. Gounand, T.F. Gallagher, W. Sandner, K.A. Safinya, and R. Kachru, *Phys. Rev. A* **27**, 1925 (1983).
17. J.P. Connerade, *Proc. Roy. Soc. (London)* **362**, 361 (1978).
18. J. Neukammer, H. Rinneberg, G. Jonsson, W.E. Cooke, H. Hieronymus, A. Konig, K. Vietzke, and H. Springer-Bolk, *Phys. Rev. Lett.* **55**, 1979 (1985).
19. S. Feneuille, S. Liberman, E. Luc-Koenig, J. Pinard, and A. Taleb, *J. Phys. B* **15**, 1205 (1985).
20. H. Friedrich and D. Wintgen, *Phys. Rev. A* **32**, 3231 (1985).
21. L.D. van Woerkom, J.G. Story, and W.E. Cooke, *Phys. Rev. A* **34**, 3457 (1986).
22. P.F. O'Mahony and C.H. Greene, *Phys. Rev. A* **31**, 250 (1985).

23. C.H. Greene and L. Kim, *Phys. Rev. A* **36**, 2706 (1987).
24. M. Aymar, *J. Phys. B* **20**, 6507 (1987).
25. M Aymar, *J. Phys. B* **23**, 2697 (1990).
26. V. Lange, U. Eichmann, and W. Sandner, *J. Phys. B* **22**, L 245 (1989).
27. C.H. Greene, private communication.
28. M. Aymar and J. M. Lecompte, *J. Phys. B.* **22**, 223 (1989).
29. V. Lange, M. Aymar, U. Eichmann, and W. Sandner, *J. Phys. B.* **24**, 91 (1991).

22

Interseries interaction in bound states

Perturbed Rydberg series

One of the most intensively studied manifestations of channel interaction in the bound states is the perturbation of the regularity of the Rydberg series, which is evident if one simply measures the energies of the atomic states. Although measurements of Rydberg energy levels by classical absorption spectroscopy show the perturbations in the series which are optically accessible from the ground state, the tunable laser has made it possible to study series which are not connected to the ground state by electric dipole transitions as well. One of the approaches which has been used widely is that used by Armstrong et al.[1] As shown in Fig. 22.1, a heat pipe oven contains Ba vapor at a pressure of ~1 Torr. Three pulsed tunable dye laser beams pass through the oven. Two are fixed in frequency, to excite the Ba atoms from the ground $6s^2\,^1S_0$ state to the $^3P_1$ state and then to the $6s7s\,^3S_1$ state. The third laser is scanned in frequency over the $6s7s\,^3S_1 \rightarrow 6snp$ transitions. The Ba atoms excited to the $6snp$ states are ionized either by collisional ionization or by the absorption of another photon. The ions produced migrate towards a negatively biased electrode inside the heat pipe. The electrode has a space charge cloud of electrons near it which limits the emission current. When an ion drifts into the space charge region it locally neutralizes the space charge, allowing many electrons to leave the region around the negative electrode, and this current pulse is detected.[2] In other words the device is a space charge limited diode. The experiments have not been confined to the resonant excitation scheme described above; off resonant multiphoton excitation has been used extensively.

Camus et al. have used an optogalvanic cell in which a glow discharge is maintained in a mixture of He and Ba.[3] In the discharge a substantial population accumulates in the metastable Ba $6s5d\,^3D_J$ levels, and from these levels it is straightforward to reach the Rydberg levels with two pulsed dye laser photons. A Ba atom in a Rydberg state is much more easily ionized than a Ba atom in a lower lying state, and as a result, whenever a Rydberg state is populated the current in the discharge increases temporarily, and this increase is detected.

Finally, thermal beams of atoms can be excited using one or more lasers, as described in Chapter 3. An interesting variant is the metastable beam used by Post et al.[4] A Ba beam from an effusive oven is used to conduct a current from a heated

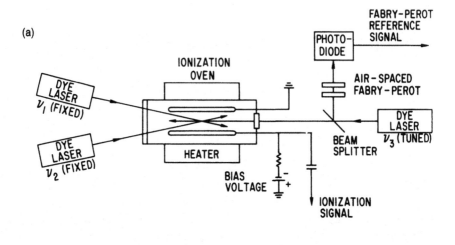

(a)

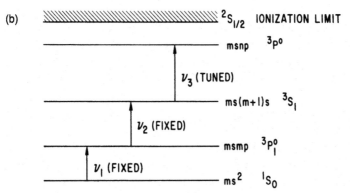

(b)

Fig. 22.1 (a) Experimental setup for three laser excitation of atomic vapors with ionization detection and reference signal for calibration. (b) Energy level diagram of a typical alkaline earth atom showing the sequence which allows easy excitation of ^{3}P° states (from ref. 1).

filament several mm from the oven orifice. A 20 V potential difference between the two produces a 400 mA current through the atomic beam, mostly carried by electrons. The electrons excite the Ba atoms, and some of the atoms decay to metastable states, which remain excited some distance downstream from the oven. The metastable Ba 6s5d ^{3}D$_J$ state is readily populated in this way. Downstream, the Ba atomic beam is crossed at 90° by the beam from a frequency doubled, single mode, cw dye laser, which excites the metastable Ba atoms to the 6snp and 6snf Rydberg states. The Rydberg atoms pass out of the excitation region and into the region between a pair of field plates, where they are field ionized, and the resulting ions detected.

All of these approaches have energy resolution limited by the laser linewidth, ~0.3 cm^{-1} for the pulsed lasers and 2 MHz for the cw laser. The intensities of the observed transitions cannot, however, be used with complete confidence, for it is

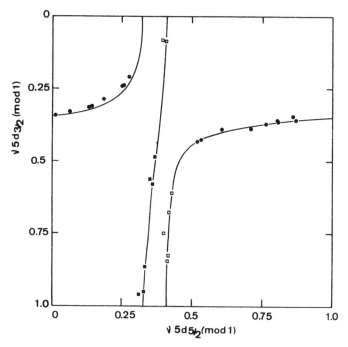

Fig. 22.2 Lu–Fano plot of the Ba $J = 4$ levels below the $5d_{3/2}$ limit: $5d_{3/2}nd_{3/2}$ (●), $5d_{5/2}nd_{5/2}$ (□), $5d_{5/2}nd_{3/2}$ (■) (from ref 5).

not clear that the efficiency of detection of all Rydberg states is the same. To the extent that collisional ionization is important, the radiative lifetimes of the states are significant, for longer lifetimes are more likely to allow ionization. As we shall see these vary dramatically in the vicinity of perturbations in the spectra. If photoionization of the Rydberg states plays a role, the cross section depends on both the wavelength and the amount of Rydberg and perturber character in the atomic states under study.

The measured energy levels cannot be expressed as

$$W = \frac{-1}{2(n - \delta)^2} \tag{22.1}$$

with a constant quantum defect. Instead, analysis of the data requires two steps. First it is important to decide how many channels are involved and how many ionization limits. Once the identity of the important limits is established each observed energy can be assigned an effective quantum number relative to each limit. For example if there are two relevant limits, 1 and 2, each observed energy is assigned the quantum numbers ν_1 and ν_2, and these can be plotted as shown in Fig. 22.2, a plot by Aymar *et al.*[5] of the effective quantum numbers corresponding to the observed term energies of the $5d_j nd_j$ $J = 4$ levels below the 5d limits of Ba^+.[6] Although all the levels studied are above the Ba^+ $6s_{1/2}$ limit, autoionization to the

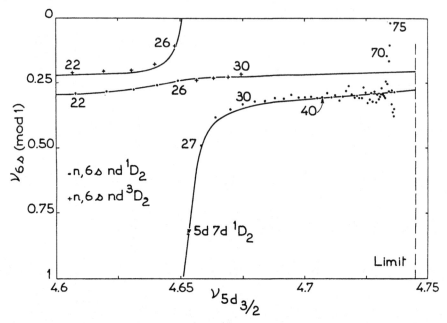

Fig. 22.3 Lu–Fano plot of the high lying Ba even parity $J=2$ levels above the 6s22d ^{3}D$_2$ level. The full curve is calculated with the QDT parameters of ref. 8: observed 6snd ^{3}D$_2$ states (+), observed 6snd^{1}D$_2$ states (●); observed 5d7d perturber (∇) (from ref. 8).

6s$\varepsilon\ell$ continua plays no role in the spectra at a resolution of 0.3 cm^{-1}. Plots such as Fig. 22.2 are called Lu–Fano plots.[7] The points represent the intersection of the quantum defect surface and the energy constraint. Using the observed points it is possible to find the parameters which generate the quantum defect surface which passes through the points. A useful observation, of Lu and Fano, is that the diagonal line from $\nu_1 = 0$, $\nu_2 = 0$ to $\nu_1 = 1$, $\nu_2 = 1$ intersects the quantum defect surface at the values of μ_α, i.e. $\nu_1 = \nu_2 = \mu_\alpha$. Fig. 22.2 is an example of a three channel, two limit problem. If there are more limits the point must be plotted in more dimensions, although if there are only more channels, but still two limits a two dimensional graph can still be used; it simply has more branches. Fig. 22.2 is in several ways an excellent example of a Lu–Fano plot. It is plain to see that the 5d$_{3/2}$nd$_{5/2}$ series has a quantum defect of 0.65 and that the 5d$_{5/2}$nd$_j$ series have quantum defects of 0.60 and 0.68 where they are not interacting and that the interaction between the series causes local deviations from these values. It is not always the case that the Lu–Fano plot consists of horizontal and vertical lines with localized avoided crossings. In their study of the ^{1}P$_1$ odd states of alkaline earth atoms Armstrong *et al.*[1] obtained Lu–Fano plots with no horizontal or vertical sections indicating that the entire series of ^{1}P$_1$ states converging to the lowest limit contained significant admixtures of doubly excited states.

The relatively well localized interaction of the Ba 5d7d ^{1}D$_2$ state with the Ba 6snd ^{1,3}D$_2$ states is shown by the section of the Lu–Fano plot of Fig. 22.3.[8] The

5d7d perturber is only mixed significantly into a few states, and only one branch of the Lu–Fano plot is significantly perturbed. However, as shown by Fig. 22.3, the branch which has $\nu_{6s} \sim 0.3$ at energies below the 5d7d perturber evolves into the branch with $\nu_{6s} \sim 0.2$ above the perturber, and as we shall see, the $^1D_2$ and $^3D_2$ states interchange character at the 5d7d $^1D_2$ state.

Properties of perturbed states

The energy level variations exhibited in the Lu–Fano plots are reflected in analogous variations in the atomic properties. One of the most apparent variations is in the lifetimes of Rydberg states in the vicinity of a perturber. Since the perturbing states tend to be relatively compact doubly excited states, they have larger matrix elements for short wavelength transitions to lower lying states than do the Rydberg states, and their lifetimes are reduced accordingly. Although the radiative lifetimes in the regions around several perturbing states have been measured,[9–12] we here focus on the variation in the lifetimes near the Ba 5d7d state perturbing the 6snd $^{1,3}D_2$ series.

The lifetimes have been measured by two techniques. The first, used by Aymar *et al.*[10] and Gallagher *et al.*[11] is to use two pulsed lasers to excite Ba atoms in a thermal atomic beam to the 6snd Rydberg states via the route 6s^2 $^1S_0 \rightarrow$ 6s6p $^1P_1 \rightarrow$ 6snd $^{1,3}D_2$. The population in the Rydberg state as a function of time after excitation is determined by applying a field ionization pulse at a variable time after the laser pulse. The second method, used by Bhatia *et al.*[12] is to use two cw lasers to excite Ba atoms by the same route. The second 6s6p $^1P_1$–6snd $^{1,3}D_2$ laser is pulse modulated, at a high repetition rate, and the temporal decay of the 6snd–6s6p fluorescence is recorded after the blue laser is pulsed off. Both methods give similar results. For example, the lifetimes reported in refs. 10–12 for the 5d7d $^1D_2$ state are 160, 217, and 170 ns, respectively. If we plot the decay rates, instead of the lifetimes of the states, we can see in a very direct way the amount of 5d7d $^1D_2$ character which exists in each state, and in Fig. 22.4 we show a plot of the decay rates as a function of term energy.[11] As shown by Fig. 22.4, roughly five states have decay rates significantly higher than the lines representing the expected n^{-3} variation in the decay rates of the 6snd $^{1,3}D_2$ Rydberg states. Note that it is the singlet states on the low energy side of the perturber which have the increased decay rates and the triplet states on the high energy side, as expected from the Lu–Fano plot of Fig. 22.3. Although Aymar *et al.*[10] have made a more sophisticated analysis of the lifetime variation, a straightforward method is to describe the *i*th state near the 5d7d state by the wavefunction[11]

$$\Psi_i = \varepsilon_i^{1/2}\Psi_{5d7d} + (1 - \varepsilon_i)^{1/2}\Psi_{6snd} , \qquad (22.2)$$

and the radiative decay rate by

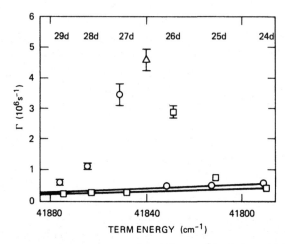

Fig. 22.4 Observed decay rates of the $6snd\,^1D_2$ ($\bigcirc$), $6snd\,^3D_2$ ($\square$), and $5d7d\,^1D_2$ ($\triangle$) states. The lines indicate the expected ν_{6s}^{-3} dependence for the unperturbed $6snd\,^1D_2$ and $6snd\,^3D_2$ Rydberg series. For points without error bars shown they are approximately the size of the symbols (from ref. 11).

$$\Gamma_i = \varepsilon_i \Gamma_{5d7d} + (1-\varepsilon_i)\Gamma_{6snd}. \qquad (22.3)$$

If we approximate Γ_{6snd} by its average, $0.35 \times 10^6\,\mathrm{s}^{-1}$, over the $24 < n < 29$ range and use the fact that

$$\sum_i \varepsilon_i = 1, \qquad (22.4)$$

i.e. that there is one 5d7d state spread among all the states shown in Fig. 22.4, then we can easily solve Eqs. (22.3) and (22.4) for the decay rate of the pure 5d7d state and the fraction ε_i of 5d7d state in the perturbed $6snd$ Rydberg states. The pure 5d7d decay rates of $1.5 \times 10^7\,\mathrm{s}^{-1}$ and $1.2 \times 10^7\,\mathrm{s}^{-1}$ were obtained by Aymar *et al.*[10] and Gallagher *et al.*[11] More interesting are the values of ε_i, the fractional perturber character in each state. In Table 22.1 we give the perturber fractions obtained from the lifetimes as well as those obtained by a three channel QDT analysis of the energies of the states. The lifetime and QDT results are certainly in reasonable agreement, and the experimental results can be read approximately from the graph of Fig. 22.4 using

$$\varepsilon_i = 8.3 \times 10^{-8}\,\mathrm{s}\,(\Gamma - 0.45 \times 10^6\,\mathrm{s}^{-1}). \qquad (22.5)$$

In the Ba $6snd\,^{1,3}D_2$ series, the higher energy levels are labelled singlets, as shown by Figs. 22.3 and 22.4. According to the QDT analysis,[8] the states on the continuous branch of the Lu–Fano plot of Fig. 22.3 evolve from singlets below the 5d7d state to triplets above the 5d7d state and contain negligible 5d7d character. The states on the broken branch change from predominantly triplet character to singlet character at the 5d7d state. This assignment of singlet and triplet character is borne out to some extent by the line strengths in absorption spectra.[13] It is

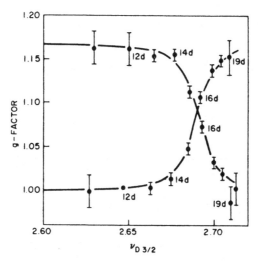

Fig. 22.5 The *g* factor of Sr 5s*nd* states as a function of $\nu_{D3/2}$, the effective principle quantum number measured relative to the 4d $^2D_{3/2}$ ionization threshold at 60488.09 cm^{-1}. The solid lines are the theoretical predictions, and the points correspond to the experimental measurements for the bound states designated by 5s*nd* (from ref. 15).

Table 22.1. *Perturber fractions obtained from the radiative decay rates, photoionization cross sections, and QDT analysis of the optical spectrum.*

	Perturber fraction ε derived from		
State	Experimental radiative decay rates[a]	Experimental photoionization cross sections[b]	QDT three channel model $6s_{1/2}$, $5d_{5/2}$ limits[a]
6s25d $^3D_2$	0.035	0.04(2)	0.02
6s26d $^3D_2$	0.219	0.11(4)	0.13
5d7d $^1D_2$	0.365	0.45(6)	0.40
6s27d $^1D_2$	0.262	0.25(5)	0.20
6s28d $^1D_2$	0.068	0.07(3)	0.05
6s29d $^1D_2$	0.028	0.05(2)	0.02

[a] (from ref. 11)
[b] (from ref. 16)

demonstrated explicitly, however, in an analogous case in Sr.[14,15] From an analysis of the energy levels of the Sr 5s*nd* $^{1,3}D_2$ states Esherick predicted that the singlet and triplet character of the Sr 5s*nd* $^{1,3}D_2$ states interchanged at $n = 15$.[14] Wynne *et al.*[15] then verified this prediction experimentally by measuring the *g* factors of these states. The singlet and triplet *g* factors are 1 and 7/6 respectively, so an accurate measurement of the g factor gives a more reliable indication of singlet and triplet composition than do the line strengths, for the composition of the initial state plays no role in the g factor measurements. The experiment was done by placing a heat pipe, such as the one shown in Fig. 22.1, between the poles of an electromagnet. A single linearly polarized laser was used to drive two photon transitions from the Sr $5s^2$ $^1S_0$ ground state to the $5snd$ states. With the laser polarization perpendicular to the magnetic field, each $5s^2$–$5snd$ resonance was split into three resonances, corresponding to the allowed $m = 0$, ± 2 final states. Measuring the relative splittings of the levels in a magnetic field of 8.3 kG yielded the relative g factors, and assigning the $5s12d$ $^1D_2$ state its calculated g factor of 1.0033 leads to the absolute values shown in Fig. 22.5.[15] Measuring the field independently could only be done to 3%, which gave an absolute value of the $5s12d$ $^1D_2$ g factor of 1.035(37), in agreement with the calculated value. In any event, it is clear from Fig. 22.5 that the singlet and triplet characters evolve as shown by the solid lines of Fig. 22.5, which are predictions from the QDT analysis of the energies only.[14]

In Ba, many of the perturbers of the bound Rydberg series are low lying states converging to the $5d_j$ limits of Ba^+ and are well described as $5dn\ell$ states of low n. Accordingly, it is reasonable to expect that they should have large cross sections for the transition to the autoionizing $6pn\ell$ state which may expected to be rather broad since n is low. The $5dn\ell \rightarrow 6pn\ell$ transitions are found in the general vicinity of the Ba^+ $5d \rightarrow 6p$ transitions, at $\lambda \sim 630$ nm. In contrast to the perturbing $5dn\ell$ states, the Ba $6sn\ell$ Rydberg states are unlikely to absorb 630 nm light. By measuring the relative photoionization cross sections at $\lambda \sim 630$ nm Mullins *et al.*[16] measured the perturber character of bound Ba $6sns$ and $6snd$ Rydberg series near several $5dn\ell$ perturbers. Their results for the Ba $6snd$ $^{1,3}D_2$ states are given in Table 22.1.

Continuity of photoexcitation across ionization limits

One of the aspects of QDT which makes it most useful is that it clearly shows the continuity of many properties across ionization limits. For example, if there are two coupled Rydberg series converging to two different ionization limits, the effective quantum numbers are ν_1 and ν_2 below the first limit. Between the two limits the spectrum is composed of a series of Beutler–Fano profiles corresponding to the Rydberg series converging to the higher limit, and plotted on a scale of ν_2 instead of energy, the spectrum repeats modulo 1 in ν_2. According to QDT the spectrum also repeats modulo 1 in ν_2 below the first limit. More precisely, since the spectrum below the first limit is composed of discrete lines, the envelope of the spectrum repeats. An excellent illustration of this notion occurs in the spectrum of the Ba $5d_jnd_j J = 4$ states as the Ba$^+$ $5d_{3/2}$ limit is traversed.[17] These are the states shown in the Lu–Fano plot of Fig. 22.2.

In Fig. 22.6(a) we show the spectrum of the transition from the Ba $5d6p$ $^1F_3$ state to the $5d_{5/2}16d_{3/2,5/2}$ states which lie just above the Ba$^+$ $5d_{3/2}$ limit, and in Fig. 22.6(b) we show the spectrum from the same initial state to the $5d_{5/2}14d_{3/2,5/2}$ states which lie just below the Ba$^+$ $5d_{3/2}$ limit. It is apparent that the envelopes of the excitations are identical. We have been discussing these states as if they were bound states, and in fact for our present purpose they might as well be. First, below the $5d_{3/2}$ limit the observed linewidths equal the laser linewidth, and second, there is no visible excitation of the $6s\varepsilon\ell$ continua below the $5d_{3/2}$ limit. It is also useful to note that a three channel quantum defect treatment of the $5d_jnd_j$ $J = 4$ series reproduces both the Lu–Fano plot of Fig. 22.2, and the spectra shown in Fig. 22.6. The experimental spectra were reproduced by QDT models using both the α and i channel dipole moment parametrizations described in Chapter 21.

Forced autoionization

Above the ionization limit interchannel interactions lead to autoionization and below the limit they lead to series perturbations. Forced autoionization allows both manifestations to be observed in the same states. The essential idea is shown in Fig. 22.7. In zero field a state converging to a higher limit lies just below the first limit and perturbs the Rydberg series converging to the first limit. If we apply an adequate static electric field we depress the lower ionization below the perturbing state so that it interacts with a Stark induced continuum of states. In this case the perturbing level appears as an autoionizing resonance. This phenomenon was termed forced autoionization by Garton *et al.*[18] who first observed it in the spectrum of shock heated Ba. Since that time it has been studied in a quantitative fashion with both static[19–22] and microwave[23] fields.

In their systematic experiments Sandner *et al.*[20,21] used two tunable dye lasers to excite ground state atoms in a Ba beam by either of two optical paths, A and B, to the vicinity of the $5d7d$ $^1D_2$ perturber, as shown in Fig. 22.8. In the absence of an electric field the Ba atoms excited to the $5d7d$ or neighboring $6snd$ Rydberg states were ionized by a field ionization pulse and the ions detected. If a static field was present it alone ionized the atoms. In Fig. 22.9 we show the excitation spectra using both paths A and B with and without a static field. Consider first the zero field spectra. Using path A, via the $6s6p$ $^1P_1$ state, the $5d7d$ state is not excited and we see weaker transitions to states near 41841 cm^{-1} which have appreciable $5d7d$ character. In contrast, using path B via the $5d6p$ $^3D_1$ state, only those states with appreciable $5d7d$ $^1D_2$ character are excited. In the E $\neq$ 0 spectra of Fig. 22.9 it is apparent that for path A we have a $q \approx 0$ Beutler–Fano profile, and for path B a

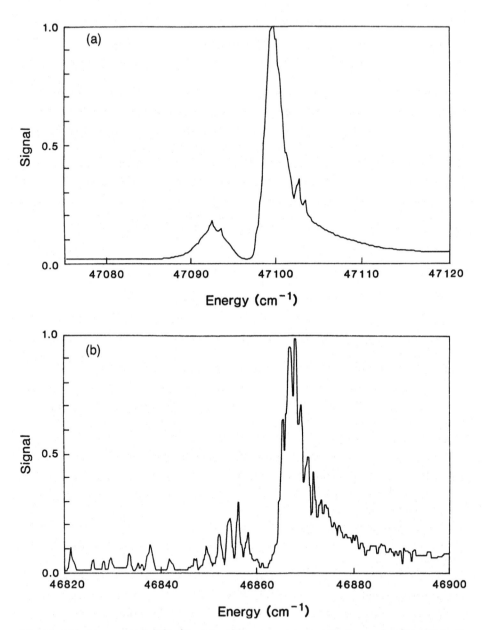

Fig. 22.6 (a) Observed Ba $5d6p\,^1F_3$–$5d_{5/2}16d_{3/2,5/2}$ spectrum. The strong doublet of broad $J = 4$ lines is located at 47093 and 47100 cm^{-1}. Two weak $J = 3$ features are visible at 47102 and 47103 cm^{-1}. (b) The transition from $5d6p\,^1F_3$ to the $5dnd\,J = 4$ levels below the $5d_{3/2}$ limit showing the same basic oscillator strength distribution to the $5d_{5/2}14d_{3/2,5/2}$ states over many essentially discrete energy levels (from ref. 17).

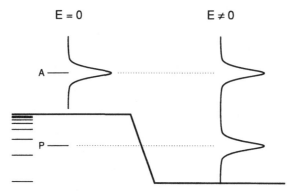

Fig. 22.7 Schematic diagram of forced autoionization. Above the ionization limit the autoionizing state A, converging to a higher limit, is manifested as an autoionization resonance. Below the limit, in zero field the interaction of the perturber P with the Rydberg series results in the perturbation of the Rydberg series. Applying an electric field E depresses the ionization limit below P, and it appears as a forced autoionization resonance.

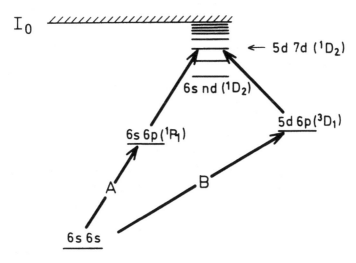

Fig. 22.8 The two different excitation schemes A and B used to observe the Ba 5d7d perturber as a forced autoionization resonance. Path A leads to a q parameter near zero, while B leads to a large q parameter (from ref. 21).

profile with large $|q|$. In both cases the field has converted the 5d7d state into an autoionizing resonance, as expected. However, even a casual inspection of Fig. 22.9 reveals that the path B zero field spectrum is not simply a discretized version of the $E = 7.5$ kV/cm spectrum; it is broader. The broadening is in fact an experimental artifact introduced by the 500 ns delay in applying the field ionization pulse.[21] When the short, ~200 ns, lifetimes of states with appreciable 5d7d character are taken into account to extrapolate the zero field spectrum of Fig. 22.9 back to zero time delay, the intensities of the central lines are increased

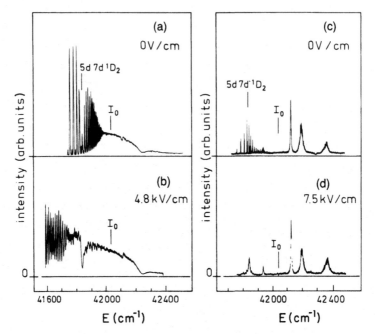

Fig. 22.9 (a) Photoabsorption spectrum observed in zero field, using excitation path A of Fig. 22.8. (b) Same as (a), but observed at an applied dc electric field of 4.8 kV/cm. (c) Zero field spectrum observed using path B of Fig. 22.8. (d) Same as (c), but with an applied field of 7.5 kV/cm (from ref. 21).

by a factor greater than 10, and the zero field path B spectrum appears to be a discretized version of the E = 7.5 kV/cm spectrum.[21]

We can compare the forced autoionization resonance to the predictions of zero field QDT. The observed width, 15.5 cm^{-1}, and the width from QDT, 15.3 cm^{-1}, are in excellent agreement. However, the energy positions are different by 5 cm^{-1}. Exactly why is not clear, but it is certainly the case that the Stark induced continuum is not in all respects like a zero field continuum. For example, with both lasers polarized parallel to the field, so as to excite $m = 0$ final states, the forced autoionization resonance analogous to the one shown in Fig. 22.9 exhibits structure due to the long lived, blue shifted, Stark states.[20,21]

References

1. J. A. Armstrong, J. J. Wynne, and P. Esherick, *J. Opt. Soc. Am.* **69**, 211 (1979).
2. K. H. Kingdon, *Phys. Rev.* **21**, 408 (1923).
3. P. Camus, M. Dieulin, and A. El Himdy, *Phys. Rev. A* **26**, 379 (1982).
4. B. H. Post, W. Vassen, W. Hogervorst, M. Aymar, and O. Robaux, *J. Phys. B* **18**, 187 (1985).
5. M. Aymar, P. Camus, and A. El Himdy, *Physica Scripta* **27**, 183 (1983).
6. P. Camus, M. Dieulin, A. El Himdy, and M. Aymar, *Physica Scripta* **27**, 125 (1983).
7. K. T. Lu and U. Fano, *Phys. Rev. A* **2**, 81 (1970).

8. M. Aymar and O. Robaux, *J. Phys. B* **12**, 531 (1979).
9. M. Aymar , P. Grafström, C. Levison, H. Lundberg and S. Svanberg, *J. Phys. B* **15**, 877 (1982).
10. M. Aymar, R. J. Champeau, C. Delsart, and J.-C. Keller, *J. Phys. B.* **14**, 4489 (1981).
11. T. F. Gallagher, W. Sandner, and K. A. Safinya, *Phys. Rev. A* **23**, 2969 (1981).
12. K. Bhatia, P. Grafström, C. Levison, H. Lundberg, L. Nelsson, and S. Svanberg, *Z. Phys. A* **303**, 1 (1981).
13. J. R. Rubbmark, S. A. Borgström, and K. Bockasten, *J. Phys. B* **10**, 421 (1977).
14. P. Esherick, *Phys. Rev. A* **15**, 1920 (1977).
15. J. J. Wynne, J. A. Armstrong, and P. Esherick, *Phys. Rev. Lett.* **39**, 1520 (1985).
16. O. C. Mullins, Y. Zhu, and T. F. Gallagher, *Phys. Rev. A* **32**, 243 (1985).
17. C. J. Dai, S. M. Jaffe, and T. F. Gallagher, *J. Opt. Soc. Am. B* **6**, 1486 (1989).
18. W. R. S. Garton, W. H. Parkinson, and E. M. Reeves, *Proc. Phys. Soc.* **80**, 860 (1962).
19. B. E. Cole, J. W. Cooper, and E. B. Salomon, *Phys. Rev. Lett.* **45**, 887 (1980).
20. W. Sandner, K. A. Safinya, and T. F. Gallagher, *Phys. Rev. A* **24**, 1647 (1981).
21. W. Sandner, K. A. Safinya, and T. F. Gallagher, *Phys. Rev. A* **33**, 1008 (1986).
22. T. F. Gallagher, F. Gounand, R. Kachru, N. H. Tran, and P. Pillet, *Phys. Rev. A* **27**, 2485 (1983).
23. R. R. Jones and T. F. Gallagher, *Phys. Rev. A* **39**, 4583 (1989).

23

Double Rydberg states

The autoionizing two electron states we have considered so far are those which can be represented sensibly by an independent electron picture. For example, an autoionizing Ba $6pnd$ state is predominantly $6pnd$ with only small admixtures of other states, and the departures from the independent electron picture can usually be described using perturbation theory or with a small number of interacting channels. In all these cases one of the electrons spends most of its time far from the core, in a coulomb potential, and the deviation of the potential from a coulomb potential occurs only within a small zone around the origin.

In contrast, in highly correlated states the noncoulomb potential seen by the outer electron is not confined to a small region. In most of its orbit the electron does not experience a coulomb potential, and an independent electron description based on $n\ell n'\ell'$ states becomes nearly useless. There are two ways in which this situation can arise. The first, and most obvious, is that the inner electron's wavefunction becomes nearly as large as that of the outer electron. If we assign the two electrons the quantum numbers $n_i\ell_i$ and $n_o\ell_o$, this requirement is met when n_i approaches n_o, which leads to what might be called radial correlation. The sizes of the two electron's orbits are related. The second way the potential seen by the outer electron can have a long range noncoulomb part is if the presence of the outer electron polarizes the inner electron states. This polarization occurs most easily if the inner electron ℓ states of the same n are degenerate, for then even a very weak field from the outer electron, equal to $1/r_o^2$, converts them to Stark states, which have permanent dipole moments. The anisotropic potential from the induced dipole produces an angular correlation in the motion of the two electrons, even if the orbit of the outer electron is many times larger than the orbit of the inner electron. Angular correlation may be found in all the doubly excited states of He since He^+ is hydrogenic. On the other hand, radial correlation is presumably only found when n_i approaches n_o. Angular correlation may also be observed in the doubly excited states converging to the nearly degenerate high ℓ states of a singly charged ion, Sr^+ for example, even when the outer electron is in a substantially larger orbit than the inner electron. On the other hand, the doubly excited states converging to low ℓ states of Sr^+, which have large quantum defects, do not exhibit angular correlation before they exhibit radial correlation. To see remarkable correlation in these states, the requirement that n_i approach n_o must be met.

Doubly excited states of He

Although the existence of autoionizing states has been known since the 1920s, it was the observation of the doubly excited states of He by Madden and Codling[1] which aroused interest in highly correlated states. Using synchrotron radiation, they recorded the absorption spectrum of He using an 83 cm path length of He at a pressure of 0.3 Torr. One might naively expect to see two series, 2snp and 2pns, both converging to the He$^+$ $n = 2$ limit. However, they observed one strong series, the + series, and a very weak series, the − series.[2] In addition to being weaker, the − series lies below the + series and has smaller autoionization widths.

More recently the measurements have been repeated and extended by Domke et al.,[3] who also used a synchrotron as a light source, but detected the photoions produced instead of the absorption of the synchrotron radiation. Their spectra are shown in Fig. 23.1. As shown by Fig. 23.1 the strong $n = 2$ + peaks appear as classic Beutler–Fano profiles on top of a small background 1sεp photoionization signal. They are roughly 100 times as intense as the − resonances. In addition, the widths of the − peaks are about 10% of the widths of the + peaks. If we examine the spectra converging to the higher He$^+$ limits, shown in Fig. 23.2, several features are apparent. First, we can see that these spectra contain only one series, the + series. In addition, in the spectra converging to higher n states of the ion the resonances become smaller relative to the background photoionization, which is hardly surprising, as there are more available continua. Stated another way, the magnitude of the Fano q parameter decreases with increasing n, although it also changes sign, which is not as simply explained. In Fig. 23.2(c) we can see that the series converging to the He$^+$ $n = 5$ and 6 and $n = 6$ and 7 states overlap, producing the same sort of pattern seen in Fig. 19.2.

The initial experiment of Madden and Codling, the observation of the states converging to the He$^+$ $n = 2$ state, was first explained by Cooper et al.[2] in the following way. Rather than use the 2snp and 2pns states as the atomic basis, they proposed using the combinations[2]

$$\Psi \pm = \frac{1}{\sqrt{2}}[u_{2snp} \pm u_{2pns}] \tag{23.1}$$

where $u(2snp)$ is a product of a 2s and an np wavefunction. The + and − states have characteristics which are different from those of angular momentum states. In the + states the radial oscillations of the two electrons are in phase, and in the − states they are out of phase. In addition, by considering the properties of the + and − states near the origin we can explain the $n = 2$ spectra, shown in Fig. 23.1. Both u_{2snp} and u_{2pns} are products of one s wavefunction and one p wavefunction. Thus at the origin they are functionally the same. Since the normalization factors of wavefunctions of principal quantum number 2 and n are given by $2^{-3/2}$ and $n^{-3/2}$ respectively, the overall normalization factors of u_{2snp} and u_{2pns} are nearly the same. Near the origin $u_{2snp} \approx u_{2pns}$, and Ψ_- nearly vanishes. As a result, the − states are only weakly excited from the ground state, and they have lower autoionization rates than the + states, as observed.

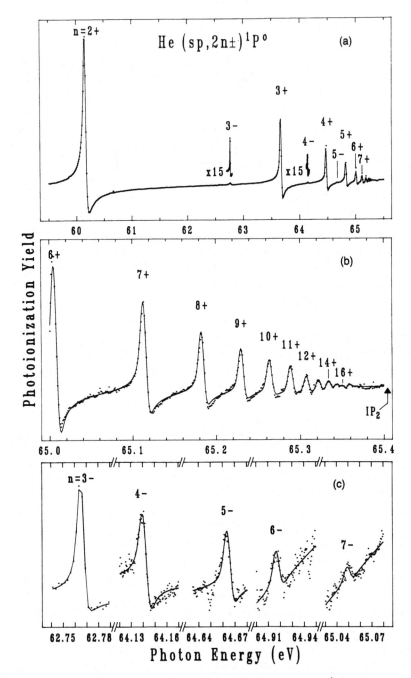

Fig. 23.1 Autoionizing states of doubly excited He below the He$^+$ $n = 2$ threshold: (a) overview, (b) magnification of the $n \geq 6$ region, and (c) $2n-$ states (from ref. 3).

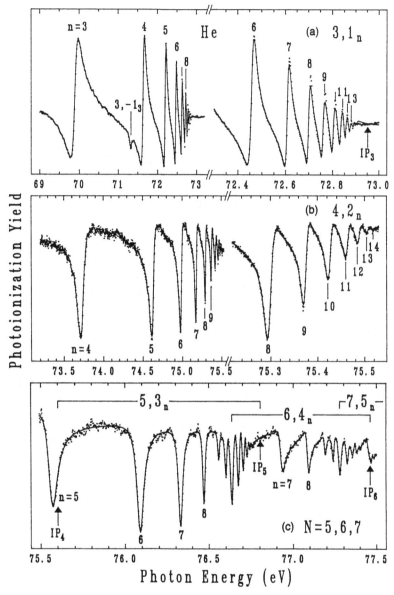

Fig. 23.2 Autoionizing states of He: (a) below the He$^+$ $n = 3$ threshold, (b) below the He$^+$ $n = 4$ threshold, and (c) below the He$^+$ $n = 5$ and 6 thresholds. The high n regions are shown magnified on the right-hand sides in (a) and (b). Note the overlapping of series in (c) (from ref. 3).

Theoretical descriptions

The work of Cooper *et al.*[2] was the beginning of a steady stream of theoretical work. Shortly after the experiments of Madden and Codling more detailed

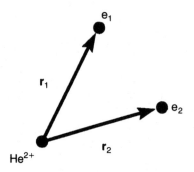

Fig. 23.3 The He atom.

calculations of the doubly excited He levels were carried out by a variety of techniques.[4-7] While these calculations showed that the description of Cooper *et al.*[2] omitted aspects of the problem, such as the contribution of the 2p*nd* states,[7] the basic feature of their explanation, the creation of superposition states, somewhat similar to Stark states, was found in all these treatments.[4-7]

While the detailed calculations gave energies and widths in reasonable agreement with the observed experimental results, they did not provide a simple physical picture of why the He spectrum was the way it was. To address this issue Macek described the He atom using hyperspherical coordinates. In this approach, the two electrons, each described by three coordinates, are replaced by an equivalent single particle in six dimensions.[8,9] In Fig. 23.3 we show the He atom in which r_1 and r_2 are the vectors from He^{2+} to each of the two electrons. The hyperradius R_h is defined by

$$R_h = \sqrt{r_1^2 + r_2^2}, \tag{23.2}$$

and the ratio of r_1 to r_2 defines the hyperangle α, i.e.

$$\tan \alpha = r_2/r_1. \tag{23.3}$$

At $\alpha = 0$ and $\pi/2$ electrons 1 and 2, respectively, are infinitely far from the He^{2+}, while at $\alpha = \pi/4$ they are equidistant from the He^{2+}. Following the approach of Macek[8] and Fano,[9] we replace the usual wavefunction ψ by Ψ, which is related to ψ by

$$\Psi(r,\alpha,\hat{\mathbf{r}}_1,\hat{\mathbf{r}}_2) = \sin \alpha \cos \alpha R_h^{5/2}\psi(R_h,\alpha,\hat{\mathbf{r}}_1,\hat{\mathbf{r}}_2), \tag{23.4}$$

where the unit vector $\hat{\mathbf{r}}_i = \mathbf{r}_i/r_i$. This substitution is roughly analogous to replacing the radial function $R(r)$ by $\rho(r)/r$ in the H atom. Using Ψ the Schroedinger equation in hyperspherical coordinates is given by[9]

$$\left[\frac{\hbar^2}{2m}\left(\frac{-\partial^2}{\partial R_h^2} + \frac{\Lambda^2}{R_h^2}\right) + V(R_h,\alpha,\hat{\mathbf{r}}_1,\hat{\mathbf{r}}_2) - E\right]\Psi(R_h,\alpha,\hat{\mathbf{r}}_1,\hat{\mathbf{r}}_2) = 0 \tag{23.5}$$

where Λ is the grand angular momentum of Smith[10] defined by

$$\Lambda^2 = \frac{-\partial^2}{\partial \alpha^2} - \frac{1}{4} + \frac{\ell_1^2}{\cos^2 \alpha} + \frac{\ell_2^2}{\sin^2 \alpha}, \tag{23.6}$$

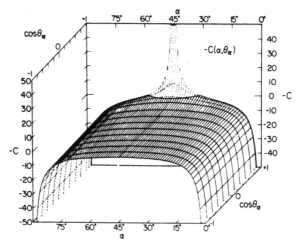

Fig. 23.4 Three dimensional plot of $-C(\alpha,\theta_{12})$ with $Z = 1$ in hyperspherical coordinates; the ordinates represent a potential surface in Rydberg units at $R_h = 1a_0$ (from ref. 11).

where ℓ_1 and ℓ_2 are the angular momenta of the two electrons. From Eqs. (23.5) and (23.6) it is apparent that Λ^2 plays the role of ℓ^2 in the H atom.

A key point of the hyperspherical approach is that the potential V can be factored;

$$V(R_h,\alpha,\hat{\mathbf{r}}_1,\hat{\mathbf{r}}_2) = \frac{e^2}{R_h} C(\alpha,\hat{\mathbf{r}}_1,\hat{\mathbf{r}}_2),\qquad(23.7)$$

where

$$C(\alpha,\hat{\mathbf{r}}_1,\hat{\mathbf{r}}_2) = \frac{-Z}{\cos\alpha} - \frac{Z}{\sin\alpha} + \frac{1}{\sqrt{1 - \hat{\mathbf{r}}_1\cdot\hat{\mathbf{r}}_2 \sin 2\alpha}}.\qquad(23.8)$$

The scaled potential $C(\alpha,\hat{\mathbf{r}}_1,\hat{\mathbf{r}}_2)$ is shown in Fig. 23.4 using $\cos\theta_{12} = \hat{\mathbf{r}}_1\cdot\hat{\mathbf{r}}_2$.[11] Inspecting it we can see that C is flat at $\alpha = 45°$, corresponding to $r_1 = r_2$, for all angles θ_{12} except $\theta_{12} = 0$. The potential valleys, at $\alpha = 0$ and $\pi/2$ correspond to one of the two electrons being far from the He^{2+} core and the other being close to the core. Inspecting Fig. 23.4 we can see that there is a saddle point in $C(\alpha,\hat{\mathbf{r}}_1,\hat{\mathbf{r}}_2)$ at $\theta_{12} = \pi$ and $\alpha = \pi/4$.

If we define the operator

$$H_R = \frac{\hbar^2\Lambda^2}{2mR} + \frac{e^2C}{R_h}(\alpha,\hat{\mathbf{r}}_1,\hat{\mathbf{r}}_2)\qquad(23.9)$$

and solve the eigenvalue equation

$$H_R\phi_\mu(\alpha,\hat{\mathbf{r}}_1,\hat{\mathbf{r}}_2) = U_\mu(R_h)\phi_\mu(\alpha,\hat{\mathbf{r}}_1,\hat{\mathbf{r}}_2)\qquad(23.10)$$

at constant R_h, we find the eigenvalue $U_\mu(R_h)$ corresponding to the solution ϕ_μ with the set of quantum numbers μ corresponding to the α, $\hat{\mathbf{r}}_1$, and $\hat{\mathbf{r}}_2$ coordinates.

An alternative approach is to use the well known eigenfunctions of Λ^2 and cope with off diagonal matrix elements due to $C(\alpha,\hat{\mathbf{r}}_1,\hat{\mathbf{r}}_2)$.

If we now assume that

$$\Psi(R_h,\alpha,\hat{\mathbf{r}}_1,\hat{\mathbf{r}}_2) = \sum_\mu F_\mu(R_h)\phi_\mu(\alpha,\hat{\mathbf{r}}_1,\hat{\mathbf{r}}_2), \tag{23.11}$$

where $F_\mu(R_h)$ is a wavefunction analogous to the vibrational wavefunction in a diatomic molecule, Eq. (23.5) becomes

$$\left[\frac{\hbar^2}{2M}\frac{\partial^2}{\partial R_h^2} + H_R\right] F_\mu(R_h)\phi_\mu R = WF_\mu(R_h)\phi_\mu(\alpha,\hat{\mathbf{r}}_1,\hat{\mathbf{r}}_2). \tag{23.12}$$

If we replace $H_R\phi_\mu$ by $V_\mu(R_h)\phi_\mu$, use the orthogonality of the ϕ_μ functions, and make the adiabatic assumption that $\partial\phi_\mu(\alpha,\hat{\mathbf{r}}_1,\hat{\mathbf{r}}_2)/\partial R_h$ is negligible, we can remove the angular variation from Eq. (23.12), leaving the radial equation

$$\left[\frac{\hbar^2}{2M}\frac{\partial^2}{\partial R_h^2} + U_\mu(R_h) - W\right] F_\mu(R_h) = 0 \tag{23.13}$$

in which $U_\mu(R_h)$ is the effective radial potential, which plays the same role as the internuclear potential curve in a diatomic molecule. In fact, the adiabatic approximation is in essence the same approximation as the Born–Oppenheimer approximation used to treat diatomic molecules. In the descriptions of a diatomic molecule the fact that the nucleii are more massive is usually given as the reason why the separation works. However, all that is really necessary is that the motion in the $\alpha,\hat{\mathbf{r}}_1,\hat{\mathbf{r}}_2$ coordinates not change rapidly with R_h.

In Fig. 23.5 we show the set of potential curves calculated by Sadegphour and Greene for H^-, which differs from He in having shallower wells.[12] Each curve corresponds to a set of μ quantum numbers. In Fig. 23.5(a) we can see that there is a family of curves which connects to H levels of different n at $R_h \to \infty$. We can also see that as R_h is reduced from infinity that the potential curves from each n split into a fan of curves, with larger spacing at higher n. This splitting is a manifestation of the Stark effect from the field of the well removed electron on the remaining H atom. It is also apparent that only the lowest lying curves have wells at small R_h allowing the formation of quasi-stable states of H^-. Each of the potential curves $U_\mu(R_h)$ of Fig. 23.5(a) supports a series of $F_\mu(R_h)$ vibrational solutions. In Fig. 23.5(b) we show the energies of the lowest lying potential wells of Fig. 23.5(a). These are the states which are accessible from the ground state of H^-, and the energies agree well with the experimental observations.

In arriving at the radial equation of Eq. (23.13) we assumed $\partial\phi/\partial R_h$ to be negligible. The neglected terms couple states on different potential curves, allowing the autoionization of He or autodetachment of H^- doubly excited states. In other words the adiabatic approximation gives excellent energies, but it does not give the decay rates of doubly excited states.

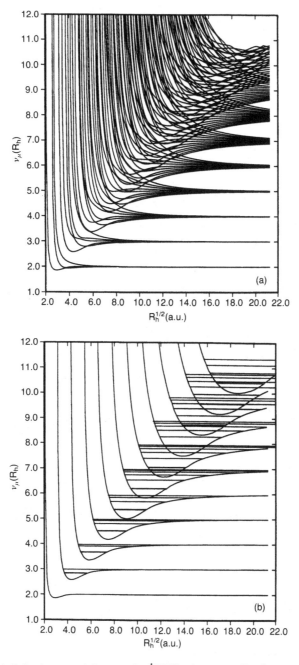

Fig. 23.5 (a) Adiabatic potential curves for $^1P^\circ$ H^- shown as effective quantum numbers vs $\sqrt{R_h}$. One should look edgewise along the the Wannier ridge line, at $\nu_\mu = 18^{-1/4}\sqrt{R_h}$, to see the lowest $+$ channels. In (b), only the lowest $+$ channels within each n manifold are plotted along with the level positions in each potential. The Wannier ridge as an imaginary straight line through the avoided crossings is clearly evident in this figure (from ref. 12).

Rau has pointed out that the doubly excited states in which both electrons are at the same distance from the He^{2+} core have $\alpha = \pi/4$ and are likely to be at the saddle point of the potential of Fig. 23.4.[13] Furthermore, if the potential V of Eq. (23.7) is expanded in the vicinity of $\theta_{12} = \pi$ and $\alpha = \pi/4$ it can be approximated by its leading term, i.e.[13]

$$V = \frac{-Z_0(Z)}{R_h},\qquad(23.14)$$

where $Z_0(Z) = 2\sqrt{2}(Z - 1/4)$. For this potential the energies relative to the double ionization limit are given by

$$W = \frac{-Z_0^2(Z)}{2(n + 5/2)^2},\qquad(23.15)$$

an expression appropriate for the six dimensional coulomb problem. To account for the higher order terms neglected in Eq. (23.14) a screening constant σ and a quantum defect δ are introduced into the expression for the energy. Explicitly,[13]

$$W = -\frac{Z_0^2(Z - \sigma)}{2(n + 5/2 - \delta)^2}\qquad(23.16)$$

A similar formula was given by Read,[14] who has, in addition, summarized much of the work on the development of Rydberg formulae for the energies of the double Rydberg states.[15] The formula of Rau, given by the expression of Eq. (23.16) gives an excellent description of the He^- $1s(ns)^2$ states observed by Buckman *et al.*[16] if $\sigma = 1/2\sqrt{2}$ and $\delta = 1.67$. An interesting aspect of this approach is that all the nodes in the wavefunction are in the R_h direction, whereas in the hyperspherical approach the ns^2 states have no nodes in the R_h direction, but are the lowest states in R_h potential curves corresponding to different values of the set of quantum numbers μ.

The hyperspherical approach has enjoyed enormous success in the calculations of energies. It does, however, have the drawback that the physical interpretation of the set of internal μ quantum numbers is not transparent. Had we used the eigenfunctions of Λ^2, it is not clear that the situation would be much improved since the individual angular momenta of the two electrons ℓ_1 and ℓ_2 appear as quantum numbers, and the $\pm$ description of Eq. (23.1) and the Stark like splittings of Fig. 23.5 suggest that ℓ_1 and ℓ_2 are not good quantum numbers. To address this issue methods of classifying states using group theory were explored by Sinanoğlu and Herrick,[17] and they made substantial progress in classifying the states.[18] One of the more interesting aspects of this approach is that the notion that the doubly excited states are related to Stark states arises naturally from the original group theory work of Park[19] relating Stark and angular momentum states in H. Molmer and Taulbjerg[20] make the point that some of the states described by Herrick are simple linear combinations of products of Stark states.

An approach which joins these two disparate points of view is the molecular orbital approach of Feagin and Briggs,[21,22] who treat the three body coulomb problem in which two of the particles have the same charge, $Z_1 = Z_2$, and mass $m_1 = m_2$, and the third has a different charge Z_3 and mass m_3. The approach is similar to the hyperspherical approach in using an adiabatic approximation. The difference lies in the choice of coordinates. If $\mathbf{r}_1$ and $\mathbf{r}_2$ are the vectors defining the locations of the two electrons relative to the He^{2+}, as shown in Fig. 23.1, the two vectors

$$\mathbf{R} = \mathbf{r}_2 - \mathbf{r}_1 \qquad (23.17a)$$

and

$$\mathbf{r} = \mathbf{r}_1 + \mathbf{r}_2 \qquad (23.17b)$$

are introduced. R plays a role comparable to the role of R_h in the hyperspherical approach. The magnitude R is the distance between the two like particles 1 and 2, the two electrons in He or H^-, and r is the distance from the center of mass of particles 1 and 2 to the third particle, He^{2+} or H^+ for He and H^-. These coordinates are precisely the coordinates commonly used to describe H_2^+. The Schroedinger equation for the three body coulomb system takes the form[22]

$$\left(\frac{-\nabla_R^2}{2\mu_{12}} - \frac{\nabla_r^2}{2\mu_{12,3}} + \frac{Z_1 Z_2}{R} + \frac{Z_1 Z_3}{r_1} - \frac{Z_2 Z_3}{r_2} \right) \psi(\mathbf{r},\mathbf{R}) = W\psi(\mathbf{r},\mathbf{R}), \qquad (23.18)$$

where μ_{12} is the reduced mass of the two like particles, $\mu_{12} = m_1 m_2/(m_1 + m_2) = 1/2$ for He or H^-, and $\mu_{12,3}$ is the reduced mass of the center of mass of the two like particles and the third particle; $\mu_{12,3} = (m_1 + m_2)m_3/(m_1 + m_2 + m_3) \cong 1/2$ for He or H^-.

The eigenfunctions of the atom must be eigenfunctions of the total angular momentum and its projection on a space fixed z axis. If we ignore the electron spins for simplicity, the total angular momentum L and its projection M on the space fixed z axis are conserved. The spatial wavefunctions $\Psi_{LM}(\mathbf{r},\mathbf{R})$ are related to the body fixed wavefunctions by the Euler transformation

$$\Psi_{LM}(\mathbf{r},\mathbf{R}) = D_{MK}^L(\psi,\theta,0) \frac{1}{R} f_{iK}(R)\phi_{iK}(\mathbf{r},R), \qquad (23.19)$$

where ψ, θ, and $\phi = 0$ are the Euler angles required to rotate the space fixed z axis into the direction of the vector $\mathbf{R}$, the body fixed frame, D_{MK}^L is a rigid top wavefunction of total angular momentum L and projections M and K on the space fixed and body fixed (along $\mathbf{R}$) axes.[22] D_{MK}^L describes the overall rotation of the atom.

$\phi_{iK}(\mathbf{r},R)$ is the molecular orbital having the projection of its angular momentum on the $\mathbf{R}$ direction equal to number K and internal quantum numbers represented by i. At a fixed separation R, ϕ_{iK} describes the motion of He^+ relative to the center of mass of the two electrons. $f_{iK}(R)$ describes the effect of separating the two electrons, which is analogous to the vibrational motion of the nuclei in a diatomic molecule. Note that the third Euler angle is zero in $D_{MK}^L(\psi,\theta,0)$ since it coincides with the azimuthal motion of $\mathbf{r}$ about $\mathbf{R}$.

We can write the Hamiltonian of Eq. (23.18) as

$$H = \frac{-\nabla_R^2}{2\mu_{12}} + h,\qquad(23.20)$$

where

$$h = \frac{-\nabla_r^2}{2\mu_{12,3}} + \frac{Z_1Z_2}{R} + \frac{Z_1Z_3}{r_1} + \frac{Z_2Z_3}{r_2}\qquad(23.21)$$

is the Hamiltonian for the motion of the third particle, He^{2+} or H^+, in the body fixed frame.

One of the aspects of the molecular orbital approach which is most interesting is that the eigenfunctions of the operator h are apparently not too different from the final atomic eigenfunctions. We shall discuss this point further, but for the moment we simply note that the eigenfunctions $\phi_{iK}(\mathbf{r},R)$ satisfy the eigenvalue equation

$$h\phi_{iK}(\mathbf{r},R) = \varepsilon_{iK}(R)\phi_{iK}(\mathbf{r},R).\qquad(23.22)$$

The ∇_R^2 term of the Hamiltonian represents the kinetic energy of the relative motion of the two electrons. It can be broken into radial and angular parts using

$$\frac{-\nabla_R^2}{2\mu_{12}} = -\frac{\partial^2/\partial R^2}{2\mu_{12}} + \frac{L_R^2}{2\mu_{12}},\qquad(23.23)$$

where $L_R = -i\mathbf{R}\times\nabla_R$.

If we now ignore the off diagonal couplings between the $\phi_{iK}(\mathbf{r},R)$ states introduced by L_R^2 and the derivatives of ϕ_{iK} with respect to R, the Schroedinger equation, Eq. (23.15), can be reduced to the ordinary differential equation

$$\left[\frac{\partial^2/\partial R^2}{2\mu_{12}} + U_{iK}^L(R) - W\right]f_{iK}(R) = 0\qquad(23.24)$$

where

$$U_{iK}^L(R) = \left\langle\phi_{iK}\left|h + \frac{L_R^2}{2\mu_{12}}\right|\phi_{iK}\right\rangle = \varepsilon_{iK}(R) + \left\langle\phi_{iK}\left|\frac{L_R^2}{2\mu_{12}}\right|\phi_{ik}\right\rangle.\qquad(23.25)$$

Eq. (23.24) leads to a series of vibrational functions in the potential of $U_{iK}{}^L(R)$ described by Eq. (23.25).

A useful feature of the molecular orbital approach is that the eigenvalue equation of Eq. (23.22) can be separated in confocal elliptic coordinates,[23] and, equally important, these eigenfunctions are apparently somewhat similar to the final atomic eigenfunctions.[22] The coordinates are given by[22]

$$\lambda = (r_1 + r_2)/R\qquad(23.26a)$$

and

$$\mu = (r_1 - r_2)/R.\qquad(23.26b)$$

The surfaces of constant μ and λ are surfaces of revolution about the interelectron axis. The eigenfunction $\phi_{iK}(\mathbf{r},R)$ of Eq. (23.22) has the form

$$\phi_{iK}(\mathbf{r},R) = \phi_\lambda(\lambda)\phi_\mu(\mu)e^{iK\phi} \tag{23.27}$$

and has the quantum numbers K, n_λ, and n_μ, the latter two specifying the number of nodes in the λ and μ directions. Since these quantum numbers are good irrespective of R, they allow the connection between the case $R = 0$ and $R = \infty$. For $R \to 0$, analogous to the united atom limit in a diatomic molecule, the problem reduces to one with spherical symmetry identical to the H atom, and $n_\lambda = n - \ell - 1$, the number of radial nodes, and $n_\mu = \ell - |m|$, the number of polar nodes. These evolutions are easily envisioned by considering Eqs. (23.26). As the two electrons are moved together the lines of constant λ become circles, and the lines of constant μ radial lines. As $R \to \infty$ n_λ and n_μ become directly related to the parabolic quantum numbers, n_1 and n_2. Explicitly, $n_\lambda = n_1$ and $n_\mu = 2n_2$ if n_μ is even and $2n_2+1$ if n_μ is odd.[22] In a real atom $R \neq 0$ so the actual atomic states have some relation to the Stark states, as mentioned in the discussion of the initial approach of Cooper et al.,[2] and as can be seen in Fig. 23.5(a).

An attractive feature of the molecular orbital approach is that the nodal lines of the atomic wavefunctions fall on surfaces of constant μ and λ. This fact seems to be true whether the wavefunctions are obtained by a diagonalization technique, a hyperspherical coordinate approach, or the molecular orbital approach. To illustrate this point we show in Fig. 23.6[24] a density plot of a wavefunction obtained in hyperspherical coordinates for $R = 80a_0$ by Sadegphour and Greene.[12] Although the figure is drawn using the hyperspherical co-ordinates θ_{12} and α, the lines of constant λ and μ are shown as well, and it is apparent that $n_\lambda = 6$ and $n_\mu = 0$. As shown by Rost et al.,[24] in most cases the nodal lines of the wavefunctions fall on surfaces of constant μ or λ, implying that the μ and λ motions are almost always separable. However, the μ and R motions are in general found to be coupled.[25]

In the discussion of both the hyperspherical method and the molecular orbital method we assumed that the motion in the R_h or R direction was slow. While in the Born–Oppenheimer approximation the slowness is generally thought to arise from the much higher mass of the nucleii, as pointed out by Feagin and Briggs,[22] it is probably the repulsive nature of the internuclear potential which allows the separability. In other words, the success of these methods in treating He is due to the interelectronic repusion of the $1/r_{12}$ potential.

Rost and Briggs adopted the opposite approach, ignoring the electron electron interaction to generate the diabatic potential $U(R)$.[26] Specifically they approximated the Hamiltonian as[26]

$$H = \frac{-\nabla_{r_1}^2}{2} - \frac{\nabla_{r_2}^2}{2} - \frac{Z}{r_1} - \frac{Z}{r_2}, \tag{23.28}$$

ignoring the $1/r_{12}$ electron electron interaction.

They used trial wavefunctions of the form[26]

$$\psi = \psi_n(\mathbf{r}_1)\psi_m(\mathbf{r}_2), \tag{23.29}$$

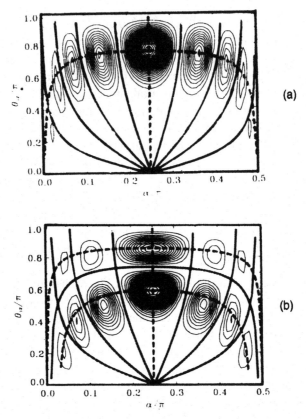

Fig. 23.6 Contour plot of adiabatic, two electron densities taken from ref. 12. The nodal (solid) and antinodal (dashed) lines of constant λ and μ are overdrawn for (a) the state with $(n_\lambda n_\mu) = (0,8)$ and (b) the state with $(n_\lambda n_\mu) = (1,6)$ (from ref. 24).

products of one electron wavefunctions with Z equal to a variational parameter α. The wavefunction of Eq. (23.29) can also be expressed in molecular orbital fashion as

$$\psi = \frac{1}{R}f(R)\phi(\mathbf{r},R),\qquad(23.30)$$

and the function $U(R)$ is calculated using

$$U(R) = \left\langle \phi \left| -\nabla_R^2 - \frac{\nabla_r^2}{4} - \frac{Z}{|\mathbf{r} - \mathbf{R}/2|} - \frac{Z}{|\mathbf{r} + \mathbf{R}/2|} \right| \phi \right\rangle,\qquad(23.31)$$

i.e. by integrating over all coordinates save R. Since $U(R)$ depends on α, both through Z in Eq. (23.31), and through ϕ, α can be varied to find the minimum energy of any state. Using nodeless 1s trial wavefunctions for ψ_n and ψ_m they found that $\alpha = 1.815$ gives the minimum He $1s^2$ energy and leads to the He potential curve of Fig. 23.7. It passes approximately through the avoided crossings of the adiabatic curves analogous to those shown in Fig. 23.5, and gives a series of

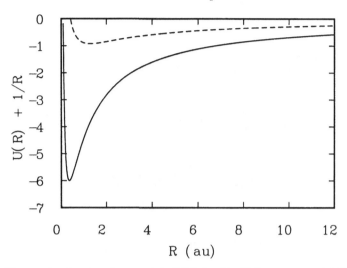

Fig. 23.7 Electronic vibrational potentials $U(R)$ for He (full curve, $\alpha = 1.8315$) and H⁻ (broken curve, $\alpha = 0.828$) (from ref. 26).

energies corresponding to increasing vibrational motion of R. The energies of the wavefunctions with even numbers of nodes correspond to the He ns^2 states and are in excellent agreement with energies calculated in other ways. It is interesting that in these wavefunctions all the nodes are in $f(R)$, as in the approach of Rau,[13] but in the molecular orbital and hyperspherical wavefunctions most often calculated, $f(R)$ is nodeless and the nodes appear in other directions. However, in either case the wavefunction has the largest amplitude at large R, so the presence or absence of an oscillating wavefunction at small R does not have much bearing on the energy, although it would make a difference in the excitation probability from an initially small atomic state. The situation is approximately the same as the H $n\ell$ states. All are degenerate, all have the same total number of nodes, and all have wavefunctions most likely to be found at similar orbital radii.

All of the above approaches are based on separating, approximately, the motion of the electrons in some coordinate system. A radically different approach is the one used by Richter and Wintgen,[27] who have rewritten the Hamiltonian using coordinate systems which are not separable but lead to equations simple in form.

One of the inherent problems in all the quantum mechanical approaches is the high density of states near a double ionization limit. As pointed out by Percival[28] and Leopold and Percival,[29] classical techniques offer an attractive alternative, and recent calculations have demonstrated the usefulness of this approach. For example, Ezra *et al.*[30] have studied the doubly excited He states of total angular momentum $L = 0$. In these states the motion of the electrons is confined to a plane. The rather surprising result is that the stable orbits correspond to an "asymmetric stretch" motion of the two electrons as shown in Fig. 23.8. The

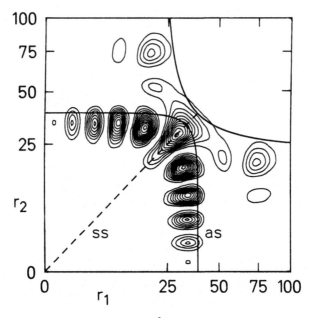

Fig. 23.8 Probability distribution $|\Psi_{Nn}(x,y,z)|^2$ for the intrashell wavefunction $N = n = 6$ in the $x = 0$ plane corresponding to the collinear arrangement $r_{12} = r_1 + r_2$. The axes have a quadratic scale to account for the wave propagation in coulombic systems, where nodal distances increase quadratically. The fundamental orbit (——) (as) as well as the symmetric stretch motion (– – –) (ss) along the Wannier ridge are overlayed on the figure (from ref. 30).

motions of the two electrons are not symmetric about the He^{2+}, i.e. they are not mirror images of each other. In other words the motion is not along the Wannier ridge but perpendicular to it. This conclusion is not really different from those drawn from quantum mechanical calculations as shown by Fig. 23.8, in which it is apparent that the asymmetric stretch classical orbit falls almost precisely on the antinodes of the quantum mechanical wavefunction.

Laser excitation

In addition to the synchrotron experiments on He, laser experiments have been done using alkaline earth atoms. The attractions of the laser are two. First, it is in principle possible to obtain better resolution with a laser than with a synchrotron, and, second, it is a straightforward matter to start from bound states other than the ground state, affording access to a wider range of doubly excited states. The drawback to laser excitation is that it is not presently possible to produce tunable radiation at 200 Å, as used in the He experiments. In fact, 2000 Å, equivalent to a photon energy of 6 eV, is a more realistic lower wavelength limit. The wavelength restriction of available lasers makes alkaline earth atoms, with their low double ionization limits, very attractive. In Ba, for example, the double ionization limit lies only 10 eV above the first limit and can be reached from a bound Rydberg state by the absorption of two 2500 Å photons, a technically manageable proposition.

The typical experimental approach is to use a thermal beam of alkaline earth atoms, which is first excited to a bound Rydberg state by one or two pulsed dye lasers. Typically, the bound Rydberg state is a high n state, although sometimes it is a high ℓ state as well. The bound Rydberg atom is excited to a high lying doubly excited state in one of two ways. First it can be excited by a two-photon ICE, for example a Ba 6snd state can be excited to the 9dnd state. The most straight-forward approach is to use a single laser operating at half the 6snd $\to$ 9dnd transition frequency.[31] This approach has the substantial attraction of simplicity, but a significant drawback as well. Absorption of a single photon leads to photoionization to the 6pεd continua. Since the virtual intermediate state of the two photon 6snd $\to$ 9dnd transition is far off resonance, driving it requires a relatively high laser intensity, and the undesired single photon ionization is not a negligible problem.

An alternative two photon excitation scheme has been used to excite the Ba 6snd $\to$ 7snd transitions.[32] Two frequencies are used, the first coincident with an overlap integral zero in the 6snd $\to$ 6pnd transitons. This approach has the attraction that the first photon produces no single photon ionization, yet the virtual intermediate state is insignificantly detuned from the real intermediate state. It has the advantage of requiring low laser powers and producing little single photon ionization. However it does have two significant drawbacks. First, it requires one more laser than does a single color, two-photon excitation, and second, it fixes the energy of the virtual state at 2.5 eV above the first ionization limit, which is rather low.

The second approach to exciting high lying doubly excited states is to make a sequence of single photon ICE transitions, for example from Ba 6snd $\to$ 6pnd $\to$ 6dnd. Eichmann *et al.*[33] have used this method to excite Ba n_ign_od states by the route 6sn_od $\to$ 6pn_od $\to$ 6dn_od $\to$ 6fn_od $\to$ n_ign_od, using four dye lasers. All four transitions are strong single photon transitions which are easily driven by lasers with 100 μJ pulse energies. The two problems are the multiple lasers and the fact that each of the several intermediate autoionizing states decays, diminishing the signal from the final doubly excited state and adding to the background ion signal.

As is by now apparent, in all these schemes for exciting high lying autoionizing states there is a substantial number of ions and electrons created having nothing to do with the final doubly excited state. As a result, simply detecting ionization is rarely adequate, although it is when studying relatively low lying states such as the Ba 7snd and 6dnd states. Detection schemes which are only sensitive to the final $n_i\ell_in_o\ell_o$ state must be used. A relatively straightforward approach is to detect only the high energy electrons which the $n_i\ell_in_o\ell_o$ state ejects in autoionization to the ground state of the ion. For example, in the two photon excitation of the Ba 6dnd states from the 6snd states such electrons have energies of $\sim$4.6 eV, while those from single photon ionization have energies of at most 2.3 eV. The success of this approach depends upon there being a favorable branching ratio for the autoionization to the continua of the ground state of the ion. As we consider doubly excited states converging to progressively higher energy states of the ion, autoionization is more likely to occur to high lying states of the ion, and detecting the high energy electrons associated with autoionization to the ground state becomes impractical. As a result, this approach has only been used on relatively low lying doubly excited states.

As mentioned above, if we excite a Ba $n_i\ell_i n_o\ell_o$ double Rydberg state, it is likely to autoionize to an excited $n_i\ell_i$ state of Ba$^+$, which is energetically nearby. The other two techniques for selectively detecting the Ba $n_i\ell_i n_o\ell_o$ states depend upon this fact. A laser, often one of the ones used to produce the Ba $n_i\ell_i n_o\ell_o$ doubly excited state, can be used to photoionize the Ba$^+ n_i\ell_i$ state to produce Ba^{2+}, which is easily separated from the copious Ba$^+$ signal to provide a clear signature of the production of the Ba $n_i\ell_i n_o\ell_o$ state. This approach assumes that the doubly excited state autoionizes and that it does not give up very much energy to the ejected electron when it does so. The photon energy of the laser photoionizing the Ba$^+$ must be high enough to ionize the ions resulting from the final doubly excited state, but low enough not to ionize background ions produced along the way.

When doubly excited states converging to Rydberg states of the ion are produced, the product of autoionization is often an ionic Rydberg state, which is easily ionized by microwave ionization[34] or pulsed field ionization,[33,35] although for field ionization to be useful, risetimes of $\sim$1 μs are required. Since either of these methods is only sensitive to Rydberg states of the ion, they provide an excellent means of detecting the high lying doubly excited states.

Experimental observations of electron correlation

The strategy of the experimental study has been to start at the autoionizing states converging to low lying ionic states, which are understood using an independent electron picture, and proceed to the doubly excited states converging to higher lying ionic states, all the while looking for clear manifestations of correlation of the motion of the two electrons.

Most of the experiments to date have been carried out on doubly excited states converging to isolated low angular momentum states of the ion. In general the spectra are well characterized as two photon ICE spectra in which the observed spectrum is characterized by the product of an overlap integral and the spectral density of the doubly excited autoionizing states. As an example we show in Fig. 23.9 the Ba 6s19d $\rightarrow$ 9dn'd spectra observed by Camus *et al.*[31] As shown in Fig.

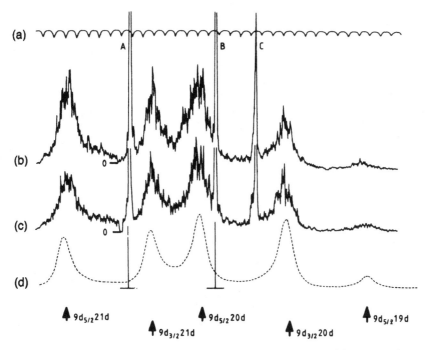

Fig. 23.9 Observed excitation spectra 6s19d → 9dn′d in neutral Ba: (b) circular, (c) linear polarization of laser beams and (d) computed spectra. The sharp resonances A, B and C correspond respectively to $6s_{1/2} \to 9d_{5/2}, 6s_{1/2} \to 9d_{3/2}$ and $5d \to 10g$ ionic transitions used as standard calibrations with the Fabry–Perot fringes (a) (from ref. 31).

23.9, there are four peaks in the spectra, corresponding to the ionic Ba$^+$ $6s_{1/2} \to 9d_{3/2}$ and $6s_{1/2} \to 9d_{5/2}$ transitions for $n = 20$ and 21, in agreement with the calculated spectrum.

In spite of the fact that the spectra can be understood as two-photon ICE spectra, implying that there is no serious departure from the independent electron picture, the quantum defects and autoionization widths of these $n_i\ell_i n_o\ell_o$ states display an interesting dependence on n_i. As shown by Bloomfield *et al.*,[36] the quantum defects of both the $n_i s n_o s$ and $n_i s n_o d$ states increase linearly with n_i, as shown by Fig. 23.10. If the quantum defect is due mostly to core penetration, as it must be when the quantum defects are so large, evaluating the n_i dependence of the expection value of the penetration energy for an outer $n_o s$ or $n_o d$ electron yields a quantum defect which increases with n_i as shown in Fig. 23.10.[36]

Not only the quantum defects of the $n_i s n_o s$ and $n_i s n_o d$ states, but also the widths change. The scaled widths $n_o^3 \Gamma$ of the $n_i s n_o s$ and $n_i s n_o d$ states are plotted in Fig. 23.11.[36] If we first consider the scaled widths of the $n_i s n_o d$ states, we can see that they rise sharply with n_i, a dependence which is not surprising if we assume that most of the autoionization rate comes from dipole autoionization to a nearby state of the ion. If the outer $n_o d$ electron remains outside the inner $n_i s$ electron, the autoionization rate Γ is given by

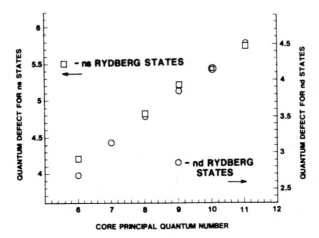

Fig. 23.10 Quantum defects of the Ba $msns$ and $msnd$ states as a function of the core principal quantum number m (from ref. 36).

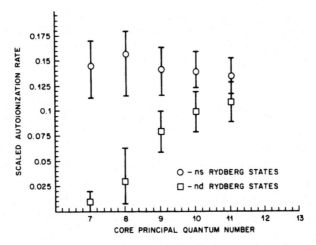

Fig. 23.11. Core scaling of autoionization rates of the Ba $msns$ and $msnd$ states as a function of the principal quantum number of the core. The nd states show increasing autoionization width with increasing core size, while the ns autoionization rates are independent of the core principal quantum number (from ref. 36).

$$\Gamma \approx 2\pi |\langle n_\text{i}s|\mu|n_\text{I}\,\text{p}\rangle|^2 \,|\langle n_0\text{d}|1/r^3|\varepsilon f\rangle|^2, \tag{23.32}$$

where the predominant dependence on n_i is through the dipole moment $\langle n_\text{i}s|\mu|n_\text{I}\text{p}\rangle$ which scales approximately as n_i^2. As a result $\Gamma \propto n_\text{i}^4$ and the rapid rise of $n_0{}^3\Gamma$ with n_i shown in Fig. 23.11 is not unexpected. Once the inner electron $n_\text{i}s$ orbit becomes large enough that the outer electron's orbit passes through most of it, so that core penetration contributes significantly to the autoionization rate, increasing the size of the inner electron's orbit does not affect the autoionization rate, because making the n_is orbit larger also makes it more dilute.

One of the ways of defining correlation is that the wavefunctions are not well represented by the designation $n_i\ell_i n_o\ell_o$ but rather as the $+$ and $-$ states of Eq. (23.1) or some other linear combination of $n_i\ell_i n_o\ell_o$ states. Excluding the configuration mixing of series converging to the two fine structure levels of an ionic state, several such cases have been observed. In any atom perturbing levels converging to higher limits which lie below the first limit often satisfy this criterion. The Ba 5d7d states are excellent examples. There have also been observations in more highly excited states. In the Ca $4s7s \rightarrow 7d7s$ transition Morita and Suzuki[37] observed clear structure due to interaction with the $5gn\ell$ states. Similarly, Jones *et al.*[38] observed structure in the Ba $5d_{3/2}5g\ J = 2 \rightarrow 4f_{5/2}7g\ J = 3$ transition due to the degenerate $6d_{5/2}nf$ and $6d_{5/2}nh$ states. In all of these cases there is evidently angular momentum exchange between the two electrons, as in Eq. (23.1); however, in both cases the Rydberg electron spends much of its time far from the ionic core where it is in a purely coulomb potential. In this sense these states are not highly correlated states.

When the series converging to high lying isolated sates of Ba$^+$ are excited it is possible to observe clear evidence of correlation. A good example occurs in the work of Camus *et al.*,[34] who observed the two photon spectra from the bound Ba $6sn_o$p states to the n_isn_op and n_idn_op states. In Fig. 23.12 we show the two photon spectrum from the Ba 6s45p state to the doubly excited states lying near the $n = 30$ states of Ba$^+$. The spectrum of Fig. 23.12 corresponds to changing the inner electron from 28s to 35s. The $6s45p \rightarrow n_id45p$ and $6s45p \rightarrow n_is45p$ transitions for $n_i < 28$ appear as simple two photon ICE transitions, which are for all intents and purposes identical to the spectrum of Fig. 23.9. As shown by Fig. 23.12, between the 27d45p resonance and the 29s45p resonance there are weak, but easily observed resonances which correspond to $28pn_o\ell$ and $25fn_o\ell$ states. As n_i is increased these p and f resonances become more pronounced, to the point that the $31pn_o\ell$ resonance is almost as large as the 30d45p resonance. At energies just above the $30pn_o\ell$ and $31pn_o\ell$ states there is a signal which is clearly nonzero, and by the time the energy of the $33pn_o\ell$, state is reached the spectrum is essentially continuous, i.e. doubly excited states are excited at any wavelength. By the time we have reached the 32d45p state the outer electron is not in a pure 45p state; however, we have used the label for simplicity. There are two important aspects of these observations; first, the appearance of the $n_ipn_o\ell$ and $n_ifn_o\ell$ resonances, and, second, which is not apparent in Fig. 23.12, the center of gravity of the resonances shifts from the ionic transition frequency as the inner electron quantum number n_i is increased. We shall return to this point shortly. Eichmann *et al.*[33] and Jones and Gallagher[35] have made observations which are similar to those shown in Fig. 23.12, so they may be regarded as representative.

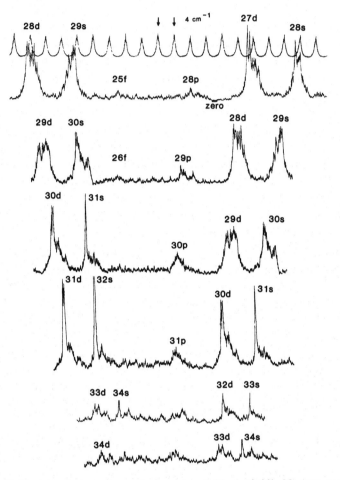

Fig. 23.12 Ba two-photon absorption spectra 6s45p→nℓn'ℓ'. Narrow resonance lines correspond to ionic resonances of Ba$^+$ due to coincidences between atomic (Ba) and ionic (Ba$^+$) resonances. They are indicated by an arrow. Note the evolution to a continuous spectrum at high photon energies (from ref. 34).

Camus *et al.*[34] explained their observations by a picture which has sometimes been called the frozen planet model. Qualitatively, the relatively slowly moving outer electron produces a quasi-static field at the inner electron given by $1/r_o^2$, and this field leads to the Stark effect in Ba$^+$. The field allows the transitions to the $n_i p n_o \ell$ and $n_i f n_o \ell$ states and leads to shifts of the ionic energies. The presence of the $n_i p n_o \ell$ and $n_i f n_o \ell$ resonances in the spectrum of Fig. 23.12 is quite evident. Camus *et al.* compared the shifts to those calculated in a fashion similar to a Born–Oppenheimer calculation. With the outer electron frozen in place at r_o they calculated the Ba$^+$ energies, $W_i(r_o)$, and wavefunctions. They then added the energy $W_o(r_o)$ to the normal screened coulomb potential seen by the outer electron. This procedure leads to a phase shift in the outer electron wavefunction

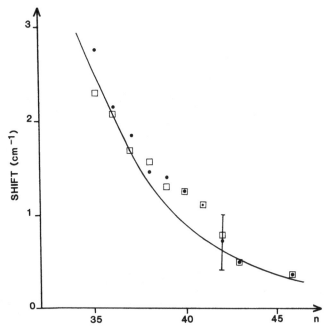

Fig. 23.13 Shifts of the centers of gravity compared to the ionic parent line positions for the observed structures for double Rydberg resonances $6snp \rightarrow 26dn'p$ ($\square$) and $6snp \rightarrow 27sn'p$ ($\bullet$) vs n^*; $n = 39$–47 and 50. The average quantum defect is 4.1 for the $6snp$ series. The full line is the theoretical curve (from ref. 34).

which varies with r_0. Using this approach they were able to compare the shifts from the ionic lines of the centers of gravity of the observed resonances.

The cross section for a two photon ICE transition is given by

$$\sigma \propto A^2(\nu)\langle \nu_b \ell^B | \nu \ell \rangle^2, \tag{23.33}$$

where we have omitted all terms which are either slowly varying or constant, and ν_b is the effective quantum number of the outer electron in the initial bound state. Camus *et al.*[34] observed that the resonances shown in Fig. 23.12 are too wide to come from a single final outer electron state. Rather, each resonance comes from a band of possible final states, in which case A^2 of Eq. (23.33) is a constant, and the cross section is determined by only the squared overlap integral. The overlap integral is centered on the ionic transition frequency. Thus by finding the center of gravity of the observed resonances we can find the transition frequency of the Ba$^+$ ion with the outer electron present, and we can compare the shifts from the bare ion frequencies to the values calculated using the procedure outlined in the previous paragraph. In Fig. 23.13 we show the excellent agreement between the measured and calculated shifts.

The frozen planet model is simple and physically appealing. In addition, it is clearly related to the treatment based on $+$ and $-$ states originally given by Cooper *et al.*[2] The most convincing demonstrations of the legitimacy of the frozen planet model is one which appears, in retrospect, obvious. Eichmann *et al.*[33] simply replaced the outer electron by a static field to check that the spectrum is the same. Using six dye lasers they excited Ba atoms first to the bound 6s78d states and then to final $n_i\ell_i$78d states of $n_i > 30$ via the route 6s78d → 6p78d → 6d78d → 6f78d → $n_i\ell_i$78d, and field ionized the Ba$^+$ ions resulting from autoionization of the final state. Their spectra are shown in Fig. 23.14. Fig. 23.14(a) shows the spectrum obtained by scanning the last laser, driving the 6f78d → $n_i\ell_i$78d transition. It shows strong background peaks at the location of the ionic transitions to the n_id and n_ig states, mostly due to excitation of Ba ions produced along the way. Fig. 23.14(c) is the spectrum obtained when Ba is photoionized by the first two lasers, instead of being excited to the 6s78d state, and then excited to produce the nd and ng Rydberg states of Ba$^+$. Fig. 23.14(d) is obtained by the same procedure as Fig. 23.14(c) except that a field of 60 V/cm is added. This field corresponds to $\langle r \rangle^{-2}$ of the 78d electron. Fig. 23.14(b) is composed of the normalized superposition of the zero field ion spectrum of Fig. 23.14(c) and the 60 V/cm ion spectrum of Fig. 23.14(d). These two components should correspond to the background ion signal and the doubly excited state signals in Fig. 23.14(a). The normalization of Figs. 23.14(c) and (d) to produce Fig. 23.14(b) effectively allows the amount of background ion signal in the synthetic spectrum of Fig. 23.14(b) to be adjusted. Finally, Fig. 23.14(e) is the computed Ba$^+$ spectrum in the 60 V/cm field. As shown, the 34g state has disappeared, being replaced by the $n = 34$ Stark manifold which is more likely to be excited on its red end than its blue end.

The good match between the actual spectrum of Fig. 23.14(a) and the spectrum of Fig. 23.14(b), synthesized from ion spectra, validates the frozen planet model. Several other aspects of this experiment are different from the other double Rydberg experiments. First, the effect of the outer electron is quite noticeable at a much lower value of n_i/n_o than in the experiments of Camus *et al.*[34] and Jones and Gallagher.[35] The reason is that the Ba$^+$ ng states are essentially degenerate with the higher ℓ states so that much smaller fields from the outer electron convert the ℓ states to Stark states. The situation is similar to He, and is a good example of angular correlation between the two electrons. Eichmann *et al.* noted that in the calculated spectrum of Fig. 23.14(e) there is almost no nd character in the Stark manifold of states and that the pronounced angular correlation shown in Fig. 23.14(a) could not have been observed if only the n_id78d states had been excited, in agreement with other results.

The frozen planet model also agrees with classical calculations of Richter and Wintgen,[39] who find a classically stable state in which both electrons are on the same side of the atom in orbits which exhibit pronounced angular correlation, even though the orbital radii of the two electrons are very different.

One of the recurring manifestations of correlation between the two valence electrons is the conversion of ℓ states to Stark states. Eichmann *et al.*[40] have studied doubly excited Sr states in which both electrons have $\ell > 2$, so that both inner and outer electrons are in approximately hydrogenic, easily polarized states

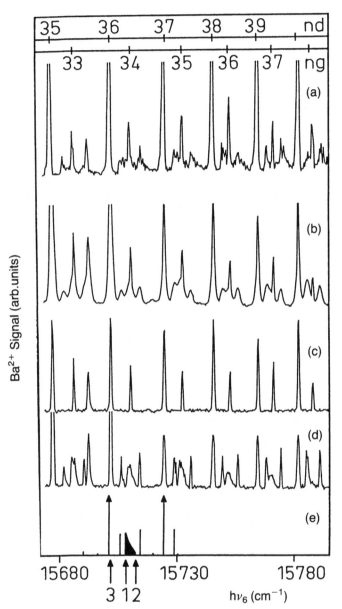

Fig. 23.14 (a) Six laser absorption spectrum of Ba $n\ell78d$ planetary states, recorded as Ba^{2+} signal vs. the photon energy, hν_6, of the laser driving the last transition, from the 6f$_{5/2}$ state to the nd78d and ng78d states. (b) Synthetic spectrum as in (a), composed from a normalized superposition of (c) and (d), including artificial autoionization broadening of (d). (c) Structure of the purely ionic component in (a), obtained by six laser excitation of ionic Ba$^+$ nd and Ba$^+$ ng states. (The extra peak between 33g and 36d appears through accidental excitation and photoionization of a 8p$_{3/2}$ state.) (d) Structure of the planetary component in (a), obtained by repeating the spectrum (c) in the presence of an external field of 60 V/cm. (e) Theoretical bar spectrum equivalent to (d), calculated around $n = 34$ (from ref. 33).

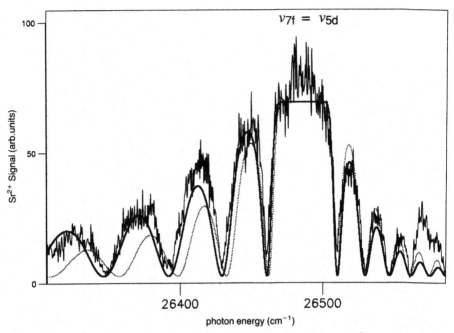

Fig. 23.15 Observed and calculated squared overlap integral $|\langle \nu_{5d}|\nu_{7f}\rangle|^2$ for the transition from the 5d20ℓ state of $\ell > 9$ to the 7f$n\ell$ states as a function of the photon energy of the laser driving the transition. The experimental spectrum (noisy trace) was taken at high laser power and shows some saturation. The smooth solid trace is the squared overlap integral calculated using as the final planetary state the wavefunction $\Psi(\nu_{7f})$, calculated using the potential of Eq. (23.34) which contains an induced dipole. The polarizability a of the Sr$^+$ 7f state is independently calculated to be $5 \times 10^5 a_0{}^3$. The dotted curve is the squared overlap integral obtained with a final state calculated using a coulomb potential ($a = 0$ in Eq. (23.34)). Note the characteristic phase shift and height asymmetry of the side maxima of the overlap integral calculated with the induced dipole potential for the final state when compared to the overlap integral calculated with a pure coulomb potential for the final state (from ref. 40).

with small quantum defects. Specifically, they produced bound Sr 5s$n_o\ell_o$ states of $n_o \sim 20$ and $\ell_o \approx 9$ using the Stark switching technique. They then further excited these states via the route 5s$n_o\ell_o \rightarrow$ 5p$n_o\ell_o \rightarrow$ 5d$n_o\ell_o \rightarrow n_ifn_o\ell_o$. The final n_if$n_o\ell_o$ state, which is a nominal description only, decays to an excited state of Sr$^+$ which is photoionized to produce Sr^{2+}. Sr^{2+} is detected as the wavelength of the laser driving the 5d$n_o\ell_o \rightarrow n_ifn_o\ell_o$ transition is scanned.

They observed two clear signatures of the Stark, or dipole, structure of the doubly excited states; first in the quantum defects and second, in the overlap integrals. They observed several Rydberg series converging to excited Sr$^+$ states. From the 5d$n_o = 17$ $\ell_o = 9$ state they observed a series with a quantum defect of 0.70(mod 1) converging to the 6g state of Sr$^+$. While the observation of a series converging to this limit alone is indicative of correlation, what is most interesting is the quantum defect. It is simply impossible for a Sr coulomb state of $\ell = 9$ to have a quantum defect of 0.70. On the other hand if the outer electron is not in a purely coulomb potential, it is possible for it to have such a quantum defect. Following Watanabe and Greene[41] they assume that the outer electron is in the potential

$$V(r_o) = \frac{-1}{r_o} + \frac{\tilde{\ell}_o(\tilde{\ell}_o + 1)}{2r_o^2} + \frac{\tilde{\alpha}}{r_o^4}, \tag{23.34}$$

where $\tilde{\ell}_o$ and $\tilde{\alpha}$ depend on r_o and are equal to the orbital angular momentum of the outer electron, ℓ_o, and the polarizability of the ion, α, as $r_o \to \infty$. Using this potential they calculate a quantum defect of 6.68 which is in excellent, perhaps fortuitous agreement with the observations.

The second manifestation of the dipole nature of the outer electron states is found in the overlap integral observed in the final ICE transition. For transitions between coulomb states of effective quantum numbers ν_1 and ν_2 the overlap integral has a $\sin(\nu_1 - \nu_2)/(\nu_1 - \nu_2)$ form. However, for the dipole states it does not, as shown by Fig. 23.15, a recording of the Sr $5dn_o\ell \to 7fn'_o\ell'_o$ transition. As shown by Fig. 23.15, the observed spectrum, taken at high laser power, is not symmetric about the ionic Sr^+ $5d \to 7f$ transition. Rather, it is more intense on the low frequency side. While the observed spectrum clearly disagrees with the assumption of the outer electron's being in a coulomb potential, it is in quite good agreement with the overlap integral assuming that in the final state the outer electron is in the potential of Eq. (23.34).

References

1. R. P. Madden and K. Codling, *Phys. Rev. Lett.* **10**, 516 (1963).
2. J. W. Cooper, U. Fano, and F. Prats, *Phys. Rev. Lett.* **10**, 518 (1963).
3. M. Domke, C. Xue, A. Puschmann, T. Mandel, E. Hudson, D. A. Shirley, G. Kaindl, C. H. Greene, H. R. Sadeghpour, and H. Peterson, *Phys. Rev. Lett.* **66**, 1306 (1991).
4. P. G. Burke and D. D. McVicar, *Proc. Phys. Soc.* **86**, 989 (1965).
5. P. L. Altick and E. N. Moore, *Phys. Rev. Lett.* **15**, 100 (1965).
6. T. F. O'Malley and S. Geltman, *Phys. Rev.* **137**, A1344 (1965).
7. L. Lipsky and A. Russek, *Phys. Rev.* **142**, 59 (1966).
8. J. M. Macek, *J. Phys. B* **1**, 831 (1968).
9. U. Fano, *Rep. Prog. Phys.* **46**, 97 (1983).
10. F. T. Smith, *Phys. Rev.* **118**, 1058 (1960).
11. C. D. Lin, *Phys. Rev. A* **10**, 1986 (1974).
12. H. R. Sadegphour and C. H. Greene, *Phys. Rev. Lett.* **65**, 313 (1990).
13. A. R. P. Rau, *J. Phys. B* **16**, L699 (1983).
14. F. H. Read, *J. Phys. B* **10**, 449 (1977).
15. F. H. Read, *J. Phys. B* **23**, 951 (1990).
16. S. J. Buckman, P. Hammond, F. H. Read, and G. C. King, *J. Phys. B* **16**, 4219 (1983).
17. O. Sinanoğlu and D. R. Herrick, *J. Chem. Phys.* **62**, 886 (1975); **65**, 850 (E) (1976).
18. D. R. Herrick, *Adv. Chem. Phys.* **52**, 1 (1988).
19. D. A. Park, *Z. Phys.* **159**, 155 (1960).
20. K. Molmer and K. Taulbjerg, *J. Phys. B* **21**, 1739 (1988).
21. J. M. Feagin and J. S. Briggs, *Phys. Rev. Lett.* **57**, 984 (1986).

22. J. M. Feagin and J. S. Briggs, *Phys. Rev. A* **37**, 4599 (1988).
23. D. R. Bates and R. G. H. Reid, in *Advances in Atomic and Molecular Physics*, Vol. 4, eds. D.R. Bates and J. Estermann (Academic Press, New York, 1968).
24. J. M. Rost, J. S. Briggs, and J. M. Feagin, *Phys. Rev. Lett.* **66**, 1642 (1991).
25. J. M. Rost, R. Gersbacher, K. Richter, J. S. Briggs, and D. Wintgen, *J. Phys. B* **24**, 2455 (1991).
26. J. M. Rost and J. S. Briggs, *J. Phys. B* **21**, L233 (1988).
27. K. Richter and D. Wintgen, *J. Phys. B* **24**, L565 (1991).
28. I. C. Percival, *Proc. Roy. Soc. London A* **353**, 189 (1967).
29. J. G. Leopold and I. C. Percival, *J. Phys. B* **13**, 1037 (1980).
30. G. S. Ezra, K. Richter, G. Tanner, and D. Wintgen, *J. Phys. B* **24**, L413 (1991).
31. P. Camus, P. Pillet, and J. Boulmer, *J. Phys. B* **18**, L481 (1985).
32. T. F. Gallagher, R. Kachru, N. H. Tran, and H. B. van Linden van den Heuvell, *Phys. Rev. Lett.* **51**, 1753 (1983).
33. U. Eichmann, V. Lange, and W. Sandner, *Phys. Rev. Lett.* **64**, 274 (1990).
34. P. Camus, T. F. Gallagher, J.-M. Lecompte, P. Pillet, L. Pruvost, and J. Boulmer, *Phys. Rev. Lett.* **62**, 2365 (1989).
35. R. R. Jones and T. F. Gallagher, *Phys. Rev. A* **42**, 2655 (1990).
36. L. A. Bloomfield, R. R. Freeman, W. E. Cooke, and J. Bokor, *Phys. Rev. Lett.* **53**, 2234 (1984).
37. N. Morita and T. Suzuki, *J. Phys. B* **21**, L439 (1988).
38. R. R. Jones, P. Fu, and T. F. Gallagher, *Phys. Rev. A* **44**, 4260 (1991).
39. K. Richter and D. Wintgen, *Phys. Rev. Lett.* **65**, 1965 (1990).
40. U. Eichmann, V. Lange, and W. Sandner, *Phys. Rev. Lett.* **68**, 21 (1992).
41. S. Watanabe and C. H. Greene, *Phys. Rev. A* **22**, 158 (1980).

Index